Modern Microwave Circuits

DISCLAIMER OF WARRANTY

For a listing of recent titles in the *Artech House Microwave Library*, turn to the back of this book.

Modern Microwave Circuits

Noyan Kinayman
M. I. Aksun

ARTECH
HOUSE

BOSTON | LONDON
artechhouse.com

Library of Congress Cataloging-in-Publication Data
A catalog record of this book is available from the Library of Congress.

British Library Cataloguing in Publication Data
Kinayman, Noyan
Modern microwave circuits.—(Artech House microwave library)
1. Microwave circuits
I. Title II. Aksun, M. I.
621.3'813

ISBN 1-58053-725-1

Cover design by Igor Valdman

Cover image is courtesy of Cascade Microtech, Inc.

The Smith® Chart is a registered trademark of Analog Instruments Co., P.O. Box 950, New Providence, NJ 07974. The Smith® Chart on page 82 has been reproduced with the courtesy of Analog Instruments Co.

International Standard Book Number: 1-58053-725-1

10 9 8 7 6 5 4 3 2 1

Contents

Preface

This book has been written to address the analysis techniques and practical applications of microwave printed passive circuits. Instead of focusing mainly on printed circuits, however, we have chosen to adopt a much broader scope so that a wide audience from research students to design engineers can benefit from it. From this respect, we tried to keep the balance between theory and applications as much as we could. Each chapter is also tailored such that they can be studied independently without losing the coherency.

Recent advances in digital computers and numerical techniques have placed computational electromagnetics in the reach of almost every electrical engineer. When used properly, computer simulations greatly increase the success in microwave circuit design. This is especially important in microwave printed circuits (i.e., microstrip circuits) where quasi-static or empirical solutions are not accurate enough at microwave- and millimeter-wave frequencies. Microstrip circuits do have an important place in microwave engineering. With continuously increasing demand for signal speed, application areas of microstrip circuits have been proliferating ever since their invention. So, it has been the philosophy of this book to promote microstrip circuits and the usage of computer simulations in microwave printed circuit design. Another feature of this book is that we tried to include as many example problems as we could. Since engineering mainly deals with the practical application of science, example problems constitute a major part of introducing engineering knowledge. Indeed, the inductive approach is arguably the best way of teaching engineering.

We would like to introduce the main features of each chapter as well. Chapter 1 provides a brief introduction to microwave circuit theory. A fundamental knowledge of vector and complex calculus is helpful to get the maximum benefit from the chapter. We include more practical topics such as network analyzer calibration methods and matching circuits in this chapter too. Another feature of this chapter is the inclusion of network synthesis. Although the importance of network synthesis in electrical engineering is not as great as it used to be, we believe that a good understanding of network synthesis methods could provide some benefit, especially in modeling passive microwave circuits. Chapter 2 sets the stage for microwave printed circuits, which is the main theme of the book.

Microstrip circuits and their production technologies are introduced here. Dielectric measurement techniques, which we thought would be useful, are also mentioned in this chapter. The next chapter, Chapter 3, talks about full-wave analysis methods of microwave printed circuits in detail, including the concept of Green's function. Although this chapter is quite theoretical, it is compact. So, a reader who follows the derivations in a rigorous manner should have no difficulty grasping it. The rest of the book is devoted to various practical applications of printed microwave circuits. Chapter 4 presents microstrip antennas, including an emphasis on bandwidth-enhancement methods. Chapter 5 contains various topics on microstrip coupled lines. We included the multiconductor transmission line analysis technique in this chapter to address the growing applications of high-frequency coupled-transmission lines such as flex-cables or unshielded twisted pair (UTP). These components can be found in many of today's electronic devices. Chapter 6 is a classical introduction to filter theory and microstrip filters. The reader will find modern filter-design techniques, such as cascaded triplets and quadruplets in this chapter, too. Chapter 7 is devoted to microwave passive elements, another growing area of microwave engineering due to advances in MMIC technology. Apart from mentioning basic lumped elements, such as spiral inductors and capacitors, a unique feature of this chapter is that it also introduces model extraction techniques for passive elements. We believe that this could be useful to designers who want to extract equivalent lumped models for their particular applications. A method which is commonly used in other areas of engineering, but not too much in electrical engineering, namely dimensional analysis, will conclude this last chapter.

Microwave circuits have a great deal of literature. It has not been our intention to provide a profound picture of the technology in this area. This would not fit into the timeline and objectives of this project. Hopefully, we will be able to provide a foundation and necessary motivation to the reader such that he or she can build on top of that foundation.

Acknowledgments

This book has not been written in a vacuum. So, we would like to thank our colleagues and friends who contributed a lot through fruitful discussions and talks. First, Dr. Kinayman is indebted to Professor M. I. Aksun, who agreed to collaborate in this project. His contributions, mentorship, and guidance were extremely valuable. He also would like to thank his colleagues in M/A-COM, namely, Dr. Tekamul Buber, Mr. Daniel G. Swanson, Jr., Dr. Ian Gresham, Dr. Alan Jenkins, Dr. Richard Anderson, Dr. Eswarappa Channabasappa, Dr. Kristi Pance, and Dr. Robert Egri. He is also grateful to Dr. Jean-Pierre Lanteri, who is director of technology at M/A-COM, and Mrs. Jackie Bennett, who is the engineering manager of strategic research and development at M/A-COM, for providing opportunity and encouragement through the development of this project. He would like to thank his former colleague Dr. Nitin Jain of Anokiwave Inc. for the valuable feedback and discussions. Finally, he is grateful to his former managers, Dr. Gerry DiPiazza and Dr. Peter Staecker, and former colleague Dr. Peter Onno for providing directions and motivation early in his career.

Professor M. I. Aksun would like to thank to Dr. N. Kinayman, who has initiated, organized, and been a main contributor to this book; without him this project would not have been completed. Moreover, Dr. Kinayman's positive energy and hard work have made this book possible, especially after Professor Aksun has been appointed as the dean of the College of Engineering at Koc University. He also would like to thank the administration of Koc University for the peaceful environment they have created.

We also would like to thank the design team at Artech House for their professional and meticulous work. We appreciate the excellent review done by the reviewer; he caught many subtle points and provided excellent feedback which helped us to improve the manuscript.

All full-wave electromagnetic simulations shown in the book have been carried out using EMPLAN. Serenade version 8.5 by Ansoft has been used in all circuit simulations.

Finally, a couple of words about how this book has been prepared are in order. The book has been written by the authors using Microsoft Word 2002 under Microsoft XP operating system. Adobe Acrobat 6.0 has been used to prepare the final camera-ready copy. All CAD figures have been drawn using AutoSketch by AutoDesk, TurboCAD by IMSI, and Microsoft Visio, depending on the requirements of individual artwork.

Chapter 1

Microwave Network Theory

Microwave frequencies refer to the frequency range starting from 300 MHz up to 300 GHz, or equivalently to the wavelength range from 1 meter down to 1 millimeter. Since the dimensions of circuit and circuit components designed to operate in microwave frequencies can easily be comparable to the wavelength, they cannot be considered as point-like objects as in the case of lumped model approximations. In lumped circuit analysis, as taught in circuit analysis and electronics courses in electrical and electronics engineering curriculum, the main assumption is that the current through a series arm and the voltage across parallel branches don't change by distance because the dimensions of the circuits are extremely small as compared to the wavelength of the signal. As a result, node voltages and loop currents become sufficient to analyze such circuits. However, if the dimensions of the circuit components become comparable to signal wavelength, the assumption no longer holds; that is, the current through a component and voltage across parallel branches vary as one moves along the circuit. This is mainly due to the finite propagation time required for an electrical disturbance, like current and voltage, to move in a circuit. Thus, the distributed nature of the circuit must be taken into account at microwave frequencies to accurately model the phase change and attenuation of signals while traveling along the circuit. To account for changes of the current and voltage along a conducting line or through a circuit component, these elements can be modeled, from a circuit point of view, by series- and parallel-connected resistance R, inductance L, capacitance C, and conductance G per unit length, distributed along the line or circuit component.

As an example, we can think of a length of coaxial line terminated with a resistor. At very low frequencies, we can assume that voltage and current magnitudes along the line are constant. In fact, this is the assumption that is used to analyze dc or very low-frequency circuits. However, as frequency increases, this assumption fails, and magnitudes of voltage and current along the line depend on position as well as the resistor value. The resistor value is important because it determines how much of the radio frequency (RF) power is reflected back. Any reflected power would cause a standing-wave pattern on the transmission line. This point will be elaborated later in detail. Another good example of

1

distinguishing lumped and distributed circuits is a simple capacitor. At low frequencies, the capacitor can very well be approximated by a single capacitance value. On the other hand, at high frequencies, parasitic series inductance of the capacitor starts to get pronounced and must be considered. In fact, after the self-resonance frequency, the lumped capacitor starts to behave like an inductor. It is important to stress that it is not the absolute value of the frequency that we use to differentiate lumped and distributed circuits. The important criterion is the ratio of the wavelength to the circuit dimensions. The more wavelength is comparable with circuit dimensions, the more the circuit becomes distributed. As a matter of fact, the theory of distributed circuits also holds for lumped circuits; we just make simplifications to make the circuit analysis easier when the frequency is low enough. As a rule of thumb, one can consider a circuit lumped if the wavelength of electrical signals passing through it is less than 1/100th of the maximum dimension.

There is a very extensive literature on the microwave network theory addressing both the theoretical as well as practical aspects. *Microwave Engineering* by D. M. Pozar is one of the classic references on microwave engineering [1]. It explains fundamental concepts of microwave engineering such as waveguides, impedance matching, *S*-parameters, filters, active microwave circuits, and the design of microwave amplifiers and oscillators. *Foundations for Microwave Engineering* by R. E. Collin is another classic reference with similar content as the one by Pozar, but it provides more details on theory [2]. *Microstrip Circuits* by F. Gardiol gives a good overview of microstrip circuits in general [3], although it presents not as much theory as the first two books but explains the underlying principles of commonly used full-wave electromagnetic simulation techniques, the indispensable tools for modern microwave engineers. Edwards et al. present very good general information on microstrip circuits in *Foundations of Interconnect and Microwave Design* [4], including models of passive MMIC elements and microstrip discontinuities. *Microwave Transistor Amplifiers* by G. Gonzales is another good reference on microwave engineering [5], and it addresses mostly active circuit design, providing a complete picture of how microstrip circuits are used with transistors to build amplifiers and oscillators.

In this chapter, we will give an introduction to microwave network theory starting with Maxwell's equations. It should be stated that an in-depth review of microwave network theory is out of the scope of this book. Therefore, interested readers should refer to the references for further reading on this matter.

1.1 REVIEW OF ELECTROMAGNETIC THEORY

Microwave engineering can be considered applied electromagnetic engineering. Therefore, to be a competent microwave engineer, one needs to understand the basic electromagnetic wave theory described by Maxwell's equations and the underlying assumptions and approximations in the analysis tools for microwave circuits. This section will outline the fundamental concepts of electromagnetic wave theory, which is built upon a group of differential equations, called Maxwell's equations.

1.1.1 Maxwell's Equations: Time-Dependent Forms

The general form of time-varying Maxwell's equations can be written in integral form as follows:

$$\oint_C \tilde{\mathbf{E}} \cdot d\mathbf{l} = -\frac{\partial}{\partial t} \iint_S \tilde{\mathbf{B}} \cdot d\mathbf{s} \tag{1.1}$$

$$\oint_C \tilde{\mathbf{H}} \cdot d\mathbf{l} = \frac{\partial}{\partial t} \iint_S \tilde{\mathbf{D}} \cdot d\mathbf{s} + \iint_S \tilde{\mathbf{J}} \cdot d\mathbf{s} \tag{1.2}$$

$$\oiint_S \tilde{\mathbf{D}} \cdot d\mathbf{s} = \iiint_V \rho \, dv \tag{1.3}$$

$$\oiint_S \tilde{\mathbf{B}} \cdot d\mathbf{s} = 0 \tag{1.4}$$

and can be converted to differential form by using the divergence and Stokes' theorems, as

$$\nabla \times \tilde{\mathbf{E}} = -\frac{\partial \tilde{\mathbf{B}}}{\partial t} \quad \text{(Faraday-Maxwell law)} \tag{1.5}$$

$$\nabla \times \tilde{\mathbf{H}} = \frac{\partial \tilde{\mathbf{D}}}{\partial t} + \tilde{\mathbf{J}} \quad \text{(Generalized Ampere's law)} \tag{1.6}$$

$$\nabla \cdot \tilde{\mathbf{D}} = \rho \quad \text{(Gauss's law)} \tag{1.7}$$

$$\nabla \cdot \tilde{\mathbf{B}} = 0 \quad \text{(Law of conservation of magnetic flux)} \tag{1.8}$$

where boldface and ~ are used throughout this text to represent vector and time-varying forms of the corresponding quantities, respectively. The time-varying vector fields in these equations are real functions of spatial coordinates x, y, z, and time coordinate t, and are defined as follows:

$$\tilde{\mathbf{E}}[\text{V/m}] \quad : \quad \text{Electric field intensity}$$

$$\tilde{\mathbf{H}}[\text{A/m}] \quad : \quad \text{Magnetic field intensity}$$

$$\tilde{\mathbf{D}}[\text{Coul/m}^2] \quad : \quad \text{Electric flux density}$$

$$\tilde{\mathbf{B}}[\text{Weber/m}^2] \quad : \quad \text{Magnetic flux density}$$

$$\tilde{\mathbf{J}}[\text{A/m}^2] \quad : \quad \text{Electric current density}$$

$$\rho[\text{Coul/m}^3] \quad : \quad \text{Electric charge density}$$

In addition to Maxwell's equations, there is another fundamental equation that describes the conservation of charges, called the continuity equation, and mathematically it can be expressed as follows:

$$\oiint_S \tilde{\mathbf{J}} \cdot d\mathbf{s} = -\frac{\partial}{\partial t} \iiint_V \rho \, dv \quad \text{(Integral form)}$$

$$\nabla \cdot \tilde{\mathbf{J}} = -\frac{\partial \rho}{\partial t} \quad \text{(Differential form)}$$

(1.9)

Note that this equation can be directly derived from Maxwell's equations above by taking the divergence of (1.6) or from the physics of the conservation of charges in a volume V enclosed by a surface S.

Most topics in microwave engineering are based on the solution of Maxwell's equations in different geometrical and material settings and for different excitation conditions. Therefore, Maxwell's equations are the only equations for which microwave engineers must be comfortable with their physical meanings and their solutions for the field quantities $\tilde{\mathbf{E}}$ and $\tilde{\mathbf{H}}$ (or current density $\tilde{\mathbf{J}}$) under different boundary conditions. In the following paragraphs, the physical meanings and significance of Maxwell's equations are discussed first, and then some important tools are introduced to help solve these equations.

Let us start with the discussion on the Faraday-Maxwell law:

$$\text{E.M.F.} = \oint_C \tilde{\mathbf{E}} \cdot d\mathbf{l} = -\frac{\partial}{\partial t} \iint_S \tilde{\mathbf{B}} \cdot d\mathbf{s} = -\frac{\partial \Phi}{dt}$$

(1.10)

where E.M.F. denotes the electromotive force and is defined as the line integral of impressed electric field, and Φ is the magnetic flux. This simple expression is of great practical importance and was referred to as Faraday's law of electromagnetic induction for a stationary loop. Faraday, in 1831, discovered experimentally that current is induced in a stationary, closed conducting loop when the magnetic flux across the surface enclosed by this loop varies in time. He, therefore, proposed the integral form of the equation and postulated that the law is valid for any conducting closed loop. However, Maxwell realized that (1.10) is valid for any closed contour in space, not only for a conducting one. In the case of a vacuum or dielectric media, the electric field with its corresponding force exists in the space, but if there is a conducting loop, the electric field induces a current along the conductor, which is the pronounced and easily observable result of the electric field.

Next, we will review the Ampere's law. The original Ampere's circuital law states that the line integral of $\tilde{\mathbf{H}}$ about any closed path is exactly equal to the direct current (time invariant or stationary) enclosed by that path:

$$\oint_C \tilde{\mathbf{H}} \cdot d\mathbf{l} = \iint_S \tilde{\mathbf{J}} \cdot d\mathbf{s} \quad \text{(Integral form)}$$

$$\nabla \times \tilde{\mathbf{H}} = \tilde{\mathbf{J}} \quad \text{(Differential form)}$$

(1.11)

After learning the Danish physicist Hans Christian Ørsted's discovery that a magnetic needle is deflected by a nearby current carrying wire, Andre-Marie Ampere formulated the law of electromagnetism in 1820, which mathematically describes the magnetic force between two electric currents. Later, Maxwell noticed that Ampere's circuital law is inconsistent with the conservation of electric charges in time-varying cases, whose mathematical representation is given by (1.9). According to the law of conservation of charges, the outward flow of charges from a closed surface S (enclosing a volume) must be accompanied by exactly the same decrease of charges in the volume, V. However, the divergence of (1.11) yields null on the left-hand side, which is a clear violation of the conservation of electric charges. To remedy this, Maxwell introduced the term $\partial \tilde{D}/\partial t$ to the right-hand side of the original Ampere's circuital law (1.11), which has resulted in the generalized Ampere's circuital law given in (1.6). The additional term on the right-hand side is called the displacement current density, as its physical nature is quite different from the conduction current density in the same expression [6]. It should be emphasized here that the conduction current occurs in material media, while the displacement current can occur in vacuum as well as in material media. Consequently, the conduction current dominates in good conductors, while the displacement current dominates in good dielectrics for time-varying fields. For example, the conduction current flowing to the plates of a capacitor is equal to the displacement current between the plates of a capacitor.

Finally, Gauss's law (1.7) and the law of conservation of magnetic flux (1.8) will be reviewed. The physical interpretation of Gauss's law is that the total electric flux emanating from a closed surface is equal to the total charge in the volume enclosed by this surface. In other words, the electric flux density (displacement vector) originates from or terminates in electric charges. Based on a similar discussion for the similar mathematical form of the law of conservation of magnetic flux, it is interpreted that there is no magnetic charge observed in nature. Therefore, the magnetic flux density lines terminate on themselves; that is, the magnetic flux densities are solenoidal fields, whose divergences are always zero.

So far, we have seen that Maxwell has contributed to Faraday's law by modifying Faraday's interpretation of (1.5), and to Ampere's circuital law by adding a term called displacement current in (1.6). These contributions of Maxwell, although they seem relatively minor as compared to the contributions of Faraday, Ampere, and Gauss, have united these four equations and turned out to be so important that, since then, they are named Maxwell's equations. At this point, we will further try to understand the importance of Maxwell's contributions, and how they pave the way for the electromagnetic wave propagation. Let us first assume that we have a conducting wire with a time-varying current flowing in it, Figure 1.1(a). According to Ampere's circuital law, whether it is in original or generalized form, this time-varying current $I(t)$ generates a time-varying magnetic field $\tilde{H}$ as depicted in Figure 1.1(a). Then, this time-varying magnetic field generates the time-varying electric field, which is made plausible by

Maxwell's interpretation of Faraday's law. Otherwise, according to the original interpretation of Faraday's law, one must have a conducting loop in place of the dashed line in Figure 1.1(b) to induce an electromotive force. In order to have a propagating wave, the time-varying electric field ought to generate a time-varying magnetic field in the absence of the time-varying current source, and the same process should repeat as described above. Without the modification by Maxwell of Ampere's circuital law, which adds the displacement current (proportional to the electric field) to Ampere's original law, the generation of magnetic field due to the time-varying electric field would be impossible to visualize. Therefore, Maxwell's contributions to those existing equations implied that there would be a propagation of electric and magnetic fields, and hence inspired Heinrich Hertz, a German professor of physics, to carry out a series of experiments to validate the existence of wave propagation.

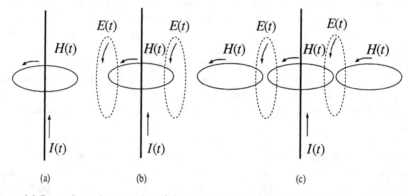

Figure 1.1 Generation and propagation of electromagnetic waves by a current carrying wire: (a) a time-varying current generates a time-varying magnetic field (Ampere's law), (b) the time-varying magnetic field generates an electric field (Maxwell's interpretation of Faraday's law), and (c) the time varying electric field generates a magnetic field (Maxwell's modification of Ampere's law).

It is important to note that for a time-varying electromagnetic field, the last two of Maxwell's equations, Gauss's law (1.7), and the conservation of magnetic flux (1.8), can be obtained from the first two of Maxwell's equations, (1.5) and (1.6). Mathematically speaking, equations (1.7) and (1.8) are linearly dependent on (1.5) and (1.6). For example, taking the divergence of (1.5) gives rise to (1.8), and taking the divergence of (1.6) and using the continuity equation (1.9) result in (1.7). Therefore, all macroscopic electromagnetic phenomena can be characterized using Faraday's law, Ampere's law, and continuity equation.

1.1.2 Maxwell's Equations: Time-Harmonic Forms

So far, we have made no assumption about time dependence of the electromagnetic sources, J and ρ, and hence Maxwell's equations are valid for any arbitrary time dependence. However, these equations, as given in (1.1)–(1.4) in integral forms and in (1.5)–(1.8) in differential forms, seem to be rather complicated to solve with the existence of both space and time dependences. There is only one exception of time dependence for which the time derivatives in Maxwell's equations can be completely eliminated: sinusoidal time dependence (equivalently, harmonic time variation). This could easily be understood by remembering the phasor notation in circuit theory, whose underlying assumption is the linearity of the circuit. As its consequence, the time variation of the signals across or through any element in the circuit is of the same functional form with a possible phase difference. For instance, for a single-frequency excitation ($e^{j\omega t}$), the time derivatives resulting from the *i-v* characteristics of the lumped components can be implemented analytically, or equivalently, the substitutions of $\partial/\partial t \rightarrow j\omega$ and $\partial^2/\partial t^2 \rightarrow -\omega^2$ can be performed either in the expressions of the *i-v* characteristics of individual components or in the governing equation of the circuit. Similarly, if the source in Maxwell's equations is assumed to be monochromatic (single frequency) and the medium is linear ($\tilde{\mathbf{D}}$ and $\tilde{\mathbf{B}}$ are linear functions of $\tilde{\mathbf{E}}$ and $\tilde{\mathbf{H}}$, respectively), then the field and source quantities can be written in phasor forms. Hence, the time derivatives are eliminated by the same substitutions as those of the circuit analysis. Of course, sinusoidal time dependence is a special case of a general time-varying excitation of electromagnetic waves, but this should not be considered too restrictive. This is mainly because any other time dependence of the source can be expressed in terms of sinusoidal signals via Fourier series expansion or transformation and because the transients involved are of little concern in many applications.

For electromagnetic waves of a particular frequency in the steady state, the fields represented in phasor forms are called time-harmonic or frequency-domain representations, while the waves are named as monochromatic or continuous waves. As stated above, the time-harmonic cases are quite important, because single-frequency sinusoidal excitation helps us to eliminate the time dependence in Maxwell's equations, thus simplifying the mathematics. For time-harmonic fields, the instantaneous fields (represented by letters with a tilde) are related to their complex forms, also called phasor forms (represented by roman letters), as

$$\tilde{\mathbf{A}}(x, y, z, t) = \text{Re}\left\{\mathbf{A}(x, y, z)e^{j\omega t}\right\} \tag{1.12}$$

where the time dependence is assumed to be cosine based, and represented by the term $e^{j\omega t}$. We should note that $e^{-i\omega t}$ representation could be adapted for the same cosine-based time dependence, as is usually the case in physic- and optics-related fields, but throughout this book, $e^{j\omega t}$ time dependence is used and suppressed wherever it is convenient. It is simple to translate any expression from one convention to the other by just changing j ($-i$) to $-i$ (j) in the expression. The magnitudes of instantaneous fields represent peak values throughout this book and are related to their corresponding root-mean-square (RMS) values by $\sqrt{2}$. For example, an instantaneous sinusoidal electric field in the x direction is written as

$$\tilde{\mathbf{E}}(x, y, z, t) = \hat{x}\, E(x, y, z)\cos(\omega t + \phi)$$

where E is the real amplitude, ω is the radian frequency, ϕ is the phase reference of the wave at $t = 0$, and $\wedge$ denotes the unit vector. This instantaneous field can be represented equivalently in the phasor form as

$$\mathbf{E}(x, y, z) = \hat{x}\, E(x, y, z)\, e^{j\phi}$$

Note that it is a trivial task to implement (1.12) to find the instantaneous field expressions when the phasor forms are available.

With the assumed time dependence and the phasor-form representation of the instantaneous fields, Maxwell's equations can be simplified to

$$\oint_C \mathbf{E} \cdot d\mathbf{l} = -j\omega \iint_S \mathbf{B} \cdot d\mathbf{s} \tag{1.13}$$

$$\oint_C \mathbf{H} \cdot d\mathbf{l} = j\omega \iint_S \mathbf{D} \cdot d\mathbf{s} + \iint_S \mathbf{J} \cdot d\mathbf{s} \tag{1.14}$$

$$\oiint_S \mathbf{D} \cdot d\mathbf{s} = \iiint_V \rho \, dv \tag{1.15}$$

$$\oiint_S \mathbf{B} \cdot d\mathbf{s} = 0 \tag{1.16}$$

in integral form, and to

$$\nabla \times \mathbf{E} = -j\omega\, \mathbf{B} \tag{1.17}$$

$$\nabla \times \mathbf{H} = j\omega\, \mathbf{D} + \mathbf{J} \tag{1.18}$$

$$\nabla \cdot \mathbf{D} = \rho \tag{1.19}$$

$$\nabla \cdot \mathbf{B} = 0 \tag{1.20}$$

in differential form. By examining the instantaneous and the phasor forms of Maxwell's equations, we see that one form can be obtained from the other by simply replacing the instantaneous field vectors with their corresponding complex forms and by replacing $\partial/\partial t$ with $j\omega$, or vice versa. For the sake of completeness, we provide the continuity equation for time-harmonic sources as follows:

$$\oiint_S \mathbf{J} \cdot d\mathbf{s} = -j\omega \iiint_V \rho \, dv \quad \text{(Integral form)} \tag{1.21}$$

$$\nabla \cdot \mathbf{J} = -j\omega\rho \quad \text{(Differential form)}$$

where S is the surface of the volume V, which contains the charges. From here on, Maxwell's equations will refer to the frequency-domain representations of Maxwell's equations, (1.13) to (1.20), unless otherwise stated explicitly.

1.1.3 Fields in Material Media: Constitutive Relations

As it is well known, Maxwell's equations are fundamental laws governing the behavior of electromagnetic fields in any medium, not only in free space. Therefore, the ultimate goal in microwave and antenna engineering is to find the solutions of Maxwell's equations for a given geometry and medium. To achieve this goal, we have simplified Maxwell's equations under sinusoidal steady-state excitation, but so far we have made no reference to material properties that provide connections to other disciplines of physics. Before getting into the influence of material media on the fields, let us first examine Maxwell's equations from the mathematical point of view to understand the need for some additional information pertaining to the material involved. Considering the differential form of Maxwell's equations, (1.17) to (1.20), there are four vector equations with four vector unknowns. However, as stated earlier, the last two of Maxwell's equations are dependent on the first two, and hence we have only two vector equations for four vector unknowns. Consequently, two more vector equations, which are to be linearly independent of the first two of Maxwell's equations, need to be introduced in order to be able to solve for the field quantities uniquely. As a conclusion, we need to incorporate the material properties into Maxwell's equations to account for the influence of the medium, and we need to introduce two more vector equations to get a unique solution. If these additional vector equations originate from the electrical characteristics of the material used, then both requirements will have been satisfied.

From the study of electric and magnetic fields in material media, it has been observed that the fields are modified by the existence of material bodies, and the macroscopic effects have been cast into mathematical forms as the relations between electric flux density (displacement vector) and electric field intensity, and between magnetic flux density and magnetic field intensity. It would be instructive to briefly discuss the nature and the physical origins of these relations for conducting, insulating (dielectric), and magnetic materials.

Conducting Materials

For conductive media, the current density $\mathbf{J}$ is proportional to the force per unit charge, $\mathbf{F}/q$, by the following relation:

$$\mathbf{J} = \sigma \mathbf{F}/q \tag{1.22}$$

where σ is an empirical constant that varies from one material to another and is called the conductivity of the material. In general, the force acting on a charge could be anything, such as chemical, gravitational, and so forth, but, for our purposes, it is the electromagnetic force that drives the charges to produce the current. Therefore, the current density can be written as

$$\mathbf{J} = \sigma(\mathbf{E} + \mathbf{v} \times \mathbf{B})$$

where the velocity of the charges **v** is sufficiently small in conducting media, leading to Ohm's law

$$\mathbf{J} = \sigma \mathbf{E} \qquad (1.23)$$

Dielectric Materials

For dielectric materials, the electric field in an electromagnetic wave polarizes individual atoms or molecules in the material, and hence lots of dipole moments are induced in the same direction as the field. From a macroscopic point of view, the material is considered to be polarized, and it is effectively quantified as the dipole moment per unit volume, called the polarization **P**. The effect of the polarization of a material appears in the definition of the displacement vector as

$$\mathbf{D} = \varepsilon_0 \mathbf{E} + \mathbf{P} \qquad (1.24)$$

and through this definition, the electrical properties of the material are incorporated into Maxwell's equations. In a vacuum, where there is no matter to polarize, the displacement vector is linearly proportional to the electric field with a constant of proportionality, ε_0, which is the permittivity of free space. As it was obvious from the physical discussion that has led to the concept of polarization, the source of polarization in a material body is the electric field in the medium and is expected to be a linear or nonlinear function of the electric field. For many materials commonly used in microwave and antenna applications, the polarization is linearly proportional to the electric field, and is given by

$$\mathbf{P} = \varepsilon_0 \chi_e \mathbf{E} \qquad (1.25)$$

where the constant of proportionality, χ_e, is called the electric susceptibility of the medium, and is dependent on the microscopic structure of the material. Note that the term ε_0 has been factored out to make χ_e dimensionless. Then, the displacement vector is written as

$$\mathbf{D} = \varepsilon_0 (1 + \chi_e) \mathbf{E} = \varepsilon_0 \varepsilon_r \mathbf{E} = \varepsilon \mathbf{E} \qquad (1.26)$$

where ε and ε_r are the permittivity and relative permittivity (or dielectric constant) of the material, respectively. The permittivity of a medium is usually a complex number, whose imaginary part accounts for the loss in the material, and is given explicitly by

$$\varepsilon = \varepsilon' - j\varepsilon'' \qquad (1.27)$$

where ε' and ε'' are positive numbers, and the negative imaginary part is used due to conservation of energy. The loss mechanism in a nonconductive dielectric material is basically due to the polarization process, where dipoles experience friction as they oscillate in a sinusoidal field, resulting in damping of vibrating dipole moments. Therefore, the polarization vector **P** will lag behind the applied electric field **E**; as a result χ_e becomes complex with a negative imaginary part, and in turn the permittivity of the medium becomes like that given in (1.27). Although the loss in a dielectric medium may be solely due to damping of vibrating dipole moments, it may be formulated as a conductor loss. If the conductivity or the resistivity of a nonperfect insulating material is known, the generalized Ampere's law can be written as

$$\nabla \times \mathbf{H} = j\omega\varepsilon\mathbf{E} + \sigma\mathbf{E} = j\omega\left(\varepsilon + \frac{\sigma}{j\omega}\right)\mathbf{E} \tag{1.28}$$

where Ohm's law is used to include the current induced in the material due to its finite conductivity. Hence, the permittivity of the material is written as a complex quantity with a negative imaginary part as

$$\hat{\varepsilon} = \varepsilon' - j\varepsilon'' - j\frac{\sigma}{\omega} = \varepsilon'\left(1 - j\frac{\sigma_{eff}}{\omega\varepsilon'}\right) \tag{1.29}$$

where the conductivity of the material appears in the imaginary part of the dielectric constant and accounts for the loss in the material. Note that different loss mechanisms in a dielectric material are mathematically modeled similarly by adding a negative imaginary part to the permittivity of the material. Therefore, by defining an effective conductivity for a material, all loss mechanisms can be combined into a single term:

$$\hat{\varepsilon} = \varepsilon'\left(1 - j\frac{\sigma_{eff}}{\omega\varepsilon'}\right) = \varepsilon_0\varepsilon_r\left(1 - j\tan\delta\right) \tag{1.30}$$

Note that $\tan\delta$ is frequency dependent according to this definition. For most substrate materials, $\sigma/\omega \ll \varepsilon''$ so that the loss tangent reduces to the well-known form of $\tan\delta = \varepsilon''/\varepsilon'$.

Magnetic Materials

After having studied the dielectric materials and incorporated their influences on the electric field into Maxwell's equation as the permittivity of the material, it is now time to discuss the effect of materials on magnetic field. While spinning and orbiting electrons can be considered as tiny currents on atomic scale, for macroscopic purposes, they can be treated as magnetic dipoles with random orientation when no magnetic field exists. With the applied magnetic field, magnetic dipole moments can be aligned to produce a magnetic polarization (equivalently called magnetization) $\mathbf{M}$. This magnetization changes the applied magnetic field, and in turn, with the change of the magnetic field in the material, the magnetization changes as well. This cyclic process goes on indefinitely until the magnetic field in the material reaches a steady state. The overall effect of the magnetization, in the steady state, on the magnetic field can be incorporated into the definition of the magnetic flux density as follows:

$$\mathbf{B} = \mu_0\left(\mathbf{H} + \mathbf{M}\right) \tag{1.31}$$

and through this definition, the magnetic properties of the material are incorporated into Maxwell's equations. In a vacuum, since there is no matter to magnetize, the magnetic flux density becomes linearly proportional to the magnetic field intensity with a constant of proportionality μ_0, which is the permeability of free space. For a linear magnetic material, the magnetization is linearly proportional to the magnetic field intensity as

$$\mathbf{M} = \chi_m \mathbf{H} \tag{1.32}$$

where χ_m is the magnetic susceptibility of the material. Hence, the magnetic flux density can be simplified to

$$\mathbf{B} = \mu_0 \left(1 + \chi_m\right)\mathbf{H} = \mu_0 \mu_r \mathbf{H} = \mu \mathbf{H} \tag{1.33}$$

where μ and μ_r are the permeability and relative permeability of the material, respectively. The loss mechanism is incorporated into the equations by allowing the permeability of the material to be complex, $\mu = \mu' - j\mu''$, whose imaginary part accounts for the loss.

To put all of these concepts into perspective, let us remember the goal of this section and what we have accomplished so far. The goal is to introduce two new equations that can account for the electrical properties of the materials involved, and that will help to solve for the field quantities uniquely together with already available and independent Maxwell's equations. This goal seems to have been accomplished with the relations between the displacement vector and the electric field (1.26), and between the magnetic flux density and the magnetic field intensity (1.33). However, because the materials were assumed to be linear in the derivations of both (1.26) and (1.33), and because not all materials are linear, these equations cannot be used for different material types. For the sake of completeness, we will briefly review the other types of materials without going into detail, by just providing the relations between **D** and **E**, and **B** and **H** in corresponding materials. It should be noted that these relations are usually called the constitutive relations or constitutive equations.

There are actually four basic categories that a material can be characterized by: (1) linear or nonlinear; (2) homogeneous or inhomogeneous; (3) isotropic or anisotropic; and (4) dispersive or nondispersive. A material or medium can be referred to as linear if the polarization and magnetization vectors are linear functions of the electric and magnetic fields, respectively. For a homogeneous medium, the electric and magnetic susceptibilities, which are the proportionality terms in the definitions of polarization and magnetization, are uniform (constant) throughout the medium. If they are varying in space, that is, functions of space coordinates, such media are called inhomogeneous. If the electric (magnetic) field induces polarization (magnetization) in a direction other than that of the electric (magnetic) field, then susceptibilities become different in different directions; such materials are commonly known as electric (magnetic) anisotropic media. Therefore, susceptibilities, and, in turn, permittivity and/or permeability, for such media can be written mathematically as tensors or matrixes rather than scalars. For a dispersive medium, the polarization and/or magnetization vectors are not dependent on the same instances of electric and/or magnetic fields, respectively. In other words, **D** and **B** not only depend on the present value of **E** and **H** but also on the time derivatives of all orders of **E** and **H**. To summarize, let us give the constitutive relations for some commonly used linear materials in microwave and antenna applications as a subset of the above mentioned characterizations:

$$\mathbf{D} = \bar{\varepsilon}(\mathbf{r}) \cdot \mathbf{E} \tag{1.34}$$

$$\mathbf{B} = \bar{\mu}(\mathbf{r}) \cdot \mathbf{H} \tag{1.35}$$

for an anisotropic and inhomogeneous medium, where $\bar{\varepsilon}(\mathbf{r})$ and $\bar{\mu}(\mathbf{r})$ are 3×3 matrixes, and their entries are functions of space coordinates.

$$\mathbf{D} = \bar{\varepsilon} \cdot \mathbf{E} \tag{1.36}$$

$$\mathbf{B} = \bar{\mu} \cdot \mathbf{H} \tag{1.37}$$

for an anisotropic and homogenous medium, where $\bar{\varepsilon}$ and $\bar{\mu}$ are 3×3 constant matrixes.

$$\mathbf{D} = \varepsilon \mathbf{E} \tag{1.38}$$

$$\mathbf{B} = \mu \mathbf{H} \tag{1.39}$$

for an isotropic and homogenous medium, where ε and μ are just constants. Note that in free space, $\mu = \mu_0$ and $\varepsilon = \varepsilon_0$, and these constants, in the international system of units (SI), are given as

$$\begin{aligned} \mu_0 &= 4\pi \times 10^{-7} \quad \text{Henry/m} \\ \varepsilon_0 &\approx 8.854187 \times 10^{-12} \quad \text{Farad/m} \\ &\approx (1/36\pi) \times 10^{-9} \text{ Farad/m} \end{aligned} \tag{1.40}$$

1.1.4 The Wave Equation

With the introduction of the constitutive relations, we now have enough independent equations to solve for the fields from Maxwell's equations. In a source-free, linear, isotropic, and homogeneous medium, Maxwell's curl equations in the frequency domain are written as

$$\nabla \times \mathbf{E} = -j\omega\mu \, \mathbf{H} \tag{1.41}$$

$$\nabla \times \mathbf{H} = j\omega\varepsilon \, \mathbf{E} \tag{1.42}$$

and can be mathematically described as first-order, coupled, partial differential equations with two unknowns. These equations can be easily cast into a form of a second-order, partial differential equation with one unknown by taking the curl of the one and substituting the other into the resulting expression, which is demonstrated using (1.41) as

$$\nabla \times \nabla \times \mathbf{E} = -j\omega\mu \nabla \times \mathbf{H} = \omega^2 \mu\varepsilon \mathbf{E}$$

$$\nabla(\nabla \cdot \mathbf{E}) - \nabla^2 \mathbf{E} = \omega^2 \mu\varepsilon \mathbf{E}$$

$$\nabla^2 \mathbf{E} + \omega^2 \mu\varepsilon \mathbf{E} = 0 \qquad (1.43)$$

where the vector identity $\nabla \times \nabla \times \mathbf{A} = \nabla(\nabla \cdot \mathbf{A}) - \nabla^2 \mathbf{A}$ and source-free condition $\nabla \cdot \mathbf{E} = 0$ are employed. Note that (1.43) is a second-order, partial differential equation with only one unknown, $\mathbf{E}$, and it is called the Helmholtz equation for the electric field. Similarly, if the same manipulations are performed on (1.42), the Helmholtz equation for the magnetic field is obtained as

$$\nabla^2 \mathbf{H} + \omega^2 \mu\varepsilon \mathbf{H} = 0 \qquad (1.44)$$

Note that the frequency-domain representations of Maxwell's equations have been used in the derivation of the above wave equations. If the time-domain representations are used, in source-free, homogeneous, isotropic media, and the same manipulations are performed, as was demonstrated in the equations leading to (1.43), the wave equations for the $\mathbf{E}$ and $\mathbf{H}$ fields in time-domain would be as follows:

$$\nabla^2 \tilde{\mathbf{E}} - \mu\varepsilon \frac{\partial^2 \tilde{\mathbf{E}}}{\partial t^2} = 0 \qquad (1.45)$$

$$\nabla^2 \tilde{\mathbf{H}} - \mu\varepsilon \frac{\partial^2 \tilde{\mathbf{H}}}{\partial t^2} = 0 \qquad (1.46)$$

This is consistent with the transformation between the time-domain and frequency-domain (phasor form) representations, as discussed in detail in Section 1.1.2, where $\partial^2/\partial t^2$ is replaced by $-\omega^2$. The mathematical form of these partial differential equations, (1.45) and (1.46) in the time domain, is named as the wave equation, and the reason for this name can only be understood when the solutions are obtained and interpreted.

To find the nature of the solutions of the wave equations, it is sufficient to begin with the simplest example, a one-dimensional solution of the time-domain wave equation in a source-free, homogeneous, and isotropic medium. By one-dimensional, we mean that the fields vary in one space coordinate (x, y, or z) and are uniform along the others. So, within this simplified setting, we can assume that $\tilde{\mathbf{E}}$ and $\tilde{\mathbf{H}}$ fields are functions of the space coordinate z only, being independent of x and y, and functions of time t. Subsequently, we have the following conditions on the derivatives of the field components:

$$\frac{\partial}{\partial x} = 0, \quad \frac{\partial}{\partial y} = 0, \quad \frac{\partial}{\partial z} \neq 0, \quad \frac{\partial}{\partial t} \neq 0$$

When these conditions are imposed on Maxwell's curl equations, together with the characteristics of the medium (lossless and source free, that is, $\sigma = 0$, $\rho = 0$, and $\tilde{\mathbf{J}} = 0$), the following six scalar differential equations are obtained:

$$\frac{\partial \tilde{E}_y}{\partial z} = \mu \frac{\partial \tilde{H}_x}{\partial t}; \quad \frac{\partial \tilde{E}_x}{\partial z} = -\mu \frac{\partial \tilde{H}_y}{\partial t}; \quad 0 = \frac{\partial \tilde{H}_z}{\partial t} \tag{1.47}$$

$$\frac{\partial \tilde{H}_x}{\partial z} = \varepsilon \frac{\partial \tilde{E}_y}{\partial t}; \quad \frac{\partial \tilde{H}_y}{\partial z} = -\varepsilon \frac{\partial \tilde{E}_x}{\partial t}; \quad 0 = \frac{\partial \tilde{E}_z}{\partial t} \tag{1.48}$$

It is obvious from the last equations in (1.47) and (1.48) that $\tilde{E}_z$ and $\tilde{H}_z$ must be equal to zero, except for dc solutions that are of no interest in the wave solutions. Therefore, we only have the x and y components of the fields, and they are functions of z and t only. With a little examination of (1.47) and (1.48), it is observed that the first equations and the second equations in (1.47) and (1.48) form coupled, first-order, partial differential equations for $\left(\tilde{E}_y, \tilde{H}_x \right)$ and $\left(\tilde{E}_x, \tilde{H}_y \right)$, respectively. As such, they become two independent sets of coupled, first-order, partial differential equations, which can be easily cast into uncoupled, second-order, partial differential equations, as demonstrated by the steps leading to (1.43). Hence, the following one-dimensional scalar wave equations are obtained:

$$\frac{\partial^2 \tilde{E}_y}{\partial z^2} - \mu\varepsilon \frac{\partial^2 \tilde{E}_y}{\partial t^2} = 0; \quad \frac{\partial^2 \tilde{H}_x}{\partial z^2} - \mu\varepsilon \frac{\partial^2 \tilde{H}_x}{\partial t^2} = 0 \tag{1.49}$$

from the first equations in (1.47) and (1.48), and

$$\frac{\partial^2 \tilde{E}_x}{\partial z^2} - \mu\varepsilon \frac{\partial^2 \tilde{E}_x}{\partial t^2} = 0; \quad \frac{\partial^2 \tilde{H}_y}{\partial z^2} - \mu\varepsilon \frac{\partial^2 \tilde{H}_y}{\partial t^2} = 0 \tag{1.50}$$

from the second equations in (1.47) and (1.48). Assuming that only one of these two sets of fields is excited by the source, for example $\left(\tilde{E}_x, \tilde{H}_y \right)$, it will be sufficient to deal with this set only. Since the goal of this section is to find the solutions of the wave equation and interpret them physically, assumption of all these simplifications will not change the physical nature of the solution, but rather circumvent the laborious part of the solution that might obscure the main purpose of this section.

For the assumed set of field components, $\left(\tilde{E}_x, \tilde{H}_y \right)$, one of the equations in (1.50) needs to be solved, and the other field component can be directly found from Maxwell's equations. So, let us try to find the solutions for

$$\frac{\partial^2 \tilde{E}_x}{\partial z^2} - \mu\varepsilon \frac{\partial^2 \tilde{E}_x}{\partial t^2} = 0 \tag{1.51}$$

Since the formal solution of this second-order, homogeneous partial differential equation (PDE) is out of the scope of this text, we will provide the solutions intuitively by investigating the properties of the differential equation. Since any solution of this differential equation must satisfy the differential equation itself, the second-order time derivative of the solution multiplied by the constant $\mu\varepsilon$ must be equal to its second-order derivative with respect to z. So, if the time dependence of the solution is assumed to be $f(t)$, then the functional form of the solutions will be $f\left(t \pm \sqrt{\mu\varepsilon}\,z\right)$, which can be easily verified by substituting it into (1.51). Since we have two independent solutions, the general solution can be written as their linear combinations as

$$\tilde{E}_x(z,t) = Af\left(t - \sqrt{\mu\varepsilon}\,z\right) + Bf\left(t + \sqrt{\mu\varepsilon}\,z\right) \tag{1.52}$$

Once the solutions are found mathematically, their physical interpretations need to be understood in order to appreciate Maxwell's equations and why we call the resulting second-order PDE a wave equation. Let us consider the first solution, $f\left(t - \sqrt{\mu\varepsilon}\,z\right)$, for which a representative view is provided in Figure 1.2 for three different time and space points.

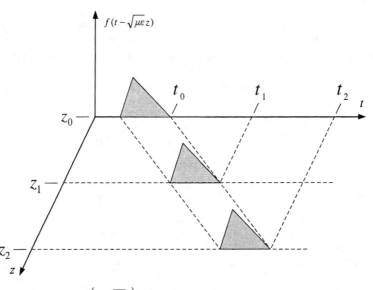

Figure 1.2 Interpretation of $f\left(t - \sqrt{\mu\varepsilon}\,z\right)$ as a wave propagating in $+z$-direction.

For the sake of argument, let's assume that the shape of the signal in the time domain is a triangular shape, as depicted in Figure 1.2. Note that any point on this waveform can be uniquely defined by providing both the time and space coordinates: for instance, the corner at the right-hand side of the waveform can be represented by the coordinate pairs (z_0, t_0), (z_1, t_1) and (z_2, t_2), and the function takes the same value at these three points; in other words, the arguments of the function at these points are the same (Figure 1.2). Since time proceeds independently, a fixed point on the waveform (equivalently, a fixed value of the argument) shifts in the $+z$ direction. To put it physically, signals whose mathematical representations are in the form of $f\left(t - \sqrt{\mu\varepsilon}\,z\right)$ propagate in the $+z$ direction as time proceeds. Therefore, functions with that given form of argument are referred to as wave functions, and the differential equations resulting in such solutions are called wave equations. Note that, using the same argument, the second solution of the wave equation, $f\left(t + \sqrt{\mu\varepsilon}\,z\right)$, represents a wave propagating in the $-z$ direction.

In this section so far, the solutions of Maxwell's equations have been obtained in a simple one-dimensional setting, and we have argued qualitatively that these solutions represent waves. So, with the knowledge of the mathematical representation of waves, the velocity of these waves can be obtained simply by tracing the fixed point of the argument of the wave function, as demonstrated below:

$$\text{Argument at } (z_0, t_0) = \text{Argument at } (z_1, t_1)$$

$$t_0 - \sqrt{\mu\varepsilon}\,z_0 = t_1 - \sqrt{\mu\varepsilon}\,z_1 \Rightarrow v = \frac{z_1 - z_0}{t_1 - t_2} = \frac{1}{\sqrt{\mu\varepsilon}} \tag{1.53}$$

which is called the phase velocity of the wave. In free space, where the permittivity and permeability are given in (1.40), the phase velocity of the wave becomes precisely the velocity of light.

Example 1.1

Find the solution of the wave equation in the frequency domain, and show that the functional form of the solutions is consistent with $f\left(t \pm \sqrt{\mu\varepsilon}\,z\right)$.

If we start with Maxwell's curl equations in the frequency domain and following the same steps leading to (1.51), the frequency-domain wave equation is obtained as

$$\frac{d^2 E_x}{dz^2} + \omega^2 \mu\varepsilon E_x = 0 \tag{1.54}$$

where E_x denotes the frequency-domain representation (phasor form, time-harmonic form) of its time-dependent (instantaneous) representation, and (1.54) is just the frequency-domain equivalent of (1.51). Since this is a simple, second-order, ordinary differential equation, its solutions can directly be written as

$$E_x(z) = Ae^{-j\omega\sqrt{\mu\varepsilon}\,z} + Be^{+j\omega\sqrt{\mu\varepsilon}\,z} \tag{1.55}$$

and the solution can be converted into the time domain via (1.12) as

$$\tilde{E}_x(z,t) = \operatorname{Re}\{E_x(z)e^{j\omega t}\}$$
$$= A\cos\left(\omega t - \omega\sqrt{\mu\varepsilon}\,z\right) + B\cos\left(\omega t + \omega\sqrt{\mu\varepsilon}\,z\right) \tag{1.56}$$

where the arguments of the solutions are similar to those of the time-domain solutions, except for a constant factor ω.

Example 1.2

Derive the wave equation in lossy media and discuss the contributions of the displacement current $\partial \tilde{\mathbf{D}}/\partial t$ ($j\omega \mathbf{D}$ in phasor form) and the conduction current $\mathbf{J}$.

Since loss in practical materials is inevitable, although it might be very small for some dielectrics, we need to understand how to account for the losses in a general medium and how losses appear in the wave equation. As we have seen in Section 1.1.3, the generalized Ampere's law in phasor form can be written in a general medium as

$$\nabla \times \mathbf{H} = j\omega \varepsilon \, \mathbf{E} + \sigma \, \mathbf{E}$$

where the first term on the right-hand side is the displacement current, while the second one is known as the conduction current. Remember that the conductivity σ that appears in the equation may well be the equivalent conductivity of the material, accounting for the damping of vibrating dipole moments, as well as the conductivity of the material. Combining this equation with the Faraday-Maxwell law, one can get the following wave equation for the electric field:

$$\nabla^2 \mathbf{E} + \omega^2 \mu \varepsilon \, \mathbf{E} - j\omega \mu \sigma \, \mathbf{E} = 0 \qquad (1.57)$$

or

$$\nabla^2 \mathbf{E} + \omega^2 \mu \hat{\varepsilon} \, \mathbf{E} = 0 \qquad (1.58)$$

where the dielectric constant of the medium is defined as a complex quantity, as discussed in Section 1.1.3:

$$\hat{\varepsilon} = \varepsilon' \left(1 - j \frac{\sigma_{eff}}{\omega \varepsilon'} \right) = \varepsilon_0 \varepsilon_r \left(1 - j \tan \delta \right)$$

Equation (1.58) implies that the solutions to the wave equation in a lossy medium are in the same functional form as those in a lossless medium, except the dielectric constant is a complex quantity in the lossy medium. If (1.57) is rearranged, we can have the displacement and conduction current explicitly in the expression as

$$\nabla^2 \mathbf{E} - j\omega \mu (j\omega \varepsilon \, \mathbf{E} + \sigma \mathbf{E}) = 0 \qquad (1.59)$$

This equation implies that if the displacement current dominates in a material ($|j\omega \varepsilon \, \mathbf{E}| \gg |\sigma \mathbf{E}|$), the material is dominantly dielectric, and the wave equation governs the behavior of the fields. However, if the conduction current dominates, the material is dominantly conductive, and hence (1.57) reduces to a diffusion equation — second derivative in space and first derivative in time. The implications of these two types of governing behavior will be clear in the discussion of the plane waves in Section 1.2.

Since the wave equations in time and frequency domains are second-order differential equations, the general solution given by (1.52) cannot be unique (A and B are unknown constants), unless some boundary and/or initial conditions are provided. Therefore, to find a unique solution, or a unique set of solutions, for the wave equation in a given geometrical setting and source conditions, one needs to know the behavior of the field components at all possible material interfaces, like dielectric-to-dielectric and dielectric-to-conductor, at infinity, and across the sources. So, in the following section, this issue will be addressed, and some conditions of the field components will be derived based on Maxwell's equations.

1.1.5 Boundary Conditions

First of all, let us emphasize that the fields and waves that are of interest in microwaves and antenna applications are always in macroscopic scales, not in microscopic scales. This point has been briefly mentioned in the topic of fields in material body, as the polarization and magnetization were defined as the average dipole moments per unit volume. Microscopic fields are given in a scale of atomic distances, while the macroscopic ones are defined as the average fields over larger regions, including many thousands of atoms. As can be easily visualized, the electric field inside matter from classical physics point of view behaves quite erratically on the microscopic level; it gets very large when observed close to electrons, and quite small and/or pointing in totally different directions when moved slightly. Therefore, throughout this book, all the quantities like fields, current and charge densities, and so forth, are macroscopic in scale.

For macroscopic quantities, the boundaries or interfaces are always considered to be geometrical surfaces. Since Maxwell's equations govern the macroscopic behavior of electric and magnetic fields everywhere, they must carry information on the behavior of the fields at geometrical surfaces. Therefore, the behavior of the field components at the boundary between two different materials, as shown in Figure 1.3, can be predicted by Maxwell's equations. Although Maxwell's equations can be written in differential and integral forms equivalently, the integral form must be employed in predicting the behavior of the fields at boundaries. This is because the differential form is valid pointwise in space, and naturally, they cannot relate fields in one layer to those in the adjacent layer. For the integral form of Maxwell's equations, one can easily choose contours and surfaces for the line and surface integrals, respectively, to cover both media, as demonstrated by the contour and the pill-box in Figure 1.3.

So, starting with Faraday's equations with a line integral over a contour C, the fields between the two media are related as follows:

$$\oint_C \mathbf{E} \cdot d\mathbf{l} = -j\omega \iint_S \mathbf{B} \cdot d\mathbf{s}$$

$$\mathbf{E}_1 \cdot \hat{\tau}\Delta l - \mathbf{E}_2 \cdot \hat{\tau}\Delta l = -j\omega \mathbf{B} \cdot \underbrace{\hat{n}_1 \Delta l\, \Delta h}_{ds}$$

where to write the left-hand side, it is assumed that Δh is already very small, and the normal component of the electric field is finite. Hence, the components of the line integrals along the perpendicular line segments of contour C are assumed to be negligible, with no loss of generality, because we are basically interested in the fields at the interface where $\Delta h \to 0$. So, for $\Delta h \to 0$, the right-hand side goes to zero as well, because **B** is the average magnetic flux density over the surface S and is always finite.

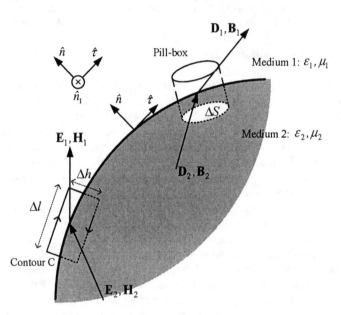

Figure 1.3 A typical interface between two different materials.

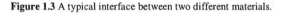

Hence, the tangential component of the electric field must be continuous across the boundary, as stated mathematically by

$$E_{1\tau} = E_{2\tau} \tag{1.60}$$

For the generalized Ampere's law, the fields between the two media for the same contour C can be related as

$$\oint_C \mathbf{H} \cdot d\mathbf{l} = j\omega \iint_S \mathbf{D} \cdot d\mathbf{s} + \iint_S \mathbf{J} \cdot d\mathbf{s}$$

using a similar argument for the evaluation of the line integral, the following equation is obtained:

$$\mathbf{H}_1 \cdot \hat{\tau}\Delta l - \mathbf{H}_2 \cdot \hat{\tau}\Delta l = j\omega \mathbf{D} \cdot \hat{n}_1 \Delta l\,\Delta h + \mathbf{J} \cdot \hat{n}_1 \Delta l\,\Delta h$$

For $\Delta h \to 0$, the first term on the right-hand side goes to zero as the average displacement vector **D** is always finite. However, for the second term, the volume current density $\mathbf{J}$ (A/m^2) may not be finite at the limit of $\Delta h \to 0$, in the case of a current sheet where the surface current density is defined as

$$\lim_{\Delta h \to 0} \mathbf{J} \cdot \hat{n}_1 \Delta h = \hat{n}_1 J_s \quad \text{A/m} \tag{1.61}$$

As a result, the tangential magnetic fields in both regions are related by

$$H_{1t} - H_{2t} = J_s \tag{1.62}$$

In other words, the equation states that the tangential component of the magnetic field intensity is discontinuous across the interface by the amount of the surface current density, if it exists.

Before going to the derivations of the other boundary conditions, it would be instructive to clarify the concept of surface current density. As mathematically defined in (1.61), the volume current density $\mathbf{J}$ must be approaching infinity in order to have a nonvanishing surface current. This is because the only way to have a product of two terms be finite when one of the terms goes to zero is to have the other term go to infinity. So, this leads us to a current sheet, where the current is thought to be distributed over a very narrow sheet of conductor. As it will be detailed in the next section, the current flow in a good conductor is practically confined to the layer next to the surface, whose thickness depends inversely on both the conductivity of the material and the frequency of the field. Therefore, for a high-frequency signal, the fields and currents tend to concentrate in an extremely thin layer at the surface of a good conductor, and the conductor is modeled as a sheet of current with a finite surface current density. This layer of finite thickness is called the skin depth of that specific conducting material.

Using the first two of Maxwell's equations, (1.13) and (1.14), at the interface along with the contour C, the boundary conditions on the tangential components of the electric and magnetic fields have been obtained. The remaining Maxwell's equations, (1.15) and (1.16), can be implemented at the same interface, now with the use of pill-box geometry covering both media (Figure 1.3). It is obvious that when there is a relation between the surface integral of a field quantity and a volume integral of a source, one needs to use a closed surface enclosing both media like a pill-box. So, evaluating Gauss's law

$$\oiint_S \mathbf{D} \cdot d\mathbf{s} = \iiint_V \rho \, dv \tag{1.63}$$

at the pill-box in Figure 1.3, the following relation is obtained:

$$\mathbf{D}_1 \cdot \hat{n} \, \Delta S - \mathbf{D}_2 \cdot \hat{n} \, \Delta S + \Psi_{sw} = \rho \, \Delta S \, \Delta h$$

where Ψ_{sw} denotes the outward electric flux through the side walls of the pill-box.

Since we are interested in the field at the interface (i.e., $\Delta h \to 0$), the surface area of the side walls goes to zero, resulting in a zero flux from the side walls with the assumption of finite $\mathbf{D}$. However, the term on the right-hand side may not go to zero if there are some surface charges right at the boundary. In mathematical terms, the surface charge density is defined by

$$\lim_{\Delta h \to 0} \rho \, \Delta h = \rho_s \quad \text{C/m}^2 \tag{1.64}$$

Hence, the evaluation of Gauss's law at the limit of $\Delta h \to 0$ becomes

$$D_{1n} - D_{2n} = \rho_s \tag{1.65}$$

When there is a surface charge density at the interface between two media, the normal component of the displacement vector is discontinuous by the amount of the surface charge density. Following the same procedure for the evaluation of the last Maxwell's equation as in the case of Gauss's law, the normal component of magnetic flux density is shown to be continuous across any boundary

$$B_{1n} = B_{2n} \tag{1.66}$$

This completes the derivation of boundary conditions. Let us summarize them here in scalar forms, as obtained above, as well as in vector forms:

Scalar form   Vector form

$$E_{1\tau} = E_{2\tau} \quad \hat{n} \times (\mathbf{E}_1 - \mathbf{E}_2) = 0$$

$$H_{1\tau} - H_{2\tau} = J_s \quad \hat{n} \times (\mathbf{H}_1 - \mathbf{H}_2) = \mathbf{J}_s$$

$$D_{1n} - D_{2n} = \rho_s \quad \hat{n} \cdot (\mathbf{D}_1 - \mathbf{D}_2) = \rho_s$$

$$B_{1n} = B_{2n} \quad \hat{n} \cdot (\mathbf{B}_1 - \mathbf{B}_2) = 0$$

The vector forms of the boundary conditions can directly be written by their scalar forms, except for the boundary condition on the tangential magnetic field intensity.

Example 1.3

Find the vector form of the boundary condition on the tangential magnetic field intensity for the geometry given in Figure 1.3.

Starting with the scalar representation $H_{1\tau} - H_{2\tau} = J_s$, the following vector equation can be written directly from Figure 1.3 as follows:

$$\hat{\tau} \cdot (\mathbf{H}_1 - \mathbf{H}_2) = \hat{n}_1 \cdot \mathbf{J}_s$$

$$\hat{\tau} \cdot (\mathbf{H}_1 - \mathbf{H}_2) = (\hat{n} \times \hat{\tau}) \cdot \mathbf{J}_s = \hat{\tau} \cdot (\mathbf{J}_s \times \hat{n})$$

where the vector identity $\mathbf{a} \cdot (\mathbf{b} \times \mathbf{c}) = \mathbf{b} \cdot (\mathbf{c} \times \mathbf{a}) = \mathbf{c} \cdot (\mathbf{a} \times \mathbf{b})$ is used. Since the orientation of the contour C, and in turn $\hat{\tau}$, is arbitrary, the terms in parentheses must be equal:

$$\mathbf{H}_1 - \mathbf{H}_2 = \mathbf{J}_s \times \hat{n}$$

Taking the cross-products of both sides from the left by $\hat{n}$ and using the vector identity $\mathbf{a} \times (\mathbf{b} \times \mathbf{c}) = \mathbf{b}(\mathbf{a} \cdot \mathbf{c}) - \mathbf{c}(\mathbf{a} \cdot \mathbf{b})$, we obtain

$$\hat{n} \times (\mathbf{H}_1 - \mathbf{H}_2) = \hat{n} \times (\mathbf{J}_s \times \hat{n}) = \mathbf{J}_s \underbrace{(\hat{n} \cdot \hat{n})}_{=1} - \hat{n} \underbrace{(\hat{n} \cdot \mathbf{J}_s)}_{=0}$$

Finally, the vector form of the boundary condition is obtained as follows:

$$\hat{n} \times \left(\mathbf{H}_1 - \mathbf{H}_2 \right) = \mathbf{J}_s$$

Note that the boundary conditions that we have derived and summarized in (1.60), (1.62), (1.65), and (1.66) are general; that is, they can be used at the boundaries of any material. It is well known that many microwave and antenna applications involve boundaries with good conductors, which are usually modeled as perfect conductors ($\sigma \to \infty$). Either considering as an exercise or as an important class of materials, boundary conditions at an interface between a dielectric material and perfect electrical conductor (PEC) are given as follows:

Scalar form	Vector form
$E_{1r} = 0$	$\hat{n} \times \mathbf{E}_1 = 0$
$H_{1r} = J_s$	$\hat{n} \times \mathbf{H}_1 = \mathbf{J}_s$
$D_{1n} = \rho_s$	$\hat{n} \cdot \mathbf{D}_1 = \rho_s$
$B_{1n} = 0$	$\hat{n} \cdot \mathbf{B}_1 = 0$

where medium 2 in Figure 1.3 is assumed to be a perfect conductor. The above boundary conditions are the result of the fact that fields in perfect conductors are zero. If it was not a perfect conductor, but rather a realistic good conductor with finite conductivity, then we would use the general boundary conditions together with the complex permittivity of the conducting medium.

With the derivation of the boundary conditions on the electric and magnetic fields, we are now equipped with all the necessary tools to be able to uniquely solve for the fields from governing differential equations (i.e., wave equations). To demonstrate the solution of the wave equations and the implementation of the boundary conditions, we provide the following example of an infinite current sheet, whose solutions are important in electromagnetic theory by their own right.

Example 1.4

Assume that an infinite sheet of electric surface current density $\mathbf{J}_s = \hat{x} J_0$ A/m is placed on a $z = 0$ plane between free space for $z < 0$ and a dielectric medium with $\varepsilon = \varepsilon_0 \varepsilon_r$ for $z > 0$, as shown in Figure 1.4. Find the resulting electric and magnetic fields in both regions.

Before getting started with the solution to this problem, let us understand the geometry and the given data first: (1) the surface current density is x-directed and uniform on the $z = 0$ plane (x-y plane); (2) the magnitude of the current density is J_0 with $e^{j\omega t}$ time dependence (i.e., it is written in phasor form in the figure); (3) we assume that layers are lossless, isotropic, homogenous, and semi-infinite in extent, and $\mu = \mu_0$ for both regions. Since the governing equation for the waves in this geometry is the wave equation, we need to solve it for the electric and magnetic fields.

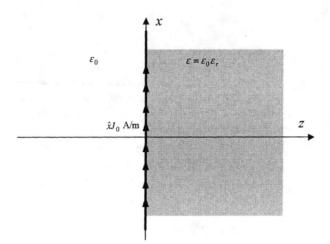

Figure 1.4 An infinite current sheet at the interface of two different media.

Perhaps the best way to solve such a differential equation in a piecewise homogeneous geometry follows the following steps: (1) find the source-free solutions with unknown coefficients in each homogeneous subregion; (2) apply the necessary boundary conditions at the interfaces to account for the boundaries between different media; and then (3) apply the boundary conditions at the sources to incorporate the influence of the sources into the solution. So, let us implement these steps one by one.

1. The frequency-domain wave equation in a source-free medium is written as follows:

$$\nabla^2 \mathbf{E} + \omega^2 \mu\varepsilon\mathbf{E} = 0$$

Then, from the source distribution and geometry, the components and the functional dependence of the fields, or better yet, the functional form of the solutions, can be predicted. For the geometry given in Figure 1.4, since the source and the boundary between the two media are uniform on the x-y plane, the solution of the differential equation must follow the same form; that is, the solution cannot be the functions of x and y. Hence, the possible electric and magnetic fields would have the components and the functional dependence of

$$\mathbf{E} = \hat{x}E_x(z) + \hat{y}E_y(z) + \hat{z}E_z(z)$$
$$\mathbf{H} = \hat{x}H_x(z) + \hat{y}H_y(z) + \hat{z}H_z(z)$$

based on the observation of the source and the boundary. However, these field components should satisfy Maxwell's equations as well, so using the fact that

$$\frac{\partial}{\partial x} = 0, \quad \frac{\partial}{\partial y} = 0, \quad \frac{\partial}{\partial z} \neq 0$$

Maxwell's curl equations result in

$$\nabla \times \mathbf{E} = -\hat{x}\frac{\partial}{\partial z}E_y + \hat{y}\frac{\partial}{\partial z}E_x = -j\omega\mu H_x\hat{x} - j\omega\mu H_y\hat{y}$$

$$\nabla \times \mathbf{H} = -\hat{x}\frac{\partial}{\partial z}H_y + \hat{y}\frac{\partial}{\partial z}H_x = j\omega\varepsilon E_x\hat{x} + j\omega\varepsilon E_y\hat{y}$$

As a result, we have obtained two decoupled sets of solutions, (E_y, H_x) and (E_x, H_y), as explained in Section 1.1.4. It should be stressed that these solutions are possible solutions; either only one of them or both with different weights can exist depending upon the boundary conditions. Since the surface current density of the source is in the x-direction, it gives rise to a discontinuity only on H_y, due to the boundary condition of the tangential magnetic field [see (1.62)]. Therefore, it can be interpreted that the source excites only one of the two possible solutions, (E_x, H_y), for which the governing equation is written as

$$\frac{d^2 E_x}{dz^2} + \underbrace{\omega^2 \mu\varepsilon}_{k^2} E_x = 0$$

where $\varepsilon = \varepsilon_0$ for $z < 0$, and $\varepsilon = \varepsilon_0\varepsilon_r$ for $z > 0$. So, the solutions for this ordinary, second-order, partial differential equation can be found as

$$E_x(z) = Ae^{-jk_0 z} + Be^{jk_0 z}, \quad \text{for } z < 0$$

$$E_x(z) = Ce^{-jkz} + De^{jkz}, \quad \text{for } z > 0$$

where A, B, C, and D are unknown coefficients to be determined by the boundary conditions. The constants $k_0 = \omega\sqrt{\mu\varepsilon_0}$ and $k = \omega\sqrt{\mu\varepsilon}$ are the wave numbers of the free space and the dielectric medium, respectively. The corresponding magnetic field H_y can easily be obtained from the Faraday-Maxwell equation, which reads as

$$\frac{\partial}{\partial z}E_x(z) = -j\omega\mu H_y(z)$$

for this specific case, and the magnetic fields in both regions are

$$H_y(z) = \frac{k_0}{\omega\mu}Ae^{-jk_0 z} - \frac{k_0}{\omega\mu}Be^{jk_0 z}, \quad \text{for } z < 0$$

$$H_y(z) = \frac{k}{\omega\mu}Ce^{-jkz} - \frac{k}{\omega\mu}De^{jkz}, \quad \text{for } z > 0$$

2. The first boundary condition, perhaps, is the fact that the source is at the $z = 0$ plane, and, therefore, the terms in the field expressions that represent $+z$-going waves for $z < 0$ and $-z$-going waves for $z > 0$ cannot exist. In other words, the

coefficients of those waves coming from $+\infty$ and $-\infty$, A and D, respectively, must be zero. Hence, the fields now would read as

$$E_x(z) = \begin{cases} B e^{jk_0 z}, & \text{for } z < 0 \\ C e^{-jkz}, & \text{for } z > 0 \end{cases}$$

$$H_y(z) = \begin{cases} -\dfrac{k_0}{\omega\mu} B e^{jk_0 z}, & \text{for } z < 0 \\ \dfrac{k}{\omega\mu} C e^{-jkz}, & \text{for } z > 0 \end{cases}$$

3. The remaining boundary conditions are at the interface between the regions and can be stated as

$$E_x(z=0^-) = E_x(z=0^+)$$
$$H_y(z=0^-) - H_y(z=0^+) = J_0$$

From the first condition, $B = C$ is found, and from the second, B is obtained as

$$B = -\frac{\omega\mu J_0}{(k_0 + k)}$$

Hence, the fields can now be written uniquely everywhere in the geometry by substituting B into the field expressions.

1.1.6 Energy Flow and the Poynting Vector

The general law of the conservation of energy states that if an object radiates electromagnetic waves (for example, light), it loses energy. Although this is an important and well-known law, it is not enough to understand the details of the mechanisms of the conservation of energy. In this section, the conservation of energy in electromagnetic fields will be detailed with a view of understanding the mechanisms how to conserve the energy, by following a similar approach to that described in *The Feynman Lectures on Physics* [7].

As the first step towards explaining the mechanism of energy conservation, we need to describe quantitatively the amount of energy in a volume element of space and the rate of energy flow. If we let a scalar function u represent the energy density in the field (the amount of energy per unit volume in space) and a vector function $\tilde{\mathbf{P}}$ represent the energy flux density of the field (the flow of energy per unit time across a unit area perpendicular to the flow), we can write the energy conservation quantitatively as

$$-\underbrace{\frac{\partial}{\partial t}\int_V u\, dV}_{\substack{\text{Decrease in the total} \\ \text{field energy}}} = \underbrace{\oint_S \tilde{\mathbf{P}} \cdot \hat{n}\, ds}_{\substack{\text{Energy flow out} \\ \text{of the volume}}} + \begin{pmatrix} +\text{Energy loss in the matter} \\ -\text{Energy gain in the matter} \end{pmatrix} \qquad (1.67)$$

where the minus sign before the time derivative represents the decrease in the following quantity, S is the surface enclosing the volume of interest, and $\hat{n}$ is the unit normal vector of the surface S. The loss or gain of the energy in the matter is actually the work being done by the field on matter or by the matter on the field, respectively. To quantify the last item in the equation (i.e., the work being done by the field on matter), we should first write the force of the fields on a charged particle as

$$\tilde{\mathbf{F}} = q\left(\tilde{\mathbf{E}} + \mathbf{v} \times \tilde{\mathbf{B}}\right) \tag{1.68}$$

Hence, the loss of energy (the rate of doing work) simply becomes $\tilde{\mathbf{F}} \cdot \mathbf{v} = q\tilde{\mathbf{E}} \cdot \mathbf{v}$. If there are N particles in the matter per unit volume, the loss of energy per unit volume can be written as $Nq\tilde{\mathbf{E}} \cdot \mathbf{v} = \tilde{\mathbf{E}} \cdot \tilde{\mathbf{J}}$, where $\tilde{\mathbf{J}}$ is the volume current density in the matter. As a result, the conservation of energy in the field can be mathematically stated as

$$-\frac{\partial}{\partial t}\int_V u\,dV = \oint_S \tilde{\mathbf{P}} \cdot \hat{n}\,ds + \int_V \tilde{\mathbf{E}} \cdot \tilde{\mathbf{J}}\,dV \tag{1.69}$$

or in differential form, as

$$-\frac{\partial u}{\partial t} = \nabla \cdot \tilde{\mathbf{P}} + \tilde{\mathbf{E}} \cdot \tilde{\mathbf{J}} \tag{1.70}$$

Now we need to find out how the energy density u and the flux density $\tilde{\mathbf{P}}$ in the formula of the conservation of energy for the electromagnetic fields (1.70) can be written in terms of the field quantities. To do that, let us rewrite the generalized Ampere's law (1.6) as

$$\tilde{\mathbf{J}} = \nabla \times \tilde{\mathbf{H}} - \varepsilon\frac{\partial \tilde{\mathbf{E}}}{\partial t}$$

and take the dot product with the electric field resulting in

$$\tilde{\mathbf{E}} \cdot \tilde{\mathbf{J}} = \tilde{\mathbf{E}} \cdot \nabla \times \tilde{\mathbf{H}} - \varepsilon\tilde{\mathbf{E}} \cdot \frac{\partial \tilde{\mathbf{E}}}{\partial t}$$

or

$$\tilde{\mathbf{E}} \cdot \tilde{\mathbf{J}} = \tilde{\mathbf{E}} \cdot \nabla \times \tilde{\mathbf{H}} - \frac{\partial}{\partial t}\left(\frac{1}{2}\varepsilon\tilde{\mathbf{E}} \cdot \tilde{\mathbf{E}}\right) \tag{1.71}$$

Comparing this equation with (1.70), it is observed that the first term on the right-hand side needs to be written as the divergence of a vector quantity. With the use of the vector identity $\nabla \cdot (\mathbf{A} \times \mathbf{B}) = \mathbf{B} \cdot (\nabla \times \mathbf{A}) - \mathbf{A} \cdot (\nabla \times \mathbf{B})$, (1.71) can be cast into the following form:

$$\tilde{\mathbf{E}} \cdot \tilde{\mathbf{J}} = -\nabla \cdot \left(\tilde{\mathbf{E}} \times \tilde{\mathbf{H}}\right) + \tilde{\mathbf{H}} \cdot \left(\nabla \times \tilde{\mathbf{E}}\right) - \frac{\partial}{\partial t}\left(\frac{1}{2}\varepsilon\tilde{\mathbf{E}} \cdot \tilde{\mathbf{E}}\right)$$

By substituting Faraday-Maxwell law (1.5), this equation is further simplified as

$$\tilde{\mathbf{E}} \cdot \tilde{\mathbf{J}} = -\nabla \cdot \underbrace{\left(\tilde{\mathbf{E}} \times \tilde{\mathbf{H}} \right)}_{\tilde{\mathbf{P}}} - \frac{\partial}{\partial t} \underbrace{\left(\frac{1}{2} \varepsilon \tilde{\mathbf{E}} \cdot \tilde{\mathbf{E}} + \frac{1}{2} \mu \tilde{\mathbf{H}} \cdot \tilde{\mathbf{H}} \right)}_{u} \qquad (1.72)$$

Now, this expression is exactly in the same form as (1.70), and the corresponding energy density u and energy flux density $\tilde{\mathbf{P}}$ are shown in the expression. Note that the energy density has turned out to be the sum of the electric and magnetic energy densities, whose forms are exactly the same as those found in static fields. Most importantly, we now have an equation for the energy flux density, $\tilde{\mathbf{P}} = \tilde{\mathbf{E}} \times \tilde{\mathbf{H}}$ [W/m^2], which is called Poynting's vector, as Poynting first derived this expression in 1884. Finally, the differential form of the conservation of energy in electromagnetic fields can be converted back to the integral form as

$$\int_V \tilde{\mathbf{E}} \cdot \tilde{\mathbf{J}} = -\oint_S \tilde{\mathbf{E}} \times \tilde{\mathbf{H}} \cdot ds - \frac{\partial}{\partial t} \int_V \left(\frac{1}{2} \varepsilon \tilde{\mathbf{E}} \cdot \tilde{\mathbf{E}} + \frac{1}{2} \mu \tilde{\mathbf{H}} \cdot \tilde{\mathbf{H}} \right) dV \qquad (1.73)$$

Remember that the term on the left-hand side was written as the loss of energy in matter inside the volume. However, if we have a source inside the volume and no loss in the volume, then the term on the left-hand side will be negative. Perhaps, the physical meaning of this expression can be better understood if we distinguish these two cases as follows:

No source exists in the volume of interest:

$$\underbrace{-\oint_S \tilde{\mathbf{E}} \times \tilde{\mathbf{H}} \cdot ds}_{\substack{\text{Inflow of energy} \\ \text{due to source} \\ \text{outside} V}} = \underbrace{\int_V \tilde{\mathbf{E}} \cdot \tilde{\mathbf{J}}}_{\substack{\text{Energy} \\ \text{dissipated} \\ \text{in volume} V}} + \underbrace{\frac{\partial}{\partial t} \int_V \left(\frac{1}{2} \varepsilon \tilde{\mathbf{E}} \cdot \tilde{\mathbf{E}} + \frac{1}{2} \mu \tilde{\mathbf{H}} \cdot \tilde{\mathbf{H}} \right) dV}_{\text{Rate of increase in stored energy}}$$

Note that energy is carried into the volume by the electromagnetic energy flux density in this case.

Source exists in the volume of interest:

$$\underbrace{-\int_V \tilde{\mathbf{E}} \cdot \tilde{\mathbf{J}}}_{\substack{\text{Source} \\ \text{contribution}}} = \underbrace{\oint_S \tilde{\mathbf{E}} \times \tilde{\mathbf{H}} \cdot ds}_{\substack{\text{Outflow of energy} \\ \text{from volume} V}} + \underbrace{\frac{\partial}{\partial t} \int_V \left(\frac{1}{2} \varepsilon \tilde{\mathbf{E}} \cdot \tilde{\mathbf{E}} + \frac{1}{2} \mu \tilde{\mathbf{H}} \cdot \tilde{\mathbf{H}} \right) dV}_{\text{Rate of increase in stored energy}}$$

So far in this section, we have only worked with arbitrarily time-dependent fields, and the terms in the mathematical expressions of the conservation of energy are time dependent (i.e., instantaneous). However, for most practical applications, the time-averaged energy and power quantities are required because most systems usually respond to the average power, rather than its instantaneous values. Perhaps this could be better explained by noting the fact that almost all physical systems have finite response times to the change of the input signal. In other words, such systems have low-pass characteristics, and hence take the average of the input signal. For periodic input signals, average quantities are defined as their instantaneous values averaged over a period. For example, using the instantaneous Poynting's vector as defined by

$$\tilde{\mathbf{P}}(\mathbf{r},t) = \tilde{\mathbf{E}}(\mathbf{r},t) \times \tilde{\mathbf{H}}(\mathbf{r},t) \tag{1.74}$$

the time-averaged Poynting's vector can be obtained from

$$\mathbf{P}(\mathbf{r}) = \frac{1}{T} \int_0^T \tilde{\mathbf{P}}(\mathbf{r},t) dt \tag{1.75}$$

where $T = 1/f$ is the period of the input signal. For time-harmonic fields, the instantaneous fields are written as

$$\tilde{\mathbf{E}}(\mathbf{r},t) = \text{Re}\{\mathbf{E}(\mathbf{r})e^{j\omega t}\} = \frac{1}{2}[\mathbf{E}(\mathbf{r})e^{j\omega t} + \mathbf{E}^*(\mathbf{r})e^{-j\omega t}]$$

$$\tilde{\mathbf{H}}(\mathbf{r},t) = \text{Re}\{\mathbf{H}(\mathbf{r})e^{j\omega t}\} = \frac{1}{2}[\mathbf{H}(\mathbf{r})e^{j\omega t} + \mathbf{H}^*(\mathbf{r})e^{-j\omega t}]$$

where we have used the fact that the sum of a complex number with its complex conjugate equals twice the real part of the complex number. Then, substituting these field representations into (1.75)

$$\mathbf{P}(\mathbf{r}) = \frac{1}{2\pi}\frac{1}{4}\left\{ (\mathbf{E}\times\mathbf{H})\underbrace{\int_0^{2\pi} e^{j2\omega t} dt}_{=0} + (\mathbf{E}^*\times\mathbf{H}^*)\underbrace{\int_0^{2\pi} e^{-j2\omega t} dt}_{=0} + \underbrace{[\mathbf{E}\times\mathbf{H}^* + \mathbf{E}^*\times\mathbf{H}]}_{2\,\text{Re}\{\mathbf{E}\times\mathbf{H}^*\}}\int_0^{2\pi} dt \right\}$$

results in the time-averaged Poynting vector as

$$\mathbf{P} = \frac{1}{2}\text{Re}\{\mathbf{E}\times\mathbf{H}^*\} \quad \text{W/m}^2 \tag{1.76}$$

Note that this expression gives us the average power flow per unit area associated with an electromagnetic wave and is valid for a sinusoidal input signal in steady state.

1.2 PLANE ELECTROMAGNETIC WAVES

Notice that in all the examples so far, we have only considered waves in the form of planes, that is, the field components of the wave are assumed to be traveling in the z direction and to have no x and y dependence. If one tries to visualize such a wave, it would look like a uniform plane of electric and magnetic fields on the x-y plane moving in the z direction, and hence such waves are called uniform plane waves. It could be easily inferred that the source of a plane wave is supposed to be uniform over an infinite plane in order to generate uniform fields over a plane parallel to the source plane. As a result, there is no actual uniform plane wave in nature. However, if one observes an incoming wave far away from a finite extent source, the constant phase surface of the fields (wavefront) becomes almost spherical. Hence, the wave looks like a uniform plane wave over a small area of a gigantic sphere of wavefront, where the observer is actually located. Perhaps it is more important for advanced studies of electromagnetic (EM) waves that an arbitrary field or wave can always be expressed in terms of plane waves, known as the plane wave spectrum representation of electromagnetic waves. Therefore, they are the building blocks of more complex waves. As a result, since plane waves play very important roles in the fields of microwaves, antenna, and propagation, they are studied briefly in the following sections to point out their salient features.

Before getting into the detailed study of general plane waves, let us review a special case, which was used to describe the wave nature of the solution of a wave equation in Section 1.1.4 and was also used to demonstrate the use of the boundary conditions in Example 1.3. In both cases, the plane waves were assumed to propagate in the z direction with uniform amplitudes of electric and magnetic fields in the plane perpendicular to the direction of propagation, the x-y plane, and the fields in time-harmonic form were written as

$$E_x = E_0 e^{-jkz}$$

$$H_y = H_0 e^{-jkz}$$

where E_0 and H_0 are constant amplitudes. So, it is rather easy to visualize these waves as they coincide with the planes of the Cartesian coordinate system. In Example 1.3, it has been demonstrated that an infinite plane of current sheet generates such waves and that resulting electric and magnetic fields are orthogonal to each other as well as to the direction of propagation. To see if this observation is true for general plane waves, rather than those conforming to the plane of a coordinate system, we need to study plane waves propagating in arbitrary directions, as shown in Figure 1.5.

1.2.1 General Uniform Plane Waves

For the sake of illustration, let us consider a uniform plane wave propagating in an arbitrary direction in a source-free, homogenous, isotropic, and lossless medium, as depicted in Figure 1.5. For such a uniform plane wave propagating in a direction $\mathbf{k}$ ($= k_x \hat{x} + k_y \hat{y} + k_z \hat{z} = k\hat{n}$), the electric and magnetic fields in time-harmonic form can be written as

$$\mathbf{E} = \mathbf{E}_0 e^{-j(k_x x + k_y y + k_z z)} = \mathbf{E}_0 e^{-j\mathbf{k}\cdot\mathbf{r}} \qquad (1.77)$$

$$\mathbf{H} = \mathbf{H}_0 e^{-j(k_x x + k_y y + k_z z)} = \mathbf{H}_0 e^{-j\mathbf{k}\cdot\mathbf{r}} \qquad (1.78)$$

where $\mathbf{E}_0$ and $\mathbf{H}_0$ are, in general, complex constant vectors of the electric and magnetic fields, respectively, $\mathbf{r}$ is the position vector (or radius vector), and $\mathbf{k}$ is a real propagation vector for the lossless medium. Note that the unit vector $\hat{n}$ is used to define the direction of propagation while k gives the propagation constant. Since the exponents are purely imaginary (for real $\mathbf{k}$ vector), they represent the phase progress of the fields (considering $e^{j\omega t}$ time dependence), and they are constant for every $\mathbf{r}$ on a plane perpendicular to the direction of propagation $\mathbf{k}$, as shown in Figure 1.5. In other words, constant-phase points constitute a plane perpendicular to $\mathbf{k}$ vector, representing a constant-phase plane propagating in the $\mathbf{k}$ direction. Therefore, this constant-phase plane is also called the phase front or wavefront of the plane wave.

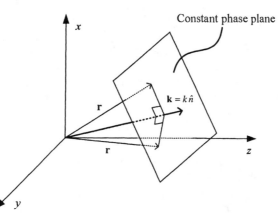

Figure 1.5 A uniform plane wave propagating in $\mathbf{k}$ direction.

So far, we have only written the mathematical forms of the electric and magnetic fields of a uniform plane wave. Now, considering that these fields have to satisfy Maxwell's equations, we may find some relations connecting the amplitudes and propagation vectors of such fields.

Implementing Gauss's law, (1.19), in a source-free, isotropic, and homogenous region (where $\mathbf{D}$ is a constant times $\mathbf{E}$) for the given electric field of a uniform plane wave, (1.77), as

$$\nabla \cdot \mathbf{D} = \varepsilon \nabla \cdot \mathbf{E} = 0 \Rightarrow \nabla \cdot \mathbf{E}_0 e^{-j\mathbf{k} \cdot \mathbf{r}} = 0$$

$$e^{-j\mathbf{k} \cdot \mathbf{r}} \underbrace{\nabla \cdot \mathbf{E}_0}_{=0} + \underbrace{\nabla e^{-j\mathbf{k} \cdot \mathbf{r}}}_{-jk\,e^{-j\mathbf{k} \cdot \mathbf{r}}} \cdot \mathbf{E}_0 = 0$$

one can get the following relation:

$$\mathbf{k} \cdot \mathbf{E}_0 = 0 \tag{1.79}$$

Note that the vector identity $\nabla \cdot f\mathbf{A} = f\nabla \cdot \mathbf{A} + \nabla f \cdot \mathbf{A}$ is used in this derivation. Additional relation can be obtained by implementing the Faraday-Maxwell law, (1.17), as follows:

$$\nabla \times \mathbf{E} = -j\omega\mu\,\mathbf{H}$$

$$\nabla \times \mathbf{E}_0 e^{-j\mathbf{k}\cdot\mathbf{r}} = e^{-j\mathbf{k}\cdot\mathbf{r}} \underbrace{\nabla \times \mathbf{E}_0}_{=0} + \nabla e^{-j\mathbf{k}\cdot\mathbf{r}} \times \mathbf{E}_0$$

$$- jk e^{-j\mathbf{k}\cdot\mathbf{r}} \times \mathbf{E}_0 = -j\omega\mu\,\mathbf{H}_0 e^{-j\mathbf{k}\cdot\mathbf{r}}$$

$$\mathbf{H}_0 = \frac{k}{\omega\mu}\hat{n} \times \mathbf{E}_0 = \frac{1}{\eta}\hat{n} \times \mathbf{E}_0 \qquad (1.80)$$

where $k = \omega\sqrt{\mu\varepsilon}$ and $\eta = \sqrt{\mu/\varepsilon}$ are real constants for lossless media, named as the wave number and the intrinsic impedance, respectively. Now, examining the equations (1.79) and (1.80) reveals that the propagation vector and the electric and magnetic field vectors are all orthogonal to each other. In other words, electric and magnetic field vectors are on the plane perpendicular to the propagation vector.

Note that the above derivation assumes the form of the electric and magnetic fields as written in (1.77) and (1.78), which were written rather intuitively, as the generalization of the special form of the plane waves we have studied earlier. The following example will demonstrate that this form actually represents physical fields.

Example 1.5

Show that the electric field of a uniform plane wave can be written as in (1.77) by solving the wave equation in Cartesian coordinates in free space via the method of separation of variables. This method is quite useful for the solution of the wave equation in many geometries in microwave and antenna engineering.

Since any physical electromagnetic field has to satisfy Maxwell's equations, it must be the solution of the wave equation. Therefore, the electric field representing a general plane wave, (1.77), can be obtained by the solution of the wave equation for the electric field (1.43). In free space, the wave equation for the electric field is written as

$$\nabla^2\mathbf{E} + k_0^2\mathbf{E} = 0$$

where $k_0 = \omega\sqrt{\mu_0\varepsilon_0}$ is the wave number of the free space, and every Cartesian component of the field must satisfy this wave equation. Since we are solving this differential equation in free space with no boundary conditions, the solution will not be unique and will provide only the functional form of the field. Moreover, there will be no difference in the functional form for different components of the field. Therefore, solving for just one component will be enough to find the rest of the components of the field. So, starting with the x-component, the wave equation can be written explicitly as

$$\left(\frac{\partial^2}{\partial x^2} + \frac{\partial^2}{\partial y^2} + \frac{\partial^2}{\partial z^2} + k_0^2\right)E_x(x, y, z) = 0 \qquad (1.81)$$

The method of separation of variables assumes that the unknown function, E_x in this case, can be written as a product of three functions, each of which is a function of a single variable. Let us point out here that this approach is only valid for geometries whose boundaries form planes in the chosen coordinate system. So, the unknown function E_x can be written as

$$E_x(x,y,z) = f(x)g(y)h(z) \tag{1.82}$$

By substituting (1.82) into (1.81) and dividing each term by fgh, the wave equation in (1.81) can be simplified to

$$\underbrace{\frac{1}{f(x)}\frac{d^2}{dx^2}f(x)}_{-k_x^2} + \underbrace{\frac{1}{g(y)}\frac{d^2}{dy^2}g(y)}_{-k_y^2} + \underbrace{\frac{1}{h(z)}\frac{d^2}{dz^2}f(z)}_{-k_z^2} + k_0^2 = 0$$

Since each term in this differential equation is a function of only one variable, each must be equal to a constant in order to satisfy the differential equation for all x, y, and z, that is $-k_x^2 - k_y^2 - k_z^2 + k_0^2 = 0$. As a result, the original second-order partial differential equation reduces to three similar second-order, ordinary differential equations, and their solutions can be directly written as

$$f(x) = A^+ e^{-jk_x x} + A^- e^{jk_x x}$$

$$g(y) = B^+ e^{-jk_y y} + B^- e^{jk_y y}$$

$$h(z) = C^+ e^{-jk_z z} + C^- e^{jk_z z}$$

For the sake of simplicity, let us assume that the wave propagates in the direction of positive x-, y-, and z-axes (i.e., in the first quadrant of the Cartesian coordinate system). Hence, the electric field E_x is obtained as

$$E_x = E_{0x} e^{-j(k_x x + k_y y + k_z z)}$$

So, since the differential equations are the same for the other vector components of the electric field, we can directly write them as

$$E_y = E_{0y} e^{-j(k_x x + k_y y + k_z z)}$$

$$E_z = E_{0z} e^{-j(k_x x + k_y y + k_z z)}$$

of course, with different constant magnitudes. When combined, the electric field can be written as

$$\mathbf{E} = \underbrace{(E_{0x}\hat{x} + E_{0y}\hat{y} + E_{0z}\hat{z})}_{\mathbf{E}_0} e^{-j\overbrace{(k_x x + k_y y + k_z z)}^{\mathbf{k}\cdot\mathbf{r}}} = \mathbf{E}_0 e^{-j\mathbf{k}\cdot\mathbf{r}}$$

which is the same expression as we have used for the fields of a plane wave propagating in a direction $\mathbf{k}$.

Throughout this section, we have discussed the plane waves in lossless media: their definitions, mathematical forms, and relations between the field components and the direction of propagation. However, to study plane waves in real media, we need to incorporate loss mechanisms into the formulations. As we have studied in Sections 1.1.3 and 1.1.4, the electrical properties of a material can be incorporated into the wave picture by providing the constituents of the complex dielectric constant of the material. In other words, all the loss mechanisms involved in a general medium can be combined into one term, the conductivity of the material. Therefore, with the use of complex dielectric constants, we will consider plane-wave solutions for lossy media in the following section, covering the plane waves in good and perfect conductors as well as poor and perfect insulators.

1.2.2 Plane Waves in Lossy Media

As we have shown in Section 1.1.4, the homogeneous wave equation in source-free lossy media is in the same form as the one in lossless media. For the completeness of the discussion, let us give the wave equation and the permittivity in a lossy medium here one more time:

$$\nabla^2 \mathbf{E} + \omega^2 \mu \hat{\varepsilon} \, \mathbf{E} = 0$$

$$\hat{\varepsilon} = \varepsilon' \left(1 - j \frac{\sigma_{eff}}{\omega \varepsilon'} \right) = \varepsilon_0 \varepsilon_r \left(1 - j \tan \delta \right)$$

where ε' $(= \varepsilon_0 \varepsilon_r)$ is the real part of the complex dielectric constant of the medium and σ_{eff} is the equivalent conductivity of the medium as described in Section 1.1.3. Hence, the wave number of a lossy medium becomes a complex number, for which a new symbol is introduced as follows:

$$\gamma = j\omega\sqrt{\mu\varepsilon} = j\omega \sqrt{\mu_0 \varepsilon_0 \varepsilon_r \left(1 - j \frac{\sigma_{eff}}{\omega \varepsilon_0 \varepsilon_r} \right)} = \alpha + j\beta \qquad (1.83)$$

where α and β are attenuation and propagation constants, respectively, $\mu = \mu_0$ and $\gamma^2 = -k^2 = -\omega^2 \mu\varepsilon$ are used. As a result, the wave equation would read

$$\nabla^2 \mathbf{E} - \gamma^2 \, \mathbf{E} = 0 \qquad (1.84)$$

and its solutions will be in the same form as those obtained for lossless cases, provided jk is replaced by γ. Therefore, electric and magnetic fields for general plane waves in lossy media can be written as

$$\mathbf{E} = \mathbf{E}_0 e^{-\gamma \hat{n} \cdot \mathbf{r}} \qquad (1.85)$$

$$\mathbf{H} = \mathbf{H}_0 e^{-\gamma \hat{n} \cdot \mathbf{r}} = \frac{1}{\eta} \hat{n} \times \mathbf{E}_0 e^{-\gamma \hat{n} \cdot \mathbf{r}} \qquad (1.86)$$

which are obtained directly from (1.77) and (1.78) with $j\mathbf{k} = \gamma\hat{n}$ substitution. Since the fields of uniform plane waves have been obtained in lossy media, now it is time to categorize materials of interest for microwave and antenna engineering and to study their influences on plane waves explicitly.

Plane Waves in Insulating or Dielectric Media

Note that this is the case of the propagation of plane waves in a vacuum, air, or any other dielectric material with no loss. The field expressions are the same as the ones obtained in the previous section for the lossless case, (1.77) and (1.78). The same field expressions can also be obtained from the fields in lossy media, (1.85) and (1.86), by just using the lossless limit of the complex propagation constant. Hence, the constants used to define the fields can be written as

$$\alpha = 0$$

$$\beta = k = \omega\sqrt{\mu_0\varepsilon'} = \omega\sqrt{\mu_0\varepsilon_0\varepsilon_r} = k_0\sqrt{\varepsilon_r} = \frac{2\pi}{\lambda_0/\sqrt{\varepsilon_r}} = \frac{2\pi}{\lambda} \tag{1.87}$$

$$\eta = \sqrt{\frac{\mu_0}{\varepsilon'}} = \sqrt{\frac{\mu_0}{\varepsilon_0\varepsilon_r}} = \frac{\eta_0}{\sqrt{\varepsilon_r}} \cong \frac{120\pi}{\sqrt{\varepsilon_r}} \tag{1.88}$$

where it was assumed that $\sigma = 0$, $\mu = \mu_0$ except for magnetic materials, λ_0 and λ are the wavelengths in free space and in the material, respectively, and, likewise η_0 and η are the wave impedances of free space and the medium, respectively.

Plane Waves in Dielectrics with Small Loss

This is a more practical situation than the completely lossless case, as almost all realistic materials demonstrate some amount of loss. For such materials, the displacement current still dominates, $\sigma/\omega\varepsilon' \ll 1$, but the complex propagation constant can be approximately written as

$$\gamma = \underbrace{j\omega\sqrt{\mu_0\varepsilon_0\varepsilon_r}}_{k_0\sqrt{\varepsilon_r}}\sqrt{1 - j\frac{\sigma}{\omega\varepsilon_0\varepsilon_r}} = \underbrace{\frac{\sigma}{2}\sqrt{\frac{\mu_0}{\varepsilon_0\varepsilon_r}}}_{\alpha} + \underbrace{jk_0\sqrt{\varepsilon_r}}_{\beta} \tag{1.89}$$

where the binomial approximation, $(1 \pm x)^{1/2} \cong (1 \pm x/2)$ for $x \ll 1$, was employed. Since the attenuation constant α is nonzero, the fields experience small exponential attenuation while propagating. The intrinsic impedance of the medium can also be approximated as

$$\eta = \sqrt{\frac{\mu_0}{\varepsilon'\left(1 - j\dfrac{\sigma}{\omega\varepsilon'}\right)}} = \sqrt{\frac{\mu_0}{\varepsilon_0\varepsilon_r}}\left(1 - j\frac{\sigma}{\omega\varepsilon_0\varepsilon_r}\right)^{-1/2} \cong \frac{\eta_0}{\sqrt{\varepsilon_r}}\left(1 + j\frac{\sigma}{2\omega\varepsilon_0\varepsilon_r}\right) \tag{1.90}$$

Since the additional reactive component due to the loss in the medium is extremely small, it can be completely ignored for most practical purposes and low-loss materials, resulting in $\eta = \eta_0/\sqrt{\varepsilon_r}$. Once these constants have been defined, then the field components can be written directly from (1.85) and (1.86).

Plane Waves in Conducting Media

In the application of microwave circuits and antennas, conductors are the indispensable parts of the system. Therefore, such systems involve loss or attenuation due to the finite conductivities of good conductors (not perfect), for which the conduction current dominates the displacement current, that is,

$$\frac{|\mathbf{J}|}{|\partial \mathbf{D}/\partial t|} = \frac{\sigma}{\omega \varepsilon'} >> 1 \tag{1.91}$$

So, ignoring the displacement current completely results in the following complex propagation constant:

$$\gamma = \alpha + j\beta \cong jk_0\sqrt{\varepsilon_r}\left(-j\frac{\sigma}{\omega\varepsilon_0\varepsilon_r}\right)^{1/2} = (1+j)\sqrt{\frac{\omega\mu\sigma}{2}} \tag{1.92}$$

where only the imaginary part is retained in the complex dielectric constant. Note that the propagation and attenuation constants have the same magnitude in good conductors. Since the fields now rapidly attenuate inside the conducting media, the distance at which the amplitudes of the fields decrease by a factor of $1/e$ is considered as the penetration depth of the field into the conductor, and is called the skin depth. So, the skin depth, denoted by δ, is found by using this definition as

$$e^{-\alpha\delta} = e^{-1} \quad \Rightarrow \quad \delta = \sqrt{\frac{2}{\omega\mu\sigma}} \tag{1.93}$$

Notice that the skin depth is inversely proportional to the frequency of operation and the conductivity of the medium. This implies that high-frequency signals cannot penetrate into good conductors as much as low-frequency signals, and especially for microwave frequencies, this distance becomes extremely small. The practical implication in microwave and antenna engineering is that the bulk of the conductors employed need not be of high quality to get low-loss components; just a thin plating of a high-quality conductor would be sufficient. By the same reasoning, the surface roughness of conductors also becomes important at microwave frequencies.

For the sake of completeness, let us give the intrinsic impedance in a good conductor as follows:

$$\eta = \sqrt{\frac{\mu_0}{\varepsilon_0\varepsilon_r}}\left(-j\frac{\sigma}{\omega\varepsilon_0\varepsilon_r}\right)^{-1/2} \cong (1+j)\sqrt{\frac{\omega\mu}{2\sigma}} \tag{1.94}$$

1.3 TRANSMISSION LINES

In general, circuits and networks are made up of some components that perform operations on the signals and of some others that only carry signals from one place to another with minimum distortion on the signals. These components that are responsible for the transmission of signals are called the transmission lines (TL), and they come in various shapes and forms [see Figure 1.6(a–c)]. It is obvious that those components performing operations on the signals vary from circuit to circuit, depending upon their intended functions, but the ones that perform transmission always exist in circuits and are indispensable. So, the transmission lines are probably one of the most basic components in any network. For microwave circuits and antennas, the characterization of transmission lines plays a much more important role than their characterization for low-frequency circuits. This is because the transmission lines are extremely short for low-frequency circuits in terms of wavelength (an exception is the ac power distribution lines, but we will not discuss them here) and can be modeled by considering only the bulk resistance, which is close to zero in most cases, between their terminals. However, for microwave and antenna applications, they are no longer short as compared to wavelengths, and hence it is necessary to include their influence on the signals passing through them.

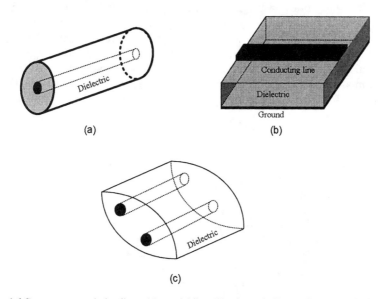

(a)

(b)

(c)

Figure 1.6 Common transmission lines: (a) coaxial line, (b) microstrip line, and (c) two-wire line.

For the characterization of transmission lines in high-frequency applications, a special method that utilizes circuit theory as well as the propagation aspect of the waves is developed and called the transmission line theory. Although the theory can be developed strictly from the circuit analysis, the major assumptions of the theory, as well as its limitations, will be missing unless its derivation via Maxwell's equations is introduced. Therefore, we will first provide a general field-analysis approach for general cylindrical waveguides, starting from Maxwell's equations, and then the transmission line theory will be developed. This is followed by a section in which the same governing equations for transmission lines will be obtained strictly from the circuit model of a general transmission line.

1.3.1 Field Analysis of General Cylindrical Waveguides

First of all, let us define the term *general cylindrical waveguide*, as illustrated in Figure 1.7, and the motivation for this section. As the name "waveguide" implies, these structures guide signals in a preferred direction, called the longitudinal direction (z-direction in Figure 1.7), while confining the fields in transverse direction (transverse to the direction of propagation, the x-y plane in Figure 1.7). A general cylindrical waveguide is assumed to be uniform in the direction of propagation and has an arbitrary cross-section in the transverse direction. As we will see later, transmission lines are made up of two conducting materials, while waveguides may consists of only one conductor or many conductors. Therefore, general cylindrical waveguides include transmission lines, multiconductor waveguides, and hollow waveguides (one conductor). Of course, this is not the only difference between the general waveguides and the transmission lines: transmission lines support a special type of waves called transverse electromagnetic (TEM) waves, while waveguides can support any type of wave configuration. To elucidate these modes of waves in a general waveguide, and to find the modes that are supported by transmission lines, we need to understand general wave solutions in a general cylindrical waveguide.

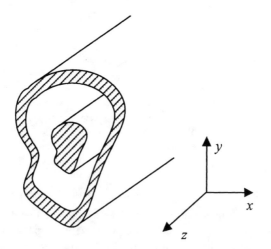

Figure 1.7 A general cylindrical TEM waveguide.

With a little study of Figure 1.7, it becomes obvious that the fields are supposed to be guided along the z-direction inside the waveguide between the two conductors. Note that the waveguide is uniform and infinite in the z-direction, the walls are assumed to be a perfect conductor, and the medium between the walls is filled with a dielectric material or air. Now, after having defined the geometry and its purpose, let us state our goal here to clarify the problem: since such geometries are used to guide the waves, our main task is to find governing equations and field expressions for such a class of geometries. Although the wave equation is the governing equation for electromagnetic waves in any geometry or media, it may be possible to find a simpler version of the wave equation for some specific class of geometries.

To achieve the goal of deriving a simpler governing equation for the wave propagation in a waveguide, we need to utilize some common features of the geometries in Maxwell's equations or in the field expressions. For example, because the geometry is uniform and infinite in the longitudinal direction (z-direction) and because it is used for the propagation of waves in the z-direction, the electric and magnetic fields should be in the following form:

$$\mathbf{E}(x, y, z) = \mathbf{E}(x, y)e^{\pm jk_z z} \tag{1.95}$$

$$\mathbf{H}(x, y, z) = \mathbf{H}(x, y)e^{\pm jk_z z} \tag{1.96}$$

where z-variations of the fields are predicted as exponentials with an unknown factor (propagation constant) k_z, and "−" and "+" on the exponents represent the fields propagating in $+z$- and $-z$-directions, respectively. Since there is no geometrical variation along the z-direction, the magnitudes of the fields cannot be forced to change by the boundary as the wave propagates along z. Therefore, only their phases change progressively with an unknown constant to be determined. Now, with this form of the fields, it would be appropriate to write vector quantities in Maxwell's curl equations in terms of their longitudinal and transverse components as follows:

$$\nabla = \hat{x}\frac{\partial}{\partial x} + \hat{y}\frac{\partial}{\partial y} + \hat{z}\frac{\partial}{\partial z} = \nabla_t + \hat{z}\frac{\partial}{\partial z} \tag{1.97}$$

$$\mathbf{E} = \hat{x}E_x + \hat{y}E_y + \hat{z}E_z = \mathbf{E}_t + \hat{z}E_z \tag{1.98}$$

$$\mathbf{H} = \hat{x}H_x + \hat{y}H_y + \hat{z}H_z = \mathbf{H}_t + \hat{z}H_z \tag{1.99}$$

where subscript t denotes transverse components of the vectors. This manipulation facilitates analytical implementation of the z-derivatives in Maxwell's equations. Once this is performed, the resulting vector components in transverse and longitudinal directions are grouped, yielding the following equations:

$$\nabla_t \times \hat{z}E_z + \frac{\partial}{\partial z}\hat{z} \times \mathbf{E}_t = -j\omega\mu\,\mathbf{H}_t \tag{1.100}$$

$$\nabla_t \times \hat{z} H_z + \frac{\partial}{\partial z} \hat{z} \times \mathbf{H}_t = j\omega\varepsilon\, \mathbf{E}_t \qquad (1.101)$$

from the transverse components, and

$$\nabla_t \times \mathbf{E}_t = -j\omega\mu H_z \hat{z} \qquad (1.102)$$

$$\nabla_t \times \mathbf{H}_t = j\omega\varepsilon E_z \hat{z} \qquad (1.103)$$

from the longitudinal components. Substituting $\mathbf{E}_t$ from (1.101) into (1.100) as

$$\nabla_t \times \hat{z} E_z + \frac{\partial}{\partial z} \hat{z} \times \left[\nabla_t \times \hat{z} H_z + \frac{\partial}{\partial z} \hat{z} \times \mathbf{H}_t \right] \frac{1}{j\omega\varepsilon} = -j\omega\mu \mathbf{H}_t$$

and making use of $\hat{z} \times (\nabla_t \times \hat{z}) = \nabla_t$, $\hat{z} \times (\hat{z} \times \mathbf{H}_t) = -\mathbf{H}_t$, and $\partial^2/\partial z^2 = -k_z^2$, the following expressions of the transverse field components in terms of their longitudinal components are obtained:

$$\mathbf{H}_t = \frac{1}{k^2 - k_z^2} \left[\frac{\partial}{\partial z} \nabla_t H_z + j\omega\varepsilon\, \nabla_t \times \hat{z} E_z \right] \qquad (1.104)$$

$$\mathbf{E}_t = \frac{1}{k^2 - k_z^2} \left[\frac{\partial}{\partial z} \nabla_t E_z - j\omega\mu\, \nabla_t \times \hat{z} H_z \right] \qquad (1.105)$$

It is evident from these two equations that if one can find the longitudinal components of the fields, (E_z, H_z), the rest of the field components can be calculated from (1.104) and (1.105). In other words, the problem of finding the electric and magnetic field vectors in a waveguide has been reduced to obtaining just the longitudinal components of the fields. In addition, with a little study of (1.104) and (1.105), and remembering that Maxwell's equations are linear in linear media, it is observed that the fields can be split into three independent modes: (1) transverse electric (TE) modes, where $E_z = 0$, $H_z \neq 0$; (2) transverse magnetic (TM) modes, where $E_z \neq 0$, $H_z = 0$; and (3) TEM modes where $E_z = 0$, $H_z = 0$. To elucidate the use of these modes, we can use an analogy from circuit analysis: since there are three independent sources that can contribute to the fields in the waveguide, $(E_z = 0,\ H_z \neq 0)$, $(E_z \neq 0,\ H_z = 0)$, and $(E_z = 0,\ H_z = 0)$, the overall response can be calculated from the responses of the individual sources with the use of the superposition principle. Therefore, any field configuration in the waveguide can be written in terms of TE, TM, and TEM mode contributions, and they are calculated independently. At this point, it is legitimate to ask the obvious question: how is it possible that one can have finite field components when both $E_z = H_z = 0$ in TEM mode as (1.104) and (1.105) become zero? Actually, (1.104) and (1.105) can be finite for $E_z = H_z = 0$ only if $k_z = k$, otherwise, there will be no TEM mode supported in the waveguide. For TE modes, also called magnetic, or H, waves in the literature, (1.105) can be reduced to the following:

$$\mathbf{E}_t = \frac{1}{k^2 - k_z^2}\left(-j\omega\mu\,\nabla_t \times \hat{z}H_z\right)$$

and substituting this into (1.102) leads to

$$\nabla_t \times \nabla_t \times \hat{z}H_z - \left(k^2 - k_z^2\right)\cdot \hat{z}H_z = 0$$

Then, using the vector identity $\nabla \times (\nabla \times A) = \nabla\nabla\cdot A - \nabla^2 A$, the following equation is obtained:

$$\nabla_t \underbrace{\left(\nabla_t \cdot \hat{z}H_z\right)}_{=0} - \nabla_t^2 \hat{z}H_z - \left(k^2 - k_z^2\right)\cdot \hat{z}H_z = 0$$

Hence, the governing equation for TE waves can be obtained as a two-dimensional scalar wave equation for H_z:

$$\nabla_t^2 H_z + \underbrace{\left(k^2 - k_z^2\right)}_{k_t^2}H_z = 0 \tag{1.106}$$

Similarly, for TM waves, also called electric, or E waves, the governing equation becomes

$$\nabla_t^2 E_z + \left(k^2 - k_z^2\right)E_z = 0 \tag{1.107}$$

Once the governing equations for TE and TM waves are solved for H_z and E_z, respectively, the rest of the field components can be obtained from (1.104) and (1.105).

For TEM waves, since the longitudinal components of the electric and magnetic fields are to be zero, $E_z = H_z = 0$, the unknown propagation constant k_z must be equal to the wave number of the medium $k = \omega\sqrt{\mu\varepsilon}$ in order to have finite field components, as can be observed from (1.104) and (1.105). This property of TEM waves has a very important practical implication: the phase velocity of a TEM wave is constant, not a function of frequency; therefore, any time waveform can be guided along a TEM-supporting waveguide without any distortion in its shape. Before going into the details of the practical advantages of TEM waves, let us first derive the governing equation for such waves. It is now clear that $k_z = k$ is a must to have a TEM wave; therefore, all field components have $e^{\pm jkz}$ dependence. In addition, we cannot use (1.104) and (1.105) as they are given, so we are left with (1.102) and (1.103) with zero longitudinal field components, rewritten as

$$\nabla_t \times \mathbf{E}_t = 0 \tag{1.108}$$

$$\nabla_t \times \mathbf{H}_t = 0 \tag{1.109}$$

It is obvious from (1.108) that the electric field is conservative in the transverse domain (i.e., its line integral over a path defined in the transverse domain is zero). This is the indication of static field behavior of TEM waves on the transverse domain. So, using a similar argument as is used for the definition of voltage in the electrostatic case, a scalar function Φ is introduced from (1.108) as

$$\nabla_t \times \mathbf{E}_t = 0 \Rightarrow \mathbf{E}_t = -\nabla_t \Phi(x, y) e^{\pm jkz} \tag{1.110}$$

where the vector identity $\nabla \times \mathbf{A} = 0 \Rightarrow \mathbf{A} = \nabla \phi$ is employed, and the scalar function Φ is called the scalar potential for the electric field due to the analogy with the electrostatic case. Since the wave propagates in a source-free region between the conductors, we obtain Laplace's equation

$$\nabla \cdot \mathbf{E}_t = 0 \Rightarrow \nabla_t^2 \Phi(x, y) = 0 \tag{1.111}$$

as the governing equation for TEM waves. In order to be able to solve for the scalar potential, and then the electric and magnetic fields, from this differential equation, one needs to have boundary conditions for the scalar potential. Writing the boundary condition of the electric field on the wall of a conducting medium leads us to

$$\hat{n} \times \mathbf{E}_t = 0 \Rightarrow \Phi = \text{constant} \tag{1.112}$$

which means that the scalar potential must be constant on the conductors. There is some additional information that we can gather from (1.100) and (1.101) for TEM waves:

$$\frac{\partial}{\partial z} \hat{z} \times \mathbf{E}_t = -j\omega\mu\, \mathbf{H}_t \tag{1.113}$$

$$\frac{\partial}{\partial z} \hat{z} \times \mathbf{H}_t = j\omega\varepsilon\, \mathbf{E}_t \tag{1.114}$$

where the coupling of the electric and magnetic fields is established. To summarize, the solution of the governing equation for TEM waves, Laplace's equation, gives the scalar potential on the transverse domain of the waveguide, and then, by the use of (1.110) and (1.113), the electric and magnetic fields are obtained, respectively.

Before getting into the details of the derivation of the transmission line equations in terms of voltage and current, it is important to understand the salient features of TEM waves: (1) fields of TEM waves are like static fields in the transverse domain, which has been evidenced by the governing equation for TEM waves, Laplace's equation, in the transverse domain; (2) there exists a unique scalar potential in cases of more than one uniform conductor in the longitudinal direction because Laplace's equation requires two boundary conditions for a unique solution; (3) the scalar potential is equivalent to voltage in electrostatic fields; and (4) although there seems to be no coupling between the electric and magnetic fields in TEM waves, which is not possible for any wave representation, the dynamic nature of the fields is taken into consideration by the propagation of the fields in the z-direction; in (1.113) and (1.114). These are the revelations of the static fields in the transverse domain, and this opens up the way of describing the fields of TEM waves in terms of voltages and currents uniquely, which will be demonstrated in the next section.

1.3.2 Transmission Line Equations Via Field Analysis

As briefly stated in Section 1.2.1, there is no distortion on the signal if it is carried by TEM-supporting waveguides, which are specifically called transmission lines. This property makes the transmission lines extremely useful for the purpose of carrying signals from one point to another. Since the governing equation for TEM waves is Laplace's equation for the scalar potential Φ, there is a unique solution for Φ if there are two independent boundaries. So, a geometry with two conductors provides two boundary conditions and, hence, can support TEM waves. Based on the same argument, N-conductor transmission lines can support $N - 1$ independent TEM modes.

It is now clear that a two-conductor line can support TEM waves and that fields of TEM waves behave like static fields in the transverse domain; as a result, voltage and current can be uniquely defined on a TEM-supporting line. Therefore, it should be possible to derive equations relating the voltage and current on the line, rather than the equations relating field quantities, (1.110), (1.113), and (1.114). Since the fields of a TEM wave on the transverse domain resemble the static fields, the electric and magnetic fields seem to be decoupled, and the line integral of the electric field between the conductors and the line integral of the magnetic field around one of the conductors give the voltage between and current through the conductors, respectively. Consequently, in order to find the relations between voltage and current on the line, one needs to start with the equations that describe the coupling of electric and magnetic fields, (1.113) and (1.114). For the sake of demonstration, let us derive the voltage-current relation on a two-wire line with an arbitrary cross-section in a homogeneous medium, as shown in Figure 1.8, starting with (1.113):

$$\frac{\partial}{\partial z} \hat{z} \times \mathbf{E}_t = -j\omega\mu\, \mathbf{H}_t$$

It is obvious that if both sides are integrated over a contour C enclosing one of the conductors, the right-hand side can be written in terms of the current I flowing in the enclosed conductor as follows:

$$\frac{\partial}{\partial z} \oint_C d\mathbf{l} \cdot \left(\hat{z} \times \mathbf{E}_t \right) = -j\omega\mu \oint_C d\mathbf{l} \cdot \mathbf{H}_t = -j\omega\mu I \tag{1.115}$$

Since the right-hand side (RHS) is written in terms of the current in the line, the left-hand side (LHS) needs to be written in terms of voltage and/or current to complete the conversion. So, substituting (1.110) into (1.115), LHS can be written in terms of the scalar potential as

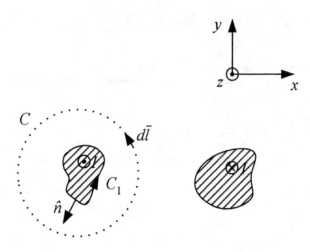

Figure 1.8 A typical two-wire line with arbitrary cross-section.

$$\text{LHS} = -\oint_C d\mathbf{l} \cdot \left(\hat{z} \times \nabla_t \Phi \right) \tag{1.116}$$

To manipulate this expression further, we should remember Stokes' theorem

$$\int_S d\mathbf{s} \cdot \nabla \times \mathbf{A} = \oint_C d\mathbf{l} \cdot \mathbf{A} \tag{1.117}$$

and make use of it in (1.116) as

$$\text{LHS} = -\int_S d\mathbf{s}\, \hat{z} \cdot \nabla_t \times \left(\hat{z} \times \nabla_t \Phi \right)$$

$$= -\int_S d\mathbf{s} \left(\nabla_t \cdot \nabla_t \Phi \right) \tag{1.118}$$

where S is the surface enclosed by contour C.

Since $\nabla_t^2 \Phi = 0$ everywhere except on the surface enclosed by C_1, the surface integral in (1.118) can be brought over the surface S_1 enclosed by the contour C_1 of the conductor. Using the divergence theorem on 2-D

$$\int_S d\mathbf{s}\, \nabla_t \cdot \mathbf{A} = \oint_C dl\, \hat{n} \cdot \mathbf{A} \tag{1.119}$$

where the contour C encloses the surface S, LHS in (1.118) can be reduced to

$$\text{LHS} = -\oint_{C_1} dl\, \hat{n} \cdot \nabla_t \Phi \tag{1.120}$$

where $\hat{n}$ is the unit normal vector to the surface enclosed by the contour C_1. Since $\mathbf{E}_t = -\nabla_t \Phi$ at $z = \text{constant}$ planes, LHS becomes an integral of the normal component of the electric field over the periphery of the conductor, which could be considered as the surface integral with a differential length in the longitudinal direction. Therefore, with the use of Gauss's law LHS becomes

$$\text{LHS} = \oint_{C_1} dl\,\hat{n} \cdot \mathbf{E}_t = \frac{Q}{\varepsilon} \tag{1.121}$$

where Q is the total charge per unit length of the line, that is, the charge enclosed in the closed surface bounded by C_1 and a differential length in the longitudinal direction. As a result, combining both sides of (1.115) results in

$$\frac{d}{dz}Q = -j\omega\mu\varepsilon\,I \tag{1.122}$$

where the charge per unit length can be written as the product of the voltage between the lines and the capacitance of the lines per unit length, $Q = CV$. Hence, the coupling between the electric and magnetic field for TEM waves, (1.115), has been represented in terms of voltage V between the lines and current I through the lines as

$$\frac{d}{dz}V = -\frac{j\omega\mu\varepsilon}{C}I \tag{1.123}$$

If a similar procedure is applied to $\hat{z} \times (1.114)$, one can get another relation between the voltage and current on the lines as

$$\frac{d}{dz}I = -j\omega CV \tag{1.124}$$

Note that the voltage and current defined between and through the lines, respectively, are in time-harmonic forms (phasor form), and uniform at $z = $ constant planes. Therefore, their variations along the line will be dictated by the coupling equations, (1.123) and (1.124). To find the governing differential equation for the voltage on the line, one can differentiate both sides of (1.123) with respect to z and substitute (1.124) into the resulting expression, yielding

$$\frac{d^2V}{dz^2} + k^2V = 0 \tag{1.125}$$

where $k^2 = \omega^2\mu\varepsilon$ is used in general. For the current along the line, a similar procedure is applied to (1.124), and the following differential equation is obtained:

$$\frac{d^2I}{dz^2} + k^2I = 0 \tag{1.126}$$

The derivation presented in this section shows that one can work with the voltage and current on TEM-supporting lines, rather than working with the field quantities. So, it is sufficient to solve one of these differential equations to find the voltage and current on the line. Since these differential equations for voltage and current are in the form of wave equations, their solutions for voltage and current represent

propagation of voltage and current along the line, and hence they are referred to as
voltage and current waves.

1.3.3 Transmission Line Equations Via Circuit Analysis

As it was detailed in the previous section, the characteristic equations for a
transmission line in terms of voltage and current on the line are derived from the
field analysis, that is, from Maxwell's equations. Although it is somewhat difficult
to follow mathematically, it provides unique understanding of TEM waves and
fields and their relations to the circuit parameters, voltage and current. The same
voltage-current relations can be obtained from the circuit model of the
transmission lines, which will be performed in this section.

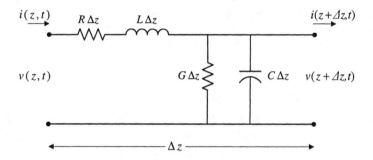

Figure 1.9 Equivalent circuit of infinitesimal length of a TEM transmission line.

To devise a circuit model for a transmission line, we need to understand the
physical mechanisms involved, as well as their modeling with the common circuit
components. Because transmission lines are always assumed to be uniform along
the longitudinal direction, the z-direction, and because voltage and current defined
on the line vary continuously along the line, it would be sufficient to model the
line over a differential length Δz. So, let us consider a transmission line with two
conductors, either in coaxial form or two-wire form as depicted in Figures 1.7 and
1.8, respectively, and try to describe the physical mechanism involved. Here are
the mechanisms and their circuit models: (1) if the conductor used in the design of
a transmission line has finite conductivity, then the current passing through the
conductor induces heat loss, which is equivalent to the loss introduced by a series
resistor; (2) since a current has to flow on both conductors in opposite directions,
provided that the two conductors are connected to complete the circuit, there must
be magnetic flux density induced between the conductors, and in turn, finite
magnetic flux linkage, which can be modeled as an inductance along the line; (3)
there are two conductors with finite surface areas and finite separation distance,
which can naturally be modeled as a capacitor between the conductors; and finally,
(4) the conductivity of the dielectric medium between the two conductors (i.e., the
loss in the dielectric media), would cause the current on one conductor to leak

through the media, which can be modeled as a conductance between the conductors. If these circuit models are put together for a differential length of a TEM-supporting transmission line, one can come up with the circuit depicted in Figure 1.9, where the circuit components are defined over a unit length.

Once the circuit has been formed to model a piece of transmission line, we intend to find the governing differential equations for the voltage and current on the transmission line based on the circuit model of the line. For this purpose, we first apply Kirchhoff's voltage and current laws (KVL and KCL) for the circuit shown in Figure 1.9, resulting in the following expressions:

$$\text{KVL}: \quad v(z+\Delta z,t) = v(z,t) - L\Delta z\frac{\partial i(z,t)}{\partial t} - R\Delta z\, i(z,t)$$

$$\text{KCL}: \quad i(z+\Delta z,t) = i(z,t) - C\Delta z\frac{\partial v(z+\Delta z,t)}{\partial t} - G\Delta z\, v(z+\Delta z,t)$$

$$(1.127)$$

Since the circuit is linear, we can introduce the time-harmonic forms of voltage and current, as the above equations can be simplified by using phasor notation as follows:

$$v(t,z) = \text{Re}\{V(z)e^{j\omega t}\}$$
$$i(t,z) = \text{Re}\{I(z)e^{j\omega t}\}$$

$$(1.128)$$

By substituting these phasor notations into (1.127), the time dependences of the KVL and KCL can be completely eliminated:

$$\text{KVL}: \quad V(z+\Delta z) = V(z) - L\Delta z\, j\omega\, I(z) - R\Delta z\, I(z)$$
$$\text{KCL}: \quad I(z+\Delta z) = I(z) - C\Delta z\, j\omega V(z+\Delta z) - G\Delta z\, V(z+\Delta z)$$

$$(1.129)$$

Now, rearranging the last two equations and taking the limit of $\Delta z \to 0$ result in the following expressions:

$$\left.\frac{V(z+\Delta z)-V(z)}{\Delta z}\right|_{\Delta z\to 0} = \frac{dV(z)}{dz} = -L\, j\omega I(z) - R\, I(z)$$

$$\left.\frac{I(z+\Delta z)-I(z)}{\Delta z}\right|_{\Delta z\to 0} = \frac{dI(z)}{dz} = -C\, j\omega V(z) - G\, V(z)$$

$$(1.130)$$

which may be written in more compact form as

$$\frac{dV(z)}{dz} = -Z\, I(z)$$

$$(1.131)$$

$$\frac{dI(z)}{dz} = -Y\, V(z)$$

$$(1.132)$$

where Z and Y are the series impedance and parallel admittance per unit length on the circuit equivalent of the transmission line, respectively. Note that these equations are actually very intuitive because the rates of changes of voltage and current along a transmission line must be due to the equivalent impedance on the series arm and equivalent admittance on the parallel arm of the circuit, respectively. Since these equations are coupled, first-order differential equations, to eliminate the coupling terms and get a single differential equation with a single unknown, we take the derivative of the first equation with respect to z and substitute the second equation into the first one, which results in a second-order differential equation for the voltage:

$$\frac{d^2V(z)}{dz^2} - \gamma^2 V(z) = 0 \tag{1.133}$$

where $\gamma^2 = ZY = (R + j\omega L)(G + j\omega C)$. Similarly, we can get another differential equation for the current on the line as

$$\frac{d^2I(z)}{dz^2} - \gamma^2 I(z) = 0 \tag{1.134}$$

Note that the equations relating the voltage and current, (1.131) and (1.132), and the governing equations on a transmission line, (1.133) and (1.134), are in the same form as (1.123)–(1.126), when no loss is assumed on the line (i.e., $R = G = 0$). Consequently, it has been demonstrated that the field-analysis approach becomes equivalent to the circuit approach in cases of TEM-supporting transmission lines.

1.3.4 Analysis of General Transmission Line Circuits

Let us first discuss what we mean by *transmission line circuits* in the title of this section and then learn how to analyze such circuits. We have already learned that transmission lines consist of conductors (at least two conductors) uniform in the intended direction of propagation of the signal and are mainly used to carry signals from one component in the circuit to another. However, this is not the only function of the transmission lines, especially in microwave circuits, where they can be designed and employed as capacitors, inductors, and matching components. So, when we refer to the analysis of transmission line circuits, we actually mean to develop mathematical tools to analyze circuits with transmission lines, with no regard to their functions, as long as they only support TEM waves or close to TEM waves.

For the sake of illustration, let us use the schematic picture in Figure 1.10 to represent a transmission line with two conductors, where z is the longitudinal direction of the transmission line. Since we have already obtained the governing differential equations for the voltage and current on a transmission line, (1.133) and (1.134), the analysis starts with the solutions of these equations, which are just second-order, ordinary differential equations. The solutions of these differential equations are quite simple and can be obtained as

$$V(z) = V^+ e^{-\gamma z} + V^- e^{\gamma z}$$
$$I(z) = I^+ e^{-\gamma z} + I^- e^{\gamma z} \tag{1.135}$$

where V^+, V^-, I^+, and I^- are the unknown amplitudes of the voltage and current waves, and the superscripts $+$ and $-$ refer to the amplitudes of the positive and negative-going waves, respectively.

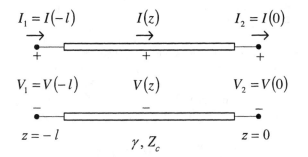

Figure 1.10 A general transmission line circuit, with port voltages and currents.

For lossy transmission lines, the variable γ is a complex quantity, as defined right after (1.133), called the complex propagation constant of the transmission line. In general, it can be decomposed into real and imaginary parts as

$$\gamma = \alpha + j\beta \tag{1.136}$$

where the real part α represents the loss in the system, while the imaginary part β is responsible for phase shift, or rather propagation, respectively. It is interesting to note that the voltage and current waveforms can be thought of as the combination of forward- and backward-traveling waves, as indicated in the signs of the exponentials.

Instead of using the solutions of governing equations for both voltage and current separately, with four unknown coefficients, one may as well solve for only one of them and use the coupling relations between the voltage and current, (1.131) and (1.132), to find the other. For example, let us start with the solutions of the voltage on the transmission line:

$$V(z) = V^+ e^{-\gamma z} + V^- e^{\gamma z} \tag{1.137}$$

and substitute (1.137) into (1.131); then the current along the line turns out to be

$$I(z) = \underbrace{\sqrt{\frac{Y}{Z}} V^+ e^{-\gamma z}}_{I^+} - \underbrace{\sqrt{\frac{Y}{Z}} V^- e^{\gamma z}}_{I^-} \tag{1.138}$$

where $\gamma = \sqrt{YZ}$ is used. As mentioned before, the first terms represent $+z$ propagating waves while the second ones represent $-z$ propagating waves. Hence, using the general definition of the impedance, the ratio of voltage to current, one can define the ratio of the $+z$ propagating voltage wave to the $+z$ propagating current wave as the characteristic impedance of the line:

$$Z_c = \frac{V^+}{I^+} = \sqrt{\frac{Z}{Y}} = \sqrt{\frac{R + j\omega L}{G + j\omega C}} \tag{1.139}$$

Note that since both series impedance and parallel admittance of the line are dependent on the geometry and electrical properties of the materials used, the impedance defined in (1.139) would be characteristic of the line with the true meaning of the word "characteristic." Since both the complex propagation constant and the characteristic impedance of a transmission line are properties of the line itself, they are specified with the given transmission line; these parameters have been included inside the schematic picture of the transmission line in Figure 1.10 as integral parts of the transmission line. With the introduction of these new parameters, the voltage and the current at any point z on the transmission line can be written as follows:

$$V(z) = V^+ e^{-\gamma z} + V^- e^{\gamma z}$$
$$I(z) = \frac{V^+}{Z_c} e^{-\gamma z} - \frac{V^-}{Z_c} e^{\gamma z} \tag{1.140}$$

where V^+ and V^- are the unknown coefficients to be determined by the source and load connections to the ports of the transmission line.

So far we have mainly talked about the solution of the governing equations and the introduction of two new parameters, namely, the complex propagation constant and the characteristic impedance of the line. Although these parameters have been defined mathematically, their interpretations, as well as their physical influences, need to be explained further to gain some insight into the functions of transmission lines in general.

Propagation Constant and Dispersion Phenomenon

Let us start with rewriting the complex propagation constant in a lossy transmission line as

$$\gamma = \sqrt{ZY} = \sqrt{(R + j\omega L)\cdot(G + j\omega C)}$$
$$= \alpha + j\beta \tag{1.141}$$

If this is substituted into the $+z$ propagating wave component of the voltage or current expression

$$e^{-\gamma z} = e^{-\alpha z} e^{-j\beta z}$$

it becomes obvious that the real part represents attenuation, while the imaginary part models the propagation in the $+z$ direction, so they are referred to as attenuation and propagation constants, respectively. If there were no loss in the transmission line (i.e., $R = G = 0$), we could have $\alpha = 0$ and $\beta = \omega\sqrt{LC}$, and the system would be called a lossless transmission line. With this in mind, let us write the time-domain representation of the $+z$ propagating exponential term as

$$\mathrm{Re}\left\{e^{-\gamma z}e^{j\omega t}\right\}= \mathrm{Re}\left\{e^{-\alpha z}e^{-j\beta z}e^{j\omega t}\right\}= e^{-\alpha z}\cos(\omega t - \beta z)$$

for which the phase velocity can be calculated, as given in (1.53), and written as

$$v_p = \lim_{\Delta t \to 0}\frac{\Delta z}{\Delta t}=\frac{\omega}{\beta} \qquad (1.142)$$

For lossless transmission lines, since the propagation constant is always linearly proportional to the frequency of the signal, $\beta = \omega\sqrt{LC}$, the phase velocity becomes independent of the frequency. However, for lossy transmission lines, the propagation constant β cannot be cast, from the definition of the complex propagation constant γ, into a form that is a linear function of frequency. Since any real signal that carries information consists of a band of frequency, transferring such signals through a lossy transmission line would result in distortion in the wave shape of the information signal at the output. This is because each frequency component propagates to the output with a different velocity; hence, the phase relations among the frequency components of the information signal change, resulting in a change in its time-domain waveform. This phenomenon is known as dispersion, and it exists in any transmission medium if the propagation constant is not a linear function of frequency. However, for a lossy transmission line, one can achieve dispersion-free transmission by selecting the loss factors appropriately, which will be detailed later in this chapter.

Characteristic Impedance
Note that the characteristic impedance of a transmission line has been defined by the ratio of the $+z$ propagating voltage and current waves. Since we have, in general, both $+z$ propagating and $-z$ propagating waves, the ratio of the total voltage to the total current at any z value on the line can give the input impedance seen from the point z forward. However, if we have no $-z$ propagating component of the waves in the line, then the input impedance will be equal to the characteristic impedance of the line. In addition to this observation, let us note that both $+z$ and $-z$ propagating wave solutions of the governing differential equation need not always exist together; their existence depends on the boundary conditions imposed by the transmission line geometry. For example, if we have a semi-infinite transmission line excited from the only port, as shown in Figure 1.11(a), there will be no $-z$ propagating voltage or current waves because there is no boundary that will cause reflection of the excited waves in the $-z$ direction. Therefore, one can equivalently define the characteristic impedance of a transmission line as the input impedance of an infinitely long transmission line, as depicted in Figure 1.11(a).

Let us also find the expression of the characteristic impedance using this alternative definition. Since the input impedance of a semi-infinite transmission line is equal to the characteristic impedance of the line by definition [Figure 1.11(a)], adding an infinitesimal segment to this line does not change the length of the line; hence, the input impedance is still the characteristic impedance of the line [Figure 1.11(b)].

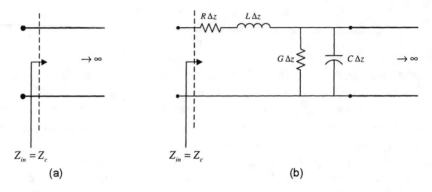

Figure 1.11 Meaning of the characteristic impedance of a transmission line: (a) input impedance of an infinitely long transmission line is the characteristic impedance of the line; and (b) addition of an infinitesimal segment to an infinitely long transmission line does not change the input impedance.

Mathematically, this can be demonstrated as follows:

$$Z_c = j\omega L\Delta z + R\Delta z + \cfrac{1}{j\omega C\Delta z + G\Delta z + \cfrac{1}{Z_c}}$$

$$= j\omega L\Delta z + R\Delta z + \frac{Z_c}{(j\omega C\Delta z + G\Delta z)Z_c + 1}$$

Then, the following quadratic equation for Z_c is obtained:

$$Z_c^2(j\omega C\Delta z + G\Delta z) = (j\omega L\Delta z + R\Delta z)(j\omega C\Delta z + G\Delta z)Z_c + j\omega L\Delta z + R\Delta z$$

$$Z_c^2(j\omega C + G)\Delta z = (j\omega L + R)(j\omega C + G)\Delta z^2 Z_c + (j\omega L + R)\Delta z$$

By taking the limit of the above quadratic equation as $\Delta z \to 0$, we find the expression for the characteristic impedance as follows:

$$Z_c = \sqrt{\frac{j\omega L + R}{j\omega C + G}} \qquad (1.143)$$

which is the same expression we obtained from the former definition of the characteristic impedance, (1.139).

So far, we have summarized the governing formulas for wave propagation along transmission lines and provided the expressions of propagation constant and characteristic impedance in terms of distributed circuit elements. It should be noted that all the expressions derived so far in this section are very general, and no assumption has been imposed in their derivation, except for the underlying assumption of the whole section that the line supports TEM waves. Since the majority of applications in microwave circuits and antennas involve either almost-lossless or low-loss transmission lines, it would be instructive to examine the propagation constant and characteristic impedance of transmission lines under different loss conditions. Therefore, this discussion is divided into lossless, low-loss, and nondispersive cases as follows:

Lossless transmission lines:

As the name implies, the conductors and dielectric materials used in the design of such transmission lines are assumed to be lossless, that is, conductors with infinite conductivity ($\sigma \to \infty$) and dielectric materials with zero conductivity ($\sigma = 0$), and no polarization loss. Therefore, the circuit components that account for the loss mechanisms of a transmission line become zero:

$$R = 0 \quad G = 0 \tag{1.144}$$

Hence, the characteristic impedance and propagation constant become pure real and imaginary, respectively, as shown below:

$$Z_c = \sqrt{\frac{L}{C}} \tag{1.145}$$

$$\gamma = j\omega\sqrt{LC} \tag{1.146}$$

Low-loss transmission lines:

This is actually the most practical case, since most materials used in transmission lines are typically low-loss materials. The criteria of low-loss in transmission lines can be stated, in terms of the circuit components, as the following inequalities:

$$\omega L \gg R \quad \omega C \gg G \tag{1.147}$$

According to these criteria, the characteristic impedance and complex propagation constant can be approximated as

$$Z_c = \sqrt{\frac{j\omega L + R}{j\omega C + G}} = \sqrt{\frac{j\omega L}{j\omega C} \frac{1 + R/j\omega L}{1 + G/j\omega C}}$$

$$\cong \sqrt{\frac{L}{C}}\left(1 + \frac{R}{2j\omega L} - \frac{G}{2j\omega C}\right) \cong \sqrt{\frac{L}{C}} \tag{1.148}$$

$$\gamma = \sqrt{(j\omega L + R)\cdot(j\omega C + G)}$$

$$\cong j\omega\sqrt{LC}\left(1 - j\frac{G}{2\omega C} - j\frac{R}{2\omega L}\right) \cong j\omega\sqrt{LC} \tag{1.149}$$

Note that the characteristic impedance and the propagation constant of a low-loss transmission line are approximately equal to their values in the lossless case. This result is particularly important because it allows using lossless approximation in most practical cases.

Nondispersive transmission lines:

As we have discussed briefly, signals carried by a lossy transmission line suffer from dispersion due to the frequency-dependent nature of the phase velocity. However, if we could choose the material properties such that the imaginary part of the complex propagation constant becomes a linear function of frequency, there will be no dispersion. This is possible if the following relation is satisfied:

$$\frac{R}{L} = \frac{G}{C} \tag{1.150}$$

Then, by substituting this relation into the expressions of the characteristic impedance and the propagation constant, these parameters are obtained as

$$Z_c = \sqrt{\frac{j\omega L + R}{j\omega C + G}}$$
$$= \sqrt{\frac{L}{C}\frac{1 + G/j\omega C}{1 + G/j\omega C}} = \sqrt{\frac{L}{C}} \tag{1.151}$$

$$\gamma = \sqrt{(j\omega L + R)\cdot(j\omega C + G)}$$
$$= \sqrt{-\omega^2 LC\left(1 - j\frac{G}{\omega C} - j\frac{G}{\omega C} - \frac{G^2}{\omega^2 C^2}\right)} \tag{1.152}$$
$$= j\omega\sqrt{LC}\left(1 - j\frac{G}{\omega C}\right) = j\omega\sqrt{LC} + R\sqrt{\frac{C}{L}}$$

where the characteristic impedance and the imaginary part of the complex propagation constant have become the same as those for the lossless line. In addition, since the propagation constant is a linear function of frequency, the imaginary part of (1.152), the transmission of a signal along this line is dispersion-free.

The above property, and the fact that attenuation on the transmission lines could be reduced by increasing the line inductance when $G \ll 1$, was first proposed by Oliver Heaviside, self-taught British engineer, mathematician, and physicist, in 1887. Later, in 1899, George Campbell of AT&T and Michael Pupin, at Columbia University, almost simultaneously discovered that increasing the line inductance could be accomplished by loading the line using discrete coils (later to be known as Pupin coils) as long as the spacing between coils is not greater than one-tenth of the wavelength. The effect of this invention on telecommunications was enormous: the maximum transmission distance was doubled and it allowed smaller radius wires to be used resulting in significant cost savings.

1.3.5 Analysis of Terminated Transmission Line Circuits

After having obtained the general solutions of a transmission line circuit, (1.140), and defined some relevant parameters, (1.139) and (1.141), now it is time to extend the analysis toward more practical circuits, namely a transmission line fed by a source at one port and terminated in a load at the other port. It is clear from (1.140) that the time-harmonic voltage and current on a transmission line depend on z, the position along the line. Therefore, the input impedance seen across a pair of terminals located at any point along the line is also dependent on the length of the line between the terminal and the load, as well as the line parameters and the termination.

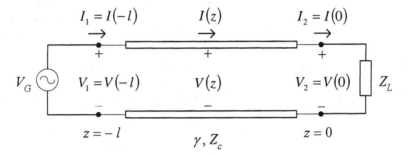

Figure 1.12 A TEM transmission line connected to a complex load.

To determine the input impedance seen from the generator, at the terminal at $z = -l$ in Figure 1.12, we need to find the voltage and current waves uniquely; that is, the unknown coefficients V^+ and V^- need to be determined from the boundary conditions at the generator and load terminals. The additional boundary conditions at the generator and load terminals are the voltage, V_G, and the load impedance Z_L, respectively. For this purpose, the total voltages and currents at both ends of the transmission line are written as follows:

$$V_1 = V(-l) = V^+ e^{\gamma l} + V^- e^{-\gamma l}$$
$$V_2 = V(0) = V^+ + V^-$$

$$I_1 = \frac{V^+}{Z_c} e^{\gamma l} - \frac{V^-}{Z_c} e^{-\gamma l}$$

$$I_2 = \frac{V^+}{Z_c} - \frac{V^-}{Z_c}$$

where V^+ and V^- are the unknown amplitudes of the $+z$ and $-z$ propagating voltage waves at the load terminals, $z = 0$, respectively. To impose a boundary condition at the load terminals, we can use the definition of the load impedance as follows:

$$Z_L = \frac{V_2}{I_2} = \frac{V^+ + V^-}{\dfrac{V^+}{Z_c} - \dfrac{V^-}{Z_c}}$$

If the load reflection coefficient is introduced as the ratio of the reflected wave amplitude to the incident wave amplitude at the load terminals, then it can be represented in terms of known impedance values as

$$\frac{V^-}{V^+} = \Gamma_L = \frac{Z_L - Z_C}{Z_L + Z_C} \tag{1.153}$$

Note that the $-z$ propagating wave can be considered a result of the reflection of the $+z$ propagating wave from the load terminals. As (1.153) manifests, the reflection is due to the impedance mismatch between the load impedance and the characteristic impedance of the line, and it provides an additional equation to be satisfied by the unknown amplitudes.

An additional equation is obtained at the terminals of the generator, as shown in Figure 1.12, where the voltage at $z = -l$ must be equal to the source voltage V_G. Without loss of generality, we can assume that the generator has 1-V peak sinusoidal excitation, and hence

$$V(-l) = V^+ e^{\gamma l} + V^- e^{-\gamma l} = 1$$

Substituting the definition of the reflection coefficient (1.153) into the above equation, V^+ and V^- can be obtained as follows:

$$V^+ = \frac{1}{e^{\gamma l} + \Gamma_L e^{-\gamma l}}$$

$$V^- = \frac{\Gamma_L}{e^{\gamma l} + \Gamma_L e^{-\gamma l}}$$

So, after having determined the amplitudes of the $+z$ propagating and $-z$ propagating waves, the input impedance of the transmission line can simply be obtained from the definition as demonstrated below:

$$Z_{in} = \frac{V_1}{I_1} = Z_c \frac{1}{V^+ e^{\gamma l} - V^- e^{-\gamma l}}$$

$$= Z_c \frac{1}{V^+ \left(e^{\gamma l} - \Gamma_L e^{-\gamma l}\right)}$$

$$= Z_c \frac{e^{\gamma l} + \Gamma_L e^{-\gamma l}}{e^{\gamma l} - \Gamma_L e^{-\gamma l}}$$

By substituting the expression for the reflection coefficient (1.153), this input impedance expression can be further simplified to, or rather rewritten for, a lossy transmission line as

$$Z_{in} = Z_c \frac{Z_L + Z_c \tanh(\gamma l)}{Z_c + Z_L \tanh(\gamma l)} \qquad (1.154)$$

and for a lossless line as

$$Z_{in} = Z_c \frac{Z_L + jZ_c \tan(\beta l)}{Z_c + jZ_L \tan(\beta l)} \qquad (1.155)$$

where $\gamma = j\beta$ is used. These equations give us the input impedance of a transmission line that is characterized by a propagation constant and a characteristic impedance terminated with a complex load. An important observation is that when a transmission line is terminated with a load that is the same as its characteristic impedance, then the reflected wave (V^-) from the load terminal becomes zero. In this case, it is said that the transmission line is matched to the load. One must note that this definition of matching is different from the definition of conjugate matching that is used in circuit theory, which states that load impedance should be equal to complex conjugate of source impedance for maximum power transfer. Therefore, in a microwave circuit, maximum power transfer to load is still possible even though there is a standing-wave pattern (i.e., V^- is nonzero) on the output transmission line. However, this is not practically useful because it makes the power transfer dependent on the length of the transmission line that is connecting load and generator. So, the common practice is to match the generator to the transmission line on the generator side, and then to match the transmission line to the load on the load side.

There are two important cases for the input impedance of a low-loss transmission line that are further studied as follows:

Quarter-wavelength-long transmission line ($\gamma l = j\pi/2$)

$$Z_{in} = \frac{Z_c^2}{Z_L} \qquad (1.156)$$

A transmission line that is a quarter-wavelength long at the operating frequency is called an impedance inverter and has a very important application in designing coupled-line microwave filters. We will visit this later in studying filter networks.

Half-wavelength-long transmission line ($\gamma l = j\pi$)

$$Z_{in} = Z_L \qquad (1.157)$$

A transmission line that is a half-wavelength long at the operating frequency transfers the load impedance unchanged to its input terminals. The periodic dips that are usually observed in the return-loss versus frequency plots of simulation and/or measurement results of transmission lines are due to this property.

1.4 CIRCUIT PARAMETERS

There are four types of circuit parameters that are commonly used in microwave engineering to uniquely characterize linear microwave circuits: scattering parameters (S-parameters), Y-parameters, Z-parameters, and $ABCD$-parameters. Among them, S-parameters are the most frequently used in microwave engineering because their derivation is based on forward- and backward-traveling waves on terminal transmission lines that are easy to measure using directional couplers. On the other hand, Z- (or Y-) and $ABCD$-parameters are based on terminal voltage and current values that are very difficult to measure directly at microwave frequencies. However, they have applications especially in the modeling of passive circuits due to the relatively simple relationship that exists between the equivalent models and relevant circuit parameters. By using Z- or Y-parameters, it is easy to extract the component values of equivalent lumped-circuit models (even by inspection in most cases), and circuit resonances are more eminent in those parameters.

In the following sections, we will provide definitions of these most commonly used circuit parameters with their applications on some simple circuits.

1.4.1 Scattering Parameters

Scattering parameters are defined using incident and reflected waves on transmission lines that connect a microwave circuit to external circuits [8, 9]. To give the definition, let us assume a two-port network whose ports are connected to transmission lines as shown in Figure 1.13. We will use voltage waves on the terminals to define the S-parameters of this circuit. Note that the transmission lines are not necessary to define the S-parameters. However, assuming a piece of transmission line connected to each port makes the visualization of traveling waves easy. From another point of view, one can think that the lengths of these transmission line segments are vanishingly small. Once we give the definitions of the S-parameters based on the two-port circuit, generalizing to n-port networks is straightforward.

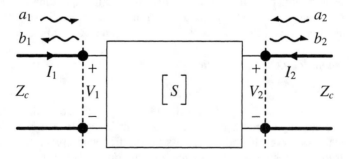

Figure 1.13 Representation of incident and reflected waves at the terminals of a linear two-port microwave network.

Let's assume that we have the port voltages and currents as shown in the figure. As demonstrated in Section 1.3.5, we can decompose the port voltages and currents into backward- and forward-traveling waves propagating on the port transmission line as follows:

$$V_1 = V_1^+ + V_1^-$$
$$I_1 = \frac{1}{Z_c}\left(V_1^+ - V_1^-\right)$$

(1.158)

$$V_2 = V_2^+ + V_2^-$$
$$I_2 = \frac{1}{Z_c}\left(V_2^+ - V_2^-\right)$$

(1.159)

where Z_c is the characteristic impedance of the port transmission lines, which is usually 50 ohms in microwave circuits. Note that it is important to have the same characteristic impedance on all ports to have symmetric S-parameters for reciprocal networks.

The above equations can be solved to determine forward- and backward-traveling voltage waves at the port terminals as

$$V_1^+ = \frac{V_1 + Z_c I_1}{2}$$

$$V_1^- = \frac{V_1 - Z_c I_1}{2}$$

$$V_2^+ = \frac{V_2 + Z_c I_2}{2}$$

$$V_2^- = \frac{V_2 - Z_c I_2}{2}$$

Then, we normalize the incident and the reflected voltages on each terminal of the network using a normalization impedance. This normalization impedance is usually selected same as the characteristic impedance of the port transmission lines:

$$a_{1,2} = \frac{V_{1,2}^+}{\sqrt{Z_c}}$$

(1.160)

$$b_{1,2} = \frac{V_{1,2}^-}{\sqrt{Z_c}}$$

(1.161)

This normalization enables that the incident and the reflected powers can always be written as follows (assuming peak voltage and current values):

$$P_{1,2}^+ = \frac{1}{2}\left|a_{1,2}\right|^2$$

(1.162)

$$P_{1,2}^- = \frac{1}{2}|b_{1,2}|^2 \tag{1.163}$$

Now, by employing these normalized voltage waveforms, we can define the scattering parameters in the following form:

$$b_1 = S_{11}a_1 + S_{12}a_2$$
$$b_2 = S_{21}a_1 + S_{22}a_2 \tag{1.164}$$

The above equations can be put in matrix form as:

$$\begin{bmatrix} b_1 \\ b_2 \end{bmatrix} = \begin{bmatrix} S_{11} & S_{12} \\ S_{21} & S_{22} \end{bmatrix} \begin{bmatrix} a_1 \\ a_2 \end{bmatrix} \tag{1.165}$$

This is the S-parameter representation of the linear two-port network shown in Figure 1.13. It fully characterizes the given linear network if the reference planes and the normalizing resistance are specified. In other words, the S-parameters of a microwave network are meaningful when they are given with the reference planes and the normalizing resistance.

Change in Reference Planes

As a next step, we will show how the shift in reference planes affects the S-parameters. To show change in S-parameters due to shift in the reference planes, let's consider the network given in Figure 1.14. Here, we are interested in finding new S-parameters when the reference planes are shifted as depicted in the figure.

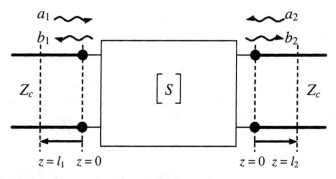

Figure 1.14 Change in reference planes for a two-port network.

When the reference planes are shifted as shown in the figure, the phase of the voltage waves is changed as shown below:

$$\tilde{b}_1 = e^{-j\beta l_1} \cdot b_1 \quad \tilde{b}_2 = e^{-j\beta l_2} \cdot b_2$$
$$\tilde{a}_1 = e^{j\beta l_1} \cdot a_1 \quad \tilde{a}_2 = e^{j\beta l_2} \cdot a_2 \tag{1.166}$$

where the parameters shown by the tilde are the modified voltage waveforms at shifted reference planes, and β is the propagation constant on the port transmission lines. Now, by inserting (1.166) into (1.164), we find the modified S-matrix:

$$\begin{bmatrix} e^{j\beta l_1} & 0 \\ 0 & e^{j\beta l_2} \end{bmatrix}\begin{bmatrix} \tilde{b}_1 \\ \tilde{b}_2 \end{bmatrix} = \begin{bmatrix} S_{11} & S_{12} \\ S_{21} & S_{22} \end{bmatrix}\begin{bmatrix} e^{-j\beta l_1} & 0 \\ 0 & e^{-j\beta l_2} \end{bmatrix}\begin{bmatrix} \tilde{a}_1 \\ \tilde{a}_2 \end{bmatrix}$$

$$\begin{bmatrix} \tilde{b}_1 \\ \tilde{b}_2 \end{bmatrix} = \begin{bmatrix} e^{-j\beta l_1} & 0 \\ 0 & e^{-j\beta l_2} \end{bmatrix}\begin{bmatrix} S_{11} & S_{12} \\ S_{21} & S_{22} \end{bmatrix}\begin{bmatrix} e^{-j\beta l_1} & 0 \\ 0 & e^{-j\beta l_2} \end{bmatrix}\begin{bmatrix} \tilde{a}_1 \\ \tilde{a}_2 \end{bmatrix}$$

The new S-parameters, due to a change in reference planes as shown in Figure 1.14, are given as follows:

$$\left[\tilde{S}\right] = \begin{bmatrix} e^{-j\beta l_1} & 0 \\ 0 & e^{-j\beta l_2} \end{bmatrix}\begin{bmatrix} S_{11} & S_{12} \\ S_{21} & S_{22} \end{bmatrix}\begin{bmatrix} e^{-j\beta l_1} & 0 \\ 0 & e^{-j\beta l_2} \end{bmatrix} \tag{1.167}$$

Change in Normalizing Impedance

As indicated before, S-parameters are specified for a given normalizing impedance; when the normalizing impedances changes, S-parameters do also change. Change in normalizing impedance can be thought of as connecting new transmission line segments with a different characteristic impedance to the terminals of the circuit. After connecting the new transmission line segments, traveling waves on the lines change, resulting in a new set of S-parameters. To find the new S-parameters, one must determine the waves on these new transmission lines. Figure 1.15 shows the interpretation of normalizing impedance change for a two-port network.

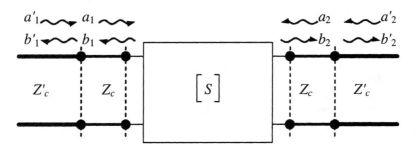

Figure 1.15 Change in normalizing impedance for a two-port network.

Let us start by writing the new S-parameters of the circuit in matrix form as follows:

$$\begin{bmatrix} b_1' \\ b_2' \end{bmatrix} = \begin{bmatrix} S_{11}' & S_{12}' \\ S_{21}' & S_{22}' \end{bmatrix}\begin{bmatrix} a_1' \\ a_2' \end{bmatrix} \tag{1.168}$$

where the primed parameters refer to the new traveling waves. To determine the new wave amplitudes, one can equalize voltages and currents at the connection points (i.e., satisfy boundary conditions) of the new and old transmission line segments:

$$V_1'^+ + V_1'^- = V_1^+ + V_1^-$$
$$\frac{V_1'^+}{Z_c'} - \frac{V_1'^-}{Z_c'} = \frac{V_1^+}{Z_c} - \frac{V_1^-}{Z_c}$$

(1.169)

$$V_2'^+ + V_2'^- = V_2^+ + V_2^-$$
$$\frac{V_2'^+}{Z_c'} - \frac{V_2'^-}{Z_c'} = \frac{V_2^+}{Z_c} - \frac{V_2^-}{Z_c}$$

(1.170)

where primed parameters and unprimed parameters represent the new and original traveling waves, respectively. After some mathematical operations:

$$V_1'^+ = mV_1^+ + nV_1^-$$
$$V_1'^- = nV_1^+ + mV_1^-$$
$$V_2'^+ = mV_2^+ + nV_2^-$$
$$V_2'^- = nV_2^+ + mV_2^-$$

where

$$m = \frac{1}{2}\left(1 + Z_c'/Z_c\right)$$
$$n = \frac{1}{2}\left(1 - Z_c'/Z_c\right)$$

Z_c and Z_c' are the original and new normalizing impedances, respectively. From the original S-parameters of the circuit, we also know that

$$V_1^- = S_{11}V_1^+ + S_{12}V_2^+$$
$$V_2^- = S_{21}V_1^+ + S_{22}V_2^+$$

Then one can show that

$$V_1'^- = mS_{12}V_2^+ + \left(n + mS_{11}\right)V_1^+$$
$$V_2'^- = mS_{21}V_1^+ + \left(n + mS_{22}\right)V_2^+$$
$$V_1'^+ = nS_{12}V_2^+ + \left(m + nS_{11}\right)V_1^+$$
$$V_2'^+ = nS_{21}V_1^+ + \left(m + nS_{22}\right)V_2^+$$

By solving the above equations simultaneously, one finds the following expressions for the wave coefficients on the inner transmission line segments in terms of wave coefficients on the outer segments:

$$V_1^+ = \frac{m + nS_{22}}{\Delta}V_1'^+ - \frac{nS_{12}}{\Delta}V_2'^+$$

$$V_2^+ = \frac{m + nS_{11}}{\Delta}V_2'^+ - \frac{nS_{21}}{\Delta}V_1'^+$$

where

$$\Delta = m^2 + mn(S_{11} + S_{22}) + n^2(S_{11}S_{22} - S_{12}S_{21})$$

The final step is to express the outgoing waves on the outer segments in terms of the incoming waves, which defines the new S-parameters:

$$V_1'^- = \frac{m^2 S_{11} + mn(1 + S_{11}S_{22} - S_{12}S_{21}) + n^2 S_{22}}{\Delta}V_1'^+ + S_{12}\frac{m^2 - n^2}{\Delta}V_2'^+$$

$$V_2'^- = S_{21}\frac{m^2 - n^2}{\Delta}V_1'^+ + \frac{m^2 S_{22} + mn(1 + S_{11}S_{22} - S_{12}S_{21}) + n^2 S_{11}}{\Delta}V_2'^+$$

The new S-parameters in matrix form, due to a change in normalization impedance as shown in Figure 1.15, are given as follows:

$$[S'] = \frac{1}{\Delta}\begin{bmatrix} m^2 S_{11} + mn \cdot K + n^2 S_{22} & S_{12}(m^2 - n^2) \\ S_{21}(m^2 - n^2) & m^2 S_{22} + mn \cdot K + n^2 S_{11} \end{bmatrix} \quad (1.171)$$

where

$$K = 1 + S_{11}S_{22} - S_{12}S_{21}$$

It can be shown that (1.171) is reduced to the original S-parameters of the circuit when $m = 1$ and $n = 0$ (i.e., $Z_c = Z_c'$). Note that the assumption of original transmission line segments with characteristic impedance Z_c is just a mathematical tool; it is not necessary to physically have these lines. However, since we are defining traveling waves on the lines, one can assume an infinitesimal length of transmission lines with characteristic impedance Z_c to start with because no reference to the line lengths is made. Since the lines have the same characteristic impedance as the normalization impedance, addition of an infinitesimal length wouldn't affect the S-parameters. A similar argument would apply to the lines with characteristic impedance Z_c'.

Note that the above formulation can be used to convert the S-parameters of any given two-port linear network to new ones with different normalization impedance. Generalization of the above analysis for an n-port network is given as follows [9]:

$$S' = A^{-1}(S - \Gamma^+) \cdot (U - \Gamma S)^{-1} A^+ \quad (1.172)$$

where S is the original scattering matrix, S' is the modified scattering matrix due to change in normalizing impedances, and U is the identity matrix. Γ and A are diagonal matrixes whose entries are given as

$$\Gamma_{ii} = \frac{Z'_i - Z_i}{Z'_i + Z^*_i}$$

$$A_{ii} = \left(1 - \Gamma^*_i\right)\frac{\sqrt{1 - \Gamma_i \Gamma^*_i}}{\left|1 - \Gamma_i\right|}$$

where Z_i and Z'_i are the original and the new characteristic impedances of the ith port, respectively.

1.4.2 Transfer Parameters

One of the disadvantages of the S-parameters is that S-parameters of cascaded networks cannot be found as easily as $ABCD$-parameters. Transfer parameters (or T-parameters) address this problem by relating traveling waves on the transmission lines in such a way that overall T-parameters of cascaded networks are given by the multiplication of individual T-parameters. This property is quite useful in the analysis of cascaded networks.

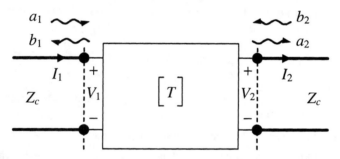

Figure 1.16 Representation of incident and reflected waves for the T-parameters. Compare this figure with the one given for S-parameters in Figure 1.13.

To demonstrate this, consider the circuit given in Figure 1.16. The definition of transfer parameters is given as follows:

$$\begin{bmatrix} a_1 \\ b_1 \end{bmatrix} = \begin{bmatrix} T_{11} & T_{12} \\ T_{21} & T_{22} \end{bmatrix} \begin{bmatrix} a_2 \\ b_2 \end{bmatrix} \tag{1.173}$$

Note that the direction of the traveling waves on the second port is reversed compared to S-parameters. Then, transfer parameters of cascaded N networks are obtained as

$$[T] = \begin{bmatrix} T^{(1)}_{11} & T^{(1)}_{12} \\ T^{(1)}_{21} & T^{(1)}_{22} \end{bmatrix} \times \begin{bmatrix} T^{(2)}_{11} & T^{(2)}_{12} \\ T^{(2)}_{21} & T^{(2)}_{22} \end{bmatrix} \times \cdots \times \begin{bmatrix} T^{(N)}_{11} & T^{(N)}_{12} \\ T^{(N)}_{21} & T^{(N)}_{22} \end{bmatrix} \tag{1.174}$$

One can easily verify that the following relationships exist between the T-parameters and S-parameters for a two-port network:

$$
\begin{bmatrix} T_{11} & T_{12} \\ T_{21} & T_{22} \end{bmatrix} = \begin{bmatrix} \dfrac{1}{S_{21}} & -\dfrac{S_{22}}{S_{21}} \\ \dfrac{S_{11}}{S_{21}} & \dfrac{S_{12}S_{21} - S_{11}S_{22}}{S_{21}} \end{bmatrix} \tag{1.175}
$$

$$
\begin{bmatrix} S_{11} & S_{12} \\ S_{21} & S_{22} \end{bmatrix} = \begin{bmatrix} \dfrac{T_{21}}{T_{11}} & -\dfrac{T_{12}T_{21} - T_{11}T_{22}}{T_{11}} \\ \dfrac{1}{T_{11}} & -\dfrac{T_{12}}{T_{11}} \end{bmatrix} \tag{1.176}
$$

Therefore, to evaluate the S-parameters of cascaded two-port networks, one can first convert the S-parameters of each network to T-parameters using (1.175). Then, overall T-parameters are found by virtue of (1.174). Finally, the resultant T-parameters are converted back to S-parameters using (1.176).

1.4.3 Z- and Y-Parameters

Z- and Y-parameters are defined by using the terminal voltages and currents shown in Figure 1.17. Note the port current convention is defined in such a way that the currents entering the circuit are taken to be positive.

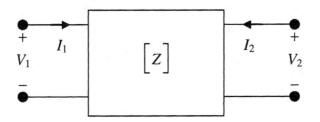

Figure 1.17 Definitions of voltages and currents for Z- and Y-parameters.

Using these terminal voltages and currents, we can write Z-parameter equations as follows:

$$
\begin{aligned}
V_1 &= Z_{11}I_1 + Z_{12}I_2 \\
V_2 &= Z_{21}I_1 + Z_{22}I_2
\end{aligned} \tag{1.177}
$$

The above equations can be put in matrix form as

$$
\begin{bmatrix} V_1 \\ V_2 \end{bmatrix} = \begin{bmatrix} Z_{11} & Z_{12} \\ Z_{21} & Z_{22} \end{bmatrix} \begin{bmatrix} I_1 \\ I_2 \end{bmatrix} \tag{1.178}
$$

Similarly, for the Y-parameters

Modern Microwave Circuits

$$I_1 = Y_{11}V_1 + Y_{12}V_2$$
$$I_2 = Y_{21}V_1 + Y_{22}V_2$$

(1.179)

and

$$\begin{bmatrix} I_1 \\ I_2 \end{bmatrix} = \begin{bmatrix} Y_{11} & Y_{12} \\ Y_{21} & Y_{22} \end{bmatrix} \begin{bmatrix} V_1 \\ V_2 \end{bmatrix}$$

(1.180)

From the above definitions, it is easy to see that

$$\begin{bmatrix} Z_{11} & Z_{12} \\ Z_{21} & Z_{22} \end{bmatrix} = \begin{bmatrix} Y_{11} & Y_{12} \\ Y_{21} & Y_{22} \end{bmatrix}^{-1}$$

(1.181)

provided that the inverse of the Y-parameters matrix exists.

The relationship between Z-parameters and S-parameters can be derived as follows [10]:

$$\begin{bmatrix} V_1^+ + V_1^- \\ V_2^+ + V_2^- \end{bmatrix} = \begin{bmatrix} Z_{11} & Z_{12} \\ Z_{21} & Z_{22} \end{bmatrix} \begin{bmatrix} \frac{1}{Z_c}(V_1^+ - V_1^-) \\ \frac{1}{Z_c}(V_2^+ - V_2^-) \end{bmatrix}$$

After some mathematical operations

$$\left(\frac{1}{Z_c}\begin{bmatrix} Z_{11} & Z_{12} \\ Z_{21} & Z_{22} \end{bmatrix} + \mathbf{U} \right) \begin{bmatrix} V_1^- \\ V_2^- \end{bmatrix} = \left(\frac{1}{Z_c}\begin{bmatrix} Z_{11} & Z_{12} \\ Z_{21} & Z_{22} \end{bmatrix} - \mathbf{U} \right) \begin{bmatrix} V_1^+ \\ V_2^+ \end{bmatrix}$$

$$\begin{bmatrix} V_1^- \\ V_2^- \end{bmatrix} = \left(\frac{1}{Z_c}\begin{bmatrix} Z_{11} & Z_{12} \\ Z_{21} & Z_{22} \end{bmatrix} + \mathbf{U} \right)^{-1} \left(\frac{1}{Z_c}\begin{bmatrix} Z_{11} & Z_{12} \\ Z_{21} & Z_{22} \end{bmatrix} - \mathbf{U} \right) \begin{bmatrix} V_1^+ \\ V_2^+ \end{bmatrix}$$

where $\mathbf{U}$ is the identity matrix. From above derivation, it is easy to see that

$$\begin{bmatrix} S_{11} & S_{12} \\ S_{21} & S_{22} \end{bmatrix} = \left(\begin{bmatrix} Z_{11} & Z_{12} \\ Z_{21} & Z_{22} \end{bmatrix} + \mathbf{Z}_c \right)^{-1} \left(\begin{bmatrix} Z_{11} & Z_{12} \\ Z_{21} & Z_{22} \end{bmatrix} - \mathbf{Z}_c \right)$$

(1.182)

and

$$\begin{bmatrix} Z_{11} & Z_{12} \\ Z_{21} & Z_{22} \end{bmatrix} = \mathbf{Z}_c \left(\mathbf{U} + \begin{bmatrix} S_{11} & S_{12} \\ S_{21} & S_{22} \end{bmatrix} \right) \left(\mathbf{U} - \begin{bmatrix} S_{11} & S_{12} \\ S_{21} & S_{22} \end{bmatrix} \right)^{-1}$$

(1.183)

where $\mathbf{Z}_c$ is the diagonal matrix whose entries are the characteristic impedances (normalization impedance) of the port transmission lines:

$$\mathbf{Z}_c = \begin{bmatrix} Z_{c1} & 0 \\ 0 & Z_{c2} \end{bmatrix}$$

1.4.4 *ABCD*-Parameters

ABCD-parameters are defined by using the terminal voltages and currents shown in Figure 1.18. Note that the direction of current on the second port is reversed compared to the definitions used for *Z*-parameters.

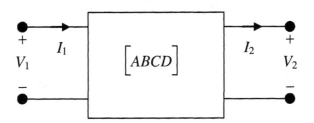

Figure 1.18 Definitions of voltages and currents for *ABCD*-parameters.

Using these terminal voltages and currents, we can write *ABCD*-parameter equations as follows:

$$V_1 = AV_2 + BI_2$$
$$I_1 = CV_2 + DI_2$$

(1.184)

The above equations can be put in matrix form as

$$\begin{bmatrix} V_1 \\ I_1 \end{bmatrix} = \begin{bmatrix} A & B \\ C & D \end{bmatrix} \begin{bmatrix} V_2 \\ I_2 \end{bmatrix}$$

(1.185)

One of the advantages of using the *ABCD*-parameters is that the overall *ABCD*-parameters of a cascaded network equal the multiplication of the *ABCD*-parameters of each individual network:

$$[ABCD] = \begin{bmatrix} A_{11}^{(1)} & B_{12}^{(1)} \\ C_{21}^{(1)} & D_{22}^{(1)} \end{bmatrix} \times \begin{bmatrix} A_{11}^{(2)} & B_{12}^{(2)} \\ C_{21}^{(2)} & D_{22}^{(2)} \end{bmatrix} \times \cdots \times \begin{bmatrix} A_{11}^{(N)} & B_{12}^{(N)} \\ C_{21}^{(N)} & D_{22}^{(N)} \end{bmatrix}$$

(1.186)

where N is the number of networks. This property is very useful in evaluating overall circuit parameters of cascaded networks.

The relationships between *ABCD*-parameters and *Z*-parameters are given as follows:

$$\begin{bmatrix} Z_{11} & Z_{12} \\ Z_{21} & Z_{22} \end{bmatrix} = \frac{1}{C} \begin{bmatrix} A & AD - BC \\ 1 & D \end{bmatrix}$$

(1.187)

$$\begin{bmatrix} A & B \\ C & D \end{bmatrix} = \frac{1}{Z_{21}} \begin{bmatrix} Z_{11} & Z_{11}Z_{22} - Z_{12}Z_{21} \\ 1 & Z_{22} \end{bmatrix}$$

(1.188)

Modern Microwave Circuits

Table 1.1
Circuit Parameters of Various Simple Networks

Network	Z- and Y-Parameters	ABCD-Parameters
Z	$Y_{11} = 1/Z$ $Y_{22} = 1/Z$ $Y_{12} = -1/Z$	$A = 1$ $B = Z$ $C = 0$ $D = 1$
Y	$Z_{11} = 1/Y$ $Z_{22} = 1/Y$ $Z_{12} = 1/Y$	$A = 1$ $B = 0$ $C = Y$ $D = 1$
Z_c l	$Z_{11} = Z_c \coth \gamma l$ $Z_{22} = Z_c \coth \gamma l$ $Z_{12} = Z_c / \sinh \gamma l$ $Y_{11} = Y_c \coth \gamma l$ $Y_{22} = Y_c \coth \gamma l$ $Z_{12} = -Y_c / \sinh \gamma l$	$A = \cosh \gamma l$ $B = Z_c \sinh \gamma l$ $C = Y_c \sinh \gamma l$ $D = \cosh \gamma l$
Y_c Y_a Y_b	$Y_{11} = Y_a + Y_c$ $Y_{22} = Y_b + Y_c$ $Y_{12} = -Y_c$ $Z_{11} = (Y_b + Y_c)/\Delta Y$ $Z_{22} = (Y_a + Y_c)/\Delta Y$ $Z_{12} = Y_c/\Delta Y$ $\Delta Y = Y_a Y_b + Y_b Y_c + Y_c Y_a$	$A = 1 + Y_b/Y_c$ $B = 1/Y_c$ $C = Y_a + Y_b + Y_a Y_b/Y_c$ $D = 1 + Y_a/Y_c$
Z_a Z_b Z_c	$Z_{11} = Z_a + Z_c$ $Z_{22} = Z_b + Z_c$ $Z_{12} = Z_c$ $Y_{11} = (Z_b + Z_c)/\Delta Z$ $Y_{22} = (Z_a + Z_c)/\Delta Z$ $Y_{12} = -Z_c/\Delta Z$ $\Delta Z = Z_a Z_b + Z_b Z_c + Z_c Z_a$	$A = 1 + Z_a/Z_c$ $B = Z_a + Z_b + Z_a Z_b/Z_c$ $C = 1/Z_c$ $D = 1 + Z_b/Z_c$

1.5 CIRCUIT PARAMETERS OF VARIOUS SIMPLE NETWORKS

In Section 1.4, we introduced the most commonly used circuit parameter definitions for microwave networks. In this section, we provide the circuit parameters of various basic networks that are frequently used in microwave engineering.

The studied networks in this section are simple TEM transmission line, T-network, and Pi-network. These simple networks are frequently used in filter circuits and modeling of passive microstrip elements. After providing the circuit parameters, we will also show mathematical derivations for *ABCD*-parameters for the sake of completeness. This will help the readers to derive the circuit parameters of other circuits that are not provided here. Circuit parameters of some of the most commonly used networks are summarized in Table 1.1.

1.5.1 TEM Transmission Line

In this section, we give the *ABCD*-parameters of a TEM transmission line characterized by a given length, characteristic impedance, and propagation constant. Voltage and current conventions that will be used for this purpose are depicted in Figure 1.19.

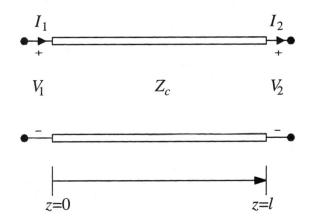

Figure 1.19 TEM transmission line and voltage and current conventions used in the *ABCD*-parameter derivations.

It was shown previously that the voltage and current on a TEM transmission line are given as follows:

$$V(z) = V^+ e^{-\gamma z} + V^- e^{\gamma z}$$

$$I(z) = \frac{V^+}{Z_c} e^{-\gamma z} - \frac{V^-}{Z_c} e^{\gamma z}$$

where Z_c and γ are the characteristic impedance and complex propagation constant, respectively. To find the $ABCD$-parameters, we will evaluate voltages and currents on the terminals of the transmission line when it is terminated by short and open circuits. Then, by using the definitions of $ABCD$-parameters, one can show that

$$A = \left.\frac{V_1}{V_2}\right|_{I_2=0} \tag{1.189}$$

$$B = \left.\frac{V_1}{I_2}\right|_{V_2=0} \tag{1.190}$$

$$C = \left.\frac{I_1}{V_2}\right|_{I_2=0} \tag{1.191}$$

$$D = \left.\frac{I_1}{I_2}\right|_{V_2=0} \tag{1.192}$$

To demonstrate the approach, we will show the derivation of the first parameter given in (1.189). By using the voltage and current expression on the transmission line, we first write

$$A = \frac{V^+ + V^-}{V^+ e^{-\gamma l} + V^- e^{\gamma l}}$$

Now, to find an expression for the above parameter, we must determine the amplitudes of forward- and backward-traveling waves. The boundary conditions on the line are used for this purpose:

$$V_1 = V^+ + V^- = 1$$

$$I_2 = \frac{V^+}{Z_c} e^{-\gamma l} - \frac{V^-}{Z_c} e^{\gamma l} = 0$$

Note that it is assumed that the line is driven by a generator with a 1-V amplitude. By using the above expressions, one can determine amplitudes of the forward- and backward-traveling waves. Then, the first parameter is found as follows:

$$A = \frac{1 + e^{2\gamma l}}{2 e^{\gamma l}} = \frac{1}{2}\left(e^{-\gamma l} + e^{\gamma l}\right)$$

By using appropriate trigonometric identities, one can show that

$$A = \cosh(\gamma l) \tag{1.193}$$

The rest of the parameters are derived using a similar approach:

$$B = Z_c \sinh(\gamma l) \tag{1.194}$$

$$C = \frac{1}{Z_c} \sinh(\gamma l) \tag{1.195}$$

$$D = \cosh(\gamma l) \tag{1.196}$$

This completes the derivation of the *ABCD*-parameters of a TEM transmission line. It is convenient to put the above equations in matrix form as follows:

$$\begin{bmatrix} A & B \\ C & D \end{bmatrix} = \begin{bmatrix} \cosh(\gamma l) & Z_c \sinh(\gamma l) \\ \dfrac{1}{Z_c} \sinh(\gamma l) & \cosh(\gamma l) \end{bmatrix} \tag{1.197}$$

For a lossless transmission line, the above matrix reduces to

$$\begin{bmatrix} A & B \\ C & D \end{bmatrix} = \begin{bmatrix} \cos(\beta l) & jZ_c \sin(\beta l) \\ j\dfrac{1}{Z_c} \sin(\beta l) & \cos(\beta l) \end{bmatrix} \tag{1.198}$$

1.5.2 Pi-Network

In this section, we will give the *ABCD*-parameters of a Pi-network, shown in Figure 1.20.

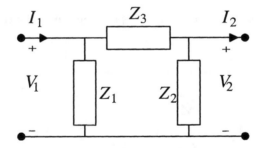

Figure 1.20 Pi-network and voltage and current conventions used in the derivations.

The *ABCD*-parameters of a Pi-network can be found using the same approach that we demonstrated for TEM transmission line:

$$A = \left.\frac{V_1}{V_2}\right|_{I_2 = 0} = \frac{V_1}{\dfrac{V_1}{Z_2 + Z_3} Z_2} = 1 + \frac{Z_3}{Z_2} \tag{1.199}$$

$$B = \left.\frac{V_1}{I_2}\right|_{V_2 = 0} = \frac{V_1}{\dfrac{V_1}{Z_3}} = Z_3 \tag{1.200}$$

$$C = \frac{I_1}{V_2}\bigg|_{I_2=0} = \frac{\dfrac{V_1}{(Z_2+Z_3)Z_1}}{\dfrac{Z_1+Z_2+Z_3}{Z_2+Z_3}Z_2} = \frac{1}{Z_1} + \frac{1}{Z_2} + \frac{Z_3}{Z_1 Z_2} \qquad (1.201)$$

$$D = \frac{I_1}{I_2}\bigg|_{V_2=0} = \frac{\dfrac{V_1}{Z_1 Z_3}}{\dfrac{Z_1+Z_3}{V_1}} = 1 + \frac{Z_3}{Z_1} \qquad (1.202)$$

The *ABCD*-parameters given above can be put into matrix form as

$$\begin{bmatrix} A & B \\ C & D \end{bmatrix} = \begin{bmatrix} 1+\dfrac{Z_3}{Z_2} & Z_3 \\ \dfrac{1}{Z_1}+\dfrac{1}{Z_2}+\dfrac{Z_3}{Z_1 Z_2} & 1+\dfrac{Z_3}{Z_1} \end{bmatrix} \qquad (1.203)$$

1.5.3 T-Network

In this section, we will give the *ABCD*-parameters of a T-network, shown in Figure 1.21.

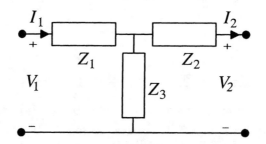

Figure 1.21 T-network and voltage and current conventions used in the derivations.

The *ABCD*-parameters of a T-network can be found using the same approach that we demonstrated for TEM transmission lines as follows:

$$A = \frac{V_1}{V_2}\bigg|_{I_2=0} = \frac{V_1}{\dfrac{V_1}{Z_1+Z_3}Z_3} = 1 + \frac{Z_1}{Z_3} \qquad (1.204)$$

$$B = \frac{V_1}{I_2}\bigg|_{V_2=0} = \frac{V_1}{Z_1 + \frac{Z_2 Z_3}{Z_2 + Z_3}} \frac{1}{Z_2} \frac{Z_2 Z_3}{Z_2 + Z_3} = \frac{Z_2 Z_1}{Z_3} + Z_1 + Z_2 \qquad (1.205)$$

$$C = \frac{I_1}{V_2}\bigg|_{I_2=0} = \frac{\dfrac{V_1}{Z_1 + Z_3}}{\dfrac{V_1}{Z_1 + Z_3} Z_3} = \frac{1}{Z_3} \qquad (1.206)$$

$$D = \frac{I_1}{I_2}\bigg|_{V_2=0} = \frac{\dfrac{V_1}{Z_1 + \dfrac{Z_2 Z_3}{Z_2 + Z_3}}}{Z_1 + \dfrac{Z_2 Z_3}{Z_2 + Z_3} \dfrac{V_1}{Z_2} \dfrac{1}{Z_2} \dfrac{Z_2 Z_3}{Z_2 + Z_3}} = 1 + \frac{Z_2}{Z_3} \qquad (1.207)$$

The above given *ABCD*-parameters can be put into matrix form as

$$\begin{bmatrix} A & B \\ C & D \end{bmatrix} = \begin{bmatrix} 1 + \dfrac{Z_1}{Z_3} & Z_1 + Z_2 + \dfrac{Z_2 Z_1}{Z_3} \\ \dfrac{1}{Z_3} & 1 + \dfrac{Z_2}{Z_3} \end{bmatrix} \qquad (1.208)$$

1.6 EQUIVALENT CIRCUIT OF A SHORT TRANSMISSION LINE

In microwave engineering, it is sometimes required to approximate electrically short transmission lines with an equivalent Pi-network. This kind of equivalent network of the short transmission line is commonly used in modeling of passive microstrip components as well as lumped approximation of distributed networks.

To find the equivalent network of an electrically short transmission line, we first assume the loss is insignificant and then find the *ABCD*-parameters of the line for $\beta l \ll 1$ by expanding trigonometric functions in series:

$$\begin{bmatrix} A & B \\ C & D \end{bmatrix} = \begin{bmatrix} 1 + \dfrac{(j\beta l)^2}{2} & Z_c \left(j\beta l + \dfrac{(j\beta l)^3}{6} \right) \\ \dfrac{1}{Z_c} \left(j\beta l + \dfrac{(j\beta l)^3}{6} \right) & 1 + \dfrac{(j\beta l)^2}{2} \end{bmatrix} \qquad (1.209)$$

Then, by comparing (1.203) and (1.209), we get

$$1+\frac{Z_3}{Z_2} = 1+\frac{Z_3}{Z_1} = 1+\frac{(j\beta l)^2}{2}$$

$$\frac{2}{Z_1}+\frac{Z_3}{Z_1 Z_2} = \frac{1}{Z_c}\left(j\beta l + \frac{(j\beta l)^3}{6}\right)$$

by assuming

$$\left|\frac{Z_3}{Z_1 Z_2}\right| \ll 1 \qquad \left|\frac{(j\beta l)^3}{6}\right| \ll 1$$

one obtains,

$$\frac{2}{Z_1} \cong \frac{j\beta l}{Z_c}$$

$$Z_3 = Z_c\left(j\beta l + \frac{(j\beta l)^3}{6}\right) \cong Z_c j\beta l$$

The above equations can be solved for the unknown impedance values by noting that $\beta = \omega\sqrt{LC}$ and $Z_c = \sqrt{L/C}$:

$$Z_1 = Z_2 = \frac{1}{j\omega C\dfrac{l}{2}} \qquad\qquad (1.210)$$

$$Z_3 = j\omega Ll \qquad\qquad (1.211)$$

where L and C are the inductance and the capacitance per unit length of the line, respectively, and l is the total length of the line. The resulting equivalent network of the electrical short transmission line is shown in Figure 1.22.

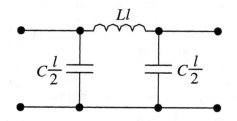

Figure 1.22 Equivalent network of an electrically short transmission line ($\beta l \ll 1$). L and C are inductance and capacitance per unit length of the line, respectively, and l is the total length of the line.

This result tells us that a short piece of transmission line can be approximated by a simple Pi-network whose component values are found from the parameters of the line. This is an important conclusion because it finds a vast application in reducing the distributed networks to equivalent lumped networks. For instance, at relatively low frequencies, microstrip branch line couplers can take up significant circuit board area. However, by using the equivalent network concept of the short transmission line demonstrated here, it is possible to design a branch line coupler using only lumped elements in a fraction of the area used by the distributed equivalent circuit (at the cost of reduced bandwidth).

Note that this approximation of transmission lines with a lumped Pi-network is theoretically good for $\beta l \ll 1$. However, it can be still used when $\beta l \leq \pi/2$ with reasonable accuracy if the bandwidth is small.

1.7 SIGNAL FLOWGRAPHS

Analysis of complex circuits by using scattering matrixes can be difficult due to involved matrix algebra. Besides, analysis using matrix algebra does not provide much insight to actual physical interactions through the circuits. Signal flowgraphs eliminate this and provide an alternative way of analyzing circuits using scattering waves. In signal flowgraphs, each node is represented by a particular wave name and lines connecting these nodes represent interactions between the relative waves as depicted in Figure 1.23. Note that the two-port circuit shown in the figure now has four wires connecting to the external world where each of them carries one of the four traveling waves. And the nodes are labeled in order to facilitate easy connection for cascading circuits.

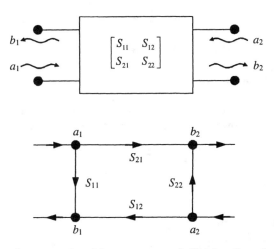

Figure 1.23 Flowgraph representation of the two-port network. Note how the nodes are labeled.

To cascade two circuits, one just needs to connect the appropriate wires of the flowgraphs as depicted in Figure 1.24. This can be easily generalized to any number of ports and circuit topologies. To find the transmission between any nodes, one traverses the existing connections between those nodes (by paying attention to the direction of arrows) and multiplies the signal by the gain factor of each connection. Of course, feedbacks should be taken into account during this process.

The resulting equation obtained using the signal flowgraphs may not be simpler than the one achieved through matrix manipulation of scattering equations, but it is usually faster and more convenient to use signal flowgraphs. The chief advantage of flowgraphs is that one can gain the physical insight even into complex circuits very easily. Similarly, flowgraphs of complex circuits can easily be generated by inspecting the physical interactions between various components. The notion of signal flowgraphs will be helpful in deriving the equivalent error models of network analyzers in the following sections.

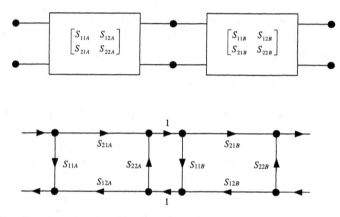

Figure 1.24 Cascade connection of two-port networks using flowgraphs.

1.7.1 Nontouching Loop Rule

The nontouching loop rule can be employed to find the ratio of an input node to an output node in signal flowgraphs [11, 12]. According to the nontouching loop rule, the ratio of the nodes is given as

$$T = \frac{\sum T_k \Delta_k}{\Delta} \tag{1.212}$$

where

T_k = Path gain on the kth forward path between the nodes

$\Delta = 1 - $ (Sum of all individual loop gains) + (Sum of loop gain products of all possible combinations of two nontouching loops) − (Sum of loop gain products of all possible combinations of three nontouching loops) + $\cdots$

Δ_k = Sum of all terms in Δ not touching the kth path.

A path must always follow the arrow directions, and a node may be passed more than once. It either forms a loop or connects the input to the output node. A first-order loop is a closed path that can be followed in the direction of the arrows. The value of the loop is the product of all path coefficients encountered during traversing. A second-order loop is the product of any two nontouching first-order loops. Similarly, a third-order loop is the product of any three nontouching first-order loops, and so forth.

Various topological rules used to simplify flowgraphs are presented in Figure 1.25.

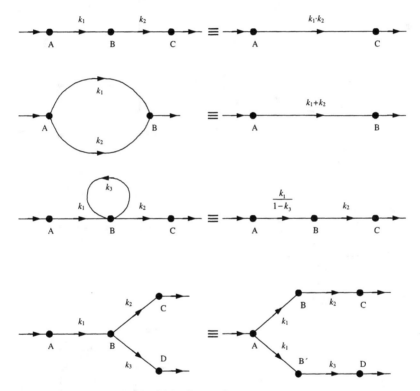

Figure 1.25 Topological rules for simplifying flowgraphs.

1.7.2 Signal Flowgraphs of Some Microwave Components

The flowgraph representation shown in Figure 1.23 is useful for general networks. It doesn't explicitly show how to model a source, a load, or a detector. Flowgraph modeling of these components is important to generate useful flowgraph models of practical devices such as power detectors and network analyzers. To facilitate this, equivalent flowgraphs of various microwave circuits are given in Figure 1.26.

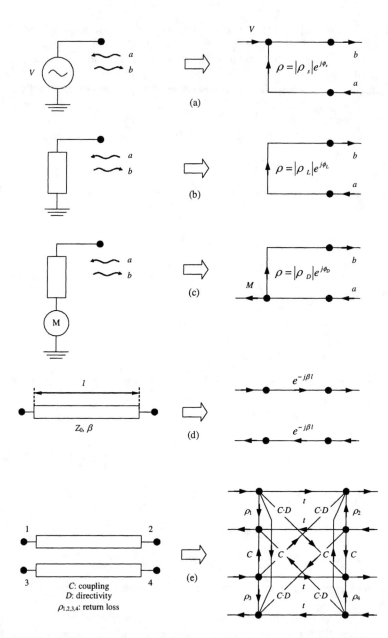

Figure 1.26 Flowgraph models of some commonly used microwave components: (a) signal source, (b) termination, (c) detector, (d) matched transmission line, and (e) microstrip directional coupler.

1.8 POWER GAIN

In this section, we will provide power gain definitions for a two-port network described by its S-parameters when it is connected between an arbitrary generator and load. There are basically three different definitions of the power gain of a two-port network: power gain, available power gain, and transducer power gain. They are given as follows [13]:

$$G = \frac{\text{Power delivered to the load}}{\text{Power input to the network}} \tag{1.213}$$

$$G_A = \frac{\text{Power available from the network}}{\text{Power available from the source}} \tag{1.214}$$

$$G_T = \frac{\text{Power delivered to the load}}{\text{Power available from the source}} \tag{1.215}$$

In this section, we will concentrate on the transducer power gain because it is the most practical definition. The transducer power gain provides a measure of the power gain when a supposedly existing passive matching circuit between the generator and load is replaced by an active two-port network (e.g., amplifier) to provide gain. Note that since the passive matching circuit can extract all the available power from the generator, the active two-port network must provide some additional gain in order to be useful. This gain is given by the transducer power gain (G_T).

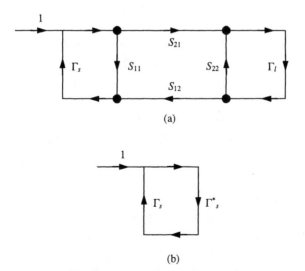

(a)

(b)

Figure 1.27 Signal flowgraph models of (a) two-port amplifier connected between an arbitrary generator and load, and (b) the generator terminated with a complex conjugate of its impedance.

If the transducer gain is unity for an amplifier circuit, then the active circuit isn't needed, and the same result could be achieved by just doing complex-conjugate matching using a passive matching circuit.

To derive the transducer power gain, we use the signal flowgraph of a two-port amplifier connected between a generator and load as shown in Figure 1.27(a). In the figure, Γ_S and Γ_L represent the generator and load reflection coefficients, respectively. We assume unity source amplitude without loss of generality. The first step is to write the power delivered to load as follows:

$$P_L = |b|^2 - |a|^2$$
$$= |b|^2 \left(1 - |\Gamma_L|^2\right) \tag{1.216}$$

where b and a are the incident and reflected waves on the load terminals, respectively. Then, by following the flowgraph model, the power delivered to the load is found as

$$P_L = \frac{|S_{21}|^2 \left(1 - |\Gamma_L|^2\right)}{|(1 - S_{22}\Gamma_L)(1 - S_{11}\Gamma_S) - S_{12}S_{21}\Gamma_S\Gamma_L|^2} \tag{1.217}$$

Note that to find the transducer power gain, we must also find the available power from the source. This can be determined with the help of Figure 1.27(b). To determine the available power from the source, we terminate the source by the complex conjugate of its impedance. This gives the available power as

$$P_{avs} = \frac{1}{1 - |\Gamma_S|^2} \tag{1.218}$$

Finally, by taking the ratio of (1.217) and (1.218), G_T is found as follows:

$$G_T = \frac{P_L}{P_{avs}} = \frac{|S_{21}|^2 \left(1 - |\Gamma_L|^2\right) \cdot \left(1 - |\Gamma_S|^2\right)}{|(1 - S_{22}\Gamma_L) \cdot (1 - S_{11}\Gamma_S) - S_{12}S_{21}\Gamma_S\Gamma_L|^2} \tag{1.219}$$

As a particular case, if $\Gamma_S = \Gamma_L = 0$, then the power gain reduces to $|S_{21}|^2$, which is not necessarily the maximum. Readers should be reminded that the condition $\Gamma_S = \Gamma_L = 0$ means only that the terminations are set to the normalization impedance so that there are no reflections (by definition). Circuit theory tells us that the maximum power transfer occurs under the complex-conjugate match conditions, which means $\Gamma_S \neq \Gamma_L \neq 0$ in general. If the source is matched ($\Gamma_S = 0$) and the amplifier is unilateral, which can be a good approximation for most practical cases, then maximum power gain occurs at $\Gamma_L = S_{22}^*$.

Note that under simultaneous complex-conjugate matching (both ports are conjugate matched), all three of the gain definitions given above provide the same result.

1.9 THE SMITH CHART

The Smith chart is a graphical tool that represents a mapping between impedance and reflection coefficient planes. It was introduced by Phil Smith of RCA in 1936. It is an extremely convenient tool because it maps the infinite Z_{in} plane to a finite Γ_{in} plane. For this reason, it greatly simplifies the impedance matching problems. Once it is understood properly, one can easily determine the type of matching network and the amount of matching required to match a particular impedance by just inspecting the location of that impedance point on the Smith chart.

To show how the chart is constructed, we first write the relationship between input impedance and the reflection coefficient:

$$Z_{in} = Z_0 \frac{1 + \Gamma_{in}}{1 - \Gamma_{in}} \tag{1.220}$$

where Z_0 is the reference impedance. Then, the above equation can be put into the following form by separating the real and imaginary parts:

$$\overline{R} + j\overline{X} = \frac{1 + \rho e^{j\theta}}{1 - \rho e^{j\theta}}$$

$$= \frac{1 + u + jv}{1 - u - jv}$$

where $u = \rho \cos\theta$ and $v = \rho \sin\theta$. $\overline{R}$ and $\overline{X}$ are normalized resistance and reactance, respectively. By equating the real and imaginary parts, the following sets of equations can be obtained:

$$\overline{R} - \overline{R}u + \overline{X}v = 1 + u$$

$$-\overline{R}v + \overline{X} - \overline{X}u = v$$

Then, by solving the above equations for $\overline{R}$ and $\overline{X}$, one gets the following:

$$\overline{X}v^2 - 2v + \overline{X}(1-u)^2 = 0$$

$$\overline{R}(1-u)^2 + v^2 + v^2\overline{R} = 1 - u^2$$

The above sets of equations can be placed into a more convenient form as follows:

$$(u-1)^2 + \left(v - \frac{1}{\overline{X}}\right)^2 = \frac{1}{\overline{X}^2} \tag{1.221}$$

$$\left(u - \frac{\overline{R}}{\overline{R}+1}\right)^2 + v^2 = \frac{1}{(\overline{R}+1)^2}$$

which represents circles in the u and v planes (i.e., reflection coefficient plane) for constant reactance, X, and constant resistance, R, respectively. Figure 1.28 shows the Smith chart constructed using circles obtained through the above equations.

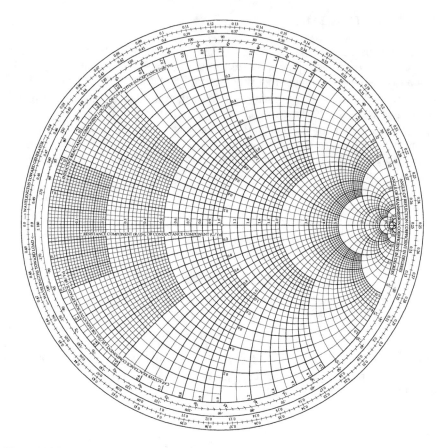

Figure 1.28 The Smith chart. "Smith" is a registered trademark of Analog Instruments Company, New Providence, New Jersey. Reprinted with permission.

At this point, it is worthwhile to discuss some properties of the Smith chart. The center of the Smith chart represents the matched condition (i.e., $\Gamma = 0$). The right ($\Gamma = 1$) and left ($\Gamma = -1$) corners of the horizontal axis are the open circuit and short circuit conditions, respectively. The top half of the impedance plane represents the inductive region and the bottom half represents the capacitive region. By using the Smith chart, one can easily determine the input impedance of a circuit if the reflection coefficient of the circuit is given, and vice versa.

Figure 1.29 shows the relation between a given impedance (Z) and the reflection coefficient (Γ) on the Smith chart. For the sake of demonstration, the impedance transformation (Z_{in}) due to a piece of transmission line, which has the same characteristic impedance as the normalization impedance, is also shown on the figure.

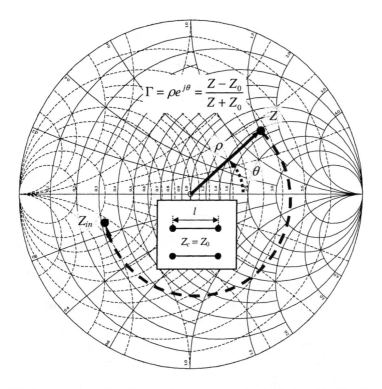

$$\Gamma = \rho e^{j\theta} = \frac{Z - Z_0}{Z + Z_0}$$

Figure 1.29 Representation of impedance and return loss on the Smith chart. Resulting impedance transformation due to a piece of transmission line is also shown. Z_0 is the reference impedance.

An important property of the Smith chart is that the input impedance locus of passive lossless circuits always follows the clockwise direction; any other behavior is not physical. The formal proof of this is given by the Foster's reactance theorem, which will be introduced at the end of this chapter. At this point, we can demonstrate this for a simple parallel *LC* resonator terminated with a resistor as follows. First, we write the input impedance of the resonator

$$Z_{in} = \frac{j\omega LR}{1 + R - \omega^2 RLC}$$

Then, reflection coefficient is determined as

$$\Gamma = \frac{Z_{in} - Z_0}{Z_{in} + Z_0} = \frac{j\omega LR - Z_0 - Z_0 R + Z_0 \omega^2 RLC}{\omega LR + Z_0 + Z_0 R - Z_0 \omega^2 RLC}$$

To show that rotation of the reflection coefficient is clockwise with respect to frequency, we should show that the frequency derivative of the phase of the reflection coefficient is always negative. To accomplish this, we write the phase of the reflection coefficient as follows:

$$\angle\Gamma = -2\tan^{-1}\left(\frac{\omega LR}{Z_0 + Z_0 R - Z_0\omega^2 RLC}\right)$$

Then, the derivative of the phase with respect to frequency is evaluated as

$$\frac{d}{d\omega}\angle\Gamma = \frac{-2}{1 + \left(\dfrac{\omega LR}{Z_0 + Z_0 R - Z_0\omega^2 RLC}\right)^2}\frac{RLZ_0\left(1 + R + \omega^2 RLC\right)}{\left(Z_0 + Z_0 R - Z_0\omega^2 RLC\right)^2}$$

The above equation is always less than zero, regardless of the value of R, L, and C, provided that they are all positive (i.e., the circuit is passive).

1.10 IMPEDANCE MATCHING

There are two different concepts of impedance matching in microwave engineering. The first one refers to terminating a transmission line with the characteristic impedance of the line. The second one refers to terminating a general source with the complex conjugate of the source impedance. The latter concept is the fundamental requirement of maximum power transfer. Note that depending on the context, one should distinguish these concepts from each other. In this section, we will investigate the problem of terminating a source with the complex conjugate of its output impedance. Note that the term *source* is used in a general sense here; it represents any network (amplifier, filter, and so forth) connected to a load resistance.

In order to terminate a general source with the complex conjugate of its output impedance, the load resistance (usually 50 ohms) must be transferred to the required value by means of an appropriate network which can be distributed or lumped. This network is called a matching network, and there are many ways of implementing it. Depending on the bandwidth requirement, the matching network can be quite complicated. Matching using stubs is the most basic form of matching networks and will be discussed first.

1.10.1 Single-Stub Matching

Single-stub matching employs a transmission line and a stub as shown in Figure 1.30. The stub can be implemented as a short- or open-circuited stub. For simple, discrete microwave amplifier designs, the short-circuited stub can be advantageous because one can feed the amplifier biases from the end of the stub while shorting the RF currents to the ground plane through a capacitor, hence increasing the isolation between the RF path and dc supply. In single-stub matching, the aim is to adjust stub and transmission line lengths in such a way that the load admittance, Y_l, is transferred to a desired admittance, Y_{in}.

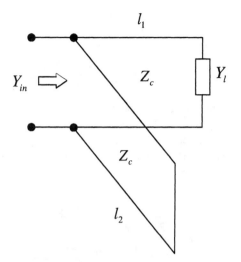

Figure 1.30 Single-stub matching. The purpose is to determine l_1 and l_2 for a given load impedance to have input matched to the generator.

To show how single-stub matching is used, we first write the input admittance of the circuit as follows:

$$Y_{in} = Y_c \frac{Y_l + jY_c \tan(\beta l_1)}{Y_c + jY_l \tan(\beta l_1)} + Y_c \frac{1}{j \tan(\beta l_2)}$$

$$Y_{in} = Y_c \frac{\tan(\beta l_2)[Y_l + jY_c \tan(\beta l_1)] - j[Y_c + jY_l \tan(\beta l_1)]}{\tan(\beta l_2)[Y_c + jY_l \tan(\beta l_1)]}$$

Now, let's assume that we want to have $Y_{in} = Y_c$ (i.e., source impedance is matched to load impedance). Having this condition satisfied and equating the real and imaginary parts separately, one can obtain the following set of expressions:

$$x(m + n) - nY_c + ymn = 0$$

$$ny + Y_c mn - Y_c + my - xmn = 0$$

where

$$m = \tan(\beta l_1) \quad n = \tan(\beta l_2)$$

$$Y_l = x + jy$$

Solution of this set of equations will give us the required transmission line and stub lengths to match the source impedance to the load impedance:

$$m = \mp \frac{(x - Y_c)}{\sqrt{\frac{\Delta}{Y_c x}}\left(x \pm y\sqrt{\frac{Y_c x}{\Delta}}\right)} \tag{1.222}$$

$$n = \pm \sqrt{\frac{Y_c x}{\Delta}} \qquad (1.223)$$

where

$$\Delta = y^2 + (x - Y_c)^2 \qquad (1.224)$$

So, for a given complex load admittance, the required transmission line lengths are found using (1.222) and (1.223) to have $Y_{in} = Y_c$. As a result, it is possible to match any complex load impedance to the source by using single-stub matching as long as it is admissible to vary the stub and transmission line lengths. Now, let us also solve the previous set of equations for the load conductance and load susceptance:

$$x = Y_c \frac{n^2(1 + m^2)}{(2mn + n^2 + m^2 + m^2 n^2)}$$

$$y = -Y_c \frac{(-n + m^2 n - m)}{(2mn + n^2 + m^2 + m^2 n^2)} \qquad (1.225)$$

From the above equations, it is seen that if the transmission line length, m, is fixed, then the load conductance and susceptance would be linked to each other to obtain a matched condition. In other words, for a fixed transmission line length, one cannot match all the possible load impedance values. Although this may not be a problem in designing matching networks, say, for an amplifier, it is a restriction for fixed tuners. To overcome this limitation, one can add a second shunt stub at the load terminals, which is the topic of the next section.

So far, we have been investigating single-stub matching analytically. It would also be extremely helpful if we could attack the matching problem using the Smith chart, and in the following paragraphs, we will show how to do this. Single-stub matching using the Smith chart is depicted in Figures 1.31 and 1.32 for an example load impedance. The first step in the matching process is to determine the length of the transmission line to carry the load impedance to the unity conductance circle. The unity conductance circle is the mirror image of the unity resistance circle with respect to the origin. Then, the type of the stub (i.e., open or short) is decided. Note that both types of stubs would yield the same result but with different lengths. So, the choice is really based on other considerations such as biasing in active circuits or availability ground via holes (to make shorted stubs) in the case of microstrip circuits. The remaining procedure is to select the stub length so that the imaginary part of the admittance is cancelled, matching the load impedance to the reference impedance. Depending on the type of the stub, the admittance moves in different directions as shown in Figures 1.31 and 1.32.

Note that the arrows and inset figures on the charts represent the location of the resulting input impedance for a given stub length and the required element for the particular transformation, respectively.

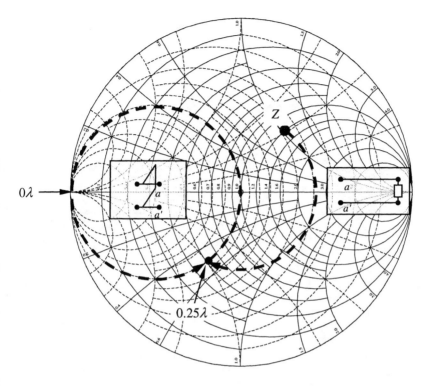

Figure 1.31 Procedure for matching different impedances using single short-stub and a transmission line. Dashed lines represent the movement of impedance value and the inset figures show the required component (transmission line and stub). Arrows show location of the input impedance versus stub length.

For example, in Figure 1.31, a zero-length shorted stub would obviously make the input impedance zero. As the stub length is increased, the input impedance starts to move counterclockwise, as indicated. At a quarter-wavelength, the stub wouldn't have any effect because the transferred impedance would approach infinity. Then, at a value greater than a quarter-wavelength, one can reach the origin and the network becomes matched at that point. On the other hand, a zero-length open-stub would not change the location of the impedance at the beginning, as shown in Figure 1.32. Now, as the stub length is increased, the input impedance starts to move clockwise in this case. At a quarter-wavelength, the stub would short-circuit the circuit because the transferred impedance would approach zero. Note that for this particular example, one would need a longer open-stub than the shorted stub to match the load impedance because it is required to travel more distance in that case. Obviously, depending on the location of the load impedance, this would change.

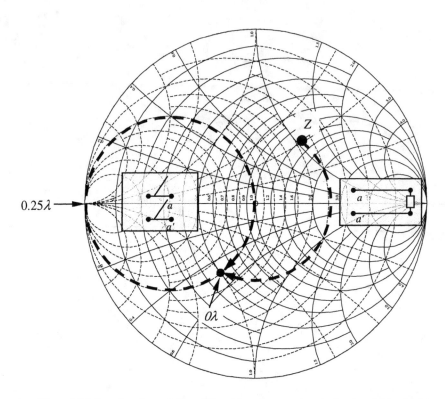

Figure 1.32 Procedure for matching different impedances using single open-stub and a transmission line. Dashed lines represent the movement of impedance value and the inset figures show the required component (transmission line and stub). Arrows show location of the input impedance versus stub length.

1.10.2 Double-Stub Matching

As indicated in the previous section, single-stub matching can match any load impedance only if it is allowed to change both stub and transmission line lengths. However, this poses practical difficulties for fixed-length tuners; it would be more advantageous to keep the transmission line length fixed yet enable matching for as wide a range of load impedances as possible. Double-stub matching shown in Figure 1.33 eliminates the problems associated with single-stub matching.

The operation of double-stub matching can be explained by examining equation (1.225). We already know that by changing the stub position on the load side in Figure 1.33, we can change the load admittance to any value we wish. This immediately suggests that in case the transmission line length, m, is fixed, we can select the second stub length, n, so that the first equation in (1.225) matches the actual load conductance. Then, we plug that n into the second equation in (1.225) and determine the required load susceptance for matching. This value would probably be different than the actual load susceptance. However, as indicated at the beginning of this paragraph, we can adjust the load stub to bring the actual load susceptance to the value given by the second equation in (1.225), which completes the matching procedure.

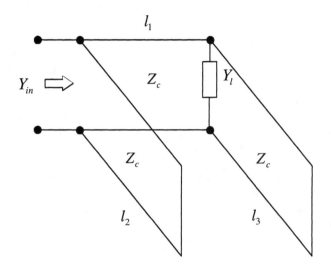

Figure 1.33 Double-stub matching. The additional stub at the load terminals is used to adjust the load susceptance so that it can be matched by keeping l_1 constant.

But, even double-stub matching cannot be used to match all possible load impedance in practice because it becomes too frequency sensitive for certain m values. Typical values for m are ∞ ($l_1 = \lambda/4$) or 1 ($l_1 = \lambda/8$). For example for $l_1 = \lambda/8$ and $l_1 = \lambda/4$, it would only be possible to match admittances whose conductance values were less than $2Y_c$ and Y_c, respectively. Although reducing l_1 close to zero would theoretically increase the maximum values of conductance that could be matched, the network becomes extremely frequency sensitive. The solution to this problem is triple-stub matching. However, its analysis will not be included here because it is similar to that presented so far. Stub matching using more than three stubs is rarely used in practice.

1.10.3 Matching with Lumped Elements

The stub matching explained in the above sections is not always practical for monolithic microwave integrated circuit (MMIC) applications due to the substrate space it consumes. Therefore, lumped element matching is usually employed in MMIC design. Commonly used lumped element matching networks are shown in Figure 1.34. Inductance terms are usually implemented using spiral inductors, and capacitive terms are usually implemented using metal-insulator-metal (MIM) capacitors. Note that although it is possible to add more elements to increase the bandwidth, the circuits shown in the figure are usually sufficient for narrow to moderate bandwidths (<10%).

It is relatively easy to find the component values of the matching networks for moderate bandwidths using the Smith chart followed by some computer optimization. Figure 1.35 summarizes the matching procedure for impedances located at four different quadrants on the Smith chart. In the figure, possible matching movements for the four different impedances are demonstrated. Each impedance requires at least two movements for matching, and required lumped elements are shown in the inset figures. For instance, to match the impedance shown on the upper right of the Smith chart in Figure 1.35, one needs to add a shunt inductance and then a series capacitance, as shown. Obviously, one can find alternative movements on the Smith chart to match shown impedances. Other circuit requirements (e.g., biasing, type of filter response) also play an important role in determining the type of the matching network. For example, a circuit with a shunt inductor and series capacitor would be ideal for matching networks where a highpass nature is required to block dc biasing and other low-frequency signals. Note that attenuation of low-frequency signals is also important in narrowband microwave amplifiers to improve stability which is automatically satisfied by selecting a highpass network. It should be noted that the designer must analyze the final layout of the lumped matching networks using a full-wave simulator because parasitic effects are important.

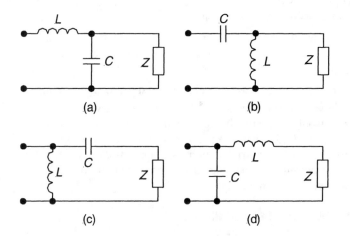

Figure 1.34 Commonly used lumped matching networks for MMIC applications: (a) and (d) lowpass networks, (b) and (c) highpass networks. The topology of the matching network is determined based on the requirements such as attenuation of low-frequency signals, biasing, and so forth.

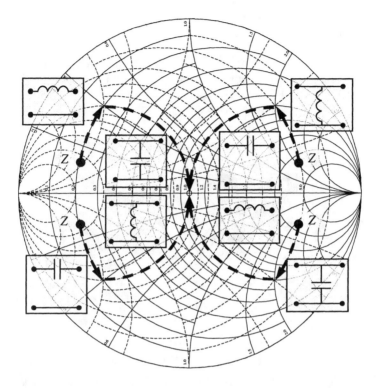

Figure 1.35 Procedure for matching different impedances using lumped components. Dashed lines represent the movement of impedance value, and the inset figures show the required lumped component (shunt or series).

1.11 NETWORK ANALYZERS

The network analyzer is perhaps the most important measurement device in a microwave laboratory [14]. The network analyzer is used to determine network parameters (S-parameters) of active or passive microwave devices in a frequency band of interest. A typical network analyzer and block diagram of the network analyzer are shown in Figures 1.36 and 1.37, respectively. Fundamentally, there are three basic blocks in a network analyzer: the receiver, the synthesizer, and the test set. The receiver is a multichannel receiver used to process signals coming from the test set. The test set contains detectors, directional couplers, and RF switches, which are necessary to detect reflected and transmitted signals and to direct RF signal to the required port. Device under test (DUT) is directly connected to the test set as shown. Finally, the synthesizer is used to generate frequency and amplitude-stabilized RF signal in a frequency band to stimulate the DUT. All these components are controlled by a central processing unit (CPU) and results are displayed on a display or transmitted to a main computer for further processing. Depending on the model and type, the fundamental blocks explained above (also shown by dotted lines in Figure 1.37) can be a separate piece of equipment or are combined into a single device (as in Figure 1.36). In case they are separate, the processor commands the instruments through a control bus called the HP-IB (or the GPIB). Implementing these blocks as separate pieces of equipment is a flexible approach. However, it usually costs more.

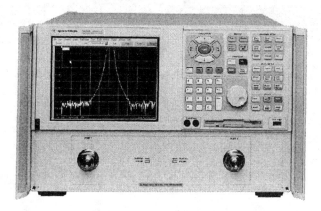

Figure 1.36 A typical network analyzer. (Courtesy of Agilent Technologies.)

There are fundamentally two types of network analyzers: scalar (e.g., Agilent 8757 series) or vector (e.g., Agilent PNA E8362 series). A scalar network analyzer can only measure the magnitudes of the S-parameters (transmission or reflection characteristics). For example, one can verify the gain of an amplifier using a scalar network analyzer, but phase information won't be available. Thus, scalar network analyzers are useful when the designer is only interested in the basic response (e.g., gain, input return loss, isolation) of the designed circuit. However, since the phase information is not extracted, scalar network analyzers cannot be used in, for instance, the modeling of passive or active circuits. Vector network analyzers (VNAs), on the other hand, provide both the magnitude and phase information of the S-parameters. Vector analyzers are therefore more capable than scalar analyzers. The disadvantage of the vector analyzers is that they are usually more expensive than the scalar analyzers, if all other specifications are the same. Apart from being scalar or vector, the other important factor that separates the network analyzers is the frequency band of operation. Due to accuracy, design issues, and cost, there are different network analyzers designed to work in different frequency bands. For example, Agilent makes network analyzers for frequencies starting from as low as 10 kHz (e.g., Agilent E5100 series) up to 110 GHz (e.g., Agilent PNA series) [15]. To change the frequency range, the test set and the synthesizer (see Figure 1.37) must be replaced.

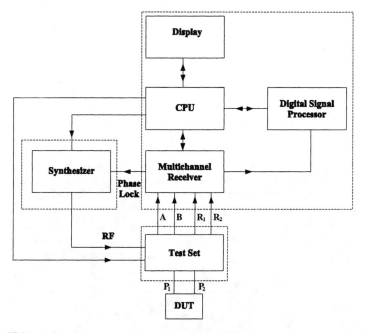

Figure 1.37 Block diagram of a network analyzer. (*After:* [14].) Sections shown by dashed lines are sometimes implemented as separate instruments for modularity and performance.

The vast majority of network analyzers are two-port analyzers meaning a two-port device (DUT) can be fully characterized without any additional processing and/or equipment. The example circuits that can be characterized with a two-port analyzer are amplifiers, filters, phase shifters, transmission lines, and transistors. There are also three-port devices, like mixers, that are commonly used in microwave engineering. While characterization of mixers is also possible with a two-port network analyzer, this will require extra connections and equipment in most cases [16]. It should be also indicated that recently introduced modern network analyzers made mixer measurements relatively easy. In recent years, true three- and four-port analyzers have also been designed (e.g., Agilent ENA 5070 series). A four-port network analyzer, for example, can make balanced and mixed-mode measurements. The application notes published by Agilent [16–24] and Rohde & Schwarz [25–27] provide a very good introduction on many aspects of the network analyzers and readers are encouraged to read them.

A microwave device is usually characterized using a network analyzer in two ways. The first way is to place the circuit in a fixture equipped with connectors, then connect the analyzer to the fixture using coaxial cables. This is the most common scenario for microstrip circuits and systems [16, 24]. Alternatively, if the device is a MMIC, then it can be directly measured on the wafer, which is called on-wafer measurement [28, 29]. In this later scenario, the circuit is first placed on a stable metal chuck (the chuck is connected to a vacuum pump so that the wafer is firmly held in place). Then, special high-frequency probes are positioned over the circuit so that the RF energy can be transferred in and out of the circuit. There are many different types of probes. The most used ones are ground-signal (GS) and ground-signal-ground (GSG) probes. The probes are actuated by the help of micrometers so that they be precisely positioned. The chuck and the micrometers are mounted on a metal frame called a probe station (see Figure 1.38). Depending on the situation, the probe station itself can also be mounted on an air-table to enable a vibration free environment. The VNA is then connected to probes through flexible coaxial cables and measurement is performed.

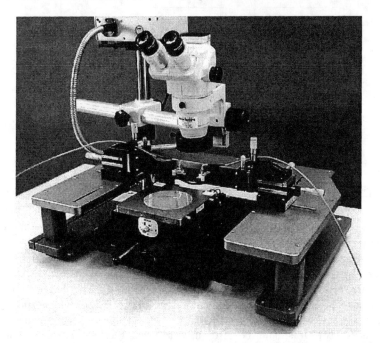

Figure 1.38 A typical probe station for on-wafer testing. Note the flexible coaxial cables connected to the two probes carried by two actuators on the left and right. The chuck (square metal piece) and the wafer are shown at the center. (Courtesy of Cascade Microtech.)

1.12 CALIBRATION OF NETWORK ANALYZERS

It is virtually impossible to design a measurement system without imperfections. Every measurement system has some number of imperfections in its construction, and depending on the frequency of the observed signals, these imperfections can be significant and therefore have to be accounted for to eliminate measurement errors. The procedure of determining the imperfections in a network analyzer is called calibration or unterminating [30]. Calibration is performed through measurement of different "known" standards. The number and type of the standards depend on the type of algorithm that is followed. After a network analyzer is calibrated, then its imperfections can be removed from the measured quantities through a procedure called de-embedding, giving the true characteristics of the DUT.

At this point, one may ask what kind of standards should be chosen in calibrating a network analyzer. After all, in order to calibrate a system, one must measure some known quantities so that the error introduced can be quantified. In fact, this concept is the central idea in different calibration procedures. Because there will also be imperfections in the standards themselves, one will never be sure how accurate the calibration is. Fortunately, it is relatively easy to design accurate calibration standards with acceptable practical accuracy in most cases. Note that it is also strongly preferable to have a calibration algorithm that requires calibration standards for which high accuracy can easily be achieved in the frequency range of interest. For example, implementing a good termination (i.e., 50-ohm load) is difficult at high frequencies due to parasitic effects. But it is very easy to accurately design a piece of microstrip transmission line. So, a calibration algorithm that relies on the characteristics of a transmission line will outperform the other algorithm which assumes the termination is purely resistive. We will elaborate this point further in the following sections.

The types of measurement errors in a network analyzer can be classified into three groups [18]: systematic errors, random errors, and drift errors. Systematic errors occur from the short-term repeatable and static imperfections in the system. For example, errors in the switches (former) and directional couplers (latter) are examples of errors belonging to this class. One can only calibrate for the systematic errors. Random errors are the errors due to electrical noise in the system. One can reduce the random errors by narrowing the intermediate frequency (IF) bandwidth (hence allowing less noise power injected into the receiver) or averaging the measurements. Note that both of these methods have the disadvantage of slowing down the measurements. Finally, drift errors are caused by changes in the mechanical and electrical components by time and/or environment. For example, as the system ages, the switches start to mechanically deteriorate, affecting contact resistance. Also, if the environment or device temperature changes, almost all of the components (electrical or mechanical) change their characteristics. The solution to drift errors is recalibrating the system with some intervals. To cope with temperature changes, the most critical components (the oscillators, for example) are placed in ovens in the instrument to keep their temperature constant.

Now, it is worthwhile to discuss the basic topologies of test sets because derivation of error models is closely related to this. Figure 1.39 shows the two most commonly used test-set topologies based on three and four samplers. A sampler can be thought of as a device which detects both the magnitude and phase of an input signal. In the three-sampler configuration [Figure 1.39(a)], the RF signal is first divided by means of a power divider. Then, one path is sent to the reference (R) sampler, while the other path is sent to a transfer switch after passing an attenuator. The transfer switch is responsible for directing the signal into the first or second port, depending on the measurement. Note that when the signal is directed into a particular port, the other port is terminated. Finally, reflected signals from the ports are detected by means of directional couplers and directed to reflection (A) and transmission (B) samplers. The network analyzer can compute the reflected and transmitted waves by evaluating the ratios of A/R and B/R. If the transfer switch is placed after the source and separate samplers are used for incident waves at each channel, then one reaches the four-sampler topology [Figure 1.39(b)]. Three-sampler architecture has the advantage of being cost effective, whereas four-sampler architecture has the capability of being calibrated with the thru-reflect-line (TRL) algorithm, which increases accuracy at microwave frequencies and makes it suitable for on-wafer measurements.

Note that it is important to place the attenuators before the directional couplers because placing them after the couplers would degrade effective directivity of the couplers [17]. This is because reflected signals from the attenuators would be seen as source leakage in the couplers. Even though this would be corrected after calibration, the accuracy of the calibration would be worse due to reduced raw directivity of the couplers. The same argument is also valid in cascading the connectors after the test ports. One should minimize cascading connectors as much as possible so as not to deter effective directivity of the couplers.

Implementation of the detectors (or samplers) also differs based on the type of network analyzer. There are two main types of detectors: diode detectors (broadband) and tuned detectors (narrowband) [17]. For a scalar network analyzer, simple diode detectors can be sufficient to convert the input RF signal to a proportional dc level. However, the phase information is lost during this process. The other way is to use a tuned receiver where a mixer is employed to convert the RF signal into an IF. Then, the IF can be sampled to extract both magnitude and phase information. This kind of detector is usually employed in commercial vector network analyzers. Note that diode receivers are broadband in nature; therefore they are convenient in characterization of frequency translating devices, such as mixers. The downside of a broadband diode receiver is the relatively low dynamic range because of the high noise floor. Diode receivers can provide a dynamic range of 75 dB, whereas tuned receivers have a dynamic range of around 100 dB [17].

There are, however, network analyzers based on six-port reflectometers, which use only scalar measurements from four power detectors to extract both magnitude and phase information [12, 31–33]. Note that to make a two-port network analyzer, one needs two six-port reflectometers thus eight (four on each reflectometer) power detectors. These kind of network analyzers are very accurate and usually employed in standards laboratories, such as the National Institute of Standards and Technology (NIST), to control quality of other equipment [34]. We will not review this kind of network analyzer here.

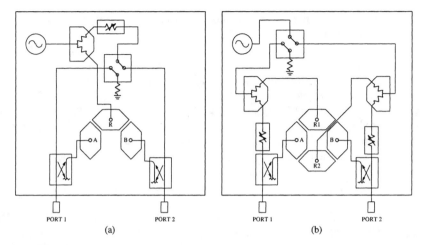

Figure 1.39 Topologies of (a) three- and (b) four-sampler vector network analyzers. Four-sampler network analyzers have the capability of being calibrated with true TRL algorithm.

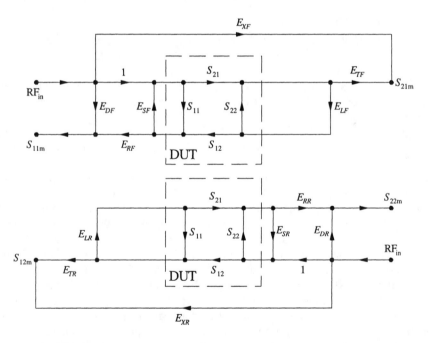

Figure 1.40 The 12-term VNA error model. This error model is the basis of most of the VNA calibration algorithms.

In the derivation of calibration algorithms, we will assume we can make *perfect* measurements using the detectors at some fictitious reference planes. The exact locations of the reference planes are not important because their effects will be included in the error terms. Since S-parameters will be based on ratios of the waves, even the absolute magnitude of the signal generator is not important as long as the signals entering the detectors have enough signal-to-noise ratio.

By studying Figure 1.39 and using signal flowgraphs, one can devise an error model as shown in Figure 1.40 for the network analyzer. Note that there are two separate error models with the same topology for forward (port 1) and reverse (port 2) paths. In each model, there are six unknowns, resulting in twelve unknowns in total. This error model is one of the fundamental error models for network analyzers, and it is called a 12-term error model. Each error term in the model has a special name that can be given as follows [18]:

$E_{DF} =$	Forward directivity	$E_{DR} =$	Reverse directivity
$E_{SF} =$	Forward source match	$E_{SR} =$	Reverse source match
$E_{RF} =$	Forward reflection tracking	$E_{RR} =$	Reverse reflection tracking
$E_{LF} =$	Forward load match	$E_{LR} =$	Reverse load match
$E_{TF} =$	Forward transmission tracking	$E_{TR} =$	Reverse transmission tracking
$E_{XF} =$	Forward isolation	$E_{XR} =$	Reverse isolation

Determination of these twelve unknowns is the duty of the calibration algorithm. There are many different calibration algorithms in the literature, and a good review can be found in [34].

Typical on-wafer calibration standards for the two frequently used calibration techniques, short-open-load-thru (SOLT) and thru-reflect-line (TRL), are shown in Figures 1.41 and 1.42. As we will see, SOLT is performed by measuring four different standards. After on-wafer SOLT calibration, the system is calibrated up to the tips of the probes. Therefore, the SOLT calibration standards are made as small as possible to reduce parasitic effects. Note that the CPW SOLT load is made by two parallel 100-ohm resistors. Also, the open circuit is achieved by raising the probes from the wafer surface and holding them in the air with a fair amount of separation to minimize crosstalk between the probe tips. It is important to indicate that SOLT does not calibrate effects of launchers that are used to carry the signal from the tips of the probes to the circuit that is being measured. In the case of GSG probes and microstrip circuit, the launcher is essentially a CPW-to-microstrip transition. Depending on the frequency, the effect of the launchers can be significant. TRL calibration, on the other hand, is based on the measurement of different lengths of transmission lines and a reflect, which is shown in Figure 1.42. The TRL standards are simpler than SOLT standards because they do not require precision resistive termination. All we need to do is implement a good reflect and a transmission line with a known length with respect to the thru. Both of these can easily be achieved with the MMIC technology. Another advantage of TRL is that the effects of the launchers can also be accounted for.

In the following sections, we will review both SOLT and TRL calibration techniques in more detail.

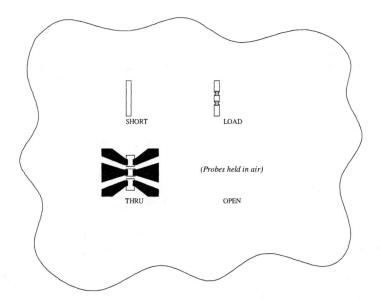

Figure 1.41 SOLT calibration set for on-wafer probe-tip calibration. Note how the probes are positioned to measure the standards (example shows the thru). After the SOLT calibration, the system is calibrated up to the probe tips. Effect of launchers must be accounted for separately.

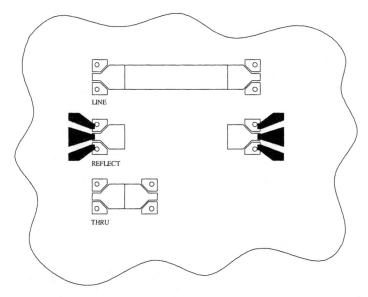

Figure 1.42 TRL calibration set for on-wafer calibration. Note how the probes are positioned to measure the standards (example shows the reflect). Length of the line is selected so that it is approximately 90° long at the center frequency. Note that TRL calibration also extracts the effect of launchers automatically.

1.12.1 SOLT Calibration

The SOLT calibration method is a widely used technique to determine the unknown error terms in network analyzers [35]. It is based on successive measurement of three different terminations on each port and one thru connection. The main drawback of the SOLT method is that the terminations (i.e., short, open, and load) or the amount of error in them, must be known precisely. Any imperfections in these standards will affect the calibration. Therefore this method is usually limited to lower microwave frequencies (i.e., less than ~10 GHz) where standards can be characterized with reasonable accuracy. In higher frequencies, different techniques (such as TRL) must be employed to calibrate network analyzers if higher accuracy is desired. In SOLT calibration, the unknowns of the 12-term error model in each flow path (i.e., forward and reverse) are determined independently for each port. In other words, by measuring the short, open, and load on the first port and a thru between the first and second ports, we can determine the six unknowns in the forward path. Then, we repeat exactly the same procedure for the second port to determine the remaining six unknowns in the reverse path. The calibration procedure is depicted in Figure 1.43(a–d) and outlined in the following paragraphs.

We start by measuring the three terminations on the first port. The measured reflection coefficient for a given load can be expressed as follows in terms of the error terms:

$$S_{11m} = E_{DF} + \frac{E_{RF}\Gamma_L}{1 - \Gamma_L E_{SF}} \qquad (1.226)$$

where

$$\Gamma_L = \begin{cases} 0, & \text{for a load} \\ 1, & \text{for a open} \\ -1, & \text{for a short} \end{cases}$$

From the above three measurements, one can obtain the following three equations:

$$S_{11m}^{(1)} = E_{DF}$$

$$S_{11m}^{(2)} = E_{DF} + \frac{E_{RF}}{1 - E_{SF}}$$

$$S_{11m}^{(3)} = E_{DF} - \frac{E_{RF}}{1 + E_{SF}}$$

where the superscript indicates the measurement number. By inspecting the above expressions, it is easy to see that E_{DF} can be determined immediately by measuring a load. E_{RF} and E_{SF} can be determined in terms of E_{DF} and measured quantities as follows:

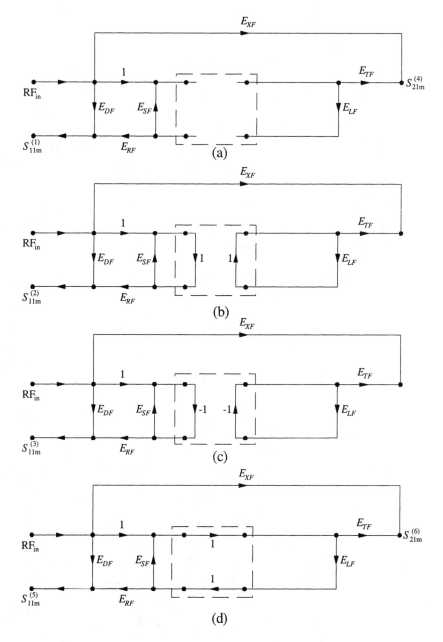

Figure 1.43 Description of the SOLT calibration measurements with flowgraphs: (a) load, (b) open, (c) short, and (d) thru. The picture depicts the calibration only for the forward path. The same procedure should be repeated for the reverse path.

$$E_{SF} = \frac{2E_{DF} - S_{11m}^{(2)} - S_{11m}^{(3)}}{S_{11m}^{(3)} - S_{11m}^{(2)}} \tag{1.227}$$

$$E_{RF} = \frac{2 \cdot \left(S_{11m}^{(2)} - E_{DF}\right) \cdot \left(S_{11m}^{(3)} - E_{DF}\right)}{S_{11m}^{(3)} - S_{11m}^{(2)}} \tag{1.228}$$

where

$$E_{DF} = S_{11m}^{(1)} \tag{1.229}$$

So far, we have determined three of the six unknowns. The forward crosstalk is determined by simply measuring the transmission without connecting the test ports:

$$E_{XF} = S_{21}^{(4)} \tag{1.230}$$

The remaining two unknowns (E_{LF} and E_{TF}) are determined using the thru connection as follows:

$$S_{11m}^{(5)} = E_{DF} + \frac{E_{RF} E_{LF}}{1 - E_{SF} E_{LF}}$$

$$S_{21m}^{(6)} = \frac{E_{TF}}{1 - E_{SF} E_{LF}} + E_{XF}$$

which results in

$$E_{LF} = \frac{S_{11m}^{(5)} - E_{DF}}{E_{RF} + \left(S_{11m}^{(5)} - E_{DF}\right)E_{SF}} \tag{1.231}$$

$$E_{TF} = \frac{S_{21m}^{(6)} - E_{XF}}{E_{RF} + \left(S_{11m}^{(5)} - E_{DF}\right)E_{SF}} E_{RF} \tag{1.232}$$

This completes the determination of first six error coefficients, and calibration of the forward path is done now. For the reverse path, the same procedure is repeated, and another set of six error coefficients is found. Then, S-parameters of the DUT are calculated in terms of the error coefficients as follows [12, 18, 36, 37]:

$$S_{11} = \frac{(1 + B \cdot E_{SR})A - C \cdot D \cdot E_{LF}}{(1 + A \cdot E_{SF})(1 + B \cdot E_{SR}) - C \cdot D \cdot E_{LF} E_{LR}} \tag{1.233}$$

$$S_{12} = \frac{[1 + A(E_{SF} - E_{LR})]D}{(1 + A \cdot E_{SF})(1 + B \cdot E_{SR}) - C \cdot D \cdot E_{LF} E_{LR}} \tag{1.234}$$

$$S_{21} = \frac{[1 + B(E_{SR} - E_{LF})]C}{(1 + A \cdot E_{SF})(1 + B \cdot E_{SR}) - C \cdot D \cdot E_{LF} E_{LR}} \tag{1.235}$$

$$S_{22} = \frac{(1 + A \cdot E_{SF})B - C \cdot D \cdot E_{LR}}{(1 + D \cdot E_{SF})(1 + C \cdot E_{SR}) - A \cdot B \cdot E_{LF} E_{LR}} \tag{1.236}$$

where

$$A = \left(S_{11m} - E_{DF}\right)\big/ E_{RF} \qquad (1.237)$$

$$B = \left(S_{22m} - E_{DR}\right)\big/ E_{RR} \qquad (1.238)$$

$$C = \left(S_{21m} - E_{XF}\right)\big/ E_{TF} \qquad (1.239)$$

$$D = \left(S_{12m} - E_{XR}\right)\big/ E_{TR} \qquad (1.240)$$

In the above equations, the S-parameters with m subscripts refer to raw measurements. An immediate observation is that the network analyzer must make a full two-port measurement to update any S-parameter of the DUT. Note that the isolation terms (E_{XF} and E_{XR}) can be assigned to zero for applications where it is not required to measure high-level isolations. This is done by omitting the isolation step during the calibration. However, if it is required to measure a device with high isolation, a band-pass filter with high stopband rejection (e.g., >80 dB) for example, then the isolation terms must be included and calibrated for.

1.12.2 TRL Calibration

The TRL calibration method eliminates the disadvantages of the SOLT calibration method. It is based on measurement of a thru, a reflect, and a known length of transmission line [38]. The most important advantage of the TRL method is that it is not necessary to know exactly the reflection coefficient of the reflect standard and the propagation constant of the line. These are obtained as byproducts of the calibration procedure. The only requirement for the reflect is that it is necessary to know the sign of the termination (i.e., short or open) because the root of an equation is determined according to this sign during the calibration. Therefore, TRL calibration is more accurate then SOLT calibration, especially at millimeter-wave frequencies.

There are two main disadvantages of the TRL calibration. The first one is the relatively narrow band of calibration. One can expect to have a good TRL calibration in a frequency band of 8:1 (there is an alternative calibration technique called line-reflect-match (LRM), which uses a matched load instead of a transmission line removing the inherent frequency band limitations of TRL). The optimum electrical line length is to be 90° at the center of the frequency band. It is also important to note that the relative length of the line must be different from the odd multiples of 180°. Second, TRL is basically an error-box removal procedure. In other words, the error model of the TRL is fixed and does not account for all the switching errors in the test set. Therefore, the 8-term error model of the TRL calibration shown in Figure 1.44 is slightly different from the 12-term error model. The error terms on the left and right will be called left error-box and right error-box, respectively. Note that SOLT copes with the problem of switching errors by determining different error coefficients both for forward and reverse paths. To do the same in TRL, the algorithm must be modified somewhat.

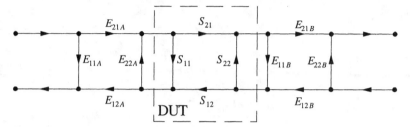

Figure 1.44 The 8-term error model of the TRL calibration.

Although the TRL error model is different from the 12-term error model, one can find the following relationships between the 12-term error model and the 8-term error model under certain assumptions:

$$E_{21A} \cdot E_{12A} = E_{RF} \qquad E_{21B} E_{12B} = E_{RR}$$
$$E_{11A} = E_{DF} \qquad E_{11B} = E_{DR}$$
$$E_{22A} = E_{SF}, E_{LR} \qquad E_{22B} = E_{SR}, E_{LF} \qquad (1.241)$$
$$E_{21A} \cdot E_{21B} = E_{TF} \qquad E_{12A} E_{12B} = E_{TR}$$

The first observation is that the isolation terms (E_{XF} and E_{RF}) do not appear in the 8-term error model and need to be determined separately. The second and more important observation is that the 8-term error model of TRL calibration cannot distinguish the effective source and load match terms, which is the result of assuming fixed error model as explained before. These errors are lumped into E_{22A} and E_{22B}. In other words, the TRL method assumes that the network analyzer presents the same termination to the test ports regardless of the excited port. This means that the transfer switch shown in Figure 1.39(a) has the same reflection coefficient in either state. However, this is only a fair assumption, and it is the reason why a three-sampler VNA cannot be accurately calibrated with the TRL algorithm, because the transfer switch is placed after the reference sampler [24]. A four-sampler VNA, on the other hand, has reference samplers after the transfer switch and can therefore be calibrated with the TRL algorithm accurately through some additional measurements [20]. Although it wouldn't be exact, calibration of the three-sampler architecture using TRL can be improved by improving the raw source match and load match of the system. This is achieved by placing high-quality fixed attenuators as closely as possible to the reference planes [24]. Note that a three-sampler VNA can be calibrated with the SOLT method accurately because in the 12-term error model of the SOLT, different error terms are found for each direction, which automatically takes care of the nonideal transfer switch. The problem arises when one fixes the error boxes, as in the 8-term error model of TRL. Nevertheless, most of the today's modern VNAs are based on the four-sampler architecture. Therefore, TRL and its variants are the primary choice for a calibration method where high accuracy is needed.

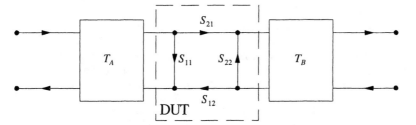

Figure 1.45 Simplified error model of the TRL calibration. Note that the error boxes are represented by their transfer matrixes.

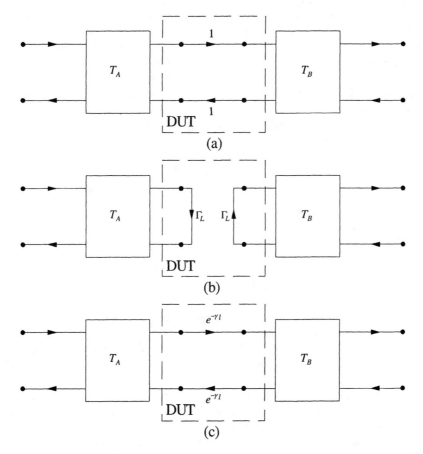

Figure 1.46 Description of the TRL calibration measurements with flowgraphs: (a) thru, (b) reflect, and (c) line.

Derivation of the TRL calibration is not as obvious as that for the SOLT algorithm; heavy physical reasoning is involved. The pivotal concept is that it is sufficient to determine only the ratios of transfer parameters of the error boxes to de-embed the DUT. Therefore, to describe the TRL calibration procedure, we first represent the error boxes by their transfer matrixes as shown in Figure 1.45. The transfer matrixes of the error boxes are defined as follows:

$$\begin{bmatrix} b_1 \\ a_1 \end{bmatrix} = \frac{1}{E_{21}} \begin{bmatrix} E_{12}E_{21} - E_{11}E_{22} & E_{11} \\ -E_{22} & 1 \end{bmatrix} \begin{bmatrix} a_2 \\ b_2 \end{bmatrix} = T_{A,B} \begin{bmatrix} a_2 \\ b_2 \end{bmatrix} \tag{1.242}$$

where a and b represent the incident and reflect waves, respectively. Once the transfer matrixes are found, the error terms can be determined, if required. One important comment at this point would be that we will not be able to distinguish the forward and reverse transmission. In other words, we will only find the product of $E_{12}E_{21}$, not the individual terms, because TRL gives us only the ratios of transfer matrix entries. In fact, it is not possible to separate these terms by any driving point impedance measurement [30]. Fortunately, it is not necessary to distinguish them either, because in de-embedding equations, we only need the product as well. The calibration procedure is depicted in Figure 1.46 and outlined as follows [38].

We first write the measured transfer matrixes of the thru and line connections in the following form:

$$R_t = T_A T_t T_B \tag{1.243}$$

$$R_l = T_A T_l T_B \tag{1.244}$$

with

$$T_t = \begin{bmatrix} 1 & 0 \\ 0 & 1 \end{bmatrix} \tag{1.245}$$

$$T_l = \begin{bmatrix} e^{-\gamma l} & 0 \\ 0 & e^{\gamma l} \end{bmatrix} \tag{1.246}$$

where T_t and T_l are the transfer matrixes of the thru and line connections, respectively, and γ is the complex propagation constant of the line. T_A and T_B are transfer matrixes of the error boxes. Raw measurements are indicated by R_t and R_l. Note that it is automatically assumed that the line is nonreflective. In fact, it really does not matter because the system will automatically be calibrated based on whatever the characteristic impedance of the line is. However, to avoid any postprocessing, the line is usually selected as 50 ohms (one can accurately design a transmission line for a given characteristic impedance using full-wave electromagnetic simulators). Then, by rearranging (1.244) and (1.243), one can obtain

$$M \cdot T_A = T_A T_l \tag{1.247}$$

where

$$M = R_i R_t^{-1} \qquad (1.248)$$

In (1.247), the only unknown quantity is the transfer matrix of the error box A. The matrix M is obtained from the measured raw quantities. By carrying out the matrix multiplication in (1.247), the following sets of equations are obtained:

$$m_{11} t_{11A} + m_{12} t_{21A} = t_{11A} e^{-\gamma l}$$

$$m_{21} t_{11A} + m_{22} t_{21A} = t_{21A} e^{-\gamma l}$$

$$m_{11} t_{12A} + m_{12} t_{22A} = t_{12A} e^{\gamma l}$$

$$m_{21} t_{12A} + m_{22} t_{22A} = t_{22A} e^{\gamma l}$$

Now, let's divide the first equation by the second and the third equation by the fourth:

$$\frac{m_{11} t_{11A} + m_{12} t_{21A}}{m_{21} t_{11A} + m_{22} t_{21A}} = \frac{t_{11A}}{t_{21A}}$$

$$\frac{m_{11} t_{12A} + m_{12} t_{22A}}{m_{21} t_{12A} + m_{22} t_{22A}} = \frac{t_{12A}}{t_{22A}}$$

Then, from the above equations, one can find the following two expressions for the ratios t_{11A}/t_{21A} and t_{12A}/t_{22A} :

$$m_{21} \left(\frac{t_{11A}}{t_{21A}} \right)^2 + \left(m_{22} - m_{11} \right) \frac{t_{11A}}{t_{21A}} - m_{12} = 0$$

$$m_{21} \left(\frac{t_{12A}}{t_{22A}} \right)^2 + \left(m_{22} - m_{11} \right) \frac{t_{12A}}{t_{22A}} - m_{12} = 0$$

Thus, both t_{11A}/t_{21A} and t_{12A}/t_{22A} are given by the solution of the same quadratic equation, which is written as follows:

$$\frac{t_{11A}}{t_{21A}} = \frac{-\left(m_{22} - m_{11}\right) \pm \sqrt{\left(m_{22} - m_{11}\right)^2 + 4 m_{21} m_{12}}}{2 m_{21}} \qquad (1.249)$$

$$\frac{t_{12A}}{t_{22A}} = \frac{-\left(m_{22} - m_{11}\right) \pm \sqrt{\left(m_{22} - m_{11}\right)^2 + 4 m_{21} m_{12}}}{2 m_{21}} \qquad (1.250)$$

Selection of the sign will be explained in a moment. One important observation is that t_{11A}/t_{21A} and t_{12A}/t_{22A} must be distinct. This can be proved with the help of (1.242) as follows: if t_{11A}/t_{21A} equals t_{12A}/t_{22A}, then this requires $E_{21} E_{12} = 0$, which can be easily shown using (1.242). However, this cannot be the case for a practical test system. Therefore, it can be concluded that the roots of the quadratic equation must be selected so that t_{11A}/t_{21A} and t_{12A}/t_{22A} are distinct. Then, the propagation constant is found from

$$e^{-2\gamma l} = \frac{m_{21}\dfrac{t_{11A}}{t_{21A}} + m_{22}}{m_{21}\dfrac{t_{12A}}{t_{22A}} + m_{22}} \qquad (1.251)$$

By inserting (1.249) and (1.250) into (1.251), we obtain

$$e^{-2\gamma l} = \frac{m_{11} + m_{22} \pm R}{m_{11} + m_{22} \mp R} \qquad (1.252)$$

where

$$R = \sqrt{(m_{22} - m_{11})^2 + 4m_{21}m_{12}} \qquad (1.253)$$

Note that the solutions of (1.252) are reciprocal to each other. This is the direct consequence of selecting distinct roots for t_{11A}/t_{21A} and t_{12A}/t_{22A}. Now, the remaining question is which sign should be selected in (1.252). Since the solutions of $e^{-2\gamma l}$ are reciprocal, the sign can be selected in a such a way that real part of γ is less than unity. A crucial point is that if the line length is electrically small, then it might be difficult to decide on the sign due to measurement noise. Therefore, it is beneficial to have the length of the line standard be sufficiently long so that the attenuation of the line is well above the measurement noise.

Up to this point, we have determined transfer parameter ratios t_{11A}/t_{21A} and t_{12A}/t_{22A} using the thru and line measurements. To be able to characterize the error box A, we need to know t_{11A}/t_{22A} as well. Now, let's return to (1.243). We multiply both side of (1.243) with the inverse of T_A, yielding:

$$\frac{1}{t_{11A}t_{22A} - t_{12A}t_{21A}} \begin{bmatrix} t_{22A} & -t_{12A} \\ -t_{21A} & t_{11A} \end{bmatrix} \begin{bmatrix} r_{11t} & r_{12t} \\ r_{21t} & r_{22t} \end{bmatrix} = \begin{bmatrix} t_{11B} & t_{12B} \\ t_{21B} & t_{22B} \end{bmatrix}$$

From the above equation, the following ratios are also determined:

$$\frac{t_{12B}}{t_{11B}} = \frac{r_{12t} - \dfrac{t_{12A}}{t_{22A}}r_{22t}}{r_{11t} - \dfrac{t_{12A}}{t_{22A}}r_{21t}} \qquad (1.254)$$

$$\frac{t_{21B}}{t_{22B}} = \frac{\dfrac{t_{11A}}{t_{21A}}r_{21t} - r_{11t}}{\dfrac{t_{11A}}{t_{21A}}r_{22t} - r_{12t}} \qquad (1.255)$$

$$\frac{t_{11B}}{t_{22B}}\frac{t_{11A}}{t_{22A}} = \frac{\dfrac{t_{12A}}{t_{22A}}r_{21t} - r_{11t}}{\dfrac{t_{21A}}{t_{11A}}r_{12t} - r_{22t}} \tag{1.256}$$

To determine t_{11A}/t_{22A}, we need one more equation relating t_{11A}/t_{22A} and t_{11B}/t_{22B}. This equation is obtained from the reflect measurements. By connecting the unknown, but same, reflect on both ports, we obtain the following equations:

$$\frac{t_{11A}}{t_{22A}} = \frac{S_{11m} - \dfrac{t_{12A}}{t_{22A}}}{\Gamma_L\left(1 - S_{11m}\dfrac{t_{21A}}{t_{11A}}\right)} \tag{1.257}$$

$$\frac{t_{11B}}{t_{22B}} = \frac{S_{22m} + \dfrac{t_{21B}}{t_{22B}}}{\Gamma_L\left(1 + S_{22m}\dfrac{t_{12B}}{t_{11B}}\right)} \tag{1.258}$$

At first glance, (1.257) seems to provide t_{11A}/t_{22A}, but we don't know Γ_L either. That is why we need (1.258) to find a relation between t_{11A}/t_{22A} and t_{11B}/t_{22B}:

$$\frac{t_{11A}}{t_{22A}}\frac{t_{22B}}{t_{11B}} = \frac{\left(S_{11m} - \dfrac{t_{12A}}{t_{22A}}\right)\left(1 + S_{22m}\dfrac{t_{12B}}{t_{11B}}\right)}{\left(1 - S_{11m}\dfrac{t_{21A}}{t_{11A}}\right)\left(S_{22m} + \dfrac{t_{21B}}{t_{22B}}\right)} \tag{1.259}$$

Finally, using (1.256) and (1.259), t_{11A}/t_{22A} is found as follows:

$$\frac{t_{11A}}{t_{22A}} = \pm\sqrt{\frac{\left(\dfrac{t_{12A}}{t_{22A}}r_{21t} - r_{11t}\right)\left(S_{11m} - \dfrac{t_{12A}}{t_{22A}}\right)\left(1 + S_{22m}\dfrac{t_{12B}}{t_{11B}}\right)}{\left(\dfrac{t_{21A}}{t_{11A}}r_{12t} - r_{22t}\right)\left(1 - S_{11m}\dfrac{t_{21A}}{t_{11A}}\right)\left(S_{22m} + \dfrac{t_{21B}}{t_{22B}}\right)}} \tag{1.260}$$

Now, it looks like we have run into another sign problem in (1.260). To resolve the sign of t_{11A}/t_{22A}, we refer to (1.257). One might recall that at the beginning of our discussion, we indicated that we don't need to know exactly the reflect standard, but it is necessary to know a nominal value of it. Now, the nominal value of the reflect (open or short) is used in (1.257) to determine the sign of t_{11A}/t_{22A} in (1.260).

At this point, all three transfer parameter ratios to characterize the error box *A* are determined using (1.249), (1.250), and (1.260). The transfer parameter ratios

t_{11B}/t_{21B} and t_{12B}/t_{22B} of error box B can be easily determined using (1.254) and (1.255). Once we determine the ratios t_{11A}/t_{21A}, t_{12A}/t_{22A}, t_{11A}/t_{22A}, t_{11B}/t_{21B}, t_{12B}/t_{22B}, and t_{11B}/t_{22B}, then the error boxes can be de-embedded from the measurements. Note that the TRL calibration can only provide us $E_{12A}E_{21A}$ and $E_{12B}E_{21B}$. This is also called phase uncertainty in TRL calibration [39]. However, separation of these error terms is not necessary to de-embed the DUT as indicated before; these terms always appear in pairs that can be verified from the 12-term de-embedding equations given in (1.233) through (1.236). In TRL calibration, we also determine reflection coefficient of the reflect standard and propagation constant of the line as a byproduct. The propagation constant can be used for further characterization of transmission lines [40, 41, 42]. A unified theory of the TRL calibration is also given in [43, 44].

1.12.3 Multiline TRL

In the previous section, we explained the fundamentals of the TRL method and stated that it is more accurate than SOLT in the sense that it does not rely on precise termination standards. One of the disadvantages of the TRL method is that the relative electrical length of the line standard cannot be an integer multiple of 180°; otherwise, singularity occurs. This limits the frequency bandwidth of the TRL calibration usually to 8:1. So, an improved TRL algorithm would be beneficial that can eliminate, or at least minimize, this restriction of the TRL. Besides, it would also be helpful if somehow redundant measurements of more than one line standard could be used to eliminate connector repeatability errors.

Multiline TRL calibration was developed with this motivation in mind [45]. The multiline TRL calibration is based on measurement of thru and multiple lines with different lengths as depicted in Figure 1.47. Then, the algorithm uses these redundant measurements to extend the bandwidth and reduce noise in the error box parameters. One might think that if all the data were equally noisy, then a simple averaging would help to reduce the noise in computed error box parameters. However, some of the measurements might be inherently noisier than the others. So, instead of applying simple averaging, the multiline method employs the Gauss-Markov theorem and tries to determine covariance matrixes to weight measurements appropriately. This results in the best unbiased linear estimate (BLUE) of error box parameters [45]. We will not go into the details of the algorithm here; interested readers can refer to the literature [45]. However, we will demonstrate how weighing can be applied in TRL for a simple case. For the sake of demonstration, we will assume the covariance matrix is diagonal (i.e., no stochastic dependence between the measurements; only variances will be considered). It should be noted that the original multiline algorithm does not make this assumption.

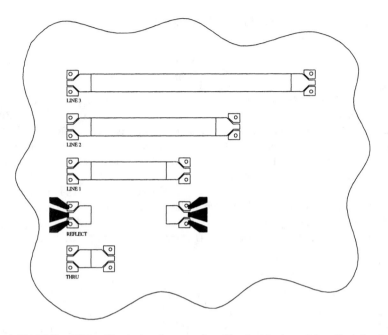

Figure 1.47 Multiline TRL calibration set for on-wafer calibration. In case of three lines (as shown), each line can be selected to be 90° long at the beginning, center, and end of the frequency band, provided that the starting frequency is not too low (otherwise an impractically long transmission line would have to be realized). Multiline TRL is the most accurate on-wafer calibration technique compared to TRL and SOLT.

Before explaining the basics of multiline TRL, let's first review the fundamentals of weighted least-squares fitting. Suppose we want to solve the following equation:

$$\mathbf{b} = \mathbf{A}x + \mathbf{r} \tag{1.261}$$

based on the condition

$$E\{\mathbf{r}\} = E\{\mathbf{b} - \mathbf{A}x\} = 0 \tag{1.262}$$

where **b** matrix represents the observations, **r** is the some error vector, x is the unknown vector to be determined, **A** is the coefficients matrix, and $E\{\}$ is the operator for mean. Now, it can be proven that BLUE of x is given by the following expression:

$$\hat{x} = \left(\mathbf{A}^T \Sigma_b^{-1} \mathbf{A}\right)^{-1} \mathbf{A}^T \Sigma_b^{-1} \mathbf{b} \tag{1.263}$$

where Σ_b is the covariance matrix for the observations. This was first proposed by Gauss, who was one of the great mathematicians known to mankind. Note that in case of unity variances and independent errors, Σ_b becomes unity and (1.263) becomes ordinary least-squares. Returning to TRL, it can be shown that the error in calculating error box coefficients is inversely proportional with the sine of the phase difference of the line standard [45–47]:

$$|E| \propto \frac{1}{2|\sin(\phi)|} \tag{1.264}$$

In particular, the error becomes infinite when the phase difference between thru and line becomes an integer multiple of 180°, a well-known disadvantage of the TRL algorithm. Equation (1.264) immediately suggests that we can use (1.264) to approximate the variances in (1.263), thus weighting the measurements based on the phase differences between particular line pairs. Since we have multiple line pairs, one can expect coefficients extracted from some pairs to be more optimal than the others because the phase difference between those pairs is close to 90° at the frequency of calculation. Now, when the phase difference becomes an integer multiple of 180°, the resulting variance becomes automatically very large, giving near-zero weight to the measurement. This can be stated mathematically in matrix form as follows:

$$\hat{t} = \left(\begin{bmatrix} 1 \\ 1 \\ \vdots \\ 1 \end{bmatrix}^T \begin{bmatrix} \sigma_1^2 & 0 & \cdots & 0 \\ 0 & \sigma_2^2 & & \vdots \\ \vdots & & \ddots & 0 \\ 0 & \cdots & 0 & \sigma_N^2 \end{bmatrix}^{-1} \begin{bmatrix} 1 \\ 1 \\ \vdots \\ 1 \end{bmatrix} \right)^{-1} \begin{bmatrix} 1 \\ 1 \\ \vdots \\ 1 \end{bmatrix}^T \begin{bmatrix} \sigma_1^2 & 0 & \cdots & 0 \\ 0 & \sigma_2^2 & & \vdots \\ \vdots & & \ddots & 0 \\ 0 & \cdots & 0 & \sigma_N^2 \end{bmatrix}^{-1} \begin{bmatrix} t_1 \\ t_2 \\ \vdots \\ t_N \end{bmatrix} \tag{1.265}$$

where

$$\sigma_i^2 = \frac{1}{\left| e^{-\gamma(l_i - l_0)} - e^{\gamma(l_i - l_0)} \right|}, \quad i = 1, 2, \ldots, N \tag{1.266}$$

In the above equations, N is the number of line pairs (there are $N + 1$ lines that needs to be measured), l_i is the length of the ith line, l_0 is the length of the thru, γ is the complex propagation constant, t_i is the calculated transfer matrix parameter based on different pairs of lines, and $\hat{t}$ is the estimate of the transfer matrix parameter. This procedure should be repeated for every computed transfer matrix parameter in the error boxes. This simple utilization of multiple line measurements provides us with a way of automatically weighting the transfer matrix parameters depending on the frequency and thus extends the usable range of TRL without worrying about ill-conditioning. Of course, numerical safeguards must be taken to prevent division by zero in calculating the variances. Note that in this demonstration, we paired the lines always with the thru line; the original multiline algorithm can also optimally select the lines to form best pairs [45].

The above equations describe the basic philosophy of the multiline TRL algorithm. We assumed there are no cross-correlations between the measurements. The multiline TRL algorithm is perhaps the most accurate on-wafer calibration method and must always be implemented where high accuracy is required.

1.12.4 Two-Tier and One-Tier Calibration

One of the disadvantages of basic TRL calibration is the lack of correcting for leakage and switching errors in the network analyzer, as discussed before. This is because the basic TRL method assumes a static error model as opposed to the dynamic error model of SOLT. By static, we mean that the error model remains the same even though the transfer switch changes its position. This is not the case for SOLT, where two different error models are used for each position of the switch. Therefore, the SOLT error model is more suitable than the error model of the basic TRL in terms of representing the physical construction of the most commonly used vector network analyzers. However, the disadvantage of SOLT is that it relies on precise terminations that are clearly difficult to achieve at microwave frequencies for on-wafer tests.

So, a natural question arises at this point: can we calibrate the network analyzer for on-wafer test while representing the internal errors of the network analyzer properly yet using on-wafer transmission line standards that can be manufactured with superior accuracy? One solution to this problem is to employ a four-sampler analyzer and use some additional measurements to make the necessary corrections in the TRL error terms [20, 23]. In this way, the whole system is calibrated in a single pass, which is advantageous.

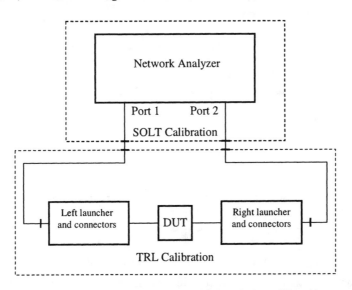

Figure 1.48 Description of two-tier calibration. Network analyzer is first calibrated by a manufacturer-suggested method (usually SOLT), then TRL is done to extract launcher and connector error boxes.

The alternative, which can be used for both three- and four-sampler analyzers, is the two-tier calibration shown in Figure 1.48. In this method, the network analyzer is first calibrated using high-quality coaxial standards on the tip of the connection cables. Note that contrary to on-wafer SOLT standards, the coaxial SOLT standard can be built with high precision. This step ensures that the internal errors of the network analyzer are accounted for properly. Then, coaxial cables are connected to probes and a second calibration is performed using TRL to extract the error boxes due to probes and connectors. Since these will remain fixed once connected, TRL can fully characterize the relevant error boxes. Finally, the DUT is measured and de-embedded.

The apparent disadvantage of the two-tier method is the necessity of calibrating the system twice; hence more labor is involved. The advantages are the proper representation of network analyzer error models and the extraction of launcher error boxes, which may be of interest.

1.13 LUMPED RESONATOR CIRCUITS

Resonator circuits are the fundamental blocks of electronic circuits. The frequency selectivity provided by resonator circuits is the key concept in filters, oscillators, and amplifiers. Figure 1.49(a–b) shows the two fundamental lumped resonator circuits. Note that the capacitor is responsible for storing electric energy and the inductance is responsible for storing magnetic energy. Resistance represents losses that may be present in the system.

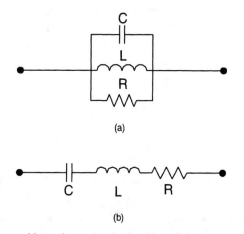

(a)

(b)

Figure 1.49 Two fundamental lumped resonator circuits: (a) parallel resonator and (b) series resonator.

In resonance, stored electric energy in the capacitor becomes equal to stored magnetic energy in the inductor, and input impedance of the resonator becomes purely resistive. The ratio of stored energy to dissipated energy in resonator circuits is called the quality factor, or simply Q.

$$Q = \omega \frac{\text{Time - average stored energy}}{\text{Energy dissipated per second}} \qquad (1.267)$$

The above equation is evaluated at resonance. Note that for lossless resonators, Q goes to infinity. But this does not happen in practice (unless superconducting materials are used). Therefore, it is one of the challenges to a microwave engineer to minimize losses in resonators to increase the quality factor. For instance, the quality factor of a resonator used in an oscillator circuit would affect the phase noise of the oscillator, which is one of the key parameters in oscillator design.

To derive Q expressions for the resonator circuits shown in the figure, we first write stored energies in the capacitor and inductor:

$$W_e = \frac{1}{4} CVV^*$$ (1.268)

$$W_m = \frac{1}{4} LII^*$$ (1.269)

where V and I are the steady-state voltage across the capacitor and current through the inductor, respectively. By using the above expressions, it can be shown that Q can be expressed for parallel and series resonators as follows:

For parallel resonators [Figure 1.49(a)]:

$$Q = \omega \frac{\frac{1}{4} C|V|^2 + \frac{1}{4} L|I|^2}{|V|^2 G} = \omega RC$$ (1.270)

For series resonators [Figure 1.49(b)]:

$$Q = \omega \frac{\frac{1}{4} C|V|^2 + \frac{1}{4} L|I|^2}{|I|^2 R} = \frac{\omega L}{R}$$ (1.271)

The above equations tells us that in a parallel resonator, one wants to have the resistor term as big as possible to maximize the Q-factor. This makes sense because the bigger the parallel resistor, the smaller the power dissipated on it. Alternatively, in a series resonator, one needs as small a resistor as possible to maximize the Q-factor. Of course, this assessment is valid with the shown circuit models. Another observation is that the Q-factor increases linearly with frequency, provided that losses are kept constant. However, in practice, this assumption is almost always violated, and losses in the circuit dominate after a certain point, causing the Q-factor to diminish afterwards. Conductor loss due to the skin depth effect is a major loss factor. Radiation and dielectric losses further lower the Q-factor as frequency increases. Note that the Q-factor is always specified at a given frequency and inductance or capacitance value. Without these two parameters, specifying the Q-factor of a resonator does not make any sense.

It would also be instructive to study the relation between the Q-factor and the input impedance of resonator circuits. For this purpose, the input impedance of a resonator circuit (series or parallel) is written in the vicinity of resonance frequency (i.e., $\omega = \omega_0 + \Delta\omega$). Then, by using the approximation $(\omega_0 + \Delta\omega)^2 \approx (\omega_0 + 2\omega_0\Delta\omega)$ for $\omega_0 \gg \Delta\omega$ (which is usually satisfied in practice), the following expressions are obtained for parallel and series resonators:

For parallel resonators:

$$Z_{in} = \frac{j\omega RL}{R + j\omega L - \omega^2 RLC} \tag{1.272}$$

$$Z_{in} = \frac{j(\omega_0 + \Delta\omega)\omega RL}{R + j(\omega_0 + \Delta\omega)L - (\omega_0 + \Delta\omega)^2 RLC}$$
$$\cong R\frac{j(\omega_0 + \Delta\omega)}{j(\omega_0 + \Delta\omega) - 2\Delta\omega Q} \cong R\frac{1}{1 + 2jQ\dfrac{\Delta\omega}{\omega_0}} \tag{1.273}$$

For series resonators:

$$Z_{in} = \frac{-\omega^2 LC + j\omega RC + 1}{j\omega C} \tag{1.274}$$

$$Z_{in} = \frac{-(\omega_0 + \Delta\omega)^2 LC + j(\omega_0 + \Delta\omega)RC + 1}{j(\omega_0 + \Delta\omega)C}$$
$$\cong R\frac{-2\Delta\omega Q + j(\omega_0 + \Delta\omega)}{j(\omega_0 + \Delta\omega)} \cong R\left(1 + 2jQ\frac{\Delta\omega}{\omega_0}\right) \tag{1.275}$$

Now, let's assume the following conditions are satisfied for the magnitude of the input impedance at some frequency points around the resonance:

For parallel resonators:

$$Z_{in}\big|_{\omega=\omega_0} = 0.707 \cdot Z_{in}\big|_{\omega=\Delta\omega}$$

For series resonators:

$$Z_{in}\big|_{\omega=\omega_0} = 1.414 \cdot Z_{in}\big|_{\omega=\Delta\omega}$$

By examining (1.273) and (1.275), the following well-known equation can be obtained for the Q-factor:

$$Q = \frac{\omega_0}{2\Delta\omega} = \frac{\omega_0}{BW} \tag{1.276}$$

where BW is called the 3-dB bandwidth of the resonator. Note that in order to use the above equation for the Q-factor, the condition $\omega_0 \gg \Delta\omega$ must be satisfied.

After studying single resonators, we will now analyze simple coupled-resonators where two resonators are coupled with each other through a mechanism. In practice, this mechanism could be a hole or a screw, as in waveguide circuits, or just the physical proximity of two resonators as in microstrip circuits. Coupled resonators are extensively used in the design of electric filter circuits. Figure 1.50 shows the two coupling mechanisms that may exist between two resonators. Note that a mixed coupling including both electric and magnetic couplings is also possible.

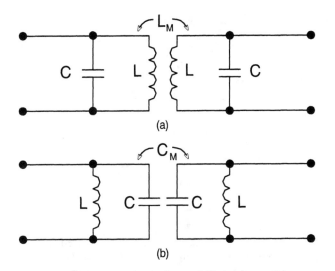

(a)

(b)

Figure 1.50 Coupled resonators: (a) magnetic coupling, and (b) electric coupling.

In the following paragraphs, we will show that a circuit that consists of two coupled resonators tuned to the same self-resonance has two different resonance frequencies because of the coupling phenomena between the circuits. The separation of those two resonance frequencies depends on the amount of coupling between the resonators. To show this, Z-parameters of the two circuits shown in Figure 1.50 are obtained as follows:

Magnetic coupling:

$$Z_{11} = Z_{22} = -j\frac{\omega^3 C\left(L_m^2 - L^2\right) + \omega L}{\omega^4 C^2\left(L_m^2 - L^2\right) + \omega^2 2CL - 1} \tag{1.277}$$

$$Z_{21} = Z_{12} = -j\frac{\omega L_m}{\omega^4 C^2\left(L_m^2 - L^2\right) + \omega^2 2CL - 1} \tag{1.278}$$

Electric coupling:

$$Z_{11} = Z_{22} = -j \frac{\omega^3 C L^2 - \omega L}{\omega^4 L^2 (C^2 - C_m^2) - \omega^2 2CL + 1} \qquad (1.279)$$

$$Z_{11} = Z_{22} = j \frac{\omega^3 C_m L^2}{\omega^4 L^2 (C^2 - C_m^2) - \omega^2 2CL + 1} \qquad (1.280)$$

where L_m and C_m are the magnetic coupling and electric coupling coefficients, respectively. Now, the poles of the above Z-parameters are the two natural frequencies described before, and they are given as follows:

Magnetic coupling:

$$\omega_1 = \pm \frac{1}{\sqrt{C(L + L_m)}}$$

$$\omega_2 = \pm \frac{1}{\sqrt{C(L - L_m)}} \qquad (1.281)$$

Electric coupling:

$$\omega_1 = \pm \frac{1}{\sqrt{L(C + C_m)}}$$

$$\omega_2 = \pm \frac{1}{\sqrt{L(C - C_m)}} \qquad (1.282)$$

For zero coupling (i.e., $L_m = 0$ and $C_m = 0$), the resonance frequencies given above reduce to the usual expressions for simple tank circuits. These expressions also suggest practical ways of determining the amount of coupling between any two resonators. For example, by measuring (or simulating) the location of two resonance frequencies, one can find the amount of coupling. This method is very powerful and yields design of filter networks through a method called filter design by coupling.

1.14 TRANSMISSION LINE RESONATORS

In the early days of filter design, transmission lines were the only means known to engineers to design high-Q resonators. Therefore, transmission lines terminated with either open- or short-circuits have been used extensively as resonators in many microwave circuits, especially in filter networks. Although this statement is probably still true, microstrip lumped resonators have started to appear in microwave circuits increasingly because of their unsurpassed advantage in miniaturization.

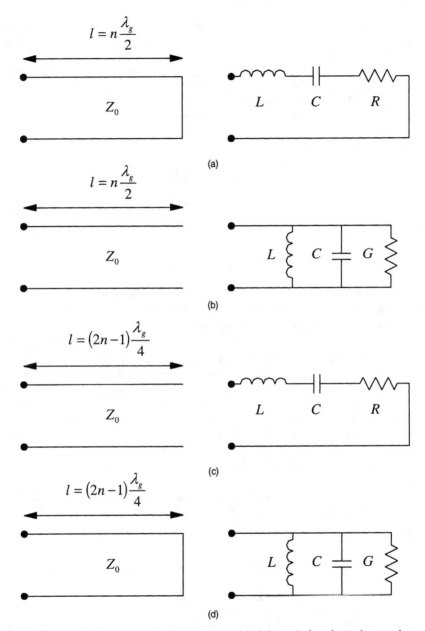

Figure 1.51 Different transmission line resonators and their equivalent lumped networks near resonance: (a) half-wavelength long short-circuited transmission line, (b) half-wavelength long open-circuited transmission line, (c) quarter-wavelength long open-circuited transmission line, and (d) quarter-wavelength long short-circuited transmission line,

Nevertheless, studying transmission line resonators will provide us with very valuable insight into general filter design. Besides, transmission line resonators are also used as intermediate components while transforming lumped prototype filters into coupled-line microstrip filters, which will be explored in detail in Chapter 6.

Figure 1.51(a–d) summarizes the four different transmission line resonators and their equivalent lumped networks near resonance. An important parameter of resonance circuits is the slope-parameter, defined as follows:

Series resonator:

$$x \equiv \omega_0 L = \frac{1}{\omega_0 C} = \frac{\omega_0}{2} \frac{dX}{d\omega}\bigg|_{\omega=\omega_0} \tag{1.283}$$

Parallel resonator:

$$b \equiv \omega_0 C = \frac{1}{\omega_0 L} = \frac{\omega_0}{2} \frac{dB}{d\omega}\bigg|_{\omega=\omega_0} \tag{1.284}$$

where X and B are the reactance and susceptance of the resonator input impedance, respectively, and ω_0 is the resonance frequency of the circuit.

This general definition of slope parameters provides a convenient way of relating the resonance properties of any circuit to an equivalent simple tank circuit. Note that slope parameters are very useful in converting lumped bandpass or bandstop filters into microstrip equivalents using transmission line resonators. For example, suppose that we have a filter circuit constructed using lumped resonators, and we want to convert the filter into a distributed one. The required parameters (characteristic impedance and length) of the transmission line resonators that will be used in the distributed equivalent can be determined from the corresponding slope parameters using (1.284) or (1.283), depending on the type of the resonator. They also provide an alternative relationship for the quality factor of narrow bandwidth resonators.

For the transmission line resonators shown in the Figure 1.51, the slope parameters and quality factors can be determined as follows:

Type A:

$$R \cong Z_0 \alpha_l l$$

$$x = \omega_0 L = \frac{1}{\omega_0 C} = \frac{n\pi Z_0}{2}\left(\frac{\lambda_g}{\lambda}\right)^2 \tag{1.285}$$

$$Q = \frac{x}{R}$$

Type B:

$$G \cong Y_0 \alpha_t l$$

$$b = \omega_0 C = \frac{1}{\omega_0 L} = \frac{n\pi Y_0}{2}\left(\frac{\lambda_g}{\lambda}\right)^2 \qquad (1.286)$$

$$Q = \frac{b}{G}$$

Type C:

$$R \cong Z_0 \alpha_t l$$

$$x = \omega_0 L = \frac{1}{\omega_0 C} = \frac{(2n-1)\pi Z_0}{4}\left(\frac{\lambda_g}{\lambda}\right)^2 \qquad (1.287)$$

$$Q = \frac{x}{R}$$

Type D:

$$G \cong Y_0 \alpha_t l$$

$$b = \omega_0 C = \frac{1}{\omega_0 L} = \frac{(2n-1)\pi Y_0}{4}\left(\frac{\lambda_g}{\lambda}\right)^2 \qquad (1.288)$$

$$Q = \frac{B}{G}$$

where α_t is the attenuation in Nepers per unit length, λ_g is the guided wavelength, and $\lambda = 1/\sqrt{\varepsilon\mu}$. The above expressions can also be used for non-TEM transmission lines (i.e., hollow waveguides) by using the modified propagation constant. That is why the $(\lambda_g/\lambda)^2$ term appears in the equations. For example, for rectangular waveguides, guided wavelength of the TE$_{mn}$ mode can be found from:

$$\lambda_g = \frac{2\pi}{\beta_{n,m}}$$

$$\beta_{n,m} = \sqrt{\left(\frac{2\pi}{\lambda_0}\right)^2 - \left(\frac{n\pi}{a}\right)^2 - \left(\frac{m\pi}{b}\right)^2}$$

where a and b are the dimensions of the rectangular waveguide. Note that for TEM transmission lines, $\lambda_g = \lambda$.

1.15 TAPPED TRANSMISSION LINE RESONATORS

Tapped transmission line resonators are commonly used in filters to feed RF
signals in and out of the circuit. Two different tapped transmission line resonators
that can be used for this purpose are shown in Figure 1.52(a–b). For the sake of
reference, the resulting current distribution is also shown in the figure by dashed
lines. It will be shown that the Q-factor is a function of the distance to the current
maximum.

In the first network, one end of the transmission line is shorted, whereas in the
second, both ends are open circuited. Therefore, the first network resonates at a
quarter-wavelength and the second network resonates at a half-wavelength. The
distance of the tapping point to the current maximum is given by l_s in both cases.
Location of the tapping determines loaded Q of the resonator, which is an
important parameter in designing the filters.

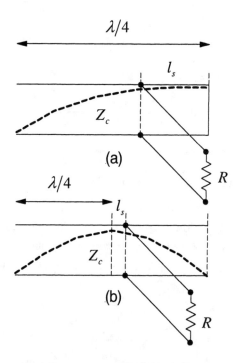

Figure 1.52 Tapped transmission line resonators: (a) quarter-wavelength long, and (b) half-wavelength
long. Distance of the tapping point to current maximum is given by l_s in both cases. Note that R
represents the load resistance. Current magnitude is represented by dashed lines.

To determine Q of the half-wavelength-long resonator [Figure 1.52(b)], we first
determine the admittance of the network as follows:

$$Y = Y_L + \frac{j}{Z_c} \left[\frac{\sin \beta l}{\cos \beta l} + \frac{\sin(\theta - \beta l)}{\cos(\theta - \beta l)} \right]$$

$$= Y_L + \frac{2j}{Z_c} \frac{\sin \theta}{\cos \beta l \cos(\theta - \beta l)}$$

where $l = \lambda/4 - l_s$ and $Y_L = 1/R$. For $\theta \to \pi$ (i.e., near resonance), the above equation can be approximated by

$$Y \cong Y_L + jY_c \frac{\sin \theta}{\sin^2 \beta l - 1}$$

The next step is to find the 3-dB bandwidth of the admittance as follows:

$$\left| Y_L + jY_c \frac{\sin \theta}{\sin^2 \beta l - 1} \right| = Y_L \sqrt{2}$$

$$\Rightarrow Y_L^2 + Y_c^2 \frac{\sin^2 \theta}{\left(\sin^2 \beta l - 1 \right)^2} = 2Y_L^2$$

$$\Rightarrow \sin \theta = \frac{Y_L}{Y_c} \cos^2 \beta l$$

$$\sin \theta \cong \pi - \theta = \pm \frac{Y_L}{Y_c} \cos^2 \beta l$$

$$\pi \pm \frac{Y_L}{Y_c} \cos^2 \beta l = \frac{2\pi}{c} f l$$

The sign change in the above equation represents the 3-dB points on both sides of the resonance point. Then, frequencies and bandwidth corresponding to the 3-dB points are written as

$$f_1 = \frac{c}{2\pi l} \left[\pi + \frac{Y_L}{Y_c} \cos^2 \beta l \right] \quad f_2 = \frac{c}{2\pi l} \left[\pi - \frac{Y_L}{Y_c} \cos^2 \beta l \right]$$

$$BW = f_1 - f_2 = \frac{c}{\pi l} \frac{Y_L}{Y_c} \cos^2 \beta l$$

Finally, the Q-factor is found as follows:

$$Q = \frac{f_0}{BW} = \frac{\pi}{2} \frac{Y_c}{Y_L} \frac{1}{\sin^2 \beta l_s} \tag{1.289}$$

The above equation tells us that Q of a tapped line resonator is inversely proportional to the characteristic impedance of the line. Also, it is a function of the tapping point; the closer the tapping point to the current maximum, the greater the quality factor. This dependency can be used to make very low-loss filters.

To determine Q of the quarter-wavelength-long resonator [Figure 1.52(a)], we follow a similar approach and first determine admittance of the network as follows:

$$Y = Y_L + \frac{j}{Z_c}\left[\frac{\cos \beta l}{\sin \beta l} + \frac{\sin(\theta - \beta l)}{\cos(\theta - \beta l)}\right]$$

$$= Y_L + \frac{2j}{Z_c}\frac{\cos \theta}{\sin \beta l \cos(\theta - \beta l)}$$

where $l = l_s$ and $Y_L = 1/R$. By following the same procedure given for the half-wavelength-long resonator, the Q-factor is found as follows:

$$Q = \frac{f_0}{BW} = \frac{\pi}{4}\frac{Y_c}{Y_L}\frac{1}{\sin^2 \beta l_s} \tag{1.290}$$

Note that Q of a quarter-wavelength-long resonator is half of the half-wavelength-long resonator, if everything else is kept constant.

1.16 SYNTHESIS OF MATCHING NETWORKS

Synthesis of matching networks, or network synthesis in general, deals with the problem of determining a network topology and related components to realize a given response under a input excitation [48–50]. Network analysis and network synthesis do represent the two different complementary processes in network theory. In the former, the network topology and excitation are given, and circuit response is found, whereas in the latter, the designer starts from the given excitation and response to find a network topology. A major distinction between synthesis and analysis of linear networks is that in analysis there is a unique solution although it may be difficult to find. On the other hand, there is no unique solution of a synthesis problem, and there may exist no solution at all [48]. Therefore, the synthesis problem is generally more difficult than the analysis. As a matter of fact, modern computer simulation techniques have made the analysis problem straightforward for many microwave circuits, yet synthesis still requires human intelligence.

It is out of the scope of this book to go into details about the network synthesis problem. There are very good references in the literature, and readers are encouraged to refer to them [48, 50]. However, it is our belief that a minimum amount of network synthesis knowledge is required to properly grasp the concepts of filtering and matching. For example, the famous Foster's reactance theorem has its roots in network synthesis. Darlington's impedance synthesis technique is used in filter synthesis, especially in coupled filters. Finally, it is very helpful for the modeling of passive components to understand the basic requirements of functions that represent circuit parameters.

1.16.1 Positive Real Functions

We will start our discussion of network analysis by giving the definition and some important properties of positive real (PR) functions since they play a very important role in network synthesis. A positive real function is an analytical function satisfying the following conditions [50]:

1. $f(s)$ is analytic in the open right-half of the s-plane (RHS).
2. $f(\bar{s}) = \bar{f}(s)$ for all s in the open RHS.
3. $\mathrm{Re}\{f(s)\} \geq 0$ whenever $\mathrm{Re}\{s\} \geq 0$

where $s = \sigma + j\omega$ is a complex variable. The above definition is valid for both rational and transcendental functions. At this point, we will state one of the important theorems in network synthesis without providing the proof: the driving point immittance (impedance or admittance) of passive one-port networks containing only passive elements (i.e., resistors, capacitors, inductors, and coupled coils) as shown in Figure 1.53 is a positive real function. The introduction of positive real functions into network synthesis is due to Otto Brune.

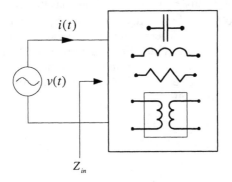

Figure 1.53 A one-port circuit containing capacitors, inductors, resistors, and transformers. Input impedance of such a circuit is a positive real function (see the text).

A rational function is defined as the ratio of two polynomials and is especially important in network synthesis because network functions associated with any linear lumped system are rational. Therefore, it would be helpful if we could define the necessary conditions for a rational function to be a PR. It can be shown that a rational function $F(s)$ is positive real if and only if the following conditions are satisfied [50]:

1. $F(s)$ is real when s is real.
2. $F(s)$ has no poles in the open RHS.

3. Poles of $F(s)$ on $j\omega$-axis, if they exist, are simple, and residues evaluated at these poles are real and positive.
4. $\text{Re}\{F(s)\} \geq 0$ for all ω, except at the poles.

For computational purposes, it would be helpful if a simpler test were proposed to check the positive realness of a rational function. To do this, we first define the rational function $F(s)$ in the following form:

$$F(s) = \frac{P(s)}{Q(s)} = \frac{m_1(s) + n_1(s)}{m_2(s) + n_2(s)} \qquad (1.291)$$

where m and n are the even and odd parts of the polynomials P and Q, respectively. Using this definition, alternative conditions for positive realness can also be given as follows [48, 52]:

1. $F(s)$ is real when s is real.
2. $P(s) + Q(s)$ is strictly Hurwitz.
3. $m_1(j\omega)m_2(j\omega) - n_1(j\omega)n_2(j\omega) \geq 0$ for al ω.

where a Hurwitz polynomial is a polynomial that has no zeroes in the open RHS. Alternatively, if a polynomial has zeros neither in the RHS nor on the $j\omega$-axis, then it is called strictly Hurwitz. The testing of the second condition can easily be accomplished with the help of the Hurwitz test: a real polynomial is strictly Hurwitz if the continued-fraction expansion of the ratio of the real part to the odd part, or the odd part to the real part, has only real and positive coefficients and does not terminate prematurely:

$$\left[\frac{m_1(s) + m_2(s)}{n_1(s) + n_2(s)} \right]^{\pm 1} = \alpha_1 s \cfrac{1}{\alpha_2 s + \cfrac{1}{\alpha_3 s + \cfrac{1}{\ddots + \cfrac{1}{\alpha_k s}}}} \qquad (1.292)$$

where α_l, $l = 1, 2, \ldots, k$, are real coefficients. Note that k must be equal to the degree of $m_1(s) + m_2(s)$ or $n_1(s) + n_2(s)$, whichever is larger.

Before applying the above test to a rational function, it would be helpful to inspect the function so that it satisfies the fundamental requirements. An inspection test for the positive realness of a rational function can be given as follows [48]:

1. All polynomials coefficients should be real and positive.
2. Degrees of numerator and denominator polynomials differ at most by one.

3. Numerator and denominator terms of the lowest degree differ at most by one.
4. Imaginary axis poles and zeroes are simple.
5. There should be no missing terms in the numerator or denominator polynomials unless all even or all odd terms are missing.

Note that these conditions are necessary but not sufficient. So, they must be used as a preliminary test to check whether or not a rational function is PR. If the function passes these, then the rigorous test should be performed as described in the previous paragraph.

1.16.2 Foster's Reactance Theorem

Foster's reactance theorem states that the frequency derivative of the reactance or the susceptance of *LC* networks is always positive [48]. To prove the theorem, let us start by expressing the input impedance of a one-port *LC* network in the following form [48]:

$$Z(s) = \frac{m_1 + n_1}{m_2 + n_2} \tag{1.293}$$

where m and n are even and odd polynomials in s as described before. Since the product of two even or two odd functions is an even function, and the product of an even and an odd function is an odd function, the input impedance can also be written as follows:

$$\mathrm{Ev}\{Z(s)\} = \frac{m_1 m_2 - n_1 n_2}{m_2^2 - n_2^2}$$

$$\mathrm{Odd}\{Z(s)\} = \frac{m_2 m_1 - m_1 n_2}{m_2^2 - n_2^2} \tag{1.294}$$

Note that substitution of $s = j\omega$ into an even polynomial gives a real number, and substitution of $s = j\omega$ into an odd polynomial gives an imaginary number. Thus, the input impedance can be written as

$$Z(s) = \mathrm{Re}\{Z(s)\} + j\,\mathrm{Im}\{Z(s)\}$$

$$= \mathrm{Ev}\{Z(s)\}\big|_{s=j\omega} + j\mathrm{Odd}\{Z(s)\}\big|_{s=j\omega} \tag{1.295}$$

Therefore, it is concluded that $\mathrm{Re}\{Z(j\omega)\}$ is an even function in ω, and $\mathrm{Im}\{Z(j\omega)\}$ is an odd function in ω. Since we have assumed a network that contains inductors and capacitors only, it is also known that $\mathrm{Re}\{Z(j\omega)\} = 0$ for all ω. Now, in order to show that $\mathrm{Re}\{Z(j\omega)\}$ is indeed equal to zero, we need to satisfy one of the following conditions: (1) $m_1 = 0$ and $n_2 = 0$, (2) $m_2 = 0$ and $n_1 = 0$, or (3) $m_1 m_2 - n_1 n_2 = 0$. Thus, the following two forms of $Z(s)$ are obtained by application any of these conditions:

$$Z(s) = \frac{m_1}{n_2} \quad \text{or} \quad Z(s) = \frac{n_1}{m_2} \tag{1.296}$$

This shows another important result in network theory that can be stated as follows: the input impedance of an LC network is always the quotient of even-to-odd or odd-to-even polynomials. Since we also know that $Z(s)$ is positive real, m_2 must be a Hurwitz, and all its zeros must be simple and pure imaginary. Similarly, n_2 is also Hurwitz and equal to s times an even polynomial, so its zeros are simple and pure imaginary as well. The same argument also applies for m_1 and n_1. In conclusion, the input impedance can be written as

$$Z(s) = \frac{H\left(s^2 + \omega_{z_1}^2\right) \cdot \left(s^2 + \omega_{z_2}^2\right) \dots \left(s^2 + \omega_{z_M}^2\right)}{s\left(s^2 + \omega_{p_1}^2\right) \cdot \left(s^2 + \omega_{p_2}^2\right) \dots \left(s^2 + \omega_{p_N}^2\right)} \tag{1.297}$$

where H is a constant term. Then, we express the above equation in the form of partial fractions as follows:

$$Z(s) = Hs + \frac{K_0}{s} + \frac{K_{p_1}}{s - j\omega_{p_1}} + \frac{K_{-p_1}}{s + j\omega_{p_1}} + \frac{K_{p_2}}{s - j\omega_{p_2}} + \frac{K_{-p_2}}{s + j\omega_{p_2}} + \dots \tag{1.298}$$

where K is the residue of the respective poles. Since $Z(s)$ is positive real, all residues at the imaginary axis poles are real and positive and $K_{p_r} = K_{-p_r} = K_{p_r} > 0$. Thus, $Z(s)$ is written as

$$Z(s) = Hs + \frac{K_0}{s} + \sum_{r=1}^{n} \frac{2K_{p_r} s}{s^2 + \omega_{p_r}^2} \tag{1.299}$$

Since we know that the even part of the impedance is zero, we can define an odd reactance function $X(\omega)$ as follows:

$$Z(j\omega) = \text{Re}\{Z(j\omega)\} + j\,\text{Im}\{Z(j\omega)\} = jX(\omega) \tag{1.300}$$

where

$$X(\omega) = H\omega - \frac{K_0}{\omega} + \sum_{r=1}^{n} \frac{2K_{p_r} \omega}{-\omega^2 + \omega_{p_r}^2} \tag{1.301}$$

$X(\omega)$ is known as the reactance function. Now, by taking the derivative of the reactance function with respect to ω, one can find that

$$\frac{dX(\omega)}{\omega} = H + \frac{K_0}{\omega^2} + \sum_{r=1}^{n} \frac{2K_{p_r}\left(\omega^2 + \omega_{p_r}^2\right)}{\left(-\omega^2 + \omega_{p_r}^2\right)^2} \tag{1.302}$$

Note that every term in the above function is positive. Thus, the frequency derivative of the reactance function is always greater than zero, which is the statement of Foster's reactance theorem:

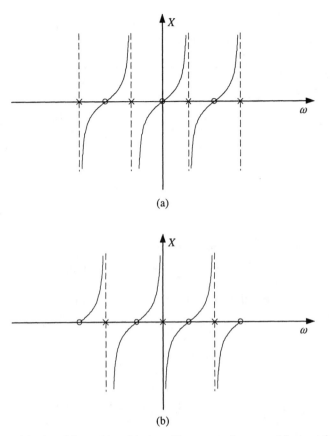

(a)

(b)

Figure 1.54 The behavior of the reactance function with respect to frequency: (a) zero at the origin and (b) pole at the origin. Note that the derivative is always positive.

$$\frac{dX(\omega)}{\omega} > 0 \quad \text{for} \quad -\infty < \omega < \infty \tag{1.303}$$

Figure 1.54 shows the two possible forms of the reactance function in the case of a simple *LC* circuit. Since the slope is positive, it is also necessary that the poles and zeros alternate along the real frequency axis:

$$0 < \omega_{z_1} < \omega_{p_1} < \omega_{z_2} < \omega_{p_2} \cdots \tag{1.304}$$

1.16.3 Darlington's Method

In this section, a brief introduction to Darlington's synthesis method will be given because it is closely related to filter and matching network synthesis. The purpose of Darlington's method is to determine the reactance network shown in Figure 1.55 from its input impedance. The input impedance of a reactive network can be extracted from the transfer function; hence, the method is suitable for synthesizing filter circuits with a specified electrical response.

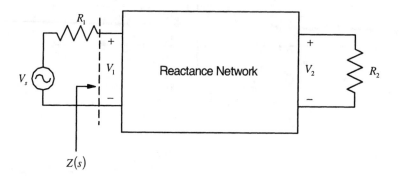

Figure 1.55 Two-port double-terminated passive network.

By simple circuit analysis, it can be shown that the input driving point impedance of the two-port network shown in Figure 1.55 is written in terms of the network parameters as follows (assuming 1-ohm termination resistance):

$$Z(s) = z_{11} \frac{1/y_{22} + 1}{z_{22} + 1} \tag{1.305}$$

$$Z(s) = \frac{1}{y_{11}} \frac{y_{22} + 1}{1/z_{22} + 1} \tag{1.306}$$

where z_{ij} and y_{ij} represent Z- and Y-parameters of the two-port network, respectively. Note that in network synthesis, small letters are typically used for the network parameters (i.e., Y- and Z-parameters). The capital letters are usually reserved for the driving point functions. Depending on the context, reader should distinguish between these two. Then, we write the $Z(s)$ in terms of the even and odd parts of its numerator and denominator polynomials:

$$Z(s) = \frac{m_1 + n_1}{m_2 + n_2} \tag{1.307}$$

where m and n refer to even and odd parts, respectively. After some algebraic manipulations, (1.307) is written as follows [48]:

$$Z(s) = \frac{m_1}{n_2} \frac{n_1/m_1 + 1}{m_2/n_2 + 1} \quad \text{(case A)} \tag{1.308}$$

$$Z(s) = \frac{n_1}{m_2} \frac{m_1/n_1 + 1}{n_2/m_2 + 1} \quad \text{(case B)} \tag{1.309}$$

By comparing (1.308) and (1.309) with (1.305) and (1.306), the network parameters can be determined as follows:

For Z-parameters:

$$z_{11} = \frac{m_1}{n_2}, \quad z_{22} = \frac{m_2}{n_2}, \quad y_{22} = \frac{m_1}{n_1}, \quad z_{12} = \frac{\sqrt{m_1 m_2 - n_1 n_2}}{n_2} \quad \text{(case A)} \qquad (1.310)$$

$$z_{11} = \frac{n_1}{m_2}, \quad z_{22} = \frac{n_2}{m_2}, \quad y_{22} = \frac{n_1}{m_1}, \quad z_{12} = \frac{\sqrt{n_1 n_2 - m_1 m_2}}{m_2} \quad \text{(case B)} \qquad (1.311)$$

For Y-parameters:

$$y_{11} = \frac{n_2}{m_1}, \quad y_{22} = \frac{n_1}{m_1}, \quad z_{22} = \frac{n_2}{m_2}, \quad y_{12} = \frac{\sqrt{n_1 n_2 - m_1 m_2}}{m_1} \quad \text{(case A)} \qquad (1.312)$$

$$y_{11} = \frac{m_2}{n_1}, \quad y_{22} = \frac{m_1}{n_1}, \quad z_{22} = \frac{m_2}{n_2}, \quad y_{12} = \frac{\sqrt{m_1 m_2 - n_1 n_2}}{n_1} \quad \text{(case B)} \qquad (1.313)$$

Now, it can be shown that z_{11}, z_{22}, y_{11}, and y_{22} are positive real functions if $Z(s)$ is positive real, which is the case. The network parameters z_{11}, z_{22}, y_{11}, y_{22}, z_{12}, and y_{12} are quotients of even-to-odd or odd-to-even polynomials. Since the functions y_{12} and z_{12} must be rational functions to represent a finite network, the expression under the radical signs must be full square. However, the positive real property of $Z(s)$ does not guarantee that those terms will be always full square. The remedy for this problem is to multiply both the numerator and denominator of the impedance function $Z(s)$ with an auxiliary Hurwitz polynomial so that y_{12} and z_{12} become rational functions. We will not go into the details of this procedure here, but interested readers can refer to literature [48]. At this point, the network parameters are determined in terms of the even and odd parts of the numerator and denominator of the input impedance.

The next natural question would be how to determine the driving point impedance $Z(s)$ from a given network function. Before giving the answer, let us first define the transmission coefficient as follows:

$$|t(j\omega)|^2 = \frac{4R_1}{R_2} \left| \frac{V_2(j\omega)}{V_s(j\omega)} \right| \qquad (1.314)$$

The relation between the transmission coefficient and the reflection coefficient can be written as

$$|t(j\omega)|^2 + |\rho(j\omega)|^2 = 1 \qquad (1.315)$$

So, once the transfer function is given, we can determine the reflection coefficient, which yields to the driving point impedance, $Z(s)$. Therefore, Darlington's method for double-terminated *LC* network synthesis is outlined as follows [48]:

1. Test the requirement $|t(j\omega)|^2 \le 1$, or do an equivalent test, such as

$$\left|\frac{V_2(j\omega)}{V_s(j\omega)}\right|^2 \le \frac{R_2}{4R_1}.$$ If this condition is not satisfied, then the network

function must be scaled.

2. Find $|\rho(j\omega)|^2 = 1 - |t(j\omega)|^2$. Form $\rho(j\omega) = \dfrac{C(-s^2)}{B(-s^2)}$ by letting $j\omega = s$.

3. Find $\rho(s) = \pm p_1(s)/q_1(s)$ by taking the left-half plane zeros of $B(-s^2)$ and either the left-half or right-half plane zeros of $C(-s^2)$. The zeros of $C(-s^2)$ must be conjugate if complex.

4. Find $\dfrac{Z(s)}{R_1} = \dfrac{q_1(s) \pm p_1(s)}{q_1(s) \mp p_1(s)}$, depending on which sign of $\rho(s)$ is selected.

5. Examine the zeros of $|t(j\omega)|^2$. If all zeros are at the origin and/or infinity, develop $Z(s)$ as a ladder. Otherwise, synthesize the network using a Cauer type of topology and remove as many ideal transformers as possible.

6. Shift the impedance levels if required.

This summarizes how to obtain the network parameters of a two-port circuit from a given transfer function specification through Darlington's synthesis method. Detailed examples will be given when we review the electrical filters.

The final step is the realization of the network topology, whose network parameters have been extracted through Darlington's method. Although there are many ways of accomplishing this, we will provide one of the fundamental ways of doing this using Cauer's method [48]. Cauer's method is based on the partial fraction expansion of network parameters and the principle that states that Z-parameters of series-connected two-port networks can be obtained by addition of Z-parameters of the individual networks when properly isolated by ideal transformers. This principle can be used to synthesize given Z-parameters through the partial fraction expansion of z_{11}, z_{22}, and z_{12}. The resulting network structure of Darlington's method according to Cauer's synthesis is given in Figure 1.56. In the figure, $Z^{(v)}$ and $a^{(v)}$ represent one-port shunt impedance of each section and transformer turn ratios, respectively. The parameters of the network are given as follows:

$$Z^{(v)} = \frac{k_{12}^{(v)}}{a^{(v)}} f_v(s) \qquad\qquad (1.316)$$

$$\left|a^{(v)}\right| = \frac{k_{22}^{(v)}}{\left|k_{12}^{(v)}\right|} = \frac{\left|k_{12}^{(v)}\right|}{k_{11}^{(v)}} \qquad\qquad (1.317)$$

$$\frac{1}{a^{(v)}}k_{12}^{(v)} \geq 0 \tag{1.318}$$

$$f_v(s) = \begin{cases} \dfrac{2s}{s^2 + \omega^2} \\[2mm] \dfrac{1}{s} \\[2mm] s \end{cases} \tag{1.319}$$

Note that the shunt impedance $Z^{(v)}$ of each section can have three different forms as given by $f_v(s)$; see (1.320). The magnitude and sign of the turn ratio of the ideal transformers, $a^{(v)}$, are selected according to (1.317) and (1.318), respectively. The residues, $k_{mn}^{(v)}$, can be determined from the partial fraction expansion of the network parameters in the following fashion:

$$z_{11} = \frac{k_{11}^{(0)}}{s} + \frac{2k_{11}^{(1)}s}{s^2 + \omega_1^2} + \frac{2k_{11}^{(2)}s}{s^2 + \omega_2^2} + \ldots + k_{11}^{(\infty)}s$$

$$z_{22} = \frac{k_{22}^{(0)}}{s} + \frac{2k_{22}^{(1)}s}{s^2 + \omega_1^2} + \frac{2k_{22}^{(2)}s}{s^2 + \omega_2^2} + \ldots + k_{22}^{(\infty)}s \tag{1.320}$$

$$z_{12} = \frac{k_{12}^{(0)}}{s} + \frac{2k_{12}^{(1)}s}{s^2 + \omega_1^2} + \frac{2k_{12}^{(2)}s}{s^2 + \omega_2^2} + \ldots + k_{12}^{(\infty)}s$$

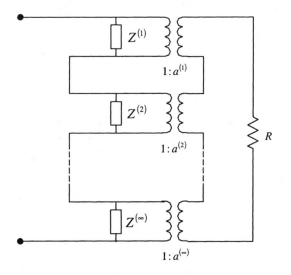

Figure 1.56 Cauer-type network structure of the Darlington synthesis method.

Therefore, once the driving point impedance of a two-port passive network is known, the network parameters of the circuit can be determined. Then, these parameters can be used to synthesize the Cauer-type network (or another suitable) topology.

1.16.4 Matching a Resistive Generator to an *RLC* Load

In the previous section, we presented Darlington's method to synthesize resistively terminated reactive networks. Although the method is general, the resulting topology may not always be desirable. Fortunately, for certain classes of transfer functions (e.g., Butterworth, Chebyshev), simple ladder network realizations are possible. Ladder networks are commonly used in practice due to their simplicity. In this section, simple closed-form expressions will be provided for different classes of ladder matching networks placed between a generator with internal resistance and a resistive load as shown in Figure 1.57.

In most of the practical cases, the load impedance can be modeled as a parallel resistor and capacitor in series with an inductor. Now, the reactive components of this load can be assumed as a part of the matching structure as depicted in Figure 1.57. Then, the matching network is derived accordingly. This simple model is usually sufficient to model input/output impedance of amplifiers over moderate bandwidths (~10%); therefore, it has wide practical applicability. Thus, simple expressions for the design of matching networks is highly desirable. It should be stated that realization of a ladder-type matching network is not always possible. For a detailed discussion of realization under those circumstances, the reader is referred to [50]. In the following paragraphs, we will provide closed-form expressions for Butterworth- and Chebyshev-type matching networks.

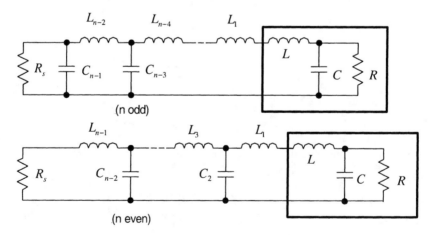

Figure 1.57 Ladder matching network to match a source resistance (R_s) to an *RLC* load.

Butterworth Matching Networks

For $RC\omega_c \geq 2\sin(\gamma_1)$ and $L_1 \geq 0$, the component values for Butterworth-type transfer function are given as follows (see Figure 1.57) [49–53]:

$$L_1 = \frac{R^2 C\omega_c \sin(\gamma_3)}{\left[(RC\omega_c - \sin(\gamma_1))^2 + \cos^2(\gamma_1)\right]\omega_c \sin(\gamma_1)} - L \tag{1.321}$$

$$C_{2m}L_{2m-1} = \frac{4\sin(\gamma_{4m-1})\sin(\gamma_{4m+1})}{\omega_c^2(1 - 2\alpha\cos(\gamma_{4m}) + \alpha^2)} \quad m \leq \frac{1}{2}(n-1) \tag{1.322}$$

$$C_{2m}L_{2m+1} = \frac{4\sin(\gamma_{4m+1})\sin(\gamma_{4m+3})}{\omega_c^2(1 - 2\alpha\cos(\gamma_{4m+2}) + \alpha^2)} \quad m < \frac{1}{2}(n-1) \tag{1.323}$$

$$C_{n-1} = \frac{2(1+\alpha^n)\sin(\gamma_1)}{R(1-\alpha^n)(1+\alpha)\omega_c} \quad n \text{ odd} \tag{1.324}$$

$$L_{n-1} = \frac{2R(1-\alpha^n)\sin(\gamma_1)}{(1+\alpha^n)(1+\alpha)\omega_c} \quad n \text{ even} \tag{1.325}$$

$$R_s = R\frac{1-\alpha^n}{1+\alpha^n} \tag{1.326}$$

where

$$m = 1,2,\ldots,\left[\frac{1}{2}(n-1)\right], \quad n > 1$$

$$\gamma_m = m\pi/2n \quad m = 1,2,\ldots \tag{1.327}$$

$$K_n = 1 - \left[1 - \frac{2\sin(\gamma_1)}{RC\omega_c}\right]^{2n} \tag{1.328}$$

$$\alpha_n = 1 - (1 - K_n)^{1/2n} \tag{1.329}$$

Note that n is the order of the matching network, and ω_c is the cutoff frequency.

Chebyshev Matching Networks

For $RC\omega_c \sinh(a) \geq 2\sin(\gamma_1)$ and $L_1 \geq 0$, the component values for Chebyshev type transfer function are given as follows (see Figure 1.57) [49–53]:

$$L_1 = \frac{R^2 C\omega_c \sin(\gamma_3)}{\left[(1 - RC\omega_c \sinh(a)\sin(\gamma_1))^2 + R^2 C^2 \omega_c^2 \cosh^2(a)\cos^2(\gamma_1)\right]\omega_c \sin(\gamma_1)} - L \tag{1.330}$$

$$C_{2m}L_{2m-1} = \frac{4\sin(\gamma_{4m-1})\sin(\gamma_{4m+1})}{\omega_c^2 f_{2m}(\sinh(a), \sinh(\hat{a}))} \quad m \leq \frac{1}{2}(n-1) \tag{1.331}$$

$$C_{2m}L_{2m+1} = \frac{4\sin(\gamma_{4m+1})\sin(\gamma_{4m+3})}{\omega_c^2 f_{2m+1}(\sinh(a),\sinh(\hat{a}))} \quad m < \frac{1}{2}(n-1) \tag{1.332}$$

$$C_{n-1} = \frac{2(\sinh(n\,a)+\sinh(n\,\hat{a}))\sin(\gamma_1)}{R\omega_c(\sinh(a)+\sinh(\hat{a}))(\sinh(n\,a)-\sinh(n\,\hat{a}))} \quad n \text{ odd} \tag{1.333}$$

$$L_{n-1} = \frac{2R(\cosh(n\,a)-\cosh(n\,\hat{a}))\sin(\gamma_1)}{\omega_c(\sinh(a)+\sinh(\hat{a}))(\cosh(n\,a)+\cosh(n\,\hat{a}))} \quad n \text{ even} \tag{1.334}$$

$$R_s = R\frac{\sinh(n\,a)-\sinh(n\,\hat{a})}{\sinh(n\,a)+\sinh(n\,\hat{a})} \quad n \text{ odd} \tag{1.335}$$

$$R_s = R\frac{\cosh(n\,a)-\cosh(n\,\hat{a})}{\cosh(n\,a)+\cosh(n\,\hat{a})} \quad n \text{ odd} \tag{1.336}$$

where

$$m = 1,2,\ldots,\left[\frac{1}{2}(n-1)\right], \quad n > 1$$

$$\gamma_m = m\pi/2n \quad m = 1,2,\ldots \tag{1.337}$$

$$K_n = 1-\varepsilon^2\sinh^2\left(n\sinh^{-1}\left[\sinh(a)-\frac{2\sin(\gamma_1)}{RC\omega_c}\right]\right) \tag{1.338}$$

$$a = \frac{1}{n}\sinh^{-1}\left(\frac{1}{\varepsilon}\right) \tag{1.339}$$

$$\hat{a} = \frac{1}{n}\sinh^{-1}\left(\frac{1}{\hat{\varepsilon}}\right) \tag{1.340}$$

$$\hat{\varepsilon} = \frac{\varepsilon}{\sqrt{1-K_n}} \tag{1.341}$$

$$f_m(\sinh(a),\sinh(\hat{a})) = \sinh^2(a)+\sinh^2(\hat{a})+\sin^2(\gamma_{2m})$$
$$-2\sinh(a)\sinh(\hat{a})\cos(\gamma_{2m}) \quad m = 1,2,\ldots,\left[\frac{1}{2}n\right] \tag{1.342}$$

Note that ε is the passband ripple of the Chebyshev response, n is the order of the matching network, and ω_c is the cutoff frequency.

References

[1]   David M. Pozar, *Microwave Engineering*, New York: John Wiley and Sons, 1998.

[2]   Robert E. Collin, *Foundations for Microwave Engineering*, IEEE Series on Electromagnetic Wave Theory, New York: IEEE Press, 2001.

[3]   F. Gardiol, *Microstrip Circuits,* New York: John Wiley and Sons, 1994.

[4]   T. C. Edwards and M. B. Steer, *Foundations of Interconnect and Microstrip Design,* New York: John Wiley and Sons, 2000.

[5]   Guillermo Gonzales, *Microwave Transistor Amplifiers,* Upper Saddle River, NJ: Prentice-Hall, 1984.

[6]   Umran S. Inan and Aziz S. Inan, *Engineering Electromagnetics,* Reading, MA: Addison-Wesley, 1999.

[7]   Richard Feynman et al., *The Feynman Lectures on Physics,* Reading, MA: Addison-Wesley, 1970.

[8]   Max W. Medley, Jr., *Microwave and RF Circuits: Analysis, Synthesis, and Design,* Norwood, MA: Artech House, 1993.

[9]   K. Kurokowa, "Power Waves and the Scattering Matrix," *IEEE Trans. on Microwave Theory and Techniques,* vol. MTT-13, pp. 194–202, Mar. 1965.

[10] Dean A. Frickey, "Conversion between *S, Z, Y, h, ABCD,* and *T* Parameters Which Are Valid for Complex Source and Load Impedances," *IEEE Trans. on Microwave Theory and Techniques,* vol. 42, pp. 205–211, Feb. 1994.

[11] J. K. Hunton, "Analysis of Microwave Measurement Techniques by Means of Signal Flow Graphs," *IRE Trans. on Microwave Theory and Techniques,* pp. 206–212, Mar. 1960.

[12] G. H. Bryant, *Principles of Microwave Measurements,* London, England: Peter Peregrinus Ltd., IEE Electrical Measurement Series, 1993.

[13] E. H. Fooks and R. A. Zakarevicius, *Microwave Engineering Using Microstrip Circuits,* Upper Saddle River, NJ: Prentice Hall, 1990.

[14] Frank K. David et al., "Design of an Enhanced Vector Network Analyzer," *Hewlett-Packard Journal,* pp. 1–13, Apr. 1997.

[15] Agilent Technologies, "Network Analyzer Selection Guide."

[16] Agilent Technologies, "10 Hints for Making Better Network Analyzer Measurements," *Application Note,* No. 1291-1B.

[17] Agilent Technologies, "Exploring the Architectures of Network Analyzers," *Application Note,* No. 1287-2.

[18] Agilent Technologies, "Applying Error Correction to Network Analyzer Measurements," *Application Note,* No. 1287-3.

[19] Agilent Technologies, "Network Analyzer Measurements: Filter and Amplifier Examples," *Application Note,* No. 1287-4.

[20] Agilent Technologies, "Applying the 8510 TRL Calibration for Non-Coaxial Measurements," *Product Note,* No. 8510-8A.

[21] Agilent Technologies, "An Introduction to Multiport and Balanced Device Measurements," *Application Note,* No. 1373-1.

[22] Agilent Technologies, "Concepts in Balanced Device Measurements," *Application Note,* No. 1373-2.

[23] Agilent Technologies, "In-Fixture Measurements Using Vector Network Analyzers," *Application Note,* No. 1287-9.

[24]   Agilent Technologies, "In-Fixture Microstrip Device Measurements Using TRL* Calibration," *Product Note*, No. 8720-2.

[25]   Hans-Gerd Krekels, "AutoKal: Automatic Calibration of Vector Network Analyzer ZVR," *Rohde & Schwarz Application Note*, No. 1EZ30, Aug. 1996.

[26]   Olaf Ostwald, "Frequently Asked Questions About Vector Network Analyzer ZVR," *Rohde & Schwarz Application Note*, No. 1EZ38, Jan. 1998.

[27]   Thilo Bednorz, "Measurement Uncertainties for Vector Network Analysis," *Rohde & Schwarz Application Note*, No. 1EZ29, Oct. 1996.

[28]   Cascade Microtech, "On-Wafer Vector Network Analyzer Calibration and Measurements," *Application Note*.

[29]   Dale E. Carlton, K. Reed Gleason, and Eric W. Strid, "Microwave Wafer Probing," *Microwave Journal*, Jan. 1985.

[30]   Ronald F. Bauer and Paul Penfield, "De-embedding and Unterminating," *IEEE Trans. Microwave Theory Tech.*, vol. 22, pp. 282–288, Mar. 1974.

[31]   Glenn F. Engen, "The Six-Port Reflectometer: An Alternative Network Analyzer," *IEEE Trans. Microwave Theory Tech.*, vol. 25, pp. 1075–1079, Dec. 1977.

[32]   Glenn F. Engen, "A (Historical) Review of the Six-Port Measurement Technique," *IEEE Trans. Microwave Theory Tech.*, vol. 45, pp. 2414–2417, Dec. 1997.

[33]   Glenn F. Engen, "A Least Squares Solution for Use in the Six-Port Measurement Technique," *IEEE Trans. Microwave Theory Tech.*, vol. 28, pp. 1473–1477, Dec. 1980.

[34]   A. J. Estin et al., "Basic RF and Microwave Measurements: A Review of Selected Programs," *Metrologia*, pp. 135–152, 1992.

[35]   Jeffrey A. Jargon, Roger B. Marks, and Doug K. Rytting, "Robust SOLT and Alternative Calibrations for Four-Sampler Vector Network Analyzers," *IEEE Trans. Microwave Theory Tech.*, vol. 47, pp. 2008–2013, Oct. 1999.

[36]   Vladimir G. Gelnovatch, "A Computer Program for the Direct Calibration of Two-Port Reflectometers for Automated Microwave Measurements," *IEEE Trans. Microwave Theory Tech.*, pp. 45–47, Jan. 1976.

[37]   Stig Rehnmark, "On the Calibration Process of Automatic Network Analyzer Systems," *IEEE Trans. Microwave Theory Tech.*, pp. 457–458, Apr. 1974.

[38]   Glenn F. Engen and Cletus A. Hoer, "Thru-Reflect-Line: An Improved Technique for Calibrating the Dual Six-Port Automatic Network Analyzer," *IEEE Trans. Microwave Theory Tech.*, vol. 27, pp. 987–993, Dec. 1979.

[39]   Ning Hua Zhu, "Phase Uncertainty in Calibrating Microwave Test Fixtures," *IEEE Trans. Microwave Theory Tech.*, vol. 47, pp. 1917–1922, Oct. 1999.

[40]   Dylan F. Williams and Roger B. Marks, "Accurate Transmission Line Characterization," IEEE *Microwave Guided Wave Lett.*, vol. 3, pp. 247–249, Aug. 1993.

[41]   Roger B. Marks and Dylan F. Williams, "Characteristic Impedance Determination Using Propagation Constant Measurement," *IEEE Microwave Guided Wave Lett.*, vol. 1, pp. 141–143, June 1991.

[42]   Raian F. Kaiser and Dylan F. Williams, "Sources of Error in Coplanar-Waveguide TRL Calibrations," *NIST Publication*.

[43] Robert A. Soares et al., "A Unified Mathematical Approach to Two-Port Calibration Techniques and Some Applications," *IEEE Trans. Microwave Theory Tech.*, vol. 37, pp. 1669–1674, Nov. 1989.

[44] Kimmo J. Silvonen, "A General Approach to Network Analyzer Calibration," *IEEE Trans. Microwave Theory Tech.*, vol. 40, pp. 754–759, Apr. 1992.

[45] Roger B. Marks, "A Multiline Method of Network Analyzer Calibration," *IEEE Trans. Microwave Theory Tech.*, vol. 39, pp. 1205–1215, July 1991.

[46] Roger B. Marks and Kurt Phillips, "Wafer-Level ANA Calibration at NIST," *NIST Publication*.

[47] Donald C. DeGroot, Jeffrey A. Jargon, and Roger B. Marks, "Multiline TRL Revealed," *NIST Publication*.

[48] Van Valkenburg, *Introduction to Modern Network Synthesis*, New York: John Wiley and Sons, 1960.

[49] Wai-Kai Chen, *Theory and Design of Broadband Matching Networks*, Oxford: Pergamon Press Ltd., 1976.

[50] Wai-Kai Chen, *Passive and Active Filters: Theory and Implementations*, New York: John Wiley and Sons, 1986.

[51] Wai-Kai Chen, "Explicit Formulas for the Synthesis of Optimum Broad-Band Impedance-Matching Networks," *IEEE Trans. on Circuits and Systems*, vol. 24, pp. 157–169, Apr. 1977.

[52] Wai-Kai Chen and T. Chaisrakeo, "Explicit Formulas for the Synthesis of Optimum Bandpass Butterworth and Chebyshev Impedance-Matching Networks," *IEEE Trans. on Circuits and Systems*, vol. 24, pp. 928–942, Oct. 1980.

[53] Yi-Sheng Zhu and Wai-Kai Chen, *Computer-Aided Design of Communication Networks*, Singapore: World Scientific, 2000.

Chapter 2

Microwave Printed Circuits

Much of the modern communication equipment has at least one form of microwave printed circuit. Besides, almost every electrical device has a printed circuit board (PCB), which can be classified as a low-frequency counterpart of a microwave printed circuit, although with ever-increasing demand for signal speed, the boundary between those two has been steadily blurring in some applications. Once devised as a simple signal transportation medium between circuit components, microwave printed circuits are continuously renewing themselves to keep up with the technology.

Microwave printed circuits, and printed circuits in general, are playing the important role of being a harness for modern electrical devices where the complex connections between the circuit components wouldn't be possible without them. Apart from transferring electrical signals from one point to another, one can also build passive components, like couplers, power dividers, filters, and so forth, using microwave printed circuit technology. Note that by saying printed circuit, we mean that circuit patterns are either etched or printed on a dielectric slab using an appropriate technology in the form of metal traces. In fact, that is the main place that all of the advantages of the technology are coming from: by transferring circuit images on dielectric slabs, multilayer and complex circuits can be built in a very cost-effective and high-volume way. The main difference between a regular printed circuit and a microwave printed circuit is the electrical and physical properties of the dielectric slab. For microwave applications, low-loss and tight control on the material specifications and physical dimensions are required.

2.1 HISTORY OF MICROWAVE PRINTED CIRCUITS

The concept of microwave printed circuits was first inspired from the flat-strip coaxial line in which the center conductor of the coaxial line was flattened and placed in a enclosure with a rectangular cross-section. According to R. M. Barrett, this configuration was first employed by V. H. Rumsey and H. W. Jamieson in an antenna and power division network during World War II in the United States [1,

2]. On the other hand, Howe indicates that Harold Wheeler was experimenting with two coplanar strips placed side by side to make a low-loss transmission line as early as 1936 [3]. Although the original inventor of the concept of planar transmission lines seems to be hidden deep in history, the concept of stripline as a means for transmitting RF energy from one point to another and for constructing various microwave components (couplers, matching circuits, filters, and so forth) was first proposed by R. M. Barrett early in 1949 [1–5]. In the following years, the Air Force Cambridge Research Center, where R. M. Barrett was a chief in the Airborne Antenna Section, provided various grants and initiated collaboration between the other research centers to work on this new transmission line medium [1, 3]. Some of the research centers that worked in this new field were Tufts University, the Airborne Instruments Laboratory (stripline), the Polytechnic Institute of Brooklyn, Sanders Associates (tri-plate — another early name for stripline), and Federal Communications Research Laboratories (microstrip). Because the printed circuits were rugged, easy, and inexpensive to produce, reproducible by simple means, and low in cross-section and weight, all of which make them very attractive for mass production and for airborne applications, they quickly gained a lot of interest. In the beginning, the circuits were either cut out from copper foil using razor blades or printed using silver ink. These initial, intense activities on printed circuits resulted in the first symposium specifically on microwave printed circuits, held at Tufts University in 1954. The papers presented at that conference were reporting striplines, microstrip lines, filters, couplers, power dividers, hybrid rings, and some theoretical aspects of this new transmission medium. In the next year, *Transactions on Microwave Theory and Techniques* published a special issue with the papers presented in that conference. That was the start of a new era in microwave engineering [3].

In the beginning, the microstrip, developed by the Federal Communications Research Laboratories [6], did not attract too much attention because the fundamental propagation mode was not TEM [4]. Stripline, developed by Airborne Instruments Laboratory and the others [7–10], was the center of attention because it did support a TEM mode. Initially, this was considered a significant problem because the phase velocity and characteristic impedance of the microstrip was changing with frequency due to the non-TEM nature of the propagation. Besides, the discontinuities of microstrip were causing undesired radiation, unlike the purely reactive discontinuities of stripline. It was, indeed, a significant problem because the first microstrip lines had electrically large cross-sections. For example, the microstrip circuits reported in [6] had a substrate thickness of 1.5 mm (~60 mils), which was approximately 10 times larger than most of today's modern substrates used for microwave circuits. Later, however, it was discovered that these disadvantages of microstrip could largely be eliminated by reducing the cross-section compared to wavelength, thus making the propagation mode very close to TEM (which would be called quasi-TEM; we will return to this concept later) [4]. After that, microstrip became the preferred transmission line medium for microwave printed circuits. In fact, this event was listed as the most dramatic inversion in the history of the microwave field according to Arthur Oliner [4].

The early theoretical treatment of striplines was based on the conformal mapping technique [11] because the propagation mode was TEM. However, the resultant analytical equations contained elliptic integrals, which were very tedious to calculate in those times because electronic computers were not widely available. Thus, a lot of work was concentrated on developing approximate and accurate equations for striplines. Microstrip lines were more difficult to analyze because of their inhomogeneous geometry [12]. In the following years, more papers were published describing theoretical aspects of microwave printed circuits and numerical analysis methods in detail [13–20]. Later, the advent of digital computers enabled fast and accurate analysis of printed structures using numerical techniques [19, 20].

Today, microstrip lines are perhaps the most commonly used form of microwave printed transmission lines because of their simple construction. Other types include striplines, coplanar waveguides (CPWs), and slotlines. Microstrip circuits are employed in the vast majority of modern microwave circuits because of their simplicity, high repeatability, ease of manufacture, small size, and extremely high suitability for incorporation with active devices. The variety of microstrip circuits, such as transmission lines, filters, matching circuits, power dividers, and directional couplers, can be implemented using microstrip technology. These components can be extremely small and low cost compared to their bulky, metallic waveguide or coaxial-line equivalents. Although from ongoing discussion it is clear that microstrip circuits are a special form of microwave printed circuits, the name "microstrip circuit" is usually used to refer to all kinds of microwave printed circuits.

The manufacturing process of microstrip circuits, and printed circuits in general, on laminate printed circuit boards and MMIC substrates is done through photolithographic techniques, which yield high dimensional accuracy and repeatability. Screen printing is another method used especially on ceramic substrates to implement printed circuits in high-volume, low-cost manufacturing. Circuits manufactured by this method are called thick-film hybrids. However, screen printing usually has a dimensional accuracy an order of magnitude lower than photolithographic techniques.

The other advantage of microstrip circuits, which was not obvious in the early years, is that accurate and very efficient analysis is possible through the use of numerical techniques. For instance, microstrip matching networks for the input and output of a transistor amplifier can be designed using electromagnetic computer simulations after replacing the active device with a two-port black box. This provides a very accurate modeling of the matching circuits. Then, the two-port circuit is replaced by the device model of the transistor in a circuit simulator and the overall circuit response of the amplifier is obtained. In essence, this is the approach that is extensively used by microwave engineers today. Note that one should be able to anticipate the overall circuit response very accurately by using this methodology provided that good models for the active circuits (i.e., transistors) are available. This happens because all interactions between the matching circuit sections, as well as radiation and other higher-order effects, are automatically taken into account during the computer simulation. Hence, the use of full-wave electromagnetic simulation tools, which is the promoted philosophy throughout this book, changed the design methodology in microwave engineering for good. However, as always repeatedly indicated, no simulation tool can replace human thought. So, learning the fundamentals is always crucial.

The literature of microstrip circuits is indeed very rich. A nonexhaustive list of books is provided in the references for the benefit of the reader [21–29]. The review papers that are also given in the references will be helpful in understanding the fundamentals.

2.2 MICROSTRIP LINES

A typical microstrip line is formed using a top conductor on a dielectric layer with a single ground plane and air above, as shown in Figure 2.1 [16]. The top conductor is basically a metallic foil shaped in the form of a narrow rectangle, and the ground plane is typically a continuous metallic plane. Note that the top foil and the ground plane form a two-conductor transmission line system; currents flowing on the top foil return back to the source through the ground plane, completing the circuit. Because the dielectric layers on the top and bottom of the foil are different, the microstrip is an inhomogeneous structure.

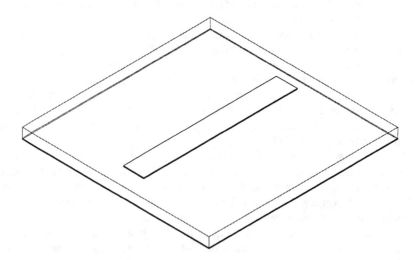

Figure 2.1 A microstrip line printed on a dielectric substrate.

The cross-section of the structure is depicted in Figure 2.2 showing the important dimensions. In the figure, w and h represent the width of the line and thickness of the substrate, respectively. The metallization thickness is given by t. Conductivity of the foil, σ, relative dielectric constant of the substrate, ε_r, and dielectric loss tangent of the substrate, $\tan\delta$, are the electrical parameters.

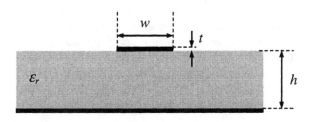

Figure 2.2 Cross-sectional view of the microstrip line.

At this point, it would be instructive to discuss the quasi-TEM propagation mode of microstrips in more detail. In a microstrip line, the wave is guided along the longitudinal direction of the structure, and the associated fields exist in the substrate, as well as in the air region of the structure, as depicted in Figure 2.3.

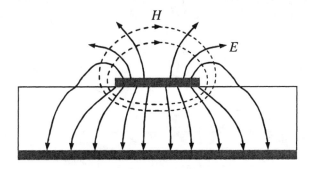

Figure 2.3 Electric and magnetic field lines for a microstrip line.

Since the microstrip line is an inhomogeneous structure, it cannot support pure TEM modes as indicated above, which can be explained qualitatively as follows: (1) the propagation constant of a TEM wave must be equal to the wave number of the medium in order to have nonzero electric and magnetic fields, as was stressed in the discussions of TEM waves in Section 1.3.1; (2) consequently, a TEM wave in an inhomogeneous structure, if it existed, would have to propagate with different velocities in different dielectric regions (i.e., in the air and substrate), as the phase velocity depends only on the material properties ε and μ; and (3) the difference in the phase velocity of the wave on both sides of the interface of two dielectric materials would have eventually led to the generation of longitudinal components of the fields, resulting in a non-TEM-like field distribution. However, for most practical microstrip lines, the field distribution is very close to the one for TEM waves; that is, the waves propagating in such structures usually have negligible longitudinal field components. Because of that, the fundamental mode of propagation of microstrip lines is called the quasi-TEM mode. Once the formation of the field distribution in a microstrip line has been established qualitatively as quasi-TEM mode with negligible longitudinal field components, it is inevitable to wonder when and how this approximation breaks down. With some very approximate mathematics, one can relate the magnitudes of the longitudinal components of the fields to their transverse components for quasi-TEM waves as follows [28]:

$$\nabla \times \mathbf{E} = -j\omega\mu\,\mathbf{H} \tag{2.1}$$

$$\hat{z}\frac{\partial}{\partial z}\times\mathbf{E}_t + \nabla_t \times \hat{z}E_z = -j\omega\mu\,\mathbf{H}_t$$

$$\nabla_t \times \hat{z}E_z = -j\omega\mu\,\mathbf{H}_t + jk_z\left(\hat{z}\times\mathbf{E}_t\right) \tag{2.2}$$

If the wave were a pure TEM, the term on the left-hand side would be zero, and the terms on the right-hand side would be equal to each other. For a quasi-TEM wave, since we have a small longitudinal field component, we can still assume that the terms on the RHS are almost equal to each other with a propagation constant less than the wave number of the dielectric medium.

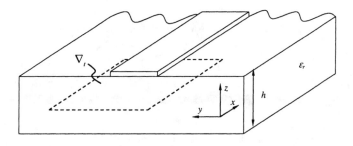

Figure 2.4 Explanation of quasi-TEM approximation in microstrip. One of the requirements for quasi-TEM assumption is that the extent of the region of field confinement in the transverse domain should be small compared to the wavelength, ($\Delta/\lambda_0 \ll 1$), which is satisfied for small thicknesses, h.

Hence, assuming the field confinement is over a finite, cross-sectional area with the shortest distance Δ in the transverse domain (see Figure 2.4), the last expression can be approximated as

$$\frac{1}{\Delta}|E_z| < k_0\sqrt{\varepsilon_r}\left|\hat{z}\times\mathbf{E}_t - \sqrt{\frac{\mu_0}{\varepsilon_0\varepsilon_r}}\mathbf{H}_t\right| \tag{2.3}$$

and the upper bound for the longitudinal field component can be written as follows:

$$|E_z| < 2\pi\frac{\Delta}{\lambda_0}\sqrt{\varepsilon_r}\left|\hat{z}\times\mathbf{E}_t - \sqrt{\frac{\mu_0}{\varepsilon_0\varepsilon_r}}\mathbf{H}_t\right| \tag{2.4}$$

where ∇_t is approximated by $1/\Delta$ for the fundamental mode, and k_z is set equal to the wave number of the dielectric medium, $k_0\sqrt{\varepsilon_r}$, to get the upper bound of the longitudinal field component. Remember that this is a very approximate expression for E_z, and the following observations are based on the fact that we have a quasi-TEM wave and wish to see how the longitudinal field component changes with the parameters of the geometry and the frequency of operation. Here are the observations: (1) If the region of field confinement in the transverse domain is homogenous or almost homogenous, then the expression in the absolute value sign will be zero or very small, resulting in a zero or small longitudinal component. Therefore, this term in (2.4) gives us the measure of homogeneity of the region where the significant portion of the field is confined. (2) If the extent of the region of field confinement in the transverse domain is small compared to the wavelength, that is $\Delta/\lambda_0 \ll 1$, the longitudinal component will be made even smaller. This observation can be understood intuitively from the fact that the fields for a TEM wave satisfy Laplace's equation in the transverse domain, as for static fields for which $\Delta/\lambda_0 \to 0$. So, if the fields are confined in a small region in the transverse domain, they will look more like static fields in the transverse domain, and hence the associated dynamic behavior will be more like a TEM wave. (3) If the contrast of the dielectric materials involved in the structure is high (i.e., $\varepsilon_r \gg 1$) for the case depicted in Figure 2.2, the longitudinal component of the electric field becomes larger, causing a departure from the TEM approximation.

As demonstrated in Chapter 1, characteristic impedance and the propagation constant are the two main parameters that describe a TEM transmission line (note that both parameters are complex in general). Therefore, in order to have a useful means for practical applications as a transmission line, one must know the characteristic impedance and propagation constant of the microstrip for a given geometry. However, exact analysis of microstrip lines is not easy because of the inhomogeneous nature of the structure. The use of effective dielectric concept as depicted in Figure 2.5 reduces the inhomogeneous microstrip problem to a homogeneous one [26]. Then, characteristic impedance can be calculated using the effective dielectric constant. In determining the effective dielectric constant, it is assumed that the inhomogeneous microstrip medium is made homogeneous by filling all of the volume with a dielectric, which has a dielectric constant of ε_{eff}, which must be determined separately. Note that both of the microstrip lines shown in Figure 2.5 have the same characteristic impedance.

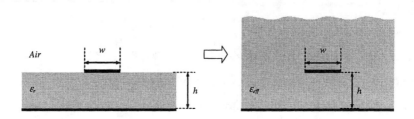

Figure 2.5 Concept of effective dielectric constant. Both of the microstrip lines have the same characteristic impedance by definition.

2.2.1 Characteristic Impedance

In this section, we demonstrate various static methods for the analysis of microstrip lines. This will provide us with a general understanding of the underlying phenomena for line capacitance and characteristic impedance.

We know from the transmission line theory that the characteristic impedance of a TEM line can be calculated once the phase velocity and line capacitance per unit length are known. Thus, once we assume that the dominant mode of propagation is quasi-TEM, it is then natural to attack the characteristic impedance problem of any printed transmission line structure by first trying to calculate its per-unit-length capacitance. Then, effective permittivity needs to be found to determine the phase velocity. From these two quantities, the characteristic impedance follows.

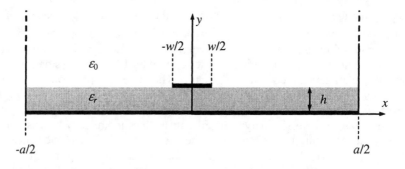

Figure 2.6 Modified microstrip structure for static analysis. The boundaries in the y-direction are terminated with perfectly conducting electrical walls.

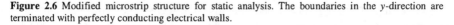

One of the most common approaches to analyzing the microstrip line is to solve Laplace's equation [21]. For this purpose, the transverse boundaries of the microstrip geometry are terminated in the y-direction using perfect electric walls as shown in Figure 2.6. This is not a significant restriction because one can select the dimension a big enough (i.e., $a > 10h$) so that the electric walls do not significantly disturb fields in the vicinity of the microstrip. Then, we determine the per-unit-length capacitance of the microstrip line. After determining the capacitance, characteristic impedance and the effective dielectric constant can be found (note that to determine the effective dielectric constant, one should also know the line capacitance in the absence of the dielectric).

Since we have assumed static case, potential distribution in the cross-section must satisfy the Laplace's equation:

$$\nabla^2 \phi(x, y) = 0 \quad \text{for } |x| \le a/2, 0 \le y < \infty \tag{2.5}$$

with boundary conditions

$$\phi(x, y) = 0, \quad |x| = \frac{a}{2}$$

$$\phi(x, y) = 0, \quad y = 0 \text{ and } y \to \infty$$

Equation (2.5) can be solved using separation of variables, yielding the following results in air and dielectric regions:

$$\phi(x, y) = \begin{cases} \sum_{n=1,3,5,\dots}^{\infty} A_n \cos\left(\frac{n\pi x}{a}\right) \sinh\left(\frac{n\pi y}{a}\right), & \text{for } 0 \le y \le h \\ \sum_{n=1,3,5,\dots}^{\infty} B_n \cos\left(\frac{n\pi x}{a}\right) e^{-n\pi y/a}, & \text{for } h \le y < \infty \end{cases} \tag{2.6}$$

As a next step, we will determine the unknown constants A_n and B_n. First, the potential across the air–dielectric interface must be continuous:

$$\sum_{n=1,3,5,\dots}^{\infty} A_n \cos\left(\frac{n\pi x}{a}\right) \sinh\left(\frac{n\pi h}{a}\right) = \sum_{n=1,3,5,\dots}^{\infty} B_n \cos\left(\frac{n\pi x}{a}\right) e^{-n\pi h/a} \tag{2.7}$$

$$\Rightarrow A_n \sinh\left(\frac{n\pi h}{a}\right) = B_n e^{-n\pi h/a}$$

Second, discontinuity in electric displacement vector, **D**, must be equal to the surface charge density on the strip:

$$\rho_s = D_y\big|_{x, y=h^+} - D_y\big|_{x, y=h^-}$$

$$= \varepsilon_0 E_y^{(2)}\big|_{x, y=h^+} - \varepsilon_0 \varepsilon_r E_y^{(1)}\big|_{x, y=h^-} \tag{2.8}$$

where

$$\mathbf{E}^{(1)} = -\nabla\phi = \hat{a}_x \sum_{n=1,3,5,\ldots}^{\infty} A_n \frac{n\pi}{a} \sin\left(\frac{n\pi x}{a}\right) \sinh\left(\frac{n\pi y}{a}\right)$$
$$-\hat{a}_y \sum_{n=1,3,5,\ldots}^{\infty} A_n \frac{n\pi}{a} \cos\left(\frac{n\pi x}{a}\right) \cosh\left(\frac{n\pi y}{a}\right) \tag{2.9}$$

$$\mathbf{E}^{(2)} = -\nabla\phi = \hat{a}_x \sum_{n=1,3,5,\ldots}^{\infty} B_n \frac{n\pi}{a} \sin\left(\frac{n\pi x}{a}\right) e^{-n\pi y/a}$$
$$+\hat{a}_y \sum_{n=1,3,5,\ldots}^{\infty} B_n \frac{n\pi}{a} \cos\left(\frac{n\pi x}{a}\right) e^{-n\pi y/a} \tag{2.10}$$

Using (2.7) and (2.8), one can obtain the following expression for the surface charge density:

$$\rho_s = \varepsilon_0 \sum_{n=1,3,5,\ldots}^{\infty} A_n \frac{n\pi}{a} \cos\left(\frac{n\pi x}{a}\right) \left[\sinh\left(\frac{n\pi h}{a}\right) + \varepsilon_r \cosh\left(\frac{n\pi h}{a}\right)\right] \tag{2.11}$$

The remaining task is to find unknown coefficients A_n, $n = 1,3,4,\ldots,\infty$. Once they are found, so is the surface charge density. Then, per-unit-length capacitance of the microstrip line can be determined. The unknown coefficients in (2.11) can be found by making an approximation to the surface charge density and then using orthogonality of trigonometric functions. Let's assume the charge density on the microstrip can be approximated by

$$\rho_s(x) = \frac{2}{\pi w\sqrt{1 - x^2/(w/2)^2}}, \quad -w/2 \le x \le w/2 \tag{2.12}$$

where w is the width of the line. Note that the charge density has singular values at the edges of the strip, as depicted in Figure 2.7. This is a classic approximation for microstrip charge density representing edge singularities.

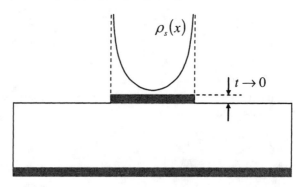

Figure 2.7 Qualitative view of the idealized charge distribution on a microstrip line. Note that the charges are concentrated more on the bottom surface in case of finite thickness microstrip.

It must be noted that this approximation is accurate only for relatively narrow strips (i.e., $w/h \ll 1$). As the strip gets wider, the accuracy of (2.12) deteriorates, because although the charge still has singularities, these singularities drop much faster to a uniform distribution for wider strips. Therefore, for wider strips (i.e., $w/h \gg 1$), uniform charge distribution provides a better approximation for characteristic impedance. One can also use linear combination of (2.12) multiplied by even powers of x to obtain a better approximation to the charge density.

Although (2.12) contains singularity, it is integrable, so the line integral of (2.12) along the strip gives total charge on the strip of

$$Q = \int_{-w/2}^{w/2} \rho_s(x)dx \tag{2.13}$$

In order to employ orthogonality to determine unknown coefficients, we first multiply (2.11) and (2.12) by $\cos(m\pi x/a)$, where m is an odd integer. Then we take the integral of both sides in the whole domain of x:

$$\int_{-a/2}^{a/2} \varepsilon_0 \sum_{n=1,3,5,...}^{\infty} A_n \frac{n\pi}{a} \cos\left(\frac{m\pi x}{a}\right) \cos\left(\frac{n\pi x}{a}\right) \left[\sinh\left(\frac{n\pi h}{a}\right) + \varepsilon_r \cosh\left(\frac{n\pi h}{a}\right) \right] dx$$

$$= \int_{-a/2}^{a/2} \frac{2}{\pi w \sqrt{1 - x^2/(w/2)^2}} \cos\left(\frac{m\pi x}{a}\right) dx, \quad m = 1,3,5,...,\infty$$

Note that the integral on the left-hand side is nonzero only when $m = n$. This yields the following equation for A_n:

$$A_n = \frac{\dfrac{1}{\pi} \displaystyle\int_{-\pi/2}^{\pi/2} \cos\left(\dfrac{n\pi w}{2a}\sin(\theta)\right) d\theta}{\varepsilon_0 \dfrac{n\pi}{2}\left[\sinh\left(\dfrac{n\pi h}{a}\right) + \varepsilon_r \cosh\left(\dfrac{n\pi h}{a}\right) \right]}$$

where the substitution $x = w/2\sin(\theta)$ was used in the surface charge density integral to handle singularity properly. Finally, per-unit-length capacitance of the line is expressed as

$$C = \frac{Q}{V} \tag{2.14}$$

where

$$Q = \frac{1}{\pi} \int_{-\pi/2}^{\pi/2} d\theta = 1$$

$$V = -\int_0^h E_y^{(1)}\Big|_{x=0} dy = \sum_{n=1,3,5}^{\infty} A_n \sinh\left(\frac{n\pi h}{a}\right)$$

This completes the determination of the line capacitance. Next, we have to find the effective dielectric constant so that the characteristic impedance can be found. The effective dielectric constant is the ratio of line capacitances with and without the dielectric (i.e., $\varepsilon_r \neq 1$ and $\varepsilon_r = 1$):

$$\varepsilon_{eff} = \frac{C\big|_{\varepsilon_r \neq 1}}{C\big|_{\varepsilon_r = 1}} \qquad (2.15)$$

Finally, the characteristic impedance is determined from

$$Z_c = \frac{1}{cC} \qquad (2.16)$$

where C is the capacitance in the presence of dielectric and c is the speed of light in a medium filled with the effective dielectric constant, which is given by $c = c_0 / \sqrt{\varepsilon_{eff}}$. So far, we have demonstrated an approximate method for determining the characteristic impedance and effective dielectric constant of a microstrip line. The method is based on the solution of Laplace's equation. Since it is a static model, the solution is an approximate one. Besides, an approximation of surface charge density is required. For that purpose, one can use a uniform or nonuniform charge distribution. Representation of edge singularities is critical for narrow microstrips (i.e., $w/h \ll 1$).

The static solution of the microstrip introduced above can be generalized to any line width by representing the charge density using multiple basis functions and employing Green's function. A spectral-domain approach was presented in the literature for this purpose [30]. It can be shown that the line capacitance of the microstrip line is found from the solution of the following linear system [30]:

$$\sum_{n=0}^{N} a_n A_{mn} = b_m, \quad m = 0,1,2,\ldots,N \qquad (2.17)$$

where

$$A_{mn} = \int_0^\infty G(u) F_m(u) F_n(u)\, du$$

$$b_m = 2\pi\varepsilon_0 \int_0^1 f_m(\xi)\, d\xi = \begin{cases} \varepsilon_0 \pi^2, & m = 0 \\ 0, & m \neq 0 \end{cases}$$

$$G(u) = \frac{1 - e^{-2\delta u}}{u \cdot (1 + \varepsilon_r) \cdot \left(1 - \dfrac{1 - \varepsilon_r}{1 + \varepsilon_r} e^{-2\delta u} \right)}$$

$$F_n(u) = \pi(-1)^n J_{2n}(u)$$

$$f_n(\xi) = \begin{cases} T_{2n}(\xi)(1-\xi^2)^{-1/2}, & |\xi| \le 1 \\ 0, & |\xi| > 1 \end{cases}$$

$$\delta = 2h/w$$

In the above equations, $T_{2n}(\xi)$ is the Chebyshev polynomial of 2nth order, $J_{2n}(u)$ is the Bessel function of 2nth order, and N is the number of basis functions. Then, the capacitance is found using [30]

$$C = \pi a_0 \tag{2.18}$$

After determining the capacitance through (2.18), the characteristic impedance can be calculated using (2.15) and (2.16). It was claimed that with a choice of only two basis functions, the C and Z found using the above formulation are both accurate to five digits for $0.1 < h/w < 0.4$. Note that the choice of the singular nature of the basis function makes the integral of Green's function, $G(u)$, slowly converge in (2.17). A common approach to addressing this problem is to subtract the slowly converging part and evaluate analytically. By noticing that

$$G(u) \approx \frac{1}{(1+\varepsilon_r)u}, \quad u \to \infty$$

we can rewrite the matrix terms as follows to improve convergence [30]:

$$A_{mn} = \pi^2 (-1)^{n+m} \left\{ \int_0^\infty \left[G(u) - \frac{1}{u(1+\varepsilon_r)} \right] J_{2m}(u) J_{2n}(u)\, du \right.$$

$$\left. + \frac{1}{1+\varepsilon_r} \int_0^\infty \frac{1}{u} J_{2m}(u) J_{2n}(u)\, du \right\}, \quad m \ne 0, n \ne 0$$

$$A_{mn} = \pi^2 \left\{ \int_0^\infty \left[G(u) - \frac{u}{u^2+1} \left(\frac{1}{1+\varepsilon_r} \right) \right] J_{2m}(u) J_{2n}(u)\, du \right.$$

$$\left. + \frac{1}{1+\varepsilon_r} \int_0^\infty \frac{u}{u^2+1} J_{2m}(u) J_{2n}(u)\, du \right\}, \quad m = 0, n = 0$$

where the second integrals in the above expressions can be evaluated analytically as follows [30]:

$$\int_0^\infty \frac{1}{u} J_{2m}(u) J_{2n}(u)\, du = \begin{cases} \dfrac{1}{4n}, & m = n \ne 0 \\ 0, & m \ne n \end{cases}$$

$$\int_0^\infty \frac{u}{u^2+1} J_0(u) J_0(u)\, du = I_0(1) K_0(1)$$

where $I_0(x)$ and $K_0(x)$ are modified Bessel functions.

Simple seminumerical formulas for the line capacitance and characteristic impedance can also be derived based on the above analysis as follows [30]:

$$
C \cong \frac{\varepsilon_0 \varepsilon_r w}{h} \left\{ 1 + \frac{2h}{\pi \varepsilon_r w} \left[\ln \frac{w}{2h} + 1.416 \varepsilon_r \right. \right.
$$
$$
\left. \left. + 1.547 + \left(1.112 - 0.028 \varepsilon_r \right) \frac{h}{w} \right] \right\}
\tag{2.19}
$$

$$
Z \cong \frac{377h}{\sqrt{\varepsilon_r} \, w} \left\{ 1 - \frac{2h}{\pi \varepsilon_r w} \left[\left(1 + \varepsilon_r \right) \ln \frac{h}{w} - 4.554 \varepsilon_r \right. \right.
$$
$$
\left. \left. - 2.230 - \left(4.464 + 3.89 \varepsilon_r \right) \frac{h}{w} \right] \right\}^{-1/2}
\tag{2.20}
$$

for h/w small, and

$$
C \cong \frac{\varepsilon_0 \varepsilon_r w}{h} \frac{\pi h \left(1 + \varepsilon_r \right)}{\varepsilon_r w} \left[\ln \frac{8h}{w} + \frac{1}{16(1 + \varepsilon_r)} \frac{w^2}{h} \right.
$$
$$
\left. + \frac{\varepsilon_r - 1}{\varepsilon_r} \cdot \left(0.041 \frac{w^2}{h^2} - 0.454 \right) \right]^{-1}
\tag{2.21}
$$

$$
Z \cong \frac{377}{\pi \sqrt{2(1 + \varepsilon_r)}} \left\{ \left[\ln \frac{8h}{w} + \frac{1}{32} \frac{w^2}{h^2} \right] \cdot \left[\ln \frac{8h}{w} + \frac{w^2}{h^2} \frac{1}{16(1 + \varepsilon_r)} \right. \right.
$$
$$
\left. \left. + \frac{\varepsilon_r - 1}{\varepsilon_r} \left(0.041 \frac{w^2}{h^2} - 0.454 \right) \right] \right\}^{1/2}
\tag{2.22}
$$

for h/w large. The reader is encouraged to study the functional forms of the equations (2.19) to (2.22). Most of the empirical and numerical expressions for the line capacitance and characteristic impedance of microstrip lines resemble similar functional forms. It would also be possible to employ pulse basis functions instead of the singular basis functions [31]. In that case, however, a relatively high number of basis functions would be required for the same accuracy.

For the sake of completeness, we will also present closed-form expressions that are commonly employed in the literature for the static effective dielectric constant and characteristic impedance. The effective dielectric constant and characteristic impedance of a zero-thickness ($t \to 0$) microstrip line for a given w, h, and ε_r can be determined by the following expressions [32]:

$$
\varepsilon_{\text{eff}} = \frac{\varepsilon_r + 1}{2} + \frac{\varepsilon_r - 1}{2} \frac{1}{\sqrt{1 + 12 h/w}}
\tag{2.23}
$$

$$
Z_c = \begin{cases} \dfrac{60}{\sqrt{\varepsilon_{\textit{eff}}}} \ln\!\left(\dfrac{8h}{w} + \dfrac{w}{4h}\right) & w/h \le 1 \\[2ex] \dfrac{120\pi}{\sqrt{\varepsilon_{\textit{eff}}}\,\big[w/h + 1.393 + 0.667\ln(w/h + 1.444)\big]} & w/h \ge 1 \end{cases} \tag{2.24}
$$

where w and h are the width of the microstrip and height of the substrate, respectively. Alternatively, for a given characteristic impedance Z_c and dielectric constant ε_r, the w/h ratio can be determined from [32]:

$$
\frac{w}{h} = \begin{cases} \dfrac{8e^A}{e^{2A} - 2} & w/h \le 2 \\[2ex] \dfrac{2}{\pi}\left[B - 1 - \ln(2B - 1) + \dfrac{\varepsilon_r - 1}{2\varepsilon_r}\left\{\ln(B - 1) + 0.39 - \dfrac{0.61}{\varepsilon_r}\right\}\right] & w/h \ge 2 \end{cases} \tag{2.25}
$$

where

$$
A = \frac{Z_c}{60}\sqrt{\frac{\varepsilon_r + 1}{2}} + \frac{\varepsilon_r - 1}{\varepsilon_r + 1}\left(0.23 + \frac{0.11}{\varepsilon_r}\right) \tag{2.26}
$$

$$
B = \frac{377\pi}{2Z_c\sqrt{\varepsilon_r}} \tag{2.27}
$$

Note that as long as the w/h ratio is the same, the static characteristic impedance of a microstrip line does not change, provided that other parameters are kept constant. These expressions are very easy to program using a digital computer.

Equations (2.23) and (2.24) are also compared with the results of the static solution obtained using (2.15) and (2.16) with one singular basis function, providing the results in Table 2.1. The substrate used in the comparisons is 10 mils thick with a relative dielectric constant of $\varepsilon_r = 2.2$. As can be seen from the table, agreement between the closed-form expressions and static solution is good for narrow strip widths.

Table 2.1

Comparison of Static Solution and Closed-Form Formulas for the Microstrip Characteristic Impedance ($h = 10$ mil, $\varepsilon_r = 2.2$)

w/h	Static Solution (2.15) and (2.16)		Closed-Form Formulas (2.23) and (2.24)	
	$\varepsilon_{\textit{eff}}$	Z_c	$\varepsilon_{\textit{eff}}$	Z_c
0.25	1.70	160.5	1.69	160.3
0.5	1.73	127.3	1.72	127.2
1	1.78	94.6	1.77	95.3
2	1.86	63.3	1.83	66.1
4	1.98	37.6	1.90	41.9
8	2.11	19.8	1.98	24.6

2.2.2 Conductor Losses

Conductor loss is the other important parameter in the characterization of microstrip lines [33–37]. Unfortunately, accurate closed-form analytical expressions for conductor losses are difficult to obtain. The difficulty arises mainly because of the nonuniform current distribution on the conductors. If the current distribution on the conductors were known, the following expression could be employed to determine the conductor losses for a microstrip line [33]:

$$\alpha_c = \frac{R_{s1}}{2Z_c} \oint_c \frac{|J_1|^2}{|I|^2} dl + \frac{R_{s2}}{2Z_c} \int_{-\infty}^{\infty} \frac{|J_2|^2}{|I|^2} dl \qquad (2.28)$$

where R_{s1} and R_{s2} are the surface resistivity for the upper and bottom conductors, respectively, $J_1(x)$ and $J_2(x)$ are the corresponding surface current densities, Z_c is the characteristic impedance, and $|I|$ is the magnitude of the total current per conductor. In the above equation, the first and second terms represent the losses due to the strip conductor and ground plane, respectively. Note that the first contour integral implies that integration must be performed on all sides of the strip conductor.

However, we do not know the exact current distribution on the conductors, which can only be obtained through numerical techniques. Thus, an alternative approach, known as Wheeler's Incremental Inductance Rule, was proposed in the literature to obtain approximate closed-form expressions for the microstrip conductor losses [13, 33, 38]. To demonstrate the method, first consider the surface impedance of a conductor due to skin effect:

$$Z_s = R + jX$$

This equation was introduced in Chapter 1. The incremental inductance rule is based on the fact that the real part of the surface impedance of good conductors is equal to its imaginary part; that is,

$$R = X = \omega L_i$$

where L_i is the inductance due to the skin-effect contribution. Therefore, once the L_i is determined, the surface resistance follows immediately. The problem is, however, in most of the inductance calculations, the effect of the skin-effect contribution is neglected compared to the external inductance L per unit length, since perfect conductors are usually assumed. In characteristic impedance calculations, this approximation does not introduce serious error for low-loss lines. To address the problem, Wheeler showed that L_i can be inferred from L as the incremental increase in L caused by an incremental recession of all metallic walls carrying a skin current [33]. The amount of recession is equal to half the skin depth. Note that this technique is useful only when the external inductance can be calculated (e.g., TEM lines). Another crucial assumption of the technique is that the conductor thickness must be greater than the skin depth. Based on Wheeler's theory, the skin-effect inductance is related to the external inductance by [33]

$$L_i = \sum_j \frac{\mu_j}{\mu_0} \frac{\partial L}{\partial n_j} \frac{\delta_j}{2} \tag{2.29}$$

and,

$$R = \frac{1}{\mu_0} \sum_j R_{sj} \frac{\partial L}{\partial n_j} \tag{2.30}$$

where

$$R_{sj} = \frac{\omega \delta_j \mu_j}{2}$$

In the above equations, $\partial L/\partial n_j$ denotes the derivative of L with respect to the incremental recession of wall j, n_j is the vector normal to the jth wall, and R_{sj} is the surface skin resistivity of wall j (see Figure 2.8). Then, the attenuation due to conductor losses can be calculated from

$$\alpha_c = \frac{1}{2\mu_0 Z_c} \sum_j R_{sj} \frac{\partial L}{\partial n_j} \tag{2.31}$$

Referring to Figure 2.8, one can obtain

$$\alpha_c = \frac{1}{2\mu_0 Z_c} \left[R_{s1} \frac{\partial L}{\partial h} - 2R_{s1} \frac{\partial L}{\partial w} - 2 \frac{\partial L}{\partial t} + R_{s2} \frac{\partial L}{\partial h} \right]$$

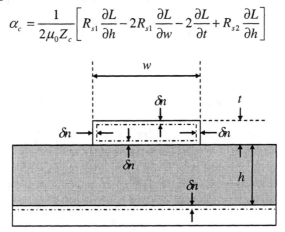

Figure 2.8 Explanation of Wheeler's method to derive conductor attenuation. (*After:* [33].)

Note that the external inductance can be obtained from the characteristic impedance expression for TEM waves. Then, it can be shown that the following equations are obtained for the attenuation constant [33]:

$$\alpha_c' = \begin{cases} \dfrac{1}{2\pi}\left\{1-\left(\dfrac{w_{eq}}{4h}\right)^2\right\}\left[1+\dfrac{h}{w_{eq}}+\dfrac{h}{\pi w_{eq}}\left\{\dfrac{t}{w}+\ln\left(\dfrac{4\pi w}{t}\right)\right\}\right], & w/h \le 1/2\pi \\[3em] \dfrac{1}{2\pi}\left\{1-\left(\dfrac{w_{eq}}{4h}\right)^2\right\}\left[1+\dfrac{h}{w_{eq}}+\dfrac{h}{\pi w_{eq}}\left\{\ln\left(\dfrac{2h}{t}\right)-\dfrac{t}{h}\right\}\right], & 1/2\pi \le w/h \le 2 \\[3em] \dfrac{\left[\dfrac{w_{eq}}{h}+\dfrac{w_{eq}/\pi h}{w_{eq}/2h+0.94}\right]\left[1+\dfrac{h}{w_{eq}}+\dfrac{h}{\pi w_{eq}}\left\{\ln\left(\dfrac{2h}{t}\right)-\dfrac{t}{h}\right\}\right]}{\left[\dfrac{w_{eq}}{h}+\dfrac{2}{\pi}\ln\left\{5.44\pi\cdot\left(\dfrac{w_{eq}}{2h}+0.94\right)\right\}\right]^2}, & 2 \le w/h \end{cases} \tag{2.32}$$

and

$$\alpha_c = 8.68\dfrac{R_s}{Z_c h}\alpha_c' \quad \text{dB/m} \tag{2.33}$$

where w_{eq} is the equivalent width of the line calculated according to

$$w_{eq} = \begin{cases} w+\dfrac{t}{\pi}\left(\ln\dfrac{4\pi w}{t}+1\right), & w/h \le 1/2\pi \\[2em] w+\dfrac{t}{\pi}\left(\ln\dfrac{2h}{t}+1\right), & w/h \ge 1/2\pi \end{cases}$$

For the sake of completeness, an approximate expression for conductor losses for wide strip widths ($w/h \to \infty$), assuming uniform current distribution, is also given here by the following expression [33, 34]:

$$\alpha_c = 8.686\dfrac{R_s}{wZ_c} \quad \text{dB/m} \tag{2.34}$$

where

$$R_s = \sqrt{\dfrac{\omega\mu}{2\sigma}} \tag{2.35}$$

In the above expression, ω is the frequency, σ is the conductivity of the line metallization, w is the line width, and Z_c is the characteristic impedance of the line. As an example, a 50-ohm microstrip line on a dielectric with $h = 5$ mils and $\varepsilon_r = 2.2$, has approximately 0.37-dB conductor loss per inch at 10 GHz. It should also be stated that this equation ignores the effect of surface roughness, which can be substantial at microwave frequencies. This is because at microwave frequencies, the skin depth is comparable to the surface roughness of PCB metallizations, causing an increase in surface resistance (see Figure 2.9).

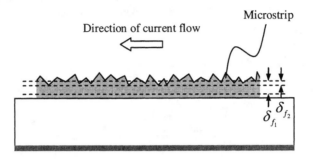

Figure 2.9 Pictorial representation of the effect of surface roughness on the surface resistance due to skin depth, δ. At higher frequencies ($f_2 \gg f_1$), the current is more confined to the surface and forced to travel an increased effective length because foil surfaces are not smooth enough. Note that the surface roughness is exaggerated for easy visualization. It also happens on both sides of the foil but in different amounts depending on how the copper foils are produced (see Section 2.5.1).

To include the effect of surface roughness, the following equation can be used, modifying R_s [32]:

$$R_s(\Delta) = R_s \left[1 + \frac{2}{\pi} \tan^{-1} \left(1.4 \frac{\Delta^2}{\delta^2} \right) \right] \qquad (2.36)$$

where Δ is the root-mean-square (RMS) surface roughness, and $R_s(\Delta)$ is the modified surface resistance.

It is important to note that (2.33) should be considered as a guideline in estimating the conductor losses. Full-wave simulations must be used to accurately predict the conductor losses in microstrip circuits. Usually, knowing the equivalent conductivity of the metallization is sufficient to model the conductor losses using simulators. Although surface roughness and skin depth are not directly modeled in planar full-wave simulators, the advantage of using a full-wave simulator would be the accurate determination of the current distribution on the metals, which affects the conductor losses. On the other hand, 3-D full-wave simulators can be instructed to solve the fields inside metallic objects to take into account skin depth accurately. But this technique is usually not resorted to in practice because it uses too much computer resources. Another point that complicates matters further is that the microstrip conductors are usually not made of a single metal but by depositing different metal layers at different steps of the process, resulting in a relatively complex expression for conductivity. Therefore, perhaps the easiest way to model conductor losses for a given circuit is to measure the S-parameters of at least two different lengths of microstrip line, and then to use a full-wave simulator to find an equivalent conductivity by trying to match simulation results with measurements. Later, this equivalent conductivity can be used in subsequent simulations of other circuits manufactured on the same substrate. Note that line standards of the thru-line-reflect (TRL) calibration can be conveniently used for this purpose. In Section 2.7.2, we will introduce a general method that can be used in extracting the loss parameters of transmission lines.

2.2.3 Dielectric Losses

An approximate expression for the dielectric losses for a microstrip line is given by the following expression [33, 34]:

$$\alpha_d = 27.3 \frac{\varepsilon_r (\varepsilon_{eff} - 1) \tan \delta}{\sqrt{\varepsilon_{eff}} (\varepsilon_r - 1) \lambda_0} \quad \text{dB/m} \quad (2.37)$$

where $\tan \delta$ and ε_{eff} are the loss tangent and effective dielectric constant of the substrate, respectively. As an example, a 50-ohm microstrip line on a dielectric with $h = 5$ mils, $\varepsilon_r = 2.2$, and $\tan \delta = 0.001$, has a 0.027-dB dielectric loss per inch at 10 GHz, which is an order of magnitude smaller than the conductor loss for the same structure.

As in the metallization losses, (2.37) should be considered as a guideline in estimating the dielectric losses. Full-wave simulations must be used to predict the dielectric losses in a microstrip circuit. Note that for most of the microstrip circuits built on low-loss dielectric materials, conductor loss is the dominating factor. It is also important to stress that determining the dielectric losses is not a trivial exercise. We will visit this important subject in Section 2.7. In most practical cases, however, the numbers provided by the substrate manufacturer may be sufficient, provided that the characterization frequency is not too far away from the frequency of interest.

2.2.4 Radiation Losses

Radiation loss is another important loss mechanism to consider, especially for microstrip lines containing many discontinuities (e.g., impedance steps, bends, open circuits) [34–36, 39]. Unwanted radiation can cause cross-talk between the circuit components.

Radiation from microstrip discontinuities can be rigorously studied by integrating the strip and polarization currents to evaluate the hertz potential from which the radiated fields may be found [34]. It was shown that the radiated power from the discontinuities can be expressed by [34]

$$P_r = 60(k_0 h)^2 F(\varepsilon_{eff}) \quad (2.38)$$

where h is the substrate thickness, and $F(\varepsilon_{eff})$ is the form factor, which can be found for different discontinuity types in Table 2.2.

Table 2.2

Radiation Form Factors for Microstrip

Discontinuity	$\varepsilon_{eff} \gg 1$	$\varepsilon_{eff} = 2.25$
Open circuit	$8/3\varepsilon_{eff}$	1.073
Short circuit	$16/15\varepsilon_{eff}^2$	0.246
90° corner	$4/3\varepsilon_{eff}$	0.610
T-junction	$2/3\varepsilon_{eff}$	0.349
Impedance step	$\dfrac{8}{3\varepsilon_{eff}}\left(\dfrac{Z_2 - Z_1}{Z_2 + Z_1}\right)^2$	0.268 (3 to 1 change)

Source: [34].

The first observation that can be made is that open circuits and 90° corners are the biggest offenders in terms of undesired radiation as far as the practical discontinuities are concerned. Another important observation is the dependency of radiated power on the square of frequency and substrate thickness, as indicated by (2.38). It is also inversely proportional to the effective dielectric constant.

Fortunately, radiation losses can also be characterized very accurately using computer simulations. Thus, (2.38) should serve as a guideline in terms of the relative levels of radiation caused by different discontinuities.

2.2.5 Higher-Order Modes and Dispersion

It has been indicated that the dominant propagation mode of a microstrip line is the quasi-TEM mode, which closely resembles the TEM mode, provided that frequency is low enough. However, when the frequency exceeds the cutoff frequency of the first higher-order mode, two modes start to propagate with different phase velocities, which can cause increased dispersion [40, 41]. As the frequency increases, more higher-order modes start to propagate, as depicted in Figure 2.10. It is important to stress that higher-order modes and surface waves (which will be introduced in the next section) are two different phenomena.

The cutoff frequency of the first higher-order mode is approximately given by the following expression [26, 36]:

$$f_c = \frac{c_0}{\sqrt{\varepsilon_r}\left(2w + 0.8h\right)} \tag{2.39}$$

Microstrip lines should be designed so that the operating frequency is always lower than the cutoff frequency given by (2.39). Fortunately, for most commonly used dielectric thicknesses of PCB materials, the f_c is quite high. For example, the cutoff frequency of the first higher-order mode for a 50-ohm microstrip line on a substrate with $h = 10$ mils and $\varepsilon_r = 2.2$ is approximately 113 GHz.

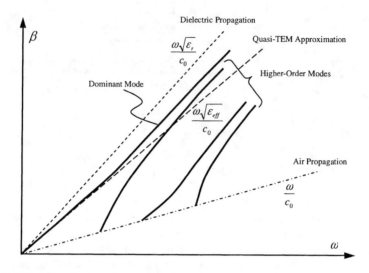

Figure 2.10 Dispersion diagram of a microstrip line showing higher-order modes.

Another important non-TEM effect that is encountered in microstrip lines is dispersion [42–44]. When the propagation constant is a nonlinear function of the frequency, the respective medium is said to be dispersive. For ideal TEM mode, the propagation constant is a linear function of the frequency. Note that the equation for effective permittivity given in (2.23) does not depend on frequency; thus, it is usually called static effective permittivity. The study of microstrip dispersion has been an important topic of research in the literature. Perhaps the most commonly employed model in practice is the model of Getsinger [42], which expresses the frequency-dependent effective dielectric constant of the dominant mode as

$$\varepsilon_{eff}(f) = \varepsilon_r - \frac{\varepsilon_r - \varepsilon_{eff}}{1 + G \cdot f^2 / f_p^2} \tag{2.40}$$

where ε_{eff} is given by (2.23). The other parameters are given by

$$f_p = \frac{Z_c}{2\mu_0 h}$$

$$G \cong 0.6 + 0.009 \frac{Z_c}{\text{ohm}}$$

It is evident from (2.40) that the dispersion is more pronounced for thicker substrates. Also, as $f \to 0$, $\varepsilon_{eff}(f) \to \varepsilon_{eff}$ as expected. Figure 2.11 shows the quantitative behavior of (2.40).

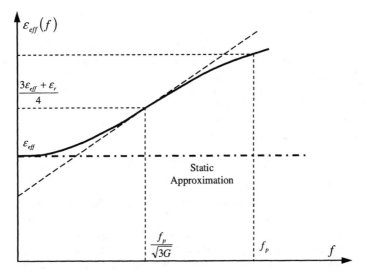

Figure 2.11 Frequency-dependent effective dielectric constant. (*After:* [42].)

A more accurate dispersion equation has also been developed covering a wide range of parameters and is given next [44]:

$$\varepsilon_{eff}(f) = \varepsilon_r - \frac{\varepsilon_r - \varepsilon_{eff}}{1 + (f/f_a)^m} \tag{2.41}$$

with

$$f_a = \frac{f_b}{0.75 + (0.75 - 0.332 \cdot \varepsilon_r^{-1.73})w/h}$$

$$f_b = \frac{c}{2\pi h\sqrt{\varepsilon_r - \varepsilon_{eff}}} \tan^{-1}\left(\varepsilon_r \sqrt{\frac{\varepsilon_{eff} - 1}{\varepsilon_r - \varepsilon_{eff}}}\right)$$

$$m = \begin{cases} m_0 m_c, & m_0 m_c \leq 2.32 \\ 2.32, & m_0 m_c > 2.32 \end{cases}$$

$$m_0 = 1 + \frac{1}{1 + \sqrt{w/h}} + 0.32\left(1 + \sqrt{w/h}\right)^{-3}$$

$$m_c = \begin{cases} 1 + \dfrac{1.4}{1 + w/h}\left(0.15 - 0.235e^{-0.45f/f_a}\right), & w/h \leq 0.7 \\ 1, & w/h > 0.7 \end{cases}$$

where f is in gigahertz. It was claimed that the error in the above dispersion equation is less than 0.6% in the range of $0.1 < w/h < 10$ and $1 < \varepsilon_r < 128$.

2.2.6 Surface Waves

Surface waves are the propagating modes of metal-dielectric-air structure, and they can travel through the dielectric-air interface with fields mostly trapped in the dielectric [23]. The lowest-order TM surface wave has zero cutoff frequency, which means that it can exist at any frequency. However, significant energy transfer occurs from the fundamental mode to the lowest-order surface wave only when the phase velocities of the two modes are close to each other. This synchronism occurs at the following frequency [26, 36]:

$$f_s = \frac{c_0 \tan^{-1}(\varepsilon_r)}{\sqrt{2\pi} \, h \sqrt{\varepsilon_r - 1}} \tag{2.42}$$

As an example, the synchronization frequency of the lowest-order surface wave mode for a 50-ohm microstrip line on a substrate with $h = 10$ mils and $\varepsilon_r = 2.2$ is approximately 278 GHz. So, like the higher-order modes, surface wave excitation is also less of an issue for most of the microwave printed circuits. One must be careful, however, in microstrip antenna applications, where substrate can be significantly thick (~ 40 mils or more). In that case, the synchronization frequency can fall well into the operating band, resulting in spurious radiation and increased coupling between the antenna elements.

2.3 STRIPLINES

Striplines are the other most commonly used type of microwave printed circuit [17, 45, 46]. As shown in Figure 2.12, a stripline structure consists of a conducting foil sandwiched between two ground planes. Contrary to microstrip lines, the main mode of propagation for stripline is TEM because the structure is homogeneous. This provides advantages over microstrip in applications where the propagation mode is critical. For example, as will be demonstrated in Chapter 5, microstrip directional couplers suffer an isolation problem because of the unequal even- and odd-mode phase velocities, a result of quasi-TEM propagation. This problem is greatly reduced in stripline couplers because the propagation mode is TEM. Another advantage of striplines is that the signal lines are shielded from the outside by means of two ground planes. An apparent disadvantage of striplines, on the other hand, is the relative difficulty in combining them with active electronics. Because it is relatively difficult to embed the electronics inside the dielectrics, any electronic component must be placed on outside surfaces of the ground planes. This creates the necessity of using transitions employing via-holes to make contact with the signal lines. Nevertheless, striplines are frequently employed when shielding and/or TEM propagation are paramount considerations.

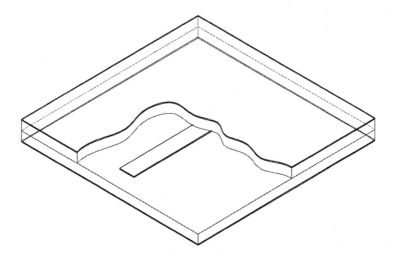

Figure 2.12 A stripline that is constructed by sandwiching a metal foil between two ground planes. Note that the top layer is cut open for easy visualization.

The cross-section of a stripline is depicted in Figure 2.13. In the figure, w and h represent the width of the center conductor and the thickness of the dielectric slab, respectively. The most commonly used striplines are the symmetric ones, as shown in the figure (i.e., center conductor is placed midway between the two ground planes, $h = (b - t)/2$), although for some applications unsymmetrical lines are employed. The best example for that kind of application is the broadside-coupled stripline coupler, where each line is placed asymmetrically with respect to ground planes, and they overlap with each other [15]. Broadside-coupled couplers are important because they allow high coupling ratios, which otherwise wouldn't be possible with edge-coupled geometry.

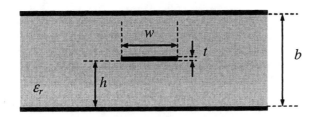

Figure 2.13 Cross-sectional view of the stripline.

2.3.1 Characteristic Impedance

To obtain an expression for the characteristic impedance of a stripline, first consider the capacitance model as shown in Figure 2.14. In the model, C_p and C_f represent parallel-plate and firing capacitances, respectively, and they are given by the following equations for a wide strip [45, 46]:

$$C_p = \varepsilon_0 \varepsilon_r \frac{w}{(b-t)/2} \quad \text{F/m} \tag{2.43}$$

$$C_f \cong \frac{\varepsilon_0 \varepsilon_r}{\pi} \left\{ \frac{2}{1-t/b} \ln\left[1 + \frac{1}{1-t/b}\right] - \left(\frac{1}{1-t/b} - 1\right) \ln\left[\frac{1}{(1-t/b)^2} - 1\right] \right\} \quad \text{F/m} \tag{2.44}$$

Then, total capacitance is found from

$$C = 2C_p + 4C_f \tag{2.45}$$

Note that (2.44) is exact for a semi-infinite center foil between two infinite ground planes. In the case of the finite-width center foil, (2.44) still provides an excellent approximation, provided that $w/(b-t) \geq 0.35$ (i.e., wide strips). For $w/(b-t) \leq 0.35$ (i.e., narrow strips), the following approximation can be used for the total capacitance [46]:

$$C = \varepsilon_0 \varepsilon_r \frac{2\pi}{\ln\dfrac{4b}{\pi d_0}} \quad \text{F/m} \tag{2.46}$$

where d_0 is the diameter of a circular cross-section equivalent to the rectangular strip, which is given by

$$d_0 = \frac{w}{2}\left[1 + \frac{t}{\pi w}\left(1 + \ln\frac{4\pi w}{t}\right)\right] \tag{2.47}$$

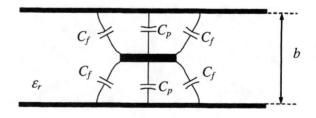

Figure 2.14 Stripline capacitance model.

Equation (2.47) is valid for $t/w \ll 1$, which is usually the case for most of today's PCB technology (i.e., the thickness of metallizations is an order of magnitude smaller than their width, even for narrower strips).

Then, the characteristic impedance can be determined using

$$Z_c = \frac{1}{cC} \tag{2.48}$$

where c is the speed of light in the medium, and C is the total line capacitance per unit length calculated using either (2.45) or (2.46).

If the center conductor is assumed to be infinitesimally thin, then an exact, analytical, closed-form solution for the characteristic impedance can be obtained through conformal mapping. The characteristic impedance of a zero-thickness stripline is given by the following expressions [17, 18]:

$$Z_c = \frac{30\pi}{\sqrt{\varepsilon_r}} \frac{K(k')}{K(k)} \tag{2.49}$$

where

$$k = \tanh\left(\frac{\pi w}{2b}\right) \tag{2.50}$$

$$k' = \sqrt{1 - k^2} \tag{2.51}$$

In the above expressions, $K(k)$ is the complete elliptic integral of the first kind defined as

$$K(k) = \int_0^1 \frac{1}{\sqrt{(1-t^2)(1-k^2 t^2)}} dt \tag{2.52}$$

The integral in (2.52) can easily be evaluated using numerical techniques. Alternatively, the following approximations can be employed, instead of resorting to numerical integration for the ratios of elliptic integrals in (2.49):

$$\frac{K(k')}{K(k)} = \begin{cases} \dfrac{1}{\pi} \ln\left(2 \dfrac{1+\sqrt{k}}{1-\sqrt{k}}\right), & 1/\sqrt{2} \le k \le 1 \\[3ex] \dfrac{\pi}{\ln\left(2 \dfrac{1+\sqrt{k'}}{1-\sqrt{k'}}\right)}, & 0 \le k \le 1/\sqrt{2} \end{cases} \tag{2.53}$$

2.3.2 Conductor Losses

By following the incremental inductance rule that was introduced in the discussion of microstrip lines, it can be shown that the conductor loss of stripline is given as follows [45, 46]:

$$\alpha_c = \frac{8.67 \cdot 4R_s Z_c \varepsilon_r}{377^2 b} \left[\frac{1}{1-t/b} + \frac{2\,w/b}{(1-t/b)^2} + \frac{1+t/b}{\pi(1-t/b)^2} \ln\left(\frac{\frac{1}{1-t/b}+1}{\frac{1}{1-t/b}-1} \right) \right] \quad \text{dB/m} \quad (2.54)$$

for $w/(b-t) > 0.35$, and

$$\alpha_c = \frac{8.67 \cdot R_s}{2\pi Z_c b} \left[1 + \frac{b}{d_0} \left\{ 0.5 + 0.669\frac{t}{w} - 0.225\frac{t}{w} + \frac{1}{2\pi}\ln\left(\frac{4\pi w}{t}\right) \right\} \right] \quad \text{dB/m} \quad (2.55)$$

for $w/(b-t) \le 0.35$. In the last equation, d_0 is the diameter of a circular cross-section equivalent to the rectangular strip, which is given in (2.47). Note that although the error between the capacitance expressions (2.45) and (2.46) in the transition region (i.e., around the point $w/(b-t) = 0.35$) is small, this is not necessarily true for their derivatives. Thus, the attenuation expressions (2.54) and (2.55) can exhibit a difference of up to 10% in the transition region. It is then appropriate to develop a transition curve to ensure a smooth transition between two formulas.

2.3.3 Dielectric Losses

Since the dielectric medium is homogeneous, the dielectric losses for a stripline can be expressed simply [46]:

$$\alpha_d = \frac{\pi}{\lambda}\sqrt{\varepsilon_r}\,\tan\delta \quad \text{nepers/m}$$
$$= \frac{27.3}{\lambda}\sqrt{\varepsilon_r}\,\tan\delta \quad \text{dB/m} \quad (2.56)$$

where λ is the free-space wavelength, ε_r is the relative dielectric constant of the medium, and $\tan\delta$ is the loss-tangent of this medium.

2.3.4 Higher-Order Modes

Unlike the microstrip, the fundamental mode of stripline is TEM. This provides a significant advantage in the design of couplers, as we will discuss in Chapter 5. Another advantage of stripline in terms of higher-order modes is that the surface waves do not exist because there is no air–dielectric interface. However, if the frequency of operation is sufficiently high, then higher-order modes do also appear in the stripline [36, 47, 48]. The nature of these higher order modes depends on the stripline structure. For instance, if the stripline is in a boxed enclosure (i.e., side walls are closed), then there is a possibility of launching rectangular waveguide modes if the cutoff frequencies are exceeded. In case of open sidewalls, transverse resonance modes can exist. The cutoff frequency of the lowest-order transverse TE mode is given by [36]:

$$f_c = \frac{30}{\sqrt{\varepsilon_r}\left(2w + \dfrac{\pi b}{2}\right)} \quad \text{GHz} \tag{2.57}$$

where w and b are in centimeters.

The other potential issue with the stripline is that it is essentially a three-conductor system, which is theoretically capable of supporting two independent TEM modes. However, since it is excited in balance mode (i.e., top and bottom ground planes are at same voltage), only one TEM mode propagates. Now, the balanced-excitation assumption would be destroyed if there were discontinuities on either ground plane. If this happens, then the top and bottom ground planes must be connected with conductive posts before and after the discontinuity to minimize excitation of higher-order modes. Sometimes, the ground planes are periodically connected along the whole length of the strip, creating a quasi-coaxial structure. This configuration is also useful to increase isolation between the adjacent striplines.

2.4 COPLANAR WAVEGUIDES

In some applications, use of microstrip lines or striplines is prohibitive due to unavailability of access to the ground plane and/or high line-to-line isolation requirements. The first situation sometimes occurs in MMIC designs because manufacturing circuits with no backside metallization is cheaper (it eliminates some steps in the wafer process). Thus, all metallizations are forced to be on the same plane. In the second, although the ground plane may exist, using a microstrip configuration may result in very wide strips if the dielectric constant of the substrate is low enough, deteriorating line-to-line isolation. Besides, if the dielectric is too lossy, then confining the E-field mostly in the dielectric won't be a good idea either. In such situations, one can use the coplanar waveguide (CPW) structure as shown in Figure 2.15 [49–52]. CPW was first reported in the literature by Wen [49].

The CPW is constructed by bringing the ground planes next to the center strip as can be seen from the figure. One of the unique advantages of CPWs is that there is a very short physical distance to the ground plane, reducing ground inductance. This is important when one would like to mount a transistor or a PIN diode on the circuit. For the same reason, it also has better heat-sink capabilities. The primary disadvantage is that for some cases it may have higher conductive loss because the center conductor is relatively narrower than the microstrip line for the same characteristic impedance. In addition to that, the current concentrates more on the edges, which tend to be rougher.

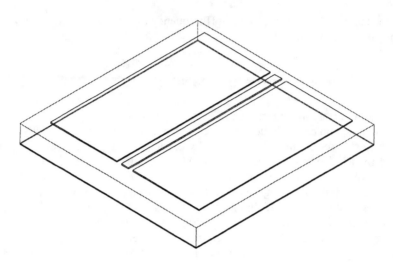

Figure 2.15 A coplanar waveguide. Note that the ground planes are brought up next to the center conductor.

Another important point is that the ground planes on both sides of the center conductor must be stitched together periodically by either wire bonds (in PCB technology) or air bridges (in MMIC technology) to prevent propagation of higher-order modes.

The cross-section of a CPW is depicted in Figure 2.16. In the figure, w, s, and h represent the width of the center conductor, the spacing between center conductor and ground planes, and the thickness of the dielectric slab, respectively. The most commonly used CPWs are the symmetrical ones as shown in the figure (i.e., the distance to the ground planes is the same on both sides).

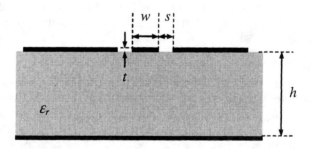

Figure 2.16 Cross-sectional view of the CPW.

A careful reader might notice that although the bottom ground plane is not necessary, it is shown in the figure. This is because most of the time the bottom ground plane is there either as backside metallization, or as a part of the packaging [52] (even though the circuits are manufactured without backside metallization, backside metal is automatically formed when the circuit is attached to a carrier, or package paddle). Thus, it makes sense to include the bottom ground plane in the calculations so that its effect is accounted for.

2.4.1 Characteristic Impedance

The capacitance per unit length, effective dielectric constant, and characteristic impedance of a symmetrical CPW with bottom and upper ground planes are given as follows [50]:

$$C(\varepsilon_r) = 2\varepsilon_0\varepsilon_r \frac{K(k_1)}{K(k_1')} + 2\varepsilon_0 \frac{K(k_2)}{K(k_2')} \quad \text{F/m} \tag{2.58}$$

where

$$k_1 = \tanh\left(\pi \frac{w}{4h}\right) \Big/ \tanh\left(\pi \frac{w+2s}{4h}\right) \tag{2.59}$$

$$k_2 = \tanh\left(\pi \frac{w}{4h_c}\right) \Big/ \tanh\left(\pi \frac{w+2s}{4h_c}\right) \tag{2.60}$$

$$k' = \sqrt{1-k^2} \tag{2.61}$$

then,

$$\varepsilon_{eff} = \frac{C(\varepsilon_r)}{C(1)} \tag{2.62}$$

$$Z_c = \frac{1}{cC(\varepsilon_r)} \tag{2.63}$$

where h_c and c are the height of upper shielding and speed of light in the medium (found using the effective dielectric constant), respectively. Note that this derivation is rigorously correct when $h = h_c$, although it was claimed to have excellent results for practical structures. In the above expressions, $K(k)$ is the complete elliptic integral of the first kind defined in (2.52). The integral in (2.52) can easily be evaluated using numerical techniques.

Additional expressions for asymmetrical (i.e., unequal gaps) coplanar waveguides with conductor backing can also be found in the literature [51].

2.4.2 Higher-Order Modes

As indicated briefly at the beginning of this section, coplanar structures, although not strictly necessary, are usually implemented with backside metallization. Note that this is a result of mechanical requirements; by no means is it intended to launch a microstrip-like mode using CPW. To complicate matters further, the dielectric thickness cannot be increased arbitrarily to minimize the effect of backside metallization because the thickness is limited by the MMIC process, although there is somewhat more flexibility in PCB circuits. Therefore, CPW circuits must be designed by considering the possible deleterious effects of backside metallization.

Depending on the structure, higher-order modes can exist due to backside metallization as depicted in Figure 2.17 [53, 54]. One way of preventing this is to connect the bottom ground to the side grounds periodically by conducting posts. This will attenuate the microstrip-like modes. If this approach is utilized, then conducting posts must be placed closer to the center strip so that the quasi-rectangular-waveguide beneath the strip remains below the cutoff frequency.

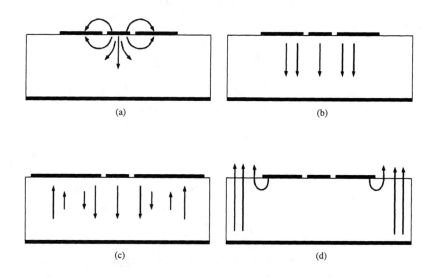

Figure 2.17 Symmetric modes of an overmoded conductor-backed coplanar waveguide: (a) CPW mode, (b) microstrip-like (MSL) mode, (c) first higher-order MSL mode (occurs if the side grounds are sufficiently wide), and (d) image-guide-like mode. (*After:* [53].)

2.5 MICROSTRIP DISCONTINUITIES

Microstrip discontinuities include open ends, gaps, steps, bends, and T-junctions. Most of the discontinuities occur because of the routing requirement of microstrip lines. For example, a microstrip circuit almost always has bends to route the transmission lines to and from the circuit components such as amplifiers and filters. Therefore, characterization of such discontinuities is important.

The results presented here are based on the work of R. Garg and I. J. Bahl and others [55–61]. Garg and Bahl devised simple equivalent models for most common discontinuities and then curve-fit the model parameters to the empirical results available in the literature. Note that since the modern microwave engineer usually employs full-wave simulation tools, the need for closed-form expressions for microstrip discontinuities is not as great as it used to be. Typically, the engineer first separately simulates anticipated T-junctions, bends, and other discontinuities and creates a library of components based on the PCB or MMIC process used. Then microstrip transmission lines are connected using these blocks to route the signals. This approach combines the accuracy of full-wave simulation with circuit simulations without increasing the overall simulation time.

However, a brief introduction to microstrip discontinuities would be useful, first to give an idea about their equivalent networks, and second, to provide the functional form of the equivalent circuit parameters in case the reader wants to refine the models using full-wave simulations to create parameterized models. For that reason, we will introduce the basic microstrip discontinues in the next sections.

2.5.1 Microstrip Open End

The effect of an open end in a microstrip line is to create extra fringe electric field at the location of the open circuit. Therefore, an open end in a microstrip line does not present an ideal open circuit at the physical reference plane. The open end can be modeled by a small extension in the physical length of the line, and then assuming no fringe fields exist (i.e., ideal open circuit). The equivalent circuit of a microstrip open end is shown in Figure 2.18(a). The length change of a microstrip line due to an open gap is given as follows [56]:

$$\frac{\Delta l}{h} = 0.412 \frac{\varepsilon_{eff} + 0.3}{\varepsilon_{eff} - 0.258} \frac{w/h + 0.264}{w/h + 0.8} \tag{2.64}$$

where h is the substrate thickness, ε_{eff} is the relative effective dielectric constant, w is the width of the line, and Δl is the extension in the line length to compensate for the effect of the open end. The error in (2.64) is less than 4% in the range of $w/h > 0.2$ and $2 < \varepsilon_r < 50$.

The open circuit capacitance shown in Figure 2.18(a) can be obtained using the value of $\Delta l / h$:

$$\frac{C_o}{w} = \frac{\Delta l}{h} \frac{1}{w/h} \frac{\sqrt{\varepsilon_{eff}}}{cZ_c} \tag{2.65}$$

where c and Z_c are the speed of light and characteristic impedance of the line, respectively.

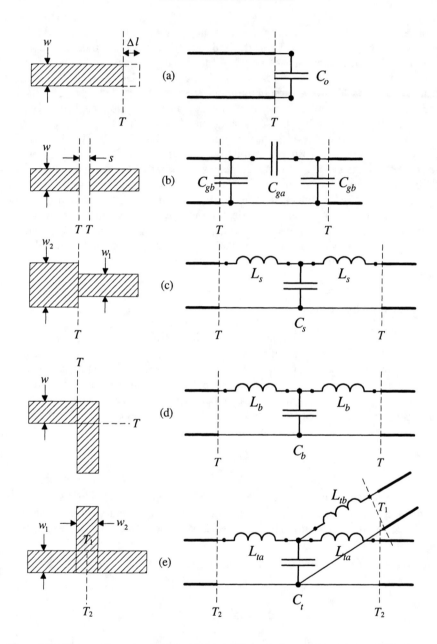

Figure 2.18 Various microstrip discontinuities and their equivalent models: (a) open end, (b) series gap, (c) change in width, (d) bend, and (e) T-junction. Reference planes are indicated by dashed lines.

2.5.2 Microstrip Series Gap

The microstrip series gap can be useful in microstrip resonators where it is used to couple RF energy from the input line to the resonator. It can also be used in dc-blocking circuits if the frequency is high enough that the small series capacitance has sufficiently low impedance. Note that a single gap is rarely enough to provide enough coupling in PCB technology. Therefore, one needs to utilize an interdigital structure to boost the coupling in that case. On the other hand, one can achieve orders-of-magnitude smaller gaps in MMIC than with PCB circuits. Thus, their actual benefit is more in MMIC than with PCB circuits.

The equivalent circuit of a microstrip gap is shown in Figure 2.18(b). The component values of the equivalent circuit that are obtained by exciting the lines symmetrically (even mode) and antisymmetrically (odd mode) are given as follows [56]:

$$C_{gb} = \frac{1}{2}C_{even} \tag{2.66}$$

$$C_{ga} = \frac{1}{2}\left(C_{odd} - \frac{1}{2}C_{even}\right) \tag{2.67}$$

where

$$\frac{C_{odd}}{w} = \left(\frac{s}{w}\right)^{m_o} e^{K_o}\left(\frac{\varepsilon_r}{9.6}\right)^{0.8} \quad \text{pF/m}$$

$$\frac{C_{even}}{w} = \left(\frac{s}{w}\right)^{m_e} e^{K_e}\left(\frac{\varepsilon_r}{9.6}\right)^{0.9} \quad \text{pF/m}$$

$$m_o = \frac{w}{h}\left(0.619\log\frac{w}{h} - 0.3853\right), \quad 0.1 \le s/w \le 1.0$$

$$m_e = \begin{cases} 0.8675, & 0.1 \le s/w \le 0.3 \\ \dfrac{1.565}{(w/h)^{0.16}} - 1, & 0.3 \le s/w \le 1.0 \end{cases}$$

$$K_o = 4.26 - 1.453\log\frac{w}{h}, \quad 0.1 \le s/w \le 1.0$$

$$K_e = \begin{cases} 2.043\cdot(w/h)^{0.12}, & 0.1 \le s/w \le 0.3 \\ 1.97 - \dfrac{0.03}{w/h}, & 0.3 \le s/w \le 1.0 \end{cases}$$

The error in (2.66) and (2.67) is less than 7% in the range of $0.5 < w/h < 2$ and $2.5 < \varepsilon_r < 15$.

2.5.3 Change in Microstrip Width

In a microstrip circuit, it is sometimes necessary to change the width of the line to alter the characteristic impedance of the line. This is typically required in designing impedance matching circuits and filters. The change in the width can be symmetrical (axes of the lines coincide) or asymmetrical (axes of the lines do not coincide). Asymmetrical change in the width is usually employed in microstrip coupled-line filters where a symmetrical change may short-circuit the two adjacent resonators and is therefore not an option. The equivalent circuit of a step in width is shown in Figure 2.18(c). The closed-form expression for the shunt capacitor is given below [56]:

$$\frac{C_s}{\sqrt{w_1 w_2}} = \left(10.1 \cdot \log \varepsilon_r + 2.33\right)\frac{w_2}{w_1} - 12.6 \cdot \log \varepsilon_r - 3.7 \quad \text{pF/m} \qquad (2.68)$$

The error in (2.68) is less than 10% in the range of $1.5 < w_2/w_1 < 3.5$ and $\varepsilon_r < 10$. The closed-form expression for the series inductors is given below [56]:

$$\frac{2L_s}{h} = 40.5\left(\frac{w_2}{w_1} - 1.0\right) - 75 \cdot \log\frac{w_2}{w_1} + 0.2\left(\frac{w_2}{w_1} - 1\right)^2 \quad \text{nH/m} \qquad (2.69)$$

The error in (2.69) is less than 5% in the range of $w_2/w_1 < 5$ and $w_1/h = 1$.

2.5.4 Microstrip Bend

Having bends in a complex microstrip circuit is almost unavoidable. Without bending the lines, it would be nearly impossible to reduce the circuit dimensions. In fact, it is one of the biggest advantages of microstrip lines that they can be shaped relatively easily; hence circuits can be placed in close proximity. However microstrip bends affect the VSWR and therefore must be properly compensated. The equivalent circuit of a right-angle bend is shown in Figure 2.18(d). The closed-form expression for the model components are given as follows [56]:

$$\frac{C_b}{w} = \begin{cases} \dfrac{\left(14\varepsilon_r + 12.5\right)w/h - \left(1.83\varepsilon_r - 2.25\right)}{\sqrt{w/h}} + \dfrac{0.02\varepsilon_r}{w/h} & \text{pF/m, \quad for } w/h < 1 \\[3mm] \left(9.5\varepsilon_r + 1.25\right)\dfrac{w}{h} + 5.2\varepsilon_r + 7.0 & \text{pF/m, \quad for } w/h \geq 1 \end{cases} \qquad (2.70)$$

$$\frac{2L_b}{h} = 100\left(4\sqrt{w/h} - 4.21\right) \quad \text{nH/m} \qquad (2.71)$$

The error in (2.70) is less than 5% in the range of $0.1 < w/h < 5$ and $2.5 < \varepsilon_r < 15$, and the error in (2.71) is less than 3% in the range of $0.5 < w/h < 2$.

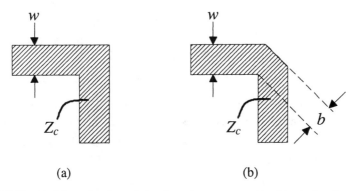

Figure 2.19 The (a) right-angle microstrip bend, can be compensated by (b) mitering the corner. The distance b is approximately chosen as $0.6w$ for optimum performance.

Fortunately, it is possible to compensate for the excessive shunt capacitance of a microstrip right-angle bend by mitering it, as shown in Figure 2.19. The amount of mitering controls the values of shunt capacitance and series inductance in different directions. In this way, the discontinuity can be matched to the characteristic impedance of the line. The exact amount of mitering depends on the line width, the effective dielectric constant, and the bend angle. Typical values of b shown in the figure are approximately $0.6w$ to $0.7w$. Exact values can easily be determined by using a full-wave simulation. The microstrip bends must always be mitered to minimize their effect on VSWR.

2.5.5 Microstrip T-Junction

Microstrip T-junction occurs when a line interacts another line at a right angle and terminates. Perhaps the most common use of a T-junction is in the microstrip stubs. Other common application is in the power dividers and branch-line couplers. The equivalent circuit of a T-junction is shown in Figure 2.18(e). The closed-form expression for the model components are given as follows when the main-line impedance is 50 ohms [56]:

$$\frac{C_t}{w_1} = \frac{100}{\tanh(0.0072 \cdot Z_o)} + 0.64Z_0 - 261 \quad \text{pF/m} \tag{2.72}$$

where Z_0 is the characteristic impedance of the stub. The error in (2.72) is less than 5% in the range of $25 < Z_0 < 100$ and $\varepsilon_r = 9.9$. The closed-form expression for the series inductors is given below [56]:

$$\frac{L_{ta}}{h} = -\frac{w_2}{h}\left[\frac{w_2}{h}\left(-0.016\frac{w_1}{h} + 0.064\right) + \frac{0.016}{w_1/h}\right]L_{w_1} \quad \text{nH/m} \tag{2.73}$$

$$\frac{L_{tb}}{h} = \left[\begin{array}{c} \left(0.12\frac{w_1}{h} - 0.47 \right)\frac{w_2}{h} + 0.195\frac{w_1}{h} - 0.357 \\ + 0.0283\sin\left(\frac{w_1}{h}\pi - 0.75\pi \right) \end{array} \right] L_{w_2} \quad \text{nH/m} \qquad (2.74)$$

The error in (2.73) is less than 5% in the range of $0.5 < \left(w_1/h, w_2/h \right) < 2$ and $\varepsilon_r = 9.9$, and the error in (2.74) is less than 5% in the range of $1 < w_1/h < 2$, $0.5 < w_2/h < 2.0$ and $\varepsilon_r = 9.9$. In the above equations, L_w is the inductance per unit length for a microstrip line of width w and is obtained from:

$$L_w = Z_c \frac{\sqrt{\varepsilon_{eff}}}{c} \qquad (2.75)$$

where Z_c and c are the characteristic impedance of the line and speed of the light, respectively.

There are mainly two ways to compensate for a T-junction, as shown in Figure 2.20. In the first method, a simple notch is placed on the main arm across the perpendicular arm. In the second method, steps are placed on the junction appropriately [57]. The dimensions of the notch and steps can be determined through full-wave simulations.

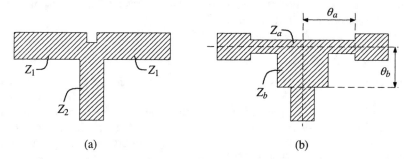

(a)   (b)

Figure 2.20 Compensation techniques for T-junctions: (a) notch, and (b) step junction.

2.6 SUBSTRATES FOR MICROWAVE PRINTED CIRCUITS

As shown so far, dielectric materials are almost always used in microwave printed circuits to support the conductors. Although special constructions are possible yielding air-dielectric microwave printed circuits to prevent dielectric losses, those won't provide the same planarity, modularity, and compactness of the structures that use substrates. Thus, the properties of substrate materials play an important role in both electrical and the mechanical performance of the microwave printed circuit. In the following paragraphs, we will briefly introduce important properties of substrate materials that are frequently used in microwave printed circuits [62–65].

The fabrication of dielectric materials is one of the important steps in microwave printed circuit technology since the requirements of microwave circuits can vary. For instance, a conformal printed antenna may require a flexible dielectric for easy application on curved surfaces. On the other hand, rigidity and planarity is required in some applications where bow and twist on the dielectric surface beyond certain limits is unacceptable. Of course, low dielectric loss is always preferable in microwave applications, yet cost issues don't always allow choosing the best material. It is one of the important steps in microwave printed circuit design to decide on the optimum material, not the best, that performs a given functionality. The thickness and electrical properties of dielectric materials must also be able to be controlled tightly for microwave applications unlike low-frequency printed circuits. Large variation in these parameters from batch to batch is not acceptable. On top of all this, the dielectric material for PCB must provide some means of adhering conductive foils or depositing metallization on its surface; otherwise, transmission lines cannot be constructed.

The thermal properties of dielectric materials are another important factor that must be considered while selecting the PCB substrate. This is especially important in manufacturing multilayer PCBs. If the coefficient of thermal expansion (CTE) of each layer in a multilayer PCB is significantly different compared to other ones, then delamination may occur, which eventually results in circuit malfunction. CTEs must also be looked at carefully when attaching other circuit components to the PCBs. If the difference between the CTE of the attached circuit component and the board is high enough, then cracking of electrical contacts may occur after thermal excursion cycles.

Metal foils used to clad dielectric slabs must also satisfy some strict requirements such as maximum thickness and surface smoothness. Dimensional resolution of printed circuits mainly depends on the thickness of the metallization on the dielectric. As a rule of thumb, the absolute minimum microstrip width that can be practically realized in PCB technology is approximately two times the thickness of the metallization. Therefore, thinner metallization means higher printing resolution, provided that the proper photolithographic technique is used. Another important point is that the yield drops in subsequent manufacturing (e.g., wire bonding) as the line widths get smaller.

Moisture absorption and resistance to chemicals are also important parameters in the selection of dielectrics. If the substrate is going to be exposed to atmosphere or placed in a nonhermetic medium, then moisture absorption could affect the RF performance because water has a high dielectric constant ($\varepsilon_r = \sim 70$). If the substrate is going be used as an enclosure for electronic circuits, then moisture absorption is again critical because oxidation will eventually occur on the circuits that are inside due to moisture diffusion through the substrate material. Substrate materials must withstand chemicals that are used in etching and cleaning processes as well.

Table 2.3

Electrical Properties of Commonly Used PCB Materials [64, 65]

Material Name	Composition	Dielectric Constant	Dielectric Loss Tangent
Rogers RO3003	PTFE/ceramic	3.0 @ 10 GHz	0.0013 @ 10 GHz
Rogers RO3006	PTFE/ceramic	6.15 @ 10 GHz	0.0025 @ 10 GHz
Rogers RO3010	PTFE/ceramic	10.2 @ 10 GHz	0.0035 @ 10 GHz
Rogers RO4003	Hydrocarbon/ceramic woven glass	3.38 @ 10 GHz	0.0027 @ 10 GHz
Rogers RO4350	Hydrocarbon/ceramic woven glass	3.48 @ 10 GHz	0.004 @ 10 GHz
RT/Duroid 5870	PTFE/glass fiber	2.33 @ 10 GHz	0.0012 @ 10 GHz
RT/Duroid 5880	PTFE/glass fiber	2.2 @ 10 GHz	0.0009 @ 10 GHz
RT/Duroid 6010	PTFE/ceramic	10.2 @ 1 GHz	0.0028 @ 1 GHz
RT/Duroid 6002	PTFE/ceramic	2.94 @ 1 GHz	0.0012 @ 1 GHz
Arlon AR320	PTFE/woven fiberglass	3.20 @ 10 GHz	0.003 @ 10 GHz
Arlon AR600	PTFE/woven fiberglass/ceramic filled	6.0 @ 10 GHz	0.0035 @ 10 GHz
Arlon 33N	Polyimide/E-glass	4.25 @ 1 MHz	0.009 @ 1 MHz
Arlon 35N	Polyimide/E-glass	4.39 @ 1 MHz	0.008 @ 1 MHz
Arlon DiClad 880	PTFE/woven fiberglass	2.2 @ 10 GHz	0.0009 @ 10 GHz
Arlon DiClad 870	PTFE/woven fiberglass	2.3 @ 10 GHz	0.0013 @ 10 GHz
Taconic TLC32	PTFE/woven glass	3.2 @ 10 GHz	0.003 @ 10 GHz
Taconic TLE95	PTFE/woven glass	2.95 @ 10 GHz	0.0028 @ 10 GHz
Taconic TLT9	PTFE/woven glass	2.5 @ 10 GHz	0.0006 @ 10 GHz
Taconic TLY5	PTFE/woven glass	2.2 @ 10 GHz	0.0009 @ 10 GHz
Taconic RF30	Glass reinforced fluoropolymer	3.0 @ 1.9 GHz	0.0014 @ 1.9 GHz
Taconic RF60	PTFE/ceramic	6.15 @ 10 GHz	0.0028 @ 10 GHz
FR4	Epoxy/glass	5.2 @ 1 MHz	0.025 @ 1 MHz
FR5	Epoxy/glass	5.2 @ 1 MHz	0.025 @ 1 MHz

Table 2.4

Electrical Properties of Commonly Used Ceramic and Amorphous Materials [64, 65]

Material Name	Composition	Dielectric Constant	Dielectric Loss Tangent
Alumina	Al_2O_3	9.6 @ 100 GHz	0.0006 @ 100GHz
Beryllia	BeO	6.7 @ 100 GHz	0.0016 @ 100 GHz
Sapphire	Al_2O_3 (single crystalline)	3.1 @ 100 GHz	0.00046 @ 100 GHz
Kyocera A493	%99.6 Al_2O_3	9.9 @ 1 MHz	0.0002 @ 1 MHz
Kyocera A494	%99.6 Al_2O_3	9.9 @ 1 MHz	0.0002 @ 1 MHz
Dupont 951	Ceramic	7.8 @ 10 GHz	0.005 @ 10 GHz
Kyocera GL550	Ceramic	5.6 @ 2 GHz	0.0009 @ 2 GHz
Kyocera GL560	Ceramic	6.0 @ 2 GHz	0.0017 @ 2 GHz
Kyocera GL660	Ceramic	9.6 @ 2 GHz	0.0017 @ 2 GHz
Quartz	SiO_2 (crystalline)	4.43 @ 35 GHz	0.00031 @ 35 GHz
Glass (7070)	SiO_2 (amorphous)	3.9 @ 25 GHz	0.0031 @ 25 GHz
Gallium Arsenide	GaAs	13 @ 10 GHz	0.006 @ 10 GHz
Silicon	Si	11.9 @ 10 GHz	0.004 @ 10 GHz
Fused Silica	SiO_2	3.78 @ 10 GHz	0.00017 @ 10 GHz
Fused Quartz	SiO_2 (crystalline)	3.8 @ 24 GHz	0.0001 @ 10 GHz

So far, we have been concentrating mostly on important mechanical and chemical properties of dielectric materials. Electrical properties of dielectric materials are also critical. The most important electrical properties of dielectrics are permittivity, loss-tangent, and isotropy. Needless to say, a good microwave substrate should have a low loss-tangent. Sapphire is one of the crystal materials that have very low dielectric loss. However, it is expensive and difficult to machine, thus, not suitable for consumer applications. Polytetrafluoroethytene (PTFE) is another low-loss material, which is extensively used in microwave printed circuit substrates. A material is said to be isotropic if permittivity and permeability are scalars. It is anisotropic if either or both the permittivity and permeability become tensors. If the permittivity becomes a tensor, for instance, then the structure presents a different dielectric constant depending on the orientation of the electric field. Sapphire is an example of an anisotropic material. Electrical properties of some of the commonly used substrate materials are given in Tables 2.3 and 2.4 for reference. In the tables, three fundamental classes are summarized: high-frequency laminates (e.g., PTFE/ceramic), low-frequency laminates (e.g., FR4), and ceramics (e.g., Alumina, LTCC). More detailed information on the dielectric properties of various materials can be found in the literature [62–65].

In the following sections, we will explain the two important subgroups of printed circuit materials in more detail.

2.6.1 Laminates

The first group of materials that is frequently used in microwave printed circuits is called laminates (or composites) [62, 63]. A laminate typically consists of a dielectric slab and copper claddings on both sides, which are bonded to the dielectric. The dielectric slab has two major elements: the fabric and the resin.

Fabrication of dielectric slabs starts with manufacturing the fabric with the desired thickness. Fabric is a flexible material, so it must be impregnated by the resin and then cured to make it harder. Note that the combination of the fabric and the resin forms the overall dielectric; therefore, the electrical properties of both materials are important in determining the dielectric constant and loss tangent, especially at high frequencies. Typical materials used for the fabric are epoxy-glass (most commonly used), quartz, and Aramid fiber (Kevlar). Typical materials used for resins are epoxies, polyimides, and PTFE. Standard epoxies are the most commonly used resins for low-frequency applications while PTFE is the choice for microwave applications. Standard epoxies provide relatively high dielectric constants ($\varepsilon_r = \sim 4$). To lower the dielectric constant, different resins must be used, such as PTFE, which is the most commonly used resin material for low-loss dielectric boards ($\varepsilon_r = \sim 2$). Note that most resins employed in laminates are thermosetting plastics, which will not remelt upon heating (the opposite is called thermoplastics, which can be melted multiple times). Although thermosetting resins will not remelt, they can soften substantially if the glass transition temperature is exceeded. Glass transition temperature is the temperature where the physical properties (including dielectric constant and CTE) of the resin start to change significantly. When the temperature exceeds the transition temperature, the substrate starts to get significantly softer. An important consequence of glass transition temperature results during drilling of the laminates. If the transition temperature is low enough, then smearing during the drilling process happens because the temperature in the vicinity of drilling area exceeds the transition temperature due to friction. In that case, the drilled holes won't be smooth because of the partially torn and partially cut substrate material. In such situations, further treatment (de-smearing) of the drilled holes might be necessary for proper plating.

After preparing the dielectric slab, copper foils with a predetermined thickness are attached to slab surfaces using appropriate epoxies under high pressure and temperature. Without going into detail, we can indicate that a good epoxy used for this purpose must withstand temperature excursions and chemicals (chiefly used during etching) to prevent delamination of the copper. The amount of copper cladding for PCB materials is usually specified in terms of the weight of the copper per square foot. For instance, 0.5-ounce copper cladding means that the copper thickness on the board will be 0.7 mils. The copper foils can be produced using rolling or electrodepositing. Electrodepositing is the standard technique and is done by electrodepositing the copper on large drums and then stripping the copper off. The bath side of the copper foil obtained through this process has more roughness than the drum side, which makes adhesion to the slab easier. In the other method, the copper is thinned until it reaches to the desired thickness by passing it through successive steel rollers. This method is expensive but results in good smoothness on both sides of the foil. However, the copper foil obtained using this latter technique requires special treatment to facilitate the attachment to the slab surface. Since the attachment is more difficult (due to the relatively smooth surface) and the process is expensive, one may wonder what the benefit of rolled copper is. The chief benefit of rolled copper is the low-insertion loss in microwave printed circuits because of the reduced surface resistance. Note that in electrodeposited copper, the foil is attached in a such a way that the grained surface is on the dielectric side. Since the current density for microstrip lines is higher on the bottom surface of the transmission line, the top, smooth side of the electrodeposited copper has little effect on improving the loss in that case. Therefore, rolled copper, which has smooth surfaces on both sides, is the best choice to reduce conductivity loss in microwave printed circuits.

CTE is a measure of the thermal expansion of the material (given in ppm/$^\circ$C) when it is subject to temperature changes. For laminate materials, it has both in-plane (x-y) and out-of-plane (z) components. Although resin materials expand almost isotropically, the expansion becomes nonisotropic for the resin-fabric combination. This is because the fabric limits the expansion of the resin in the x-y plane. Note that this results in more expansion in the z-direction because the resin expansion is practically incompressible. This is important for the reliability of via-holes.

2.6.2 Ceramics

The second group of materials that is used in microwave printed circuits is ceramics. Aluminum oxide (Al_2O_3), or Alumina, is one of the mostly commonly used ceramic materials for ceramic microwave printed circuits. It has a higher dielectric constant ($\varepsilon_r = \sim 9.6$) than PTFE. The advantages of Alumina are high dimensional resolution and high dielectric constant. Since the dielectric constant is relatively higher, circuits dimensions can be reduced. Besides, ceramic materials are very rigid and therefore can be made extremely planar. One of the distinguishing differences between ceramic and laminate printed circuits is the forming of metallizations. Metallizations are usually formed on the ceramic surface through thin film deposition by evaporation or sputtering (contrary to bonding metal foils on dielectric surfaces as in laminates). Hence the metal thickness can be very small compared to laminate boards, yielding high dimensional resolution. Another unique feature of the ceramic process is that the via-holes are completely filled unlike the plating in the laminate process.

Ceramic materials are also classified into more subgroups. Two main subgroups of ceramic materials are low-temperature cofired (LTCC) and high-temperature cofired (HTCC) ceramics. Cofiring means sintering multiple layers of ceramics at the same time with metallizations. This makes multilayer construction possible. In these groups, the substrate is in green, or soft, form to begin with, unlike the Alumina, which is already sintered before the deposition of metallizations. Then, via-holes and cavities are opened. The next step is to fill the via-holes using the screen-printing method. In screen printing, a stencil, which has exactly the same pattern as the features that need to be filled, is placed on the substrate. Then, a blade pushes a conductive paste through the openings on the stencil, filling the via-holes in the ceramic. Typically, via-holes with diameter of 4 mils can be realized with this approach. After this step, metallization patterns are formed again, using the screen-printing method. If desired, resistor and capacitor pastes can also be applied to create passive components. The whole process is repeated until the multilayer construction is completed. Then, the circuitry is fired to make it rigid. Depending on the type (LTCC or HTCC), the firing temperature differs; therefore different metal types must be used to comply with the temperature elevation. For example, LTCC is sintered below 900°C, which allows cofiring with gold and silver conductors (both of them melt above 900°C). However, in HTCC, the sintering temperature is around 1600°C, which only allows cofiring with metals like tungsten, which melts at 3370°C. Therefore, LTCC technology is very important for microwave printed circuits because it allows the use of high-conductivity metals. It is also extensively used in electronic packaging because it allows forming massive numbers of via-holes and cutouts much more easily than Alumina or laminates. A disadvantage of both HTCC and LTCC processes is that during firing the ceramic material shrinks slightly. However, the amount of shrinkage is controlled and therefore can be accounted for. The tooling costs for LTCC is more expensive than the PCB process. Another important point is that

HTCC has a higher thermal conductivity than the LTCC, which is an important consideration for circuits that dissipate high dc power.

2.7 MANUFACTURING TECHNIQUES FOR PRINTED CIRCUITS

In this section, we will briefly review laminate manufacturing techniques for microwave printed circuit boards. The laminate PCB manufacturing can broadly be classified into two groups: pattern plating and panel plating. In the pattern plating technique, the only areas that are electroplated are the circuit traces. The panel plating, on the other hand, reverses the sequence of plating and imaging. In other words, the whole panel is electroplated prior to imaging. In the pattern plating, the tin-lead plating serves as the etch resist, whereas in the panel plating, the photoresist does the same job. Note that these methods use different polarity circuit film for imaging. In the following paragraphs, we will explain pattern plating in more detail. Interested readers can refer to the literature for more information [62, 63].

A typical pattern-plating PCB manufacturing process is depicted in Figure 2.21. The process starts with a substrate with copper foil on both sides [Figure 2.21(a)] whose thickness and dielectric constant are selected based on the requirements. Typical board thickness for microwave circuits is 5 mils (~125 μm) to 20 mils (~500 μm). Here we will assume that the circuit pattern will be printed only on one side of the circuit board. Although it should be noted that multiple layer boards are possible, having different circuit patterns on each layer, the main manufacturing steps are similar. In the figure shown, top and bottom layers will serve as the signal layer and ground layer of a microstrip circuit, respectively. The next step is to drill the via-holes of the circuit, if there are any [Figure 2.21(b)]. Typically, via-holes serve the basic purpose of reaching to ground in a microstrip circuit. Since drilling simply removes the material, via-holes would be nonconductive without plating.

To plate the holes, a very thin layer (~2.5 μm) of copper is first deposited on the surfaces, including via-holes, using the electroless technique [Figure 2.21(c)]. Electroless deposition is a chemical process, and it is too slow to achieve thicker (~25 μm) layers of depositions under normal conditions. Thus, a secondary deposition through electrolytic means is necessary to achieve the required plating thickness for printed circuits. Note that electroless deposition is required because the electrolytic method needs a conductive path for every surface, which is missing inside the via-hole surfaces in the beginning. In other words, electroless deposition forms the seed layer for the subsequent electrolytic deposition. Note that depending on the application, via-hole plating can be skipped.

After electroless plating, a photoresist layer is deposited [Figure 2.21(d)]. The photoresist is a special chemical sensitive to light that makes photolithography possible. Then, the photoresist is exposed and etched, revealing the positive image of the circuit traces [Figure 2.21(e)]. Note that to expose the resist, the circuit pattern must be previously printed on a transparent medium (e.g., film) through a special equipment called a photo-plotter. Then, the film is positioned on the substrate and everything is exposed to ultraviolet (UV) light. The UV light causes chemical changes in exposed areas of the resist, making it vulnerable to certain chemicals, which will eventually be used to wash away the exposed sections. After imaging and resist etching, the substrate now has the image of the circuit traces.

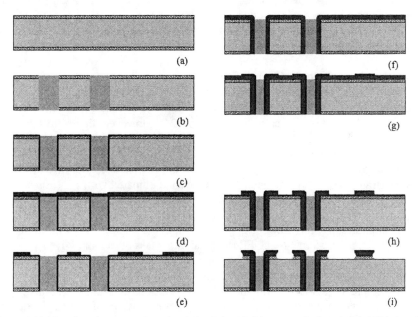

Figure 2.21 Manufacturing steps of a printed circuit board: (a) prepare the board, (b) drill holes, (c) electroless copper plating, (d) apply photoresist, (e) print and develop, (f) electrolytic copper plate, (g) Sn-Pb plate, (h) resist removal, and (i) etch unwanted copper.

At this step, although an image of the traces was transferred to the surface of the copper foil, all traces are still connected with each other because the copper foil is not etched yet. The next step is to electroplate the metallization with copper [Figure 2.21(f)] and then with tin-lead [Figure 2.21(g)]. The copper and tin-lead will deposit on all surfaces not covered by the unetched resist. Note that the copper plating will increase the thickness of metallization in the via-holes as indicated before. The purpose of the tin-lead plating will be clear in a moment. After copper and tin-lead plating, the remaining resist is etched away [Figure 2.21(h)] and the circuit pattern appears as tin-lead plated traces. Now, the top copper cladding of the substrate is still continuous and must be etched properly to isolate the traces. Fortunately, this can be done without etching the trace images because the deposited tin-lead makes traces resistant to etching. Therefore, suitable chemicals are used to etch the unwanted copper, leaving tin-lead covered traces. After etching the copper cladding, the circuit is ready electrically [Figure 2.21(i)]. There will be some additional steps to treat the circuit further (e.g., gold-plating to prevent corrosion and facilitate wire bonding).

As one might notice, the final profiles of the traces look like an aircraft carrier due to chemical etching trying get inside the soft sections of metals [Figure 2.21(i)]. This profile is very typical in PCB manufacturing. This is why the minimum line width that can be realized on printed circuits is a function of metal thickness. If the trace width is too small, then it may not carry itself due to this underetching and may collapse during assembly, which has deleterious effects. As a rule of thumb, the trace widths must be at least four times greater than the amount of underetching.

2.7.1 Multilayer Printed Circuits

Construction of multilayer printed circuits follows a similar procedure to that described for single-layer circuits in Section 2.7. However, there are additional steps to facilitate lamination and blind via-holes (a via-hole is called blind when it does not go through all the layers).

Examples of typical multilayer RF boards are depicted in Figure 2.22. In the first configuration, a high-frequency RF board is laminated with FR4 for physical stiffness and dc function. This is necessary because the laminate materials that are used for RF functions are usually thin and flexible. Therefore, if the RF board will not be supported by a metal housing, then it is a common practice to laminate it with a stronger material (e.g., FR4) containing glass fibers or woven glass, as shown, to make the whole assembly rigid enough. The second configuration shows a four-layer circuit for stripline/microstrip applications. In this configuration, the first two dielectrics are used for the stripline. Note that the first board has copper foils on both sides, but the second board has foil only on one side. The center conductor of stripline is then formed by the bottom foil of the first board. The third board has copper foil again on one side only and can be used for microstrip and/or low-frequency (e.g., biasing) functions.

The material that is used to bond the two boards together is called prepreg, meaning previously impregnated. It has a similar resin to the other boards used in the construction, but it is partially cured. The prepreg is also called B-stage. B-stage is the name of the intermediate stage in the reaction of thermosetting resins in which the material can be remelted and/or dissolved in certain chemicals (the first and last stages are called A-stage and C-stage, in which the material is still fusible and no longer fusible, respectively). Therefore, when the laminate boards are stacked together using prepreg as intermediate layers under elevated temperature and pressure, the prepreg gets cured, bonding the boards together. The electrical properties of prepreg can be important for some applications. For example, in the configuration shown in Figure 2.22(b), the prepreg is, in fact, part of the inner dielectric layers. Therefore, its thickness, permittivity, and loss-tangent should be taken into account during the design. A bad prepreg not only ruins the microwave performance because of poor electrical characteristics, but delamination may occur after environmental extremities. The designer should check with the manufacturing house to use the thinnest and lowest-loss prepregs for stripline circuits where RF fields significantly penetrate into the prepreg. Note that for the configuration given in Figure 2.22(a), the prepreg does not affect the RF performance because the RF ground is placed above it.

Another way of bonding two or more substrates is to use a bond film. The main difference between a bond film and prepreg is that thickness of the bond film is generally in one mil range, whereas prepreg comes in many thicknesses. Bonding films are mostly acrylic materials.

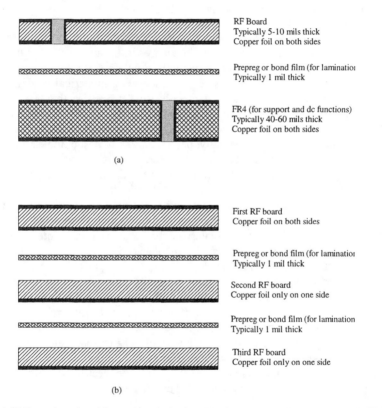

RF Board
Typically 5-10 mils thick
Copper foil on both sides

Prepreg or bond film (for lamination
Typically 1 mil thick

FR4 (for support and dc functions)
Typically 40-60 mils thick
Copper foil on both sides

(a)

First RF board
Copper foil on both sides

Prepreg or bond film (for lamination
Typically 1 mil thick

Second RF board
Copper foil only on one side

Prepreg or bond film (for lamination
Typically 1 mil thick

Third RF board
Copper foil only on one side

(b)

Figure 2.22 Examples of multilayer printed circuits: (a) microstrip circuit supported with FR4 for stiffness and dc functions (shown with blind via-holes), and (b) three-layer printed circuit for stripline/microstrip applications.

In a multilayer construction, if the boards have blind via-holes, then those must be drilled and plated before the lamination. Then, after lamination, through holes (if any) are drilled and plated. Registration between the layers is also important in a multilayer construction.

2.8 MEASUREMENT OF SUBSTRATE MATERIALS

Accurate characterization of electrical properties (e.g., permittivity, electric loss-tangent) of substrate materials is important because they are crucial parameters in the design of microwave and millimeter-wave printed circuits. Although in many cases these parameters are supplied by the manufacturer, the supplied data are usually limited to low frequencies. Thus, for accurate modeling, the designer may need first to characterize the employed substrate material in the required frequency of operation. In this section, we review some of the techniques available to engineers that can be used for this purpose. Note that some of these techniques can also be used to determine the metallization losses of the traces.

Why are we so concerned about the accurate determination of the electrical properties of the substrate materials? The answer is, because we will have to enter those numbers into full-wave electromagnetic simulators. It is important to realize that the full-wave simulations, besides the underlying numerical model, can only be as accurate as the substrate and metallization properties that we have been using. The weakest point of full-wave simulations is perhaps predicting the resulting loss of the microwave circuits. That is why one needs to obtain these parameters, preferably experimentally, before starting to design the circuits.

There are many experimental ways of determining the electrical properties of the substrate materials. Materials can also have a very wide range of electrical and physical properties such as isotropic or nonisotropic, linear or nonlinear, magnetic or nonmagnetic, and so forth. So, depending on the material type, frequency range, and accuracy required, a proper characterization method is selected. One of the important features of substrate materials that are used in printed microwave circuits is that the samples are usually in planar (i.e., slab) form. Therefore, a technique that can give very accurate results, but requires that the material be shaped as a cylindrical rod and placed in a metallic cavity, may not be too suitable for printed circuit substrates. Thus, although there are many different methods that are at our disposal in determining the properties of substrate materials, we will limit our discussion to methods that can be applied to slab forms or, at least, to bricks, which can be shaped relatively easily. Literature references will be provided for other techniques for interested readers.

Before starting our discussion, let's first review the electric-field wave equation in the presence of dielectric, magnetic, and ohmic losses:

$$\nabla^2 \mathbf{E} + \omega^2 \hat{\mu}\left(\hat{\varepsilon} - \frac{j\sigma}{\omega}\right)\mathbf{E} = 0 \qquad (2.76)$$

In the above equation, the quantities represented by the hat symbol $\wedge$ (i.e., permittivity and permeability) are complex quantities in general:

$$\hat{\varepsilon} = \varepsilon' - j\varepsilon'' \qquad (2.77)$$

$$\hat{\mu} = \mu' - j\mu'' \qquad (2.78)$$

Note that the imaginary part of the complex permittivity can be expanded to include the ohmic losses as well. However, it is important to realize that ohmic losses and the imaginary part of the permittivity are, in fact, results of different physical phenomena. The former is merely the result of conduction of free charges in the material, whereas the latter is the result of damping effects on vibrating polar molecules and/or charges that try to align themselves with the alternating electric field. One can define the loss tangent and propagation constant in the presence of these losses as follows [66, 67]:

$$\tan(\delta) = \frac{\varepsilon'' + \sigma/\omega}{\varepsilon'} \qquad (2.79)$$

$$\gamma = \omega \sqrt{\hat{\mu}\left[\varepsilon' - j\left(\varepsilon'' + \frac{\sigma}{\omega}\right)\right]} \qquad (2.80)$$

Dielectrics, by definition, do not have free charges. Therefore ohmic conductivity is zero for dielectrics. In the following discussion, we will also assume nonmagnetic materials (i.e., $\hat{\mu} = \mu_0$). Fortunately, most of the printed circuit board materials (e.g., PTFE, Alumina), electronic packaging materials (e.g., LTCC), thin- and thick-film materials, and GaAs have negligible ohmic conductivities in the microwave and millimeter-wave regions, so the loss-tangent conductivity term can safely be dropped for such materials. They are also nonmagnetic. However, there are some materials (SiGe, for example) that are used in MMIC design that can have a significant substrate conductivity (0.1 to 5 S/m). Therefore, one should be aware that different frequency dependency of loss results in such cases if a significant ohmic conductivity term is included in the imaginary part of the permittivity. What makes things more complicated is the separation of losses due to conductivity and damping of oscillating dipoles from measurement data. This may not be a problem if we are only interested in substrate loss figures at discrete frequencies in a given frequency band. The measured permittivity would contain a real part, which accounts for the dispersion, and an imaginary part, which accounts for the loss, which combines two major loss mechanisms together. The difficulty arises when one wants to separate the functional forms of those loss terms from measurements. Fortunately, in most practical applications, it is sufficient to represent substrate losses by just using the loss-tangent figure. Note that in the foregoing discussion, we have been assuming no conductor losses exist in the measured data. If conductor losses have a significant effect on the employed characterization technique, then they must be taken into account as well.

The measurement techniques for the electrical properties of materials can be broadly classified into four groups [68–74]: transmission line methods [75–84], microwave cavity resonator methods [85–94], open resonator methods [95–97], and free-space methods [98–104]. In the following sections we will go into detail about some of these methods. But, it would be beneficial to give an introduction to each method at this point. Transmission line methods require the sample to be inserted in a transmission line, waveguide or coaxial line completely. Then, from measured circuit parameters (S-parameters, for example) of the modified line, one can extract the complex permittivity and permeability of the sample. Note that these methods can also be applied to microstrip circuits. In microwave cavity resonator methods, the sample is placed in a metallic cavity and permittivity and loss tangent are extracted from the resonance frequency shift and Q of the cavity, respectively, in the presence of the sample. The open resonator methods are very similar to cavity methods in that both are based on the change of resonance behavior of a resonance structure. However, the open resonators can achieve much higher unloaded Q-factors than metallic cavities, thus enabling more precise measurements in principle. For example, Q-factors of TE_{01}-mode cylindrical cavities can reach to 50,000, whereas a Fabry-Perot resonator can achieve a Q of 150,000 in the same frequency range [73]. Unfortunately, open resonator methods also need relatively complicated measurement techniques. The advantage of transmission line methods over the cavity approach is that they can provide broadband data, whereas cavity methods provide data only at the designed resonance frequency. On the other hand, cavity methods (whether they are closed or open) are much more accurate than the transmission line methods in principle due to low conductor losses.

Table 2.5

Comparison of Dielectric Measurement Techniques

Technique	Field	Advantages	Disadvantages	$\Delta\varepsilon_r'$	$\Delta\tan\delta$
Full-sheet resonance	TE_{10}	ε_r'	Poor ε_r''	±2%	
Transmission line	TEM, TE_{10}	Broadband	Machining of the specimen	±1–10%	$\pm 1 \times 10^{-3}$
Capacitor	E field	Normal E field	Air gap between the specimen and contacts	±1%	$\pm 1 \times 10^{-4}$
Cavity	TE_{01}	Very accurate	Narrow bandwidth	±0.2%	$\pm 5 \times 10^{-5}$
Cavity	TM_{0m}	ε_{rz}'	Air gap between the specimen and cavity	±0.5%	$\pm 5 \times 10^{-4}$
Dielectric resonator	TE_{01}	Very accurate	Low-loss region	±0.2%	$\pm 5 \times 10^{-5}$
Open-ended probe	TEM, TM_{01}	Nondestructive	Contact problem with the specimen	±10%	±0.002
Fabry-Perot resonator	TEM	High frequency	Low-loss materials	±2%	$\pm 5 \times 10^{-4}$

Source: [70].

The free-space method that we will briefly mention here is the dispersive Fourier transform spectroscopy (DTFS) technique [104, 105]. This is an interesting technique that combines both the broadband characterization and accuracy into a single method. It is based on measurement of the complex transfer function of a specimen placed in an interferometer using a broadband source. It is regarded as the most accurate method to characterize material properties [104].

However, it is extremely complicated compared to the transmission line or cavity methods. It is also employed at the high end of the millimeter-wave band and near optical frequencies.

A summary of some of the methods used in dielectric measurements is provided in Table 2.5. In the next section, an introduction to the dielectric properties of materials will be given.

2.8.1 Dielectric Properties of Materials

Study of the dielectric properties of materials has been an active field in physics during the last century [67]. We will not attempt to give a detailed treatment of the subject here; only an introduction will be provided. Interested readers can refer to the literature [66, 67].

The effect of an applied electric field on a dielectric is the displacement of charged particles in the material. Under an electric field, the positive and negative charges move in opposite directions to align themselves with the field, creating a polarization in the material. The real part of the permittivity is a measure of this polarization: the larger the real part, the stronger the polarization is. When the applied field is removed, the charges return to their equilibrium states. However, the movements of the charges cannot happen instantaneously. The two main mechanisms that prevent instantaneous movement are the inertia of the particles and the damping effect due to interactions between the charged particles and surrounding matter. The imaginary part of the permittivity measures the net loss due to these two mechanisms (provided that the ohmic conductivity is zero). To model this motion, a relaxation time constant, τ, can be introduced that defines the amount of time required for the charges to return to $1/e$ of their initial states.

This concept was studied in depth by Debye, who first derived the following equations for the real and imaginary part of the permittivity when the material exhibits a single relaxation time constant [67]:

$$\varepsilon' = \varepsilon_\infty + \frac{\varepsilon_s - \varepsilon_\infty}{\left(1 + \omega^2 \tau^2\right)} \tag{2.81}$$

$$\varepsilon'' = \frac{\left(\varepsilon_s - \varepsilon_\infty\right)\omega\tau}{\left(1 + \omega^2 \tau^2\right)} \tag{2.82}$$

where ε_s is the permittivity at dc, ε_∞ is the permittivity when frequency approaches infinity, and τ is the relaxation time constant, all of which are dependent on the material being investigated. Typical values for the relaxation times are in the order of 10^{-11} seconds. The main assumptions that Debye made while deriving the above equations are twofold [67]: first, the polarization is due to dipoles rotating synchronously with the applied field while being dragged by the viscosity of the medium, and, second, the dipoles do not have any inertia. Although they look simple, it was proved that Debye's equations modeled the dielectric phenomenon for most solids and liquids, especially at frequencies lower than the inverse of the relaxation time. A qualitative plot of Debye's equations is given in Figure 2.23.

There are many observations that one can make by inspecting the given plot. First of all, the dielectric loss is not a monotonic function of the frequency. It increases when the frequency is lower than the inverse of the relaxation time. After the relaxation frequency is passed, it starts to decrease again. This behavior can be explained as follows: When the frequency is sufficiently lower, the dipoles have enough time to align themselves with the alternating frequency, thus fully developing the polarization. Therefore, in this region, the increasing frequency also increases the losses due to the friction. However, when the relaxation frequency is passed, the dipoles now start to fail in keeping up with the electric field. Since polarization is not fully developed, the resultant loss also becomes low. The same logic can also be used to explain the drop in the real part of the permittivity.

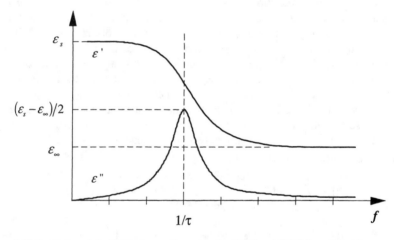

Figure 2.23 Qualitative plot of Debye's equations given by (2.81) and (2.82) for a dielectric.

It is important to note that the high-frequency, asymptotic behavior of Debye's equations is wrong because of the crucial assumption that was made by ignoring the inertias of the dipoles. The practical result of this is the asymptotic limit reached by the attenuation constant as the frequency goes to infinity (note that the attenuation is proportional by $\omega \varepsilon''$). This is called Debye plateau, which would necessitate that all polar liquids be opaque throughout the optical frequencies in contradiction with the measurements [67]. That is why Debye's model, in its original form, has been used mostly at frequencies less than the inverse of the relaxation times. Another discrepancy between real dielectrics and Debye's model is the assumption of a single relaxation frequency. Real systems tend to have a spread of relaxation times, hence smoothing the ideal Lorentzian shape of the ε''.

To correct this, many researchers modified Debye's equations. One of those equations is called the Cole-Cole equation and is given as follows [67]:

$$\varepsilon' - j\varepsilon'' = \varepsilon_\infty + \frac{\varepsilon_s - \varepsilon_\infty}{1 + (j\omega\tau)^{1-\alpha}}, \quad 0 < \alpha < 1 \qquad (2.83)$$

where α is a parameter that should be determined experimentally. The Cole-Cole equation provides a better approximation of the permittivity by adjusting the α parameter.

There are fundamentally four different polarization mechanisms in a dielectric [66]: electronic, atomic, dipole, and space-charge polarizations. In the first mechanism, electrons of an atom are displaced slightly under the applied electric field. When the atoms that form molecules are displaced from their equilibrium positions in the molecule, then the second polarization mechanism takes place. The third mechanism is because of the asymmetric charge distribution in the molecule, which also creates permanent dipole moments (e.g., water molecule). Note that such moments tend to orient themselves with the field direction. All three mechanisms introduced so far are due to bound charges in atoms or molecules. The fourth and last mechanism, on the other hand, occurs because of the small number of free charges, which are injected into the dielectric before they get trapped in the material.

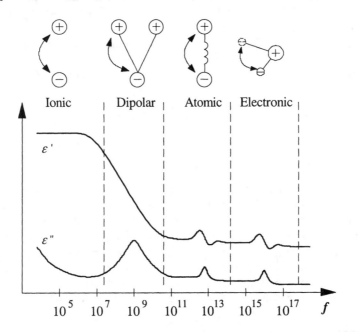

Figure 2.24 Typical multiple relaxation and resonance mechanisms in a dielectric. (*After:* [68].)

A resonant effect is usually related to atomic or electronic polarization, whereas relaxation is related to polarization orientation [68]. As the frequency increases, the slower mechanisms, such as vibration of molecules, drop out, and relatively faster mechanisms start to contribute. A typical plot showing different relaxation and resonance mechanisms is shown in Figure 2.24.

At this point one might ask the following natural question: is it, then, possible to determine the permittivity function of a dielectric by performing dielectric measurements at some discrete frequencies to curve-fit Debye's or Cole-Cole's equations? The answer strongly depends on the material and frequency band. For low-loss and low-permittivity materials that are commonly used in microwave engineering, such as PTFE, fitting to Debye's model with insufficient data points could result in unreal relaxation phenomena. Unfortunately, Debye's theory is a simplification of resonance and relaxation phenomena that can exist in a material and, therefore, one must be extremely careful in extrapolating and/or interpolating (2.81) and (2.82), or (2.83). A better approach would be to express the permittivity in terms of poles and zeros based on causality, which will be demonstrated later.

Kramers-Kronig Relations

So far, we have briefly explained the relaxation and resonance phenomena in dielectrics and introduced Debye's model. One can approach the problem using the causality principle and find that, in fact, ε' and ε'' are related by an integral transformation. The causality principle fundamentally states that causes must always precede their effects. For any causal transfer function, the real and imaginary components are related by the Hilbert transform. Using this fact, one can show that the following relationship exists between ε' and ε'' [67]:

$$\varepsilon'(\omega_0) = \varepsilon_\infty + \frac{2}{\pi} P \int_0^\infty \frac{\omega \varepsilon''(\omega)}{\omega^2 - \omega_0^2} d\omega \qquad (2.84)$$

$$\varepsilon''(\omega_0) = -\frac{2\omega_0}{\pi} P \int_0^\infty \frac{\varepsilon'(\omega)}{\omega^2 - \omega_0^2} d\omega \qquad (2.85)$$

where P denotes the Cauchy principal value, and ω_0 is the frequency of interest. These equations are called Kramers-Kronig relations. This treatment of dielectrics is very beautiful in that if one accurately knows one of the parameters, for example ε', in a extended range of frequencies, then the ε'' is already defined through the Kramers-Kronig relations! However, practical application of (2.85) to microwave engineering might be limited due to the necessity of knowing the real part of the permittivity for an extended range of frequencies. Another immediate consequence of these relations is that, for any lossy dielectric, there must be an associated dispersion (i.e., change of ε' with frequency).

The Kramers-Kronig equations, which establish a relation between the dispersive and absorptive properties of a dielectric medium, are very general. They only depend on the causality and the inertial limitations of the polarization induced at very high frequencies. Note that the inertial limitations of the dielectric at very high frequencies are mathematically expressed by $\varepsilon' = \varepsilon_\infty$ as $f \to \infty$. It should also be stated that causality requires that $\hat{\varepsilon}(\omega)$ be a single valued and zeroless function everywhere in the upper half-plane (i.e., $\mathrm{Im}\{\omega\} > 0$). In the lower half-plane ($\mathrm{Im}\{\omega\} < 0$), on the other hand, the permittivity function can have zeros or poles [74]. In general, the causality demands the following symmetry on the permittivity:

$$\hat{\varepsilon}(-\omega^*) = \varepsilon^*(\omega) \tag{2.86}$$

Therefore, one can express the permittivity function as a ratio of two polynomials as follows [74]:

$$\hat{\varepsilon}(\omega) = \varepsilon_\infty + \frac{(\omega - \zeta_1)(\omega - \zeta_2)...(\omega - \zeta_M)}{(\omega - \pi_1)(\omega - \pi_2)...(\omega - \pi_N)} \tag{2.87}$$

Since $N > M$, we can do a partial fraction expansion of (2.87). By using (2.86) and employing a partial fraction expansion, one can also have an alternative form for the permittivity [74]:

$$\hat{\varepsilon}(\omega) = \varepsilon_\infty + \sum_{k=1}^{K} \frac{A_k}{(\omega - \pi_k)(\omega + \pi_k^*)} \tag{2.88}$$

Note that without making any specific assumptions, we derived a functional form for the permittivity using only causality and the inertia concept. Equation (2.87) or (2.88) can be very convenient in curve-fitting to the measurement data.

2.8.2 Transmission/Reflection Method

Because of its simplicity, the transmission/reflection (TR) method is commonly employed for broadband characterization of dielectrics. In this method, a precisely machined specimen is inserted into a waveguide as depicted in Figure 2.25. One can also use a coaxial line instead of a waveguide. Then, the S-parameters of the waveguide/sample combination are measured using a vector network analyzer (VNA) and the dielectric properties of the sample are extracted from the measurement data. The VNA should be calibrated using a multiline transmission-reflect-line (TRL) method for best accuracy. Note that L_1 and L_2 represent a shift in the reference planes and L_s is the length of the sample.

One of the disadvantages of the TR method is the difficulty of determining the precise location of the sample in the waveguide. That is why L_1 and L_2 are included as parameters in the model shown. However, we will show that L_1 and L_2 can be eliminated through some mathematical manipulations obtaining reference-plane invariant equations. The same measurement accuracy issue is also valid for the sample length, L_s, but, unlike with the reference plane locations, measurement of the sample length can be done quite accurately (± 2.5 μm, typically) by simple means. Another disadvantage of the method is that air gaps might be inadvertently left between the waveguide walls and the sample during placement of the sample. To minimize this, the sample must be machined precisely to fit into the waveguide. As we will show, the effect of the air gaps is more severe for high-permittivity materials.

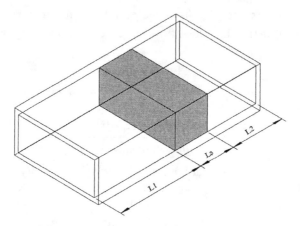

Figure 2.25 Placement of the specimen in a waveguide for the transmission/reflection method.

It should also be noted that the sample length should be different from half of the guided wavelength for optimum accuracy. Finally, the TR method assumes only a single mode of propagation in the waveguide and sample. The main advantages of the TR method are that it is straightforward so that it can be readily employed in a laboratory equipped with a network analyzer, and it provides broadband data.

It can be shown that the scattering matrix of the structure given in Figure 2.25 can be written as follows, assuming a single mode of propagation [79–81]:

$$S_{11} = R_1^2 \left[\frac{\Gamma\left(1 - T^2\right)}{1 - \Gamma^2 T^2} \right] \tag{2.89}$$

$$S_{22} = R_2^2 \left[\frac{\Gamma\left(1 - T^2\right)}{1 - \Gamma^2 T^2} \right] \tag{2.90}$$

$$S_{12} = S_{21} = R_1 R_2 \left[\frac{T\left(1 - \Gamma^2\right)}{1 - \Gamma^2 T^2} \right] \tag{2.91}$$

where

$$R_1 = e^{-\gamma_0 L_1}$$
$$R_2 = e^{-\gamma_0 L_2}$$

$$\Gamma = \frac{\gamma_0/\mu_0 - \gamma/\mu}{\gamma_0/\mu_0 + \gamma/\mu}$$

$$T = e^{-\gamma L_s}$$

with

$$\gamma = \sqrt{\left(\frac{2\pi}{\lambda_c}\right)^2 - \frac{\omega^2 \hat{\mu}_R \hat{\varepsilon}_R}{c^2}} \tag{2.92}$$

$$\gamma_0 = \sqrt{\left(\frac{2\pi}{\lambda_c}\right)^2 - \frac{\omega^2}{c^2}} \tag{2.93}$$

$$\lambda_c = \frac{2\pi}{\sqrt{\left(\frac{m\pi}{a}\right)^2 + \left(\frac{n\pi}{b}\right)^2}} \tag{2.94}$$

where c is the speed of the light in free space, a and b are the dimensions of the waveguide, and $\hat{\mu}_R$ and $\hat{\varepsilon}_R$ are the relative complex permeability and permittivity of the material, respectively. In the following sections, we will assume $\hat{\mu}_R = 1$, which is valid for most of the substrate materials used in microwave printed circuits. The complex propagation constants given above are for a rectangular waveguide. For coaxial lines, they must be modified accordingly. Now, (2.89), (2.90), and (2.91) can be recast as follows:

$$\frac{S_{11}S_{22}}{S_{21}S_{12}} = \frac{\Gamma^2}{\left(1-\Gamma^2\right)^2} \frac{\left(1-T^2\right)^2}{T^2} \tag{2.95}$$

$$S_{21}S_{12} - S_{11}S_{22} = e^{-2\gamma_0 \left(L_{air}-L_s\right)} \frac{T^2 - \Gamma^2}{1 - T^2\Gamma^2} \tag{2.96}$$

where L_{air} is the length of the waveguide. Note that both L_{air} and L_s can be measured accurately, so we will assume they are known quantities. The above equations can be solved for Γ and T as follows:

$$\Gamma = \sqrt{\frac{-b \pm \sqrt{b^2 - 4ac}}{2a}} \tag{2.97}$$

$$T = \sqrt{\frac{B + \Gamma^2}{1 + B\Gamma^2}} \tag{2.98}$$

where

$$a = \frac{AB}{\left(1-B\right)^2} \qquad b = \frac{A\left(B^2 + 1\right)}{\left(1-B\right)^2} - 1 \qquad c = \frac{AB}{\left(1-B\right)^2}$$

$$A = \frac{S_{11}S_{22}}{S_{21}S_{12}}$$

$$B = \frac{S_{21}S_{12} - S_{11}S_{22}}{e^{-2\gamma_0(L_{air}-L_s)}}$$

Note that the sign of the square root in the Γ expression must be selected so that its magnitude will be less than unity. After determining T, the propagation constant can be found from:

$$\gamma = -\frac{1}{L_s}\left[\ln|T| + j(2\pi n + \theta)\right], \quad n = 0,1,2,\ldots \qquad (2.99)$$

where θ is the angle of T. The parameter n is used to select the proper Riemann sheet for the logarithm function. Its value can be determined by comparing the group velocities of measurement and the model [81]. Once γ is determined, the permittivity can be found from (2.92) by assuming $\mu_R = 1$.

As indicated before, air gaps that might exist between the sample and waveguide walls will effect the measured permittivity. A correction for this is given as follows [77]:

$$\varepsilon_r^{(m)} = \frac{\varepsilon_r}{1+(\varepsilon_r - 1)2t/b} \qquad (2.100)$$

where superscript m represents the value determined from measurements, ε_r is the actual relative permittivity, b is the narrow dimension of the waveguide, and t is the gap left between the sample and wide wall of the waveguide (see Figure 2.26).

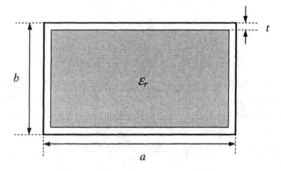

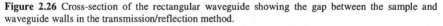

Figure 2.26 Cross-section of the rectangular waveguide showing the gap between the sample and waveguide walls in the transmission/reflection method.

Note that only the gaps left at the top and bottom of the sample are important because the electric field is normal at those interfaces. For small permittivity values, the effect of the gap is usually insignificant, provided that the sample is tightly fit in the waveguide. For example, for a sample placed in a WR-28 waveguide, the error in the permittivity will be less than 3% for $\varepsilon_r < 3$, provided that the gap between the sample and waveguide walls is no more than 1 mil.

Nonlinear Least-Squares Approach

The TR method summarized above is a point-by-point approach, which means that a new complex permittivity has to be computed at every frequency. The procedure is repeated as many times as the number of frequency samples. Then, a curve-fitting technique can be employed to obtain broadband permittivity. In this approach, the main issue that can significantly affect the accuracy of the results is the measurement noise. Because of this, the resulting ε' and ε'' may not satisfy the Kramers-Kronig relations since the procedure doesn't make any assumptions about causality. Therefore, it would be better if we could develop an alternative approach that causes causality through a model and obtains an optimum fit between the model and measurement, using an iterative least-squares algorithm [80]. Note that the procedure must be iterative because the equations connecting permittivity to scattering matrixes are nonlinear.

For this purpose, the Gauss-Newton technique can be used, which is represented by the following iterative equation:

$$\mathbf{x}_1 = \mathbf{x}_0 - \left[\mathbf{J}^T \mathbf{J}\right]^{-1} \mathbf{J}^T \left[\mathbf{f} - \mathbf{F}\right] \tag{2.101}$$

where $\mathbf{x}_0$ is the initial estimate vector of unknowns, $\mathbf{x}_1$ is the updated estimate vector of unknowns, the superscript T is the complex conjugate transpose operator, and $\mathbf{J}$ is the Jacobian matrix. The vectors $\mathbf{f}$ and $\mathbf{F}$ represent the dielectric model and measurement data, respectively. A typical model for the complex relative permittivity can be written as follows [80]:

$$\hat{\varepsilon}_r = \varepsilon_\infty + \frac{(j\omega - z_1)}{(j\omega - p_1)} \cdot \frac{(j\omega - z_2)}{(j\omega - p_2)} \cdots \frac{(j\omega - z_M)}{(j\omega - p_N)} \tag{2.102}$$

where N and M are the number of poles and zeros, respectively. Note that this function for permittivity satisfies the causality, provided that the poles are confined to the left half-plane and are in conjugate pairs if they are complex. In the above equation, one can assume that $\varepsilon_\infty = 1$. Then, the vectors $\mathbf{x}$, $\mathbf{F}$, $\mathbf{f}$, and the Jacobian matrix $\mathbf{J}$ are written as:

$$\mathbf{x} = \begin{bmatrix} z_1 & \cdots & z_M & p_1 & \cdots & p_N \end{bmatrix}^T \tag{2.103}$$

$$F_{m,1} = \left. \frac{S_{21}S_{12} - S_{11}S_{22}}{e^{-2\gamma_0(L_{air} - L_s)}} \right|_{\omega = \omega_m} , \quad m = 1, 2, \ldots, K \tag{2.104}$$

$$f_{m,1} = \left. \frac{T^2(\mathbf{x}_0, \omega) - \Gamma^2(\mathbf{x}_0, \omega)}{1 - T^2(\mathbf{x}_0, \omega)\Gamma^2(\mathbf{x}_0, \omega)} \right|_{\omega = \omega_m} , \quad m = 1, 2, \ldots, K \tag{2.105}$$

$$
\mathbf{J} = \begin{bmatrix}
\dfrac{\partial f(\mathbf{x}_0,\omega_1)}{\partial z_1} & \cdots & \dfrac{\partial f(\mathbf{x}_0,\omega_1)}{\partial z_M} & \dfrac{\partial f(\mathbf{x}_0,\omega_1)}{\partial p_1} & \cdots & \dfrac{\partial f(\mathbf{x}_0,\omega_1)}{\partial p_N} \\[3mm]
\dfrac{\partial f(\mathbf{x}_0,\omega_2)}{\partial z_1} & \cdots & \dfrac{\partial f(\mathbf{x}_0,\omega_2)}{\partial z_M} & \dfrac{\partial f(\mathbf{x}_0,\omega_2)}{\partial p_1} & \cdots & \dfrac{\partial f(\mathbf{x}_0,\omega_2)}{\partial p_N} \\[3mm]
\vdots & \ddots & \vdots & \vdots & \ddots & \vdots \\[3mm]
\dfrac{\partial f(\mathbf{x}_0,\omega_K)}{\partial z_1} & \cdots & \dfrac{\partial f(\mathbf{x}_0,\omega_K)}{\partial z_M} & \dfrac{\partial f(\mathbf{x}_0,\omega_K)}{\partial p_1} & \cdots & \dfrac{\partial f\mathbf{x}_0(x_0,\omega_K)}{\partial p_N}
\end{bmatrix} \quad (2.106)
$$

where K is the number of frequency samples. The derivatives in the Jacobian matrix can be evaluated analytically or numerically. The algorithm starts with an initial vector $\mathbf{x}_0$; then an updated vector $\mathbf{x}_1$ is calculated by the virtue of (2.101). If the difference between $\mathbf{x}_0$ and $\mathbf{x}_1$ is not acceptable, $\mathbf{x}_0$ is replaced by $\mathbf{x}_1$, and a new updated vector, $\mathbf{x}_2$, is calculated. This procedure is repeated until the difference between $\mathbf{x}_n$ and $\mathbf{x}_{n+1}$ is acceptable.

In the above algorithm, we only allowed the zeros and poles of the permittivity function to vary. One could also let the algorithm optimize the lengths of the sample and waveguide, if desired. However, one must be cautious about the nonphysical solution if too many parameters are optimized without proper bounds. In this respect, it is important to force the complex poles, if any, to appear in conjugate pairs in the algorithm. It has been reported in [80] that a couple of poles per frequency decade is sufficient.

Eigenvalue Approach

So far, we have been dealing only with measurements of a single waveguide filled with a specimen and demonstrating different approaches to extract the permittivity of the sample. One might ask if we can gain some additional information by measuring two samples with different lengths instead of one. In fact, an elegant solution is possible through an eigenvalue equation if one utilizes measurements of two samples as depicted in Figure 2.27 [78]. Although it is demonstrated with waveguides here, the method is also applicable to microstrip lines.

To demonstrate the approach, let's assume that the measured transfer matrixes of the samples are given as follows:

$$
\mathbf{M}^{(1)} = \mathbf{X}\,\mathbf{T}^{(1)}\,\mathbf{Y}
$$

$$
\mathbf{M}^{(2)} = \mathbf{X}\,\mathbf{T}^{(2)}\,\mathbf{Y}
$$

where superscripts indicate the sample number, $\mathbf{M}$ is the measurement matrix, and, $\mathbf{X}$ and $\mathbf{Y}$ are the generalized error box models, including waveguide transitions, launchers, and so forth. It is assumed that $\mathbf{X}$ and $\mathbf{Y}$ are the same for each line measurement, which can be achieved if the sample is made flush with the waveguide flanges. Note that although calibration of the network analyzer is not required in principle for the application of this method, it is advised to perform a standard SOLT calibration at the coaxial reference planes to minimize measurement noise.

Then, the transfer matrix of the waveguide filled with specimen can be written as follows:

$$\mathbf{T}^{(i)} = \begin{bmatrix} e^{-\gamma l_i} & 0 \\ 0 & e^{\gamma l_i} \end{bmatrix}, \quad i = 1,2 \qquad (2.107)$$

where l_i is the length of the ith sample. Now, the two measurements can be combined into an eigenvalue equation:

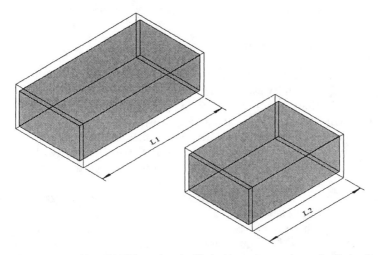

Figure 2.27 Two waveguides with different lengths filled with specimen to be used with the eigenvalue approach (waveguide flanges are not shown).

$$\mathbf{M}\,\mathbf{X} = \mathbf{X}\,\mathbf{T} \qquad (2.108)$$

where

$$\mathbf{M} = \mathbf{M}^{(1)} \left[\mathbf{M}^{(2)} \right]^{-1}$$

$$\mathbf{T} = \mathbf{T}^{(1)} \left[\mathbf{T}^{(2)} \right]^{-1}$$

It is immediately noticed that the matrix equation given by (2.108) is a similarity transformation. Therefore, the eigenvalues of $\mathbf{M}$ must be equal to the eigenvalues of $\mathbf{T}$ provided that $\mathbf{M}$ and $\mathbf{T}$ are invertible:

$$\lambda_{1,2}^{M} = \frac{\mathbf{M}_{11} + \mathbf{M}_{22} \pm \sqrt{(\mathbf{M}_{11} - \mathbf{M}_{22})^2 + 4\mathbf{M}_{12}\mathbf{M}_{21}}}{2} \qquad (2.109)$$

$$\lambda_{1,2}^{T} = e^{\pm \gamma (l_1 - l_2)} \qquad (2.110)$$

Then, the propagation constant is extracted as follows:

$$\gamma = \frac{\ln(\lambda)}{l_1 - l_2} \qquad (2.111)$$

where

$$\lambda = \frac{1}{2}\left(\lambda_1^M + \frac{1}{\lambda_2^M}\right)$$

Note that average of the eigenvalues is taken to minimize the error. Once γ is determined, the permittivity can be found from (2.92) by assuming $\hat{\mu}_R = 1$.

2.8.3 Split-Cylinder Resonator Method

Perhaps the most precise means of determining the complex permittivity of low-loss dielectrics in the microwave and low millimeter-wave region are the resonator methods. Commonly used resonators for this purpose are the dielectric-post resonator, cylindrical cavity resonator, and whispering-cavity resonator [71]. The main difficulty in all those methods, however, is that they need specially machined samples, such as in the shape of a rod, to be inserted into the cavity (these are also called destructive methods). This might be a disadvantage in characterizing the substrate materials because these materials almost always exist in slab form, and it may not be easy to obtain the bulk material for machining. Therefore, a resonator method that can use slab dielectrics as specimens without additional machining would simplify the sample preparation phase extremely. Kent developed a nondestructive cavity technique known as the split-cavity resonator technique, which can characterize the complex permittivity of sheet dielectrics [92, 94]. Later, Janezic and Baker-Jarvis refined the method further [93].

In the split-cavity resonator technique, the sample is placed between two shorted cylindrical waveguides as shown in Figure 2.28. A TE_{011} mode is excited in the sample, and from the measurements of the resonance frequency and quality factor, one can determine the complex permittivity of the sample. One of the assumptions made is that the sample is infinite in radial direction. Therefore, it is important to have the sample and cavity flanges extend as shown in the figure. Since the fields in these radial waveguide sections are rapidly evanescent, it is not necessary that the radius of the samples be excessively large. As a rule of thumb, the sample radius should be approximately 50% greater than the radius of the cavity [93]. Another important assumption is that the TE_{0n} mode is above cutoff in the sample, but it is below cutoff (evanescent) in the cylindrical cavity regions.

The procedure for determining the real part of the permittivity requires the solution of the following resonance equation [93]:

$$|X| = 0 \qquad (2.112)$$

where X is a matrix whose entries are given as follows:

$$X_{mn} = -k_m \cos(k_m L)\delta_{mn} + \sin(k_n L)\int_0^\infty k_s \tan\left(k_s \frac{d}{2}\right)\tilde{R}_m(\zeta)\tilde{R}_n(\zeta)\zeta\, d\zeta \quad (2.113)$$

with

$$k_n^2 = \omega\mu_0\varepsilon_0\varepsilon'_{ra} - h_n^2$$

$$k_s^2 = \omega\mu_0\varepsilon_0\varepsilon'_{rs} - \zeta^2$$

$$h_n = \frac{j_{1,n}}{a}$$

$$\tilde{R}_n(\zeta) = \frac{\sqrt{2}}{\zeta^2 - h_n^2} h_n J_1(\zeta a)$$

$$\delta_{mn} = \begin{cases} 1, & \text{if } m = n \\ 0, & \text{otherwise} \end{cases}$$

In the above expressions, J_1 is the Bessel function of the first kind, and $j_{1,n}$ is the nth zero of J_1. The real part of the permittivity of the material filling the cavity (usually air) and sample are given by ε'_{ra} and ε'_{rs}, respectively. Since the sample thickness d, cavity diameter $2a$, length L, and the resonance frequency ω, are known (from the measurement), the sample permittivity is the only unknown in (2.112).

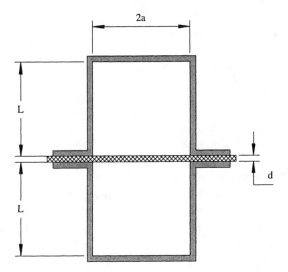

Figure 2.28 Cross-sectional diagram of the split-cylinder resonator that is used in permittivity measurements.

Note that (2.112) cannot be solved explicitly for the permittivity, but a numerical technique, such as Newton-Raphson, could be employed to determine the solution.

Determination of the loss-tangent is done through the Q-factor measurement of the cavity. An expression for this purpose can be written as follows [93]:

$$Q = \omega \frac{W_c + W_s}{P_c + P_l + P_s} \tag{2.114}$$

where W_c and W_s are the average energy stored in the cavity sections and the sample, respectively, and P_c, P_l, and P_s are the power dissipated in the cavity walls, coupling loops, and specimen, respectively. At this point, several approximations are made to simplify (2.114). First, it is assumed that the losses in the coupling loops are insignificant; therefore, P_l is ignored. Second, for the samples under study, the TE_{0n} modes in the cylindrical sections are evanescent; therefore, W_c and P_c are also ignored. This reduces (2.114) to the following form where the loss-tangent is easily extracted from the measured Q-factor:

$$Q \cong \omega \frac{W_s}{P_s} = \frac{1}{\tan(\delta_s)} \tag{2.115}$$

References

[1]   Robert M. Barrett, "Microwave Printed Circuits — The Early Years," *IEEE Trans. on Microwave Theory Tech.*, vol. 32, pp. 983–990, Sept. 1984.

[2]   Robert M. Barrett, "Microwave Printed Circuits — A Historical Survey," *IEEE Trans. on Microwave Theory Tech.*, vol. 3, pp. 1–9, Mar. 1955.

[3]   Harlan Howe, Jr., "Microwave Integrated Circuits — An Historical Perspective," *IEEE Trans. on Microwave Theory Tech.*, vol. 32, pp. 991–996, Sept. 1984.

[4]   Arthur A. Oliner, "Historical Perspectives on Microwave Field Theory," *IEEE Trans. on Microwave Theory Tech.*, vol. 32, pp. 1022–1045, Sept. 1984.

[5]   Edward C. Niehenke, Robert A. Pucel, and Inder J. Bahl, "Microwave and Millimeter-Wave Integrated Circuits," *IEEE Trans. on Microwave Theory Tech.*, vol. 50, pp. 846–857, Mar. 2002.

[6]   M. Arditi, "Characteristics and Applications of Microstrip for Microwave Wiring," *IEEE Trans. on Microwave Theory Tech.*, vol. 3, pp. 31–56, Mar. 1955.

[7]   W. E. Fromm, "Characteristics and Some Applications of Stripline Components," *IEEE Trans. on Microwave Theory Tech.*, vol. 3, pp. 13–20, Mar. 1955.

[8]   Albert D. Frost and Charles R. Mingins, "Microwave Strip Circuits Research at Tufts College," *IEEE Trans. on Microwave Theory Tech.*, vol. 3, pp. 10–12, Mar. 1955.

[9]   Norman R. Wild, "Photo-Etched Microwave Transmission Lines," *IEEE Trans. on Microwave Theory Tech.*, vol. 3, pp. 21–30, Mar. 1955.

[10]   E. H. Bradley and D. R. J. White, "Bandpass Filters Using Strip Line Techniques," *IEEE Trans. on Microwave Theory Tech.*, vol. 3, pp. 163–169, Mar. 1955.

[11] Roland Schinzinger and Patricio A. A. Laura, *Conformal Mapping: Methods and Applications*, New York: Elsevier, 1991.

[12] Kenneth G. Black and Thomas J. Higgins, "Rigorous Determination of the Parameters of Microstrip Transmission Lines," *IEEE Trans. on Microwave Theory Tech.*, vol. 3, pp. 93–113, Mar. 1955.

[13] H. Guckel, "Characteristic Impedances of Generalized Rectangular Transmission Lines," *IEEE Trans. on Microwave Theory Tech.*, vol. 13, pp. 270–274, May 1965.

[14] Thomas G. Bryant and Jerald A. Weiss, "Parameters of Microstrip Transmission Lines and of Coupled Pairs of Microstrip Lines," *IEEE Trans. on Microwave Theory Tech.*, vol. 16, pp. 1021–1027, Dec. 1968.

[15] Harold A. Wheeler, "Transmission-Line Properties of Parallel Strips Separated by a Dielectric Sheet," *IEEE Trans. on Microwave Theory Tech.*, vol. 13, pp. 172–185, Mar. 1965.

[16] Harold A. Wheeler, "Transmission-Line Properties of a Strip on a Dielectric Sheet on a Plane," *IEEE Trans. on Microwave Theory Tech.*, vol. 25, pp. 631–647, Aug. 1977.

[17] Harold A. Wheeler, "Transmission-Line Properties of a Strip Line Between Parallel Planes," *IEEE Trans. on Microwave Theory Tech.*, vol. 26, pp. 866–876, Nov. 1978.

[18] Wolfgang Hilberg, "From Approximations to Exact Relations for Characteristic Impedances," *IEEE Trans. on Microwave Theory Tech.*, vol. 17, pp. 259–265, May 1969.

[19] R. F. Harrington, "Matrix Methods for Field Problems," *IEEE Proceedings*, vol. 55, pp. 136–149, Feb. 1967.

[20] J. R. Mosig, "Arbitrarily Shaped Microstrip Structures and Their Analysis with a Mixed Potential Integral Equation," *IEEE Trans. on Microwave Theory Tech.*, vol. MTT-36, pp. 314–323, Feb. 1988.

[21] David M. Pozar, *Microwave Engineering*, New York: John Wiley and Sons, 1998.

[22] Robert E. Collin, *Foundations for Microwave Engineering*, 2nd ed., IEEE Series on Electromagnetic Wave Theory, 2001.

[23] Robert E. Collin, *Field Theory of Guided Waves*, 2nd ed., IEEE Series on Electromagnetic Wave Theory, New York: IEEE Press, 1991.

[24] K. C. Gupta et al., *Microstrip Lines and Slotlines*, Norwood, MA: Artech House, 1996.

[25] Harlan Howe, Jr., *Stripline Circuit Design*, Dedham MA: Artech House, 1974.

[26] F. Gardiol, *Microstrip Circuits*, New York: John Wiley and Sons, 1994.

[27] R. F. Harrington, *Field Computations by Moment Methods*, New York: Krieger Publishing, 1983.

[28] Frank Olyslager, *Electromagnetic Waveguides and Transmission Lines*, Oxford Science Publications, 1999.

[29] E. H. Fooks and R. A. Zakarevicius, *Microwave Engineering Using Microstrip Circuits*, Prentice Hall Australia, 1990.

[30] Soon Yun Poh, Weng Cho Chew, and Jin Au Kong, "Approximate Formulas for Line Capacitance and Characteristic Impedance of Microstrip Line," *IEEE Trans. on Microwave Theory Tech.*, vol. 29, pp. 135–142, Feb. 1981 (Corrections: Oct. 1981).

[31] K. K. M. Cheng, and J. K. A. Everard, "Accurate Formulas for Efficient Calculation of the Characteristic Impedance of Microstrip Line," *IEEE Trans. on Microwave Theory Tech.*, vol. 39, pp. 1658–1661, Feb. 1991.

[32] E. Hammerstad and O. Jensen, "Accurate Models for Microstrip Computer-Aided Design," *IEEE MTT-S Digest*, vol. 80, pp. 407–409, May 1980.

[33] Robert A. Pucel, D. J. Masse, C. P. Hartwig, "Losses in Microstrip," *IEEE Trans. on Microwave Theory Tech.*, vol. 16, pp. 342–350, June 1968 (Corrections: Dec. 1968).

[34] Edgar J. Denlinger, "Losses of Microstrip Lines," *IEEE Trans. on Microwave Theory Tech.*, vol. 28, pp. 513–522, June 1980.

[35] J. H. C. Van Heuven, "Conduction and Radiation Losses in Microstrip," *IEEE Trans. on Microwave Theory Tech.*, vol. 22, pp. 841–844, Sept. 1974.

[36] George D. Vendelin, "Limitations on Stripline Q," *Microwave Journal*, pp. 63–69, May 1970.

[37] M. E. Goldfarb and A. Platzker, "Losses in GaAs Microstrip," *IEEE Trans. on Microwave Theory Tech.*, vol. 38, pp. 1957–1963, Dec. 1990.

[38] H. A. Wheeler, "Formulas for the Skin Effect," *IRE Proceedings*, vol. 30, pp. 412–424, Sept. 1942.

[39] Mohamad D. Abouzahra, "On the Radiation from Microstrip Discontinuities," *IEEE Trans. on Microwave Theory Tech.*, vol. 29, pp. 666–668, July 1981.

[40] John F. Whitaker et al., "Pulse Dispersion and Shaping in Microstrip Lines," *IEEE Trans. on Microwave Theory Tech.*, vol. 35, pp. 41–47, Jan. 1987.

[41] Douglas G. Corr and J. Brian Davies, "Computer Analysis of the Fundamental and Higher Order Modes in Single and Coupled Microstrip," *IEEE Trans. on Microwave Theory Tech.*, vol. 20, pp. 669–678, Oct. 1972.

[42] William J. Getsinger, "Microstrip Dispersion Model," *IEEE Trans. on Microwave Theory Tech.*, vol. 21, pp. 34–39, Jan. 1973.

[43] R. H. Jansen, "High-Speed Computation of Single and Coupled Microstrip Parameters Including Dispersion, High-Order Modes, Loss and Finite Strip Thickness," *IEEE Trans. on Microwave Theory Tech.*, vol. 26, pp. 75–82, Feb. 1978 (Corrections: Mar. 1978).

[44] Masanori Kobayashi, "A Dispersion Formula Satisfying Recent Requirements in Microstrip CAD," *IEEE Trans. on Microwave Theory Tech.*, vol. 36, pp. 1246–1250, Aug. 1988.

[45] Seymour B. Cohn, "Characteristic Impedance of the Shielded-Strip Transmission Line," *IEEE Trans. on Microwave Theory Tech.*, vol. 3, pp. 52–57, July 1954.

[46] Seymour B. Cohn, "Problems in Strip Transmission Lines," *IEEE Trans. on Microwave Theory Tech.*, vol. 3, pp. 119–126, Mar. 1955.

[47] Stelios Tsitsos, Andrew A. P. Gibson, and Alan H. I. McCormick, "Higher Order Modes in Coupled Striplines: Prediction and Measurement," *IEEE Trans. on Microwave Theory Tech.*, vol. 42, pp. 2071–2077, Nov. 1994.

[48] Claude M. Weil and Lucian Gruner, "High-Order Mode Cutoff in Rectangular Striplines," *IEEE Trans. on Microwave Theory Tech.*, vol. 32, pp. 638–641, June 1984.

[49] Cheng P. Wen, "Coplanar Waveguide: A Surface Strip Transmission Line Suitable for Nonreciprocal Gyromagnetic Device Applications," *IEEE Trans. on Microwave Theory Tech.*, vol. 17, pp. 1087–1090, Dec. 1969.

[50] Giovanni Ghione and Carlu U. Naldi, "Coplanar Waveguide for MMIC Applications: Effect of Upper Shielding, Conductor Backing, Finite-Extend Ground Planes, and Line-to-Line Coupling," *IEEE Trans. on Microwave Theory Tech.*, vol. 35, pp. 260–267, Mar. 1987.

[51] Shao-Jun Fang and Bai-Suo Wang, "Analysis of Asymmetric Coplanar Waveguide with Conductor Backing," *IEEE Trans. on Microwave Theory Tech.*, vol. 47, pp. 238–240, Feb. 1999.

[52] Robert W. Jackson, "Considerations in the Use of Coplanar Waveguide for Millimeter-Wave Integrated Circuits," *IEEE Trans. on Microwave Theory Tech.*, vol. 34, pp. 1450–1456, Dec. 1986.

[53] Ching-Cheng Tien et al., "Transmission Characteristics of Finite-Width Conductor-Backed Coplanar Waveguide," *IEEE Trans. on Microwave Theory Tech.*, vol. 41, pp. 1616–1624, Sept. 1993.

[54] Ke Wu et al., "The Influence of Finite Conductor Thickness and Conductivity on Fundamental and Higher-Order Modes in Miniature Hybrid MIC's (HMIC's) and MMIC's," *IEEE Trans. on Microwave Theory Tech.*, vol. 41, pp. 421–430, Mar. 1993.

[55] E. Hammerstad, "Computer-Aided Design of Microstrip Couplers with Accurate Discontinuity Models," *IEEE MTT-S Digest*, vol. 81, pp. 54–56, June 1981.

[56] Ramesh Garg and I. J. Bahl, "Microstrip Discontinuities," *Int. J. Electronics*, vol. 45, no. 1, pp. 81–87, 1978.

[57] M. Dydyk, "Master the T-Junction and Sharpen Your MIC Designs," *Microwaves*, pp. 184–186, May 1977.

[58] P. Silvester and P. Benedek, "Equivalent Capacitance of Microstrip Open Circuits," *IEEE Trans. on Microwave Theory Tech.*, vol. 20, pp. 511–516, Aug. 1972.

[59] P. Silvester and P. Benedek, "Equivalent Capacitance of Microstrip Gaps and Steps," *IEEE Trans. on Microwave Theory Tech.*, vol. 20, pp. 729–733, Nov. 1972.

[60] Peter Silvester and Peter Benedek, "Microstrip Discontinuity Capacitances for Right-Angle Bends, T-Junctions, and Crossings," *IEEE Trans. on Microwave Theory Tech.*, vol. 21, pp. 341–346, May 1973 (Corrections: May 1975).

[61] Wolfgang Menzel and Ingo Wolff, "A Method for Calculating the Frequency-Dependent Properties of Microstrip Discontinuities," *IEEE Trans. on Microwave Theory Tech.*, vol. 25, pp. 107–112, Feb. 1977.

[62] Charles A. Harper (Ed.), *Electronic Packaging and Interconnection Handbook*, 3rd ed., New York: McGraw-Hill, 2000.

[63] ASM International, *Electronic Materials Handbook Volume 1 — Packaging*, Materials Park, OH: ASM International, 1989.

[64] Thomas S. Laverghetta, *Microwave Materials and Fabrication Techniques*, Norwood, MA: Artech House, 2000.

[65] James W. Lamb, *Miscellaneous Data on Materials for Millimeter and Submillimeter Optics*, 2004.

[66] Arthur von Hippel (Ed.), *Dielectric Materials and Applications*, 2nd ed., Norwood, MA: Artech House, 1995.

[67] G. W. Chantry, "Properties of Dielectric Materials," in *Infrared and Millimeter Waves*, vol. 8, pp. 1–49, New York: Academic Press, 1983.

[68]   Hewlett-Packard Application Note, *Basics of Measuring the Dielectric Properties of Materials*, AN 1217-1, 1992.

[69]   James Baker-Jarvis et al., "Dielectric and Conductor-Loss Characterization and Measurements on Electronics Packaging Materials," *NIST Technical Note*, July 2001.

[70]   James Baker-Jarvis, Bill Riddle, and Michael D. Janezic, "Dielectric and Magnetic Properties of Printed Wiring Boards and Other Substrate Parameters," *NIST Technical Note*, Mar. 1999.

[71]   James Baker-Jarvis et al., "Dielectric Characterization of Low-Loss Materials — A Comparison of Techniques," *IEEE Trans. on Dielectric. Electrical Ins.*, vol. 5, pp. 571–577, Aug. 1998.

[72]   James Baker-Jarvis and Chriss A. Jones, "Dielectric Measurements on Printed Wiring and Circuit Boards, Thin Films, and Substrates: An Overview," *Mat. Res. Soc. Symp. Proceedings*, pp. 153–164, 1995.

[73]   Mohammed Nurul Afsar, James R. Birch, and R. N. Clarke, "The Measurements of the Properties of Materials," IEEE Proceedings, vol. 47, pp. 183–199, Jan. 1986.

[74]   Rodolfo E. Diaz and Nicolaos G. Alexopoulos, "An Analytic Continuation Method for the Analysis and Design of Dispersive Materials," *IEEE Trans. on Antennas Prop.*, vol. 45, pp. 1602–1610, Nov. 1997.

[75]   Michael D. Janezic and Dylan F. Williams, "Permittivity Characterization From Transmission-Line Measurement," *IEEE MTT-S Digest*, pp. 1343–1345, 1997.

[76]   William A. Davis, Charles F. Bunting, and Steven E. Bucca, "Measurement and Analysis for Stripline Material Parameters Using Network Analyzers," *IEEE Trans. on Instrum. Meas.*, vol. 41, pp. 286–290, Apr. 1992.

[77]   K. S. Champlin and G. H. Glover, "Gap Effect in Measurement of Large Permittivities," *IEEE Trans. on Microwave Theory Tech.*, pp. 397–398, Aug. 1966.

[78]   Michael D. Janezic and Jeffey A. Jargon, "Complex Permittivity Determination From Propagation Constant Measurements," *IEEE Microwave Guided Wave Lett.*, pp. 76–78, Feb. 1999.

[79]   James Baker-Jarvis, Eric J. Vanzura, and William A. Kissick, "Improved Technique for Determining Complex Permittivity with the Transmission/Reflection Method," *IEEE Trans. on Microwave Theory Tech.*, vol. 38, pp. 1096–1102, Aug. 1990.

[80]   James Baker-Jarvis, Richard G. Geyer, and Paul D. Domich, "A Nonlinear Least-Squares Solution with Causality Constraints Applied to Transmission Line Permittivity and Permeability Determination," *IEEE Trans. on Instrum. Meas.*, vol. 41, pp. 646–652, Oct. 1992.

[81]   James Baker-Jarvis, "Transmission/Reflection and Short-Circuit Line Permittivity Measurements," *NIST Publication*, July 1990.

[82]   Hiroyuki Tanaka and Fumiaki Okada, "Precise Measurements of Dissipation Factor in Microwave Printed Circuit Boards," *IEEE Trans. on Instrum. Meas.*, vol. 38, pp. 509–514, Apr. 1989.

[83]   Charles B. Sharpe, "A Graphical Method for Measuring Dielectric Constants at Microwave Frequencies," *IRE Trans. on Microwave Theory Tech.*, pp. 155–159, Mar. 1960.

[84]   Nirod K. Das, Susanne M. Voda, and David M. Pozar, "Two Methods for the Measurement of Substrate Dielectric Constant," *IEEE Trans. on Microwave Theory Tech.*, vol. MTT-35, pp. 636–642, July 1987.

[85]   Darko Kajfez and Eugene J. Hwan, "Q-Factor Measurement with Network Analyzer," *IEEE Trans. on Microwave Theory Tech.*, vol. MTT-32, pp. 666–670, July 1984.

[86] Anand Parkash, J. K. Vaid, and Abhai Mansingh, "Measurement of Dielectric Parameters at Microwave Frequencies by Cavity-Perturbation Technique," *IEEE Trans. on Microwave Theory Tech.*, vol. MTT-27, pp. 791–795, Sept. 1979.

[87] Shihe Li, Cevdet Akyel, and Renato G. Bosisio, "Precise Calculations and Measurements on the Complex Dielectric Constant of Lossy Materials Using TM_{010} Cavity Perturbation Techniques," *IEEE Trans. on Microwave Theory Tech.*, vol. MTT-29, pp. 1041–1047, Oct. 1981.

[88] William E. Courtney, "Analysis and Evaluation of a Method of Measuring the Complex Permittivity and Permeability of Microwave Insulators," *IEEE Trans. on Microwave Theory Tech.*, vol. MTT-18, pp. 476–484, Aug. 1970.

[89] Xiaolu Zhao, Ce Liu, and Liang C. Shen, "Numerical Analysis of a TM_{010} Cavity for Dielectric Measurements," *IEEE Trans. on Microwave Theory Tech.*, vol. 40, pp. 1951–1959, Oct. 1992.

[90] John Q. Howell, "A Quick Accurate Method to Measure the Dielectric Constant of Microwave Integrated-Circuit Substrates," *IEEE Trans. on Microwave Theory Tech.*, pp. 142–143, Mar. 1973.

[91] Mohammed Nurul Afsar and Hanyi Ding, "A Novel Open-Resonator System for Precise Measurement of Permittivity and Loss-Tangent," *IEEE Trans. on Instrum. Meas.*, vol. 50, pp. 402–405, Apr. 2001.

[92] Gordon Kent, "An Evanescent-Mode Tester for Ceramic Dielectric Substrates," *IEEE Trans. on Microwave Theory Tech.*, vol. 36, pp. 1451–1454, Oct. 1988.

[93] Michael D. Janezic and James Baker-Jarvis, "Full-Wave Analysis of a Split-Cylinder Resonator for Nondestructive Permittivity Measurements," *IEEE Trans. on Microwave Theory Tech.*, vol. 47, pp. 2014–2020, Oct. 1999.

[94] Gordon Kent, "Nondestructive Permittivity Measurement of Substrates," *IEEE Trans. on Instrum. Meas.*, vol. 45, pp. 102–106, Feb. 1996.

[95] Rajendra K. Arora, Sheel Aditya, and Xinzhong Xu, "Computer-Aided Measurement of Q-Factor with Application to Quasi-Optical Open Resonators," *IEEE Trans. on Instrum. Meas.*, vol. 40, pp. 863–866, Oct. 1991.

[96] Mohammed Nurul Afsar, Xiaohui Li, and Hua Chi, "An Automated 60 GHz Open Resonator System for Precision Dielectric Measurement," *IEEE Trans. on Microwave Theory Tech.*, vol. 38, pp. 1845–1853, Dec. 1990.

[97] B. Komiyama, M. Kiyokawa, and T. Matsui, "Open Resonator for Precision Dielectric Measurements in the 100 GHz Band," *IEEE Trans. on Microwave Theory Tech.*, vol. 39, pp. 1792–1796, Oct. 1991.

[98] Vasundara V. Varadan et al., "Free-Space, Broadband Measurements of High-Temperature, Complex Dielectric Properties at Microwave Frequencies," *IEEE Trans. on Instrum. Meas.*, vol. 40, pp. 842–846, Oct. 1991.

[99] Deepak K. Ghodaonkar, Vasundra V. Varadan, and Vijay K. Varadan, "A Free-Space Method for Measurement of Dielectric Constants and Loss Tangents at Microwave Frequencies," *IEEE Trans. on Instrum. Meas.*, vol. 37, pp. 789–793, June 1989.

[100] Weiming Ou, Gerald Gardner, and Stuart A. Long, "Nondestructive Measurement of a Dielectric Layer Using Surface Electromagnetic Waves," *IEEE Trans. on Microwave Theory Tech.*, vol. MTT-31, pp. 255–261, Mar. 1983.

[101] Mohammed Nurul Afsar and G. W. Chantry, "Precise Dielectric Measurements of Low-Loss Materials at Millimeter and Submillimeter Wavelengths," *IEEE Trans. on Microwave Theory Tech.*, vol. MTT-25, pp. 509–511, June 1977.

[102] Mohammed Nurul Afsar, "Dielectric Measurements of Millimeter-Wave Materials," *IEEE Trans. on Microwave Theory Tech.*, vol. MTT-32, pp. 1598–1609, Dec. 1984.

[103] Mohammed Nurul Afsar, Igor I. Tkachov, and Karen N. Kocharyan, "A Novel W-Band Spectrometer for Dielectric Measurements," *IEEE Trans. on Microwave Theory Tech.*, vol. 48, pp. 2637–2643, Dec. 2000.

[104] Mohammed Nurul Afsar and Kenneth J. Button, "Millimeter-Wave Dielectric Properties of Materials," in *Infrared and Millimeter Waves*, vol. 12, pp. 1–41, New York: Academic Press, 1983.

[105] J. R. Birch and T. J. Parker, "Dispersive Fourier Transform Spectrometry," in *Infrared and Millimeter Waves*, vol. 2, pp. 137–271, New York: Academic Press, 1983.

Chapter 3

Full-Wave Analysis of Printed Circuits

In addition to the more visible advantages of printed structures, which have already been stated in the previous chapter, there is another one that was not so obvious in the early years; accurate and time-efficient designing of such circuits has become possible through the use of full-wave electromagnetic simulation tools. For instance, a microstrip matching network for input and output of a transistor amplifier can be simulated by replacing the active devices with suitable signal terminals. This would provide a very accurate model of the matching circuit. Then, the device model of the transistor is inserted between the terminals in a circuit simulator and overall circuit response of the amplifier is obtained. In essence, this is the approach that is used by a microwave engineer today. One should be able to anticipate the overall response very accurately by using this approach provided that good models for the active circuits are available. Note that all interactions between the matching circuit sections, as well as radiation and other higher-order effects, are automatically taken into account during the simulation process. Hence, use of full-wave electromagnetic simulation tools changed the design methodology in microwave engineering for good. Once regarded as black magic, microwave engineering has been placed in extremely well-defined design principles, if not simplified, by the help of computer simulation tools, which is the promoted philosophy through this book.

Full-wave computer simulations significantly reduce the trial and error in microwave engineering. However, as is always repeatedly indicated, no simulation tool can replace human thought. So, learning the fundamentals is crucial. In this chapter, we will introduce full-wave electromagnetic analysis methods for planar printed circuits.

3.1 REVIEW OF ANALYSIS TECHNIQUES FOR PRINTED CIRCUITS

With the increased use of printed structures in microwave and antenna applications, research for the development of simulation tools for such structures has significantly increased. As a result, plenty of analysis methods have been

211

proposed; some have been tuned for a specific class of printed geometries, some are rigorous enough but not computationally efficient, and others are not rigorous but are very efficient. These methods can be mainly classified into two groups: (1) approximate but numerically efficient methods like quasi-static methods [1–4]; and (2) full-wave methods, which are accurate but computationally expensive methods, such as the method of moments (MoM) [5, 6], the finite-difference time-domain (FDTD) method [7, 8] and the finite element method (FEM) [9]. Note that with the ever-increasing speed of computers, full-wave methods have already become the preferred choice for the analysis of printed structures, not only for researchers in the field but also for practitioners in industry. In addition, among these commonly known and widely used full-wave methods, the MoM has proved to be the most suitable numerical method for the rigorous analysis of printed geometries in multilayer planar environments. Therefore, this section is mainly devoted to the analysis of printed geometries in planar media via the MoM. However, this section would not be complete unless a few words were written for the salient features about the other two very common and popular full-wave methods, namely the FEM and FDTD methods.

The FEM is a general numerical technique to find the approximate solutions to the boundary value problems. In the application of FEM, the entire volume is divided into subvolumes in which the unknown function is represented by simple interpolating functions. Then, a set of algebraic equations for the solution of the system is obtained by applying the Rayleigh-Ritz procedure. The main advantage of the FEM is that it is very versatile and can be applied to almost any geometry. However, the number of unknowns grows very rapidly with the fine features of the geometry, rendering the method computationally inefficient for such structures.

The FDTD method is based on the discretization of Maxwell's equations both in the space and time domains, and the derivatives involved in Maxwell's equations are implemented by finite differencing. The method is relatively easy to implement, is made to converge almost by brute force, and, with enough high-speed computer resources, any electromagnetic (EM) problem can be solved no matter how complicated the geometry is. Moreover, since the FDTD is a time-domain method, contrary to the conventional MoM and FEM, the frequency response of a circuit over a band of frequency can be extracted in one step from the response of the circuit to a narrow Gaussian pulse excitation in the time domain. One of the disadvantages of the FDTD is that analysis of resonance structures can take a relatively long time.

Because the FEM and FDTD methods are quite general and versatile, meaning applicable to any arbitrary geometry, and because the resulting matrix equations are sparse, they have attracted much attention with the development of high-speed computers. From a computational point of view, the sparsity of the resulting matrixes in these methods leads to efficiency in memory usage and in the solution time of the resulting matrix equations. However, both methods have difficulties when open geometries are analyzed, such as radiating structures and unshielded microwave and millimeter-wave circuit components. Since these techniques require discretization of the whole volume of interest, the difficulties arise from the discretization of an infinite region in cases of open structures. In other words, both the FEM and FDTD methods require finite domain to set up a mesh (i.e., the computational space must be of limited size or must be approximated by a limited region by introducing some artificial boundaries, of course without altering the electromagnetic characteristics of the geometry).

After having stated the pros and cons of the FEM and FDTD methods, and before going into the details of the MoM, it is now time to give a brief overview of the MoM as applied to EM problems to justify the ordering of the following subsections. Although the MoM is a numerical technique that can be used to solve differential and integral equations [5], it is mainly used for the solution of integral equations in EM problems. The main motivation for this use comes from the fact that the integral equations formulated for EM problems involve current densities as their unknowns, while the fields are the unknowns for the differential equations. As it is obvious, the fields in an open geometry extend to infinity, whereas the current densities are either bounded in a volume or on a surface; that is, the domain of an integral equation is finite while the domain of a differential equation is infinite. Therefore, as the first step of the MoM, one needs to write the governing equation of the problem at hand as an integral equation. Note that the integral equations for EM problems can be written either in the spatial domain or spectral domain, whose precise definitions will be provided in the following subsections. Since the integral equations involve Green's functions, which are specifically obtained for the underlying media in which a specific geometry is printed, the concept of Green's functions and their derivations in multilayer planar media need to be given in detail. Once the integral equation is set up in terms of known Green's functions and unknown current densities, the MoM procedure can now be applied to this integral equation to solve for the unknown current densities. Consequently, the subsections will follow the steps of the application procedure of the MoM.

Since it is established that the MoM is the preferred choice for the solution of integral equations for printed geometries, at least for the authors of this book, this chapter is completely devoted to the derivation of Green's functions (for fields and potentials) in planar layered media and to the introduction and implementation of the MoM for such structures. Therefore, we start by introducing the general concept of Green's functions in Section 3.2, which also includes the derivation of the Green's function of the scalar wave equation. In addition, in the same section, it is demonstrated that the Green's function of the vector wave equation for the electric field, or for any other field or potential, can be obtained from the Green's function of the scalar wave equation. This section is followed by the introduction of impulse sources in the electromagnetic field in Section 3.3, as they are the sources to be used for the derivation of Green's functions. Moreover, the solution methods of the scalar wave equation with such source terms are reviewed, and, as a result, it is demonstrated that the same solution (spherical wave in nature) can be represented as the integral sum of plane waves or as the integral sum of cylindrical waves. Then, in Section 3.4, it is demonstrated that the vector wave equations for electric and magnetic fields can be reduced down to two scalar equations for a source-free and planar medium, with decoupled solutions referred to as transverse electric (TE) and transverse magnetic (TM). Subsequently, in the same section, by extending these solutions to the cases with point sources in a multilayer environment, Green's functions for fields and potentials are derived. Once the Green's functions are derived, the MoM procedure, which is used for the numerical solution of integral equations, is discussed in Section 3.5.

3.2 GENERAL REVIEW OF GREEN'S FUNCTIONS

The concept of Green's function in electromagnetic problems is equivalent to the impulse response in the circuit and system problems. Let us briefly remember the use of the impulse response in the analysis of linear systems. For a linear, time-invariant system, if one knows the impulse response of the system, the response due to an arbitrary input signal can be simply calculated as the convolution of the impulse response with the input signal. In addition, since convolution integrals become simple multiplications in the frequency domain, the frequency-domain output of a linear time invariant system due to an arbitrary input signal is just the multiplication of the Fourier transforms of the input signal and the impulse response of the system. Of course, this approach makes sense if we can get the impulse response of the system, either in the time domain or frequency domain, analytically. Note that the impulse input signal for circuit and system problems is defined in time domain, $\delta(t)$, as the input signals in such systems are usually functions of time. However, in EM problems, since we are usually interested in the field distribution due to an arbitrary source distribution in space, the impulse source in space will be the source for the Green's function. In other words, since Green's functions in EM problems play the role of the impulse response in system theory, point source, which is an impulse source in space, plays the role of the impulse function in time.

Although the analogy between Green's functions and impulse responses has been established, we cannot utilize the concept of convolution integral to find the field distribution due to an arbitrary source distribution unless the linearity of Maxwell's equations is demonstrated. Since Maxwell's equations consist of curl and divergence operations and time derivatives, all of which are linear operations, the only nonlinearity may arise from the electrical properties of the medium. If the medium is linear, Maxwell's equations satisfy the homogeneity and superposition principles of the linearity in terms of the magnitude and location of the sources. That is, one can mathematically show these as

$$\begin{matrix} \mathbf{J}_1(\mathbf{r}=\mathbf{r}_1) \rightarrow \mathbf{E}_1(\mathbf{r}) \\ \mathbf{J}_2(\mathbf{r}=\mathbf{r}_2) \rightarrow \mathbf{E}_2(\mathbf{r}) \end{matrix} \Rightarrow \alpha\mathbf{J}_1 + \beta\mathbf{J}_2 \rightarrow \alpha\mathbf{E}_1 + \beta\mathbf{E}_2 \qquad (3.1)$$

Now, we are equipped with the tool that would provide us with the field distribution due to an arbitrary source distribution in space via the convolution integral (superposition integral in cases of nonshift invariance), provided that we know the Green's functions.

Although Green's function has been defined as the field distribution due to a point source, it is not clear yet what type of point source and which field distribution it is associated with. The point source could be a scalar or vector quantity, that is, a charge or current source, respectively, and the field could be an electric or magnetic field, or perhaps scalar or vector potential. Note that there is no ambiguity in the choices of the source and field quantity for a Green's function, because a Green's function can only be defined and calculated when a governing differential equation for a problem is obtained. For example, if we are looking for the electric field in a homogenous and isotropic medium due to a given electric current source, $\mathbf{J}(\mathbf{r})$, we can derive the following governing vector wave equation for the electric field from Maxwell's equations as

$$\nabla \times \nabla \times \mathbf{E}(\mathbf{r}) - k^2 \mathbf{E}(\mathbf{r}) = -j\omega\mu\, \mathbf{J}(\mathbf{r}) \tag{3.2}$$

Once the governing equation is obtained as a linear differential equation, one may prefer to solve it for an impulse-like source term on the right-hand side. Then, the actual field due to the actual source can be calculated via the convolution integral. In that case, the Green's function is nothing but the electric field due to a vector point source (electric dipole); that is,

$$\nabla \times \nabla \times \overline{\mathbf{G}}^E(\mathbf{r}) - k^2 \overline{\mathbf{G}}^E(\mathbf{r}) = -\hat{a}\,\delta(\mathbf{r}) \tag{3.3}$$

where $\overline{\mathbf{G}}^E$ $(= G_{xx}^E \hat{x}\hat{x} + G_{yy}^E \hat{y}\hat{y} + G_{zz}^E \hat{z}\hat{z} + G_{xy}^E \hat{x}\hat{y} + G_{yx}^E \hat{y}\hat{x} + \dots)$ is a dyadic Green's function for the electric field, $\hat{\alpha}$ is an arbitrary unit vector for the electric dipole, and $k^2 = \omega^2 \mu\varepsilon$. It is obvious that if one wants to solve (3.2) by first solving (3.3), the solution of (3.3) must be simple enough. Hence, the electric field due to the actual electric current source can be obtained from the superposition integral as

$$\mathbf{E}(\mathbf{r}) = j\omega\mu \int_V \overline{\mathbf{G}}^E(\mathbf{r}, \mathbf{r}') \cdot \mathbf{J}(\mathbf{r}')\, d\mathbf{r}' \tag{3.4}$$

where the convolution operator involves a vector dot product, and the integration is over the source domain, volume in cases of volume current sources and surface in cases of surface current sources. Note that the dyadic form is just a compact way of writing all the possible components of the electric field due to an arbitrarily oriented electric dipole.

From the above exercise, it has been demonstrated that there is no ambiguity in the choices of the point source and the associated field quantity for a given governing equation. However, there is no word yet how to obtain Green's function, as it should be easier when compared to the direct solution of the governing equation. From this point of view, (3.3) does not seem to be any simpler than the original governing equation for the electric field, (3.2). Let us try to simplify the governing equation by the use of a vector identity for the first term on the left-hand side of (3.2) as

$$\nabla \times \nabla \times \mathbf{E}(\mathbf{r}) = \nabla \underbrace{\nabla \cdot \mathbf{E}(\mathbf{r})}_{\dfrac{\rho}{\varepsilon} \; \dfrac{\nabla \cdot \mathbf{J}(\mathbf{r})}{-j\omega\varepsilon}} - \nabla^2 \mathbf{E}(\mathbf{r}) \qquad (3.5)$$

where Gauss's law and the continuity equation are used to get the following wave equation for the electric field:

$$\nabla^2 \mathbf{E}(\mathbf{r}) + k^2 \mathbf{E}(\mathbf{r}) = j\omega\mu \left[\bar{\mathbf{I}} + \frac{\nabla \nabla}{k^2} \right] \cdot \mathbf{J}(\mathbf{r}) \qquad (3.6)$$

where $\bar{\mathbf{I}} = \hat{x}\hat{x} + \hat{y}\hat{y} + \hat{z}\hat{z}$ is called the idem factor or idem dyad. Therefore, the original vector wave equation, (3.2), has been reduced to a more manageable differential equation with a more complex source term on the right-hand side. Hence, the differential equation to be solved for Green's function becomes simpler, as the term on the right-hand side and the electric field vector are replaced by an impulse current source and the dyadic Green's function for the electric field, respectively, in (3.6). It should be noted that Green's functions obtained from (3.3) and (3.6) are different, but the electric fields obtained by the convolutions of these Green's functions with the associated source terms must be the same.

To find Green's function for the Helmholtz equation, (3.6), we should realize that the Helmholtz equation consists of three scalar wave equations,

$$\left(\nabla^2 + k^2 \right)\psi(\mathbf{r}) = 0 \quad \text{where} \quad \psi = E_x, E_y, E_z \qquad (3.7)$$

and its solution would give the components of the electric field. Note that this separation of the vector electric field into three scalar equations in the directions of the coordinate axes can only be done in the Cartesian coordinate system; it cannot be done in the cylindrical or spherical coordinate systems. This is due to the fact that

$$\nabla^2 \mathbf{A} = \hat{x}\nabla^2 A_x + \hat{y}\nabla^2 A_y + \hat{z}\nabla^2 A_z$$

where $\nabla^2 \hat{x} A_x = \hat{x}\nabla^2 A_x$, $\nabla^2 \hat{y} A_y = \hat{y}\nabla^2 A_y$, and $\nabla^2 \hat{z} A_z = \hat{z}\nabla^2 A_z$ are used, and it is not possible in other curvilinear coordinate systems.

3.2.1 Green's Function of Scalar Wave Equation

As was demonstrated in the previous section, Green's function for the electric field can be obtained from Green's function of a scalar wave equation. In addition, it will be shown later that the vector and scalar potentials satisfy the same differential equation as the electric field with different right-hand sides. Therefore, getting Green's function of a scalar wave equation helps to find Green's functions for other field quantities. So, let us write a scalar wave equation with a general right-hand side as

$$\left(\nabla^2 + k^2\right)\psi(\mathbf{r}) = s(\mathbf{r}) \tag{3.8}$$

where $s(\mathbf{r})$ is a source function defined in a volume V'. Remember that if the scalar function $\psi(\mathbf{r})$ is obtained for the point source at $\mathbf{r} = \mathbf{r}'$ [i.e., $s(\mathbf{r}) = -\delta(\mathbf{r} - \mathbf{r}')$], then it is called the Green's function of the scalar wave equation, and denoted by $g(\mathbf{r},\mathbf{r}')$; that is,

$$\left(\nabla^2 + k^2\right)g(\mathbf{r},\mathbf{r}') = -\delta(\mathbf{r} - \mathbf{r}') \tag{3.9}$$

Note that once Green's function is obtained, the scalar function $\psi(\mathbf{r})$, due to the actual source $s(\mathbf{r})$, is calculated by the superposition integral as

$$\psi(\mathbf{r}) = -\int_{V'} d\mathbf{r}'\, g(\mathbf{r},\mathbf{r}')s(\mathbf{r}') \tag{3.10}$$

Now, since the only unknown is the Green's function of the scalar wave equation, let us find it as the solution of (3.9) in an unbounded, homogenous medium, with the point source located at the origin, $\mathbf{r}' = 0$. It is obvious that, with the location of the point source at the origin, the solution should be spherically symmetrical, resulting in a solution independent of θ and ϕ coordinates in the spherical coordinate system. Therefore, it would be more convenient to solve this differential equation in spherical coordinates:

$$\left(\nabla^2 + k^2\right)g(\mathbf{r}) = -\delta(\mathbf{r}) = -\delta(x)\delta(y)\delta(z) \tag{3.11}$$

where the Laplacian operator (∇^2) in spherical coordinates is written as

$$\nabla^2\Phi(r,\theta,\phi) = \frac{1}{r^2}\frac{\partial}{\partial r}\left(r^2\frac{\partial\Phi}{\partial r}\right) + \frac{1}{r^2\sin\theta}\frac{\partial}{\partial\theta}\left(\sin\theta\frac{\partial\Phi}{\partial\theta}\right) + \frac{1}{r^2\sin^2\theta}\frac{\partial^2\Phi}{\partial\phi^2}$$

Due to the spherical symmetry of the source, as mention above, $g(\mathbf{r})$ cannot be the function of θ and ϕ coordinates; therefore, only the first term in the Laplacian operator needs to be retained. Hence, the differential equation for Green's function (3.11) can be written explicitly as

$$\frac{1}{r}\frac{d^2}{dr^2}[r\,g(r)] + k^2 g(r) = -\delta(x)\delta(y)\delta(z)$$

where the following equality is used:

$$\frac{1}{r^2}\frac{\partial}{\partial r}\left(r^2\frac{\partial\Phi}{\partial r}\right) = \frac{1}{r}\frac{d^2}{dr^2}[r\,g(r)]$$

In addition, as is usually the case in the solution of a differential equation, the homogeneous part of (3.11) is solved first, and then the boundary condition at the source is implemented to find the general solution, with, of course, one more boundary condition at infinity. So, the solution of the homogenous part can be obtained as follows:

$$\frac{1}{r}\frac{d^2}{dr^2}[r\,g(r)]+k^2 g(r)=0$$

$$\frac{d^2}{dr^2}[r\,g(r)]+k^2[r\,g(r)]=0$$

$$r\,g(r)=C\,e^{-jkr}+D\,e^{jkr}$$

For $r\neq 0$, the solution becomes

$$g(r)=C\frac{e^{-jkr}}{r}+D\frac{e^{jkr}}{r} \qquad (3.12)$$

where the unknown coefficients C and D can be found by imposing the boundary conditions onto the solution. Since sources are absent at infinity, physical intuition dictates that only an outgoing wave solution can exist, which is one of the boundary conditions, and hence

$$g(r)=C\frac{e^{-jkr}}{r} \qquad (3.13)$$

where the constant C can be obtained from the other boundary condition at the source point. Substituting the solution (3.13) into the scalar wave equation (3.11) and integrating both sides in a small spherical volume around the origin as

$$\int_{\Delta V} dV\left\{(\nabla^2+k^2)C\frac{e^{-jkr}}{r}=-\delta(r)\right\}$$

results in

$$\underbrace{\int_{\Delta V} dV\,\nabla\cdot\nabla C\frac{e^{-jkr}}{r}}_{\text{Use divergence theorem}}\quad+\quad\underbrace{\int_{\Delta V} dV\,k^2\,C\frac{e^{-jkr}}{r}}_{\to 0 \text{ when } \Delta V\to 0 \text{ because } dV=4\pi r^2 dr}\quad=-1$$

The first integral is further simplified by the use of the divergence theorem, while the second integral goes to zero in the limit of the radius of the sphere converging to zero. As a result, the first integral can be reduced to

$$\oint_{\Delta S} ds\,\hat{r}\cdot\nabla C\frac{e^{-jkr}}{r}=-1\Rightarrow 4\pi r^2\frac{d}{dr}C\frac{e^{-jkr}}{r}=-1$$

where ΔS is the surface enclosing the spherical volume ΔV. Hence, the unknown constant C can be obtained by

$$\lim_{r\to 0}\left(-C\,4\pi\,e^{-jkr}\right)=-1\Rightarrow C=\frac{1}{4\pi}$$

Finally, Green's function for the scalar wave equation can be written as

$$g(r) = \frac{e^{-jkr}}{4\pi r} \tag{3.14}$$

or, for a general point source located at $\mathbf{r'}$, it can be written as

$$g(\mathbf{r},\mathbf{r'}) = g(\mathbf{r} - \mathbf{r'}) = \frac{e^{-jk|\mathbf{r}-\mathbf{r'}|}}{4\pi|\mathbf{r} - \mathbf{r'}|} \tag{3.15}$$

This implies that $g(\mathbf{r},\mathbf{r'})$ is translationally invariant for unbounded, homogenous media.

3.2.2 Green's Function of Vector Wave Equation

Once Green's function of the scalar wave equation has been obtained in unbounded and homogenous media, Green's function for the electric field due to a general source term can be obtained as an extension simply from the wave equation (3.6).

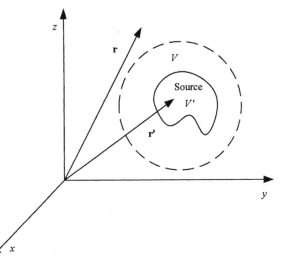

Figure 3.1 A general geometry defining the source and observation in an electromagnetic problem.

Because the vector wave equation has been simplified to the Helmholtz equation with more complex source term, as given in (3.6), and because Green's function of (3.6) for each component of the electric field in Cartesian coordinates is the scalar Green's function in an unbounded medium, the electric field is simply obtained as the convolution of the scalar Green's function and the source term:

$$\mathbf{E}(\mathbf{r}) = -j\omega\mu \int_V d\mathbf{r'}\, g(\mathbf{r} - \mathbf{r'}) \left[\bar{\mathbf{I}} + \frac{\nabla'\nabla'}{k^2} \right] \cdot \mathbf{J}(\mathbf{r'}) \tag{3.16}$$

where

$$\nabla' = \frac{\partial}{\partial x'}\hat{x} + \frac{\partial}{\partial y'}\hat{y} + \frac{\partial}{\partial z'}\hat{z}$$

and the integral is taken over the volume V' where the source is defined (see Figure 3.1). Remember that the goal is to find Green's function of the vector wave equation for the electric field. Therefore, the convolution integral in (3.16) should be rewritten as the convolution integral of the current density and Green's function for the electric field, that is, in the form of

$$\mathbf{E}(\mathbf{r}) = -j\omega\mu\int_{V'} d\mathbf{r}' \overline{\mathbf{G}}^E(\mathbf{r},\mathbf{r}') \cdot \mathbf{J}(\mathbf{r}') \qquad (3.17)$$

where Green's function is written as a dyadic-function, double-vector notation. After some vector manipulations and using some integral identities, Green's function of the vector wave equation for the electric field is obtained in terms of the scalar Green's function as

$$\overline{\mathbf{G}}^E(\mathbf{r},\mathbf{r}') = \left[\overline{\mathbf{I}} + \frac{\nabla\nabla}{k^2}\right]g(\mathbf{r}-\mathbf{r}') \qquad (3.18)$$

in an unbounded, homogenous, and isotropic medium.

Example 3.1

Obtain Green's function for the electric field given in (3.18) from the integral representation (3.16).

Using the geometry provided in Figure 3.1, let us start rewriting (3.16) as

$$\mathbf{E}(\mathbf{r}) = \underbrace{-j\omega\mu\int_{V'} d\mathbf{r}' g(\mathbf{r}-\mathbf{r}')\overline{\mathbf{I}} \cdot \mathbf{J}(\mathbf{r}')}_{\mathbf{I}_1} - \underbrace{j\omega\mu\int_{V'} d\mathbf{r}' g(\mathbf{r}-\mathbf{r}')\frac{\nabla'\nabla'}{k^2} \cdot \mathbf{J}(\mathbf{r}')}_{\mathbf{I}_2}$$

where the first integral provides the idem dyad in (3.18) and hence is already in its final form. However, the del operators operating on the prime coordinates in the second integral need to be moved in front of the scalar Green's function as operating on the unprimed coordinates. Renaming the divergence of the current density as a scalar function $f(\mathbf{r}')$,

$$\mathbf{I}_2 = \frac{1}{k^2}\int_{V'} d\mathbf{r}' g(\mathbf{r}-\mathbf{r}')\nabla'\underbrace{(\nabla' \cdot \mathbf{J}(\mathbf{r}'))}_{f(\mathbf{r}')}$$

and using the product rule of gradient

$$g(\mathbf{r}-\mathbf{r}')\nabla'f(\mathbf{r}') = \nabla'[g(\mathbf{r}-\mathbf{r}')f(\mathbf{r}')] - [\nabla'g(\mathbf{r}-\mathbf{r}')]f(\mathbf{r}')$$

the integral can be cast into the following form:

$$\mathbf{I}_2 = \frac{1}{k^2}\int_{V'} d\mathbf{r}'\nabla'\big[g(\mathbf{r}-\mathbf{r}')f(\mathbf{r}')\big] - \frac{1}{k^2}\int_{V'} d\mathbf{r}'\underbrace{\big[\nabla'g(\mathbf{r}-\mathbf{r}')\big]}_{-\nabla g(\mathbf{r}-\mathbf{r}')}f(\mathbf{r}')$$

Since $\mathbf{J}(\mathbf{r}')$, and hence $f(\mathbf{r}')$, is only defined in V' and is zero everywhere else, the volume can be arbitrarily expanded without altering the value of the integral (see Figure 3.1). Therefore, with the use of the gradient theorem for the first integral, the volume integral can be transformed into a surface integral over the surface S of the expanded volume V, where the current density is zero, and in turn the integral becomes zero:

$$\mathbf{I}_2 = \underbrace{\frac{1}{k^2}\oint_S ds\,\hat{n}\,g(\mathbf{r}-\mathbf{r}')f(\mathbf{r}')}_{=0} + \frac{\nabla}{k^2}\int_V d\mathbf{r}'g(\mathbf{r}-\mathbf{r}')\nabla'\cdot\mathbf{J}(\mathbf{r}')$$

For the second integral, using the product rule of divergence

$$g(\mathbf{r}-\mathbf{r}')\nabla'\cdot\mathbf{J}(\mathbf{r}') = \nabla'\cdot\big[g(\mathbf{r}-\mathbf{r}')\mathbf{J}(\mathbf{r}')\big] - \big[\nabla'g(\mathbf{r}-\mathbf{r}')\big]\cdot\mathbf{J}(\mathbf{r}')$$

and $\nabla'g(\mathbf{r}-\mathbf{r}') = -\nabla g(\mathbf{r}-\mathbf{r}')$, the following integral is obtained:

$$\mathbf{I}_2 = \frac{\nabla}{k^2}\underbrace{\int_V d\mathbf{r}'\nabla'\cdot\big[g(\mathbf{r}-\mathbf{r}')\mathbf{J}(\mathbf{r}')\big]}_{\oint_S ds\,\hat{n}\cdot g(\mathbf{r}-\mathbf{r}')\mathbf{J}(\mathbf{r}')=0} + \frac{\nabla\nabla}{k^2}\cdot\int_V d\mathbf{r}'g(\mathbf{r}-\mathbf{r}')\mathbf{J}(\mathbf{r}')$$

For the first integral, remembering that the volume is an expanded one, with the use of the divergence theorem, the volume integral is converted into a surface integral where the current density is zero. In addition, the order of the two gradient operators and the volume integral is interchanged with the assumption that the volume integral is convergent with a proper differentiability condition. As a result, combining the two terms $\mathbf{I}_1$ and $\mathbf{I}_2$, the electric field can be written as

$$\mathbf{E}(\mathbf{r}) = -j\omega\mu\int_V d\mathbf{r}'\underbrace{\left[\overline{\mathbf{I}}+\frac{\nabla\nabla}{k^2}\right]g(\mathbf{r}-\mathbf{r}')}_{\overline{\mathbf{G}}^E(\mathbf{r},\mathbf{r}')}\cdot\mathbf{J}(\mathbf{r}')$$

Hence, Green's function for the electric field is obtained as (3.18).

Another approach to derive the dyadic Green's function for the electric field is based on using scalar and vector potentials. The importance of this approach stems from the fact that the vector and scalar potentials are defined in the process of derivation, and that the electric field is written in terms of these potentials, resulting in a well-known mixed-potential integral equation (MPIE). Therefore, it would be instructive to give the derivation of the dyadic Green's function for the electric field in a homogenous, isotropic region with a current source $\mathbf{J}(\mathbf{r})$ defined in a volume V'.

Derivation starts with exploiting the divergence-free nature of the magnetic flux density **B**, that is, the law of conservation of magnetic flux (1.20) of the four Maxwell's equations,

$$\nabla \cdot \mathbf{B}(\mathbf{r}) = \nabla \cdot \mu \mathbf{H}(\mathbf{r}) = 0$$

with the help of the vector identity $\nabla \cdot \nabla \times \mathbf{F}(\mathbf{r}) = 0$. Thus, the magnetic flux density can be written as the curl of a vector field,

$$\mu \mathbf{H}(\mathbf{r}) = \nabla \times \mathbf{A}(\mathbf{r}) \qquad (3.19)$$

where $\mathbf{A}(\mathbf{r})$ is introduced as a new vector function, called vector potential hereafter. Note that the vector potential is not uniquely defined yet, because according to the Helmholtz theorem, a vector can be uniquely defined if its curl and divergence are defined. Upon substituting (3.19) into Faraday's law of Maxwell's equations,

$$\nabla \times \mathbf{E}(\mathbf{r}) = -j\omega\mu \mathbf{H}(\mathbf{r}) \qquad (3.20)$$

and rearranging it results in

$$\nabla \times [\mathbf{E}(\mathbf{r}) + j\omega \mathbf{A}(\mathbf{r})] = 0$$

Recognizing that a curl-free vector field can be written as the gradient of a scalar function, due to the vector identity $\nabla \times \mathbf{F}(\mathbf{r}) = 0 \Rightarrow \mathbf{F}(\mathbf{r}) = -\nabla\psi(\mathbf{r})$, $\psi(\mathbf{r})$ being a scalar function, the electric field can be written in terms of these newly introduced vector and scalar functions as

$$\mathbf{E}(\mathbf{r}) = -j\omega \mathbf{A}(\mathbf{r}) - \nabla\phi(\mathbf{r}) \qquad (3.21)$$

The scalar function that was used here, $\phi(\mathbf{r})$, is hereafter called the scalar potential, because in the static case, it reduces down to the electrostatic potential. Therefore, this representation of the electric field, (3.21), is referred to as a mixed-potential representation. Now, the question is how to find the vector and scalar potentials, and what are the governing equations for these potentials. To find the governing equations, substitute (3.19) and (3.21) into the generalized Ampere's law of Maxwell's equations

$$\nabla \times \mathbf{H}(\mathbf{r}) = j\omega\varepsilon\mathbf{E}(\mathbf{r}) + \mathbf{J}(\mathbf{r})$$

to get

$$\underbrace{\nabla \times \nabla \times \mathbf{A}(\mathbf{r})}_{\nabla(\nabla \cdot \mathbf{A}) - \nabla^2 \mathbf{A}} = j\omega\varepsilon\mu[-j\omega\mathbf{A}(\mathbf{r}) - \nabla\phi(\mathbf{r})] + \mu\mathbf{J}(\mathbf{r}) \qquad (3.22)$$

Remember that the curl of the vector potential has been defined, but the divergence has not. In addition, the Helmholtz theorem has been cited, and we have concluded that the vector potential **A** has not been uniquely defined yet. Now, to simplify (3.22), one can define the divergence of **A** as

$$\nabla \cdot \mathbf{A}(\mathbf{r}) = -j\omega\varepsilon\mu\,\phi(\mathbf{r}) \qquad (3.23)$$

which is called the Lorentz gauge, by which the definition of the vector potential has been completed, and the governing equation for the vector potential is obtained as

$$\nabla^2 \mathbf{A}(\mathbf{r}) + k^2 \mathbf{A}(\mathbf{r}) = -\mu \mathbf{J}(\mathbf{r}) \tag{3.24}$$

To find the governing equation for the scalar potential, the only Maxwell's equation not used so far in the derivation, namely Gauss's law, needs to be used as the starting equation:

$$\nabla \cdot \mathbf{D}(\mathbf{r}) = \rho(\mathbf{r}) \Rightarrow \nabla \cdot \mathbf{E}(\mathbf{r}) = \frac{\rho(\mathbf{r})}{\varepsilon} \tag{3.25}$$

Substituting the expression for the electric field in terms of both vector and scalar potentials, (3.21), into (3.25) results in

$$-j\omega \nabla \cdot \mathbf{A}(\mathbf{r}) - \nabla^2 \phi(\mathbf{r}) = \frac{\rho(\mathbf{r})}{\varepsilon} \tag{3.26}$$

and with the use of the Lorentz gauge, the following governing equation for the scalar potential is obtained:

$$\nabla^2 \phi(\mathbf{r}) + k^2 \phi(\mathbf{r}) = -\frac{\rho(\mathbf{r})}{\varepsilon} \tag{3.27}$$

Note that the governing equations for both vector and scalar potentials are in the form of the wave equation, and they can be written as the convolution integral of the scalar Green's function and the source terms at the right-hand sides of the governing equations:

$$\mathbf{A}(\mathbf{r}) = \mu \int_{V'} d\mathbf{r}' \, g(\mathbf{r} - \mathbf{r}') \mathbf{J}(\mathbf{r}') \tag{3.28}$$

$$\phi(\mathbf{r}) = \frac{1}{\varepsilon} \int_{V'} d\mathbf{r}' \, g(\mathbf{r} - \mathbf{r}') \rho(\mathbf{r}') \tag{3.29}$$

Substituting (3.28) and (3.29) into the definition of the electric field in terms of the potentials, (3.21),

$$\mathbf{E}(\mathbf{r}) = -j\omega\mu \int_{V'} d\mathbf{r}' \, g(\mathbf{r} - \mathbf{r}') \mathbf{J}(\mathbf{r}') - \frac{\nabla}{\varepsilon} \int_{V'} d\mathbf{r}' \, g(\mathbf{r} - \mathbf{r}') \rho(\mathbf{r}')$$

and using the continuity equation

$$\rho(\mathbf{r}) = -\frac{\nabla \cdot \mathbf{J}(\mathbf{r})}{j\omega}$$

result in the following expression:

$$\mathbf{E}(\mathbf{r}) = -j\omega\mu \int_{V'} d\mathbf{r}' \, g(\mathbf{r} - \mathbf{r}') \mathbf{J}(\mathbf{r}') + \frac{\nabla}{\varepsilon} \int_{V'} d\mathbf{r}' \, g(\mathbf{r} - \mathbf{r}') \frac{\nabla' \cdot \mathbf{J}(\mathbf{r}')}{j\omega}$$

This is further simplified with the use of the integral identity described in Example 3.1, and the electric field is written as

$$\mathbf{E}(\mathbf{r}) = -j\omega\mu \int_{V'} d\mathbf{r}'\, g(\mathbf{r}-\mathbf{r}')\mathbf{J}(\mathbf{r}') + \frac{\nabla\nabla\cdot}{j\omega\varepsilon} \int_{V'} d\mathbf{r}'\, g(\mathbf{r}-\mathbf{r}')\mathbf{J}(\mathbf{r}')$$

It is a simple matter to cast this into the form of (3.17); hence the dyadic Green's function for the electric field is obtained as

$$\overline{\mathbf{G}}^{E}(\mathbf{r}-\mathbf{r}') = \left[\overline{\mathbf{I}} + \frac{\nabla\nabla}{j\omega\varepsilon}\right] g(\mathbf{r}-\mathbf{r}') \qquad (3.30)$$

Note that the introduction of two gradient operators onto the scalar Green's function $g(\mathbf{r},\mathbf{r}')$ increases the order of the already available singularity of $g(\mathbf{r},\mathbf{r}')$ from $O(1/|\mathbf{r}-\mathbf{r}'|)$ to $O(1/|\mathbf{r}-\mathbf{r}'|^{3})$. In other words, if $\mathbf{r}$, the point where the field $\mathbf{E}(\mathbf{r})$ is observed, is in the source region V', the volume integral in the electric field expression in (3.17) may not converge for some current density distributions. Therefore, if the electric field is to be represented via the superposition integral of the electric field Green's function with the current density, the choice of the mathematical description of current density would be quite critical; that is, it has to be chosen from a class of functions that are smooth enough to make the superposition integral convergent [10].

3.3 POINT SOURCES AND THEIR SPECTRAL REPRESENTATIONS

With the introduction of the concept of Green's functions as the solution of an operator equation due to an impulse source function, we realize that such a source needs to be defined mathematically for electromagnetic problems in different coordinate systems, such as the Cartesian, cylindrical, and spherical coordinate systems. In addition, depending upon the dimensionality of the problem, such sources can take the physical form of line or point sources in EM problems, corresponding to two-dimensional or three-dimensional geometries, respectively. Therefore, in this section, EM sources that could be represented mathematically as impulse functions will be introduced in two- and three-dimensional coordinate systems, namely in Cartesian, cylindrical, and spherical coordinate systems [11]. Then, the solutions of the scalar wave equation due to such sources are obtained in an unbounded-material medium, as they are the basic Green's functions that would provide any field components and potential functions through the superposition integral. Moreover, it will be demonstrated that the solutions of the scalar wave equation can be obtained using different approaches, resulting in different representations for the same solution. It is important to note that these different representations of the solution will provide plane wave representations of the scalar Green's functions due to line and point sources in EM. Therefore, it will be concluded that one can represent the fields and/or potentials due to a finite (in space) source in terms of the integral of plane waves.

3.3.1 Impulse Function Representations of Point Sources

To facilitate the solution of the scalar wave equation due to a point source in different coordinate systems, one needs to review the representation of such sources in different coordinate systems. Therefore, let us start with a line source, as shown in Figure 3.2, which can be considered a point source in two-dimensional space (*xy*-plane). As the source is located at (x',y') and is not a function of *z*, it can be represented mathematically as $\delta(x - x')\delta(y - y')$. If it is assumed to be a line electrical current source, then of course the volume current density is represented as

$$\mathbf{J} = \hat{z} I_0 \delta(x - x')\delta(y - y') \tag{3.31}$$

where I_0 is the constant amplitude in amps, and the unit of current density is amps/m^2, as the unit of each impulse function in space is 1/length.

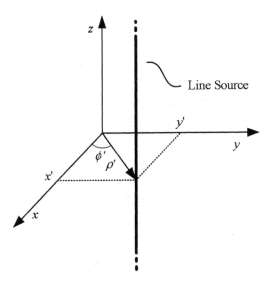

Figure 3.2 A point source in two dimensions, or equivalently, a line source at (x',y').

If this current density needs to be defined in cylindrical coordinates, it is necessary to introduce an additional factor to accommodate the Jacobian resulting from the coordinate transformation; that is,

$$\delta(x - x')\delta(y - y') = f(\rho, \phi)\delta(\rho - \rho')\delta(\phi - \phi')$$

Since the defining feature of an impulse function is its integral over a domain where the impulse exists, the integrals of both sides over the entire xy-plane should provide the same result, which is unity:

$$\int_0^{2\pi}\int_0^\infty f(\rho,\phi)\delta(\rho-\rho')\delta(\phi-\phi')\rho\,d\rho\,d\phi = 1$$

where it is obvious that $f(\rho,\phi)=1/\rho$ is to be selected to make the integral equal to unity. Hence, the current density in (3.40) is represented in cylindrical coordinates as

$$\mathbf{J} = \hat{z}I_0\frac{1}{\rho}\delta(\rho-\rho')\delta(\phi-\phi') \tag{3.32}$$

It is natural to ask if the same representation in cylindrical coordinates would be valid if the line source is positioned at the origin. To answer this, it should be noted that the origin in cylindrical coordinates can only be defined by setting ρ to zero and is independent of ϕ. Therefore, the impulses in Cartesian coordinates can be equivalently written in cylindrical coordinates as

$$\delta(x)\delta(y)= f(\rho)\delta(\rho)$$

and integrating both sides over the entire xy-plane with

$$\int_0^{2\pi}\int_0^\infty f(\rho)\delta(\rho)\rho\,d\rho\,d\phi = 1$$

results in $f(\rho)=1/2\pi\rho$. As a result, the current density of a line source at the origin can be written in cylindrical coordinates as

$$\mathbf{J} = \hat{z}I_0\frac{1}{2\pi\rho}\delta(\rho) \tag{3.33}$$

For the point source in three-dimensional space, apart from the unit vector denoting the polarization, the current density can be written as

$$J = I_0 \cdot l \cdot \delta(\mathbf{r}-\mathbf{r}')= I_0 \cdot l \cdot \delta(x-x')\delta(y-y')\delta(z-z') \tag{3.34}$$

where $I_0 l$ is the current moment of the short dipole with the unit of (amp·m). Since the z-axis is the same for both Cartesian and cylindrical coordinate systems, the representation of such a source in cylindrical coordinates can be inferred from the above discussions of line source. That is, the current density becomes, for a point source at an arbitrary point (x',y',z'),

$$J = \frac{I_0 l}{\rho}\delta(\rho-\rho')\delta(\phi-\phi')\delta(z-z') \tag{3.35}$$

or, for a point source at the origin,

$$J = \frac{I_0 l}{2\pi\rho}\delta(\rho)\delta(z) \tag{3.36}$$

in cylindrical coordinates. To represent the point source (3.34), positioned in an arbitrary point (x', y', z'), in the spherical coordinate system, the impulse functions can be written as

$$\delta(x-x')\delta(y-y')\delta(z-z') = f(r,\theta,\phi)\delta(r-r')\delta(\theta-\theta')\delta(\phi-\phi')$$

where, as before, $f(r,\theta,\phi)$ accounts for the Jacobian in the coordinate transformation from the Cartesian to the spherical coordinate system. Then, integrating both sides over the entire volume provides the following relation:

$$\int_0^{2\pi}\int_0^\pi\int_0^\infty f(r,\theta,\phi)\delta(r-r')\delta(\theta-\theta')\delta(\phi-\phi')r^2\sin\theta\,dr\,d\theta\,d\phi = 1$$

where $f(r,\theta,\phi) = 1/r^2\sin\theta$ satisfies the relation. As a result, the current density can be written in the spherical coordinates as

$$J = \frac{I_0 l}{r^2\sin\theta}\delta(r-r')\delta(\theta-\theta')\delta(\phi-\phi') \tag{3.37}$$

Note that there are two special cases for the location of the point source: on the z-axis ($r' = z$, $\theta = 0$, independent of ϕ) and at the origin ($r' = 0$, independent of θ and ϕ). Following a similar procedure, the current density for the former can be obtained as

$$J = \frac{I_0 l}{2\pi r^2\sin\theta}\delta(r-r')\delta(\theta) \tag{3.38}$$

and, for the latter, as

$$J = \frac{I_0 l}{4\pi r^2}\delta(r) \tag{3.39}$$

3.3.2 Scalar Green's Function for a Line Source

For the sake of illustration, let us consider a line source extending to infinity along z in an unbounded medium and located at the origin on the xy-plane, as shown in Figure 3.3. As the source and geometry are uniform along z, the problem can be formulated as a two-dimensional problem, where fields or potentials due to the line source can vary only in the xy-plane, and the line source appears to be a point source in this plane. Therefore, the line source at the origin is written in Cartesian coordinates as

$$\mathbf{J}(x, y) = \hat{z}I_0\delta(x)\delta(y) \tag{3.40}$$

where I_0 is the amplitude of the current in amps.

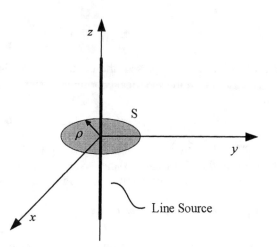

Figure 3.3 A line source at the origin.

To find the fields and potentials due to such sources, it should be remembered that the governing equations for all the fields and potentials are wave equations; that is,

$$\nabla^2 \mathbf{E} + k^2 \mathbf{E} = \left(\bar{\mathbf{I}} + \frac{\nabla \nabla}{k^2} \right) \cdot \mathbf{J} \quad \text{for electric field}$$

$$\nabla^2 \mathbf{H} + k^2 \mathbf{H} = -\nabla \times \mathbf{J} \quad \text{for magnetic field}$$

$$\nabla^2 \mathbf{A} + k^2 \mathbf{A} = -\mu \mathbf{J} \quad \text{for vector potential}$$

$$\nabla^2 \phi + k^2 \phi = -\frac{\rho}{\varepsilon} \quad \text{for scalar potential}$$

Also, these wave equations can be reduced down to scalar wave equations by splitting the vectors into their components in Cartesian coordinates. Therefore, solving the scalar wave equation due to a line source would be sufficient to find the fields and potentials via the superposition integral. As defined earlier, the solution to the scalar wave equation due to a line source is defined as the scalar Green's function, so the wave equation to be solved is written as

$$\nabla^2 g(\mathbf{r}) + k^2 g(\mathbf{r}) = -\delta(x)\delta(y) \tag{3.41}$$

and can be written in cylindrical coordinates as

$$\frac{1}{\rho} \frac{\partial}{\partial \rho} \left[\rho \frac{\partial g(\rho)}{\partial \rho} \right] + k^2 g(\rho) = -\frac{\delta(\rho)}{2\pi\rho} \tag{3.42}$$

where the solution $g(\mathbf{r})$ for a line source at the origin (z and ϕ independent) has to be z and ϕ independent in an unbounded medium, and the point source is written in cylindrical coordinates as in (3.33).

Once the governing equation for the problem is stated, its solution can be obtained using different approaches. In this section, the main purpose is to find the solution of the governing equation, (3.41) or (3.42), as well as to review the three methods used to find the solution. Therefore, through the introduction of these methods, equivalent but different representations of the same scalar Green's function are obtained.

Direct Solution in the Spatial Domain

It is recognized that the partial differential equation in (3.42) is Bessel's equation of order zero, and its homogenous solution can be written as

$$g(\rho) = A H_0^{(2)}(k\rho) + B H_0^{(1)}(k\rho) \qquad (3.43)$$

where $H_0^{(1)}$ and $H_0^{(2)}$ are the first and second kinds of Hankel functions of zero order, respectively. Remembering from Section 3.2.1 (i.e., from the solution of the scalar wave equation due to a point source in three-dimensional space), one can eliminate one of the terms in (3.43) by using the fact that the source is at the origin and the wave generated by the source propagates in an outward direction only. This condition is known as the radiation condition, and it is equivalent to imposing a boundary condition that the scalar Green's function must be zero at infinity. In addition to this, one must impose the effect of the source term in (3.42) onto the homogenous solution to get a unique solution for the scalar Green's function, as demonstrated in Section 3.2.1 for a point source. Therefore, it is necessary to examine the Hankel functions of zero order for small and large arguments; that is, as $k\rho \to 0$ and $k\rho \to \infty$, respectively:

$$H_0^{(1)}(k\rho) \approx \frac{j2}{\pi} \ln(k\rho) \quad \text{for } k\rho \to 0 \qquad (3.44)$$

$$H_0^{(2)}(k\rho) \approx -\frac{j2}{\pi} \ln(k\rho) \quad \text{for } k\rho \to 0 \qquad (3.45)$$

$$H_0^{(1)}(k\rho) \approx \sqrt{\frac{2}{j\pi k \rho}} e^{jk\rho} \quad \text{for } k\rho \to \infty \qquad (3.46)$$

$$H_0^{(2)}(k\rho) \approx \sqrt{\frac{2}{j\pi k \rho}} e^{-jk\rho} \quad \text{for } k\rho \to \infty \qquad (3.47)$$

Since the $\exp(j\omega t)$ time dependence used throughout the book for the time-harmonic representation of fields and potentials, the large argument limits of the two solutions $H_0^{(1)}$ and $H_0^{(2)}$ represent incoming and outgoing cylindrical waves, respectively. Therefore, the coefficient B in (3.43) must be zero to allow the solution to be an outgoing wave only, because the line source is at the origin in the xy-plane and because there is no source at infinity. With the implementation of the source condition, as is detailed in Example 3.3, the scalar Green's function for the scalar wave equation due to a line source can be uniquely obtained as

$$g(\rho) = \frac{1}{j4} H_0^{(2)}(k\rho)$$

(3.48)

where the unknown coefficient A was found to be $1/j4$ from the boundary condition at the source; that is, as $\rho \to 0$.

Example 3.2

Find the governing equation for the electric field due to a line source in an unbounded medium (see Figure 3.3) in cylindrical coordinates.

Because the current source is independent of ϕ and z, and because there is no boundary that would give rise to ϕ- and z-dependent field behavior,

$$\frac{\partial}{\partial \phi} = \frac{\partial}{\partial z} = 0$$

(3.49)

can be imposed on Maxwell's equations. Therefore, writing the curl operator in the generalized Faraday's law and Ampere's circuital law in cylindrical coordinates, as well as imposing (3.49), result in

$$-\hat{\phi} \frac{\partial E_z}{\partial \rho} + \hat{z} \frac{1}{\rho} \frac{\partial (\rho E_\phi)}{\partial \rho} = -j\omega \mu \mathbf{H}$$

(3.50)

$$-\hat{\phi} \frac{\partial H_z}{\partial \rho} + \hat{z} \frac{1}{\rho} \frac{\partial (\rho H_\phi)}{\partial \rho} = j\omega \varepsilon \mathbf{E} + \mathbf{J}$$

(3.51)

from which $E_\rho = H_\rho = 0$ can be easily deduced. In addition, equating the same vector components on both sides of (3.50) and (3.51) yields the following two sets of coupled differential equations:

$$\frac{\partial H_z}{\partial \rho} = -j\omega \varepsilon E_\phi; \qquad \frac{1}{\rho} \frac{\partial (\rho E_\phi)}{\partial \rho} = -j\omega \mu H_z$$

(3.52)

$$\frac{\partial E_z}{\partial \rho} = j\omega \mu H_\phi; \qquad \frac{1}{\rho} \frac{\partial (\rho H_\phi)}{\partial \rho} = j\omega \varepsilon E_z + J_z$$

(3.53)

where these sets are decoupled. With a little study, it can be observed that there is no source term in the first set, (3.52), and therefore (E_ϕ, H_z) field components must be zero. However, from the second set, (3.53), one can easily obtain the following wave equation for E_z:

$$\frac{1}{\rho} \frac{\partial}{\partial \rho} \left(\rho \frac{\partial E_z}{\partial \rho} \right) + k^2 E_z = j\omega \mu J_z$$

(3.54)

where the first equation of (3.53) is multiplied by ρ, then differentiated with respect to ρ, and then substituted into the second equation. Hence, substituting the mathematical description of the line source in cylindrical coordinates into (3.54) provides the governing equation for the electric field in cylindrical coordinates as follows:

$$\frac{1}{\rho}\frac{\partial}{\partial\rho}\left(\rho\frac{\partial E_z}{\partial\rho}\right) + k^2 E_z = j\omega\mu\frac{I_0\delta(\rho)}{2\pi\rho} \tag{3.55}$$

Example 3.3

Show that the coefficient of Green's function of a scalar wave equation for a line source [A in (3.43)] is equal to $1/j4$.

As was done for a point source in three-dimensional space (see Section 3.2.1) the coefficient A in (3.43) can be obtained by imposing a condition at the source. The general solution of the scalar wave equation due to a line source was obtained as

$$g(\rho) = A H_0^{(2)}(k\rho)$$

The source condition can be implemented by substituting this solution into the scalar wave equation for Green's function (3.42) as

$$\left(\nabla_t^2 + k^2\right)A H_0^{(2)}(k\rho) = -\frac{\delta(\rho)}{2\pi\rho}$$

where ∇_t^2 is used in place of its representation in the cylindrical coordinates in (3.42). Rewriting the operator ∇_t^2 as the divergence of a gradient,

$$A\nabla_t \cdot \nabla_t H_0^{(2)}(k\rho) + A k^2 H_0^{(2)}(k\rho) = -\frac{\delta(\rho)}{2\pi\rho}$$

and integrating over the surface of a circle around the origin (Figure 3.3) as

$$A\int_S \nabla_t \cdot \nabla_t H_0^{(2)}(k\rho)ds + A k^2 \int_S H_0^{(2)}(k\rho)ds = -\int_S \frac{\delta(\rho)}{2\pi\rho}\rho\,d\rho\,d\phi$$

result in

$$A\int_S \nabla_t H_0^{(2)}(k\rho)\cdot\hat{\rho}\,\rho\,d\phi + A k^2 \int_S H_0^{(2)}(k\rho)2\pi\,\rho\,d\rho = -1$$

where the divergence theorem for two-dimensional space is employed for the first integral. Since the line source is a point source defined at the origin in the xy-plane, the above equation holds for any surface including the origin, such as a disc, as shown in Figure 3.3. Therefore, to get closed-form expressions for the integrals involved, one can use the limiting surface of integration as ρ goes to zero. So, remembering the limiting value of the Hankel function of the second kind, (3.45), and stating the derivative of it with respect to its argument as

$$\frac{\partial}{\partial \rho} H_0^{(2)}(k\rho) = -kH_1^{(2)}(k\rho)$$

(3.56)

results in

$$\lim_{\rho \to 0} \left\{ - A2\pi\rho \, kH_1^{(2)}(k\rho) - Ak^2 \, j4 \int_S \ln(k\rho)\rho \, d\rho \right\} = -1$$

Hence, the coefficient A is obtained as $1/j4$ by using the limiting value of $H_0^{(2)}$ for the small argument

$$H_1^{(2)}(k\rho) \approx j\frac{1}{\pi}\left(\frac{2}{k\rho}\right)$$

(3.57)

and by setting the second term to zero as $\lim_{x \to 0} x \ln(x) = 0$.

Fourier Transform Technique

An alternative approach for the solution of the scalar wave equation with a line source, (3.41), originates from the assumption that the Fourier transform of the unknown function of the differential equation exists. Hence, the Fourier-transform pair can be employed in the solution of the differential equation.

Before getting into the details of the method, let's rewrite the differential equation to be solved as

$$\left(\frac{\partial^2}{\partial x^2} + \frac{\partial^2}{\partial y^2} + k^2\right) g(x,y) = -\delta(x)\delta(y)$$

(3.58)

where $\partial/\partial z = 0$ is assumed as before, because the line source is independent of z and so are the fields generated by such a source in an unbounded medium. The method starts with the assumption that the Fourier transform of the scalar Green's function $g(x, y)$ exists, with the following definition of the Fourier-transform pair:

$$\tilde{G}(k_x, k_y) = \int_{-\infty}^{\infty}\int_{-\infty}^{\infty} dx \, dy \, g(x,y) e^{jk_x x + jk_y y}$$

(3.59)

$$g(x,y) = \frac{1}{(2\pi)^2} \int_{-\infty}^{\infty}\int_{-\infty}^{\infty} dk_x \, dk_y \, \tilde{G}(k_x, k_y) e^{-jk_x x - jk_y y}$$

(3.60)

Note that $\tilde{G}(k_x, k_y)$ is also referred to as the spectral representation of $g(x, y)$ and that the Fourier transform is written between the space variables x, y and the propagation constants (k_x, k_y). It should be clarified here that the word "spectrum" does not only mean the frequency content of a time signal as is often the case in communications, but also refers to the plane wave spectrum. In other words, if the inverse Fourier transform (3.60) is examined a little, it may be interpreted that $g(x, y)$ is written as the integral sum of the plane waves with amplitudes and phase propagation denoted by $\tilde{G}(k_x, k_y)$ and the exponential term, respectively. Therefore, the spectrum in this context means the collection of plane waves propagating in different directions denoted by the pair (k_x, k_y).

Since the scalar Green's function in the spatial domain has been written as the integral of its spectral-domain counterpart as in (3.60), with explicit and differentiable x and y dependence, substituting this into the differential equation (3.58) facilitates the analytical implementation of the derivatives. However, to eliminate the resulting integrals on the left-hand side, one may use the spectral representation of the source term,

$$\delta(x)\delta(y) = \frac{1}{(2\pi)^2} \int\limits_{-\infty}^{\infty}\int\limits_{-\infty}^{\infty} dk_x dk_y\, e^{-j(k_x x + k_y y)} \tag{3.61}$$

After having introduced the spectral representation of $g(x, y)$ into the scalar wave equation, the order of integration and differentiation are interchanged because $\tilde{G}(k_x, k_y)$ exists, and is absolutely convergent. Hence, the differential equation (3.58) can be transformed to

$$\int\limits_{-\infty}^{\infty}\int\limits_{-\infty}^{\infty} dk_x dk_y\, \left(k^2 - k_x^2 - k_y^2\right)\tilde{G}(k_x, k_y) e^{-jk_x x - jk_y y} = -\int\limits_{-\infty}^{\infty}\int\limits_{-\infty}^{\infty} dk_x dk_y\, e^{-jk_x x - jk_y y}$$

Since this must be true for every x and y, the integrands of both sides must be equal to each other; that is, the spectral-domain representation of Green's function can be obtained as

$$\tilde{G}(k_x, k_y) = -\frac{1}{\left(k^2 - k_x^2 - k_y^2\right)} \tag{3.62}$$

and, consequently, the scalar Green's function for a line source can be obtained from the inverse Fourier transform as

$$g(x, y) = \frac{-1}{(2\pi)^2} \int\limits_{-\infty}^{\infty}\int\limits_{-\infty}^{\infty} dk_x\, dk_y\, \frac{1}{\left(k^2 - k_x^2 - k_y^2\right)} e^{-jk_x x - jk_y y} \tag{3.63}$$

This is the spectral-domain representation of the scalar Green's function for a line source, which is an alternative representation to the one obtained in (3.48). Since the solution of the scalar wave equation with a boundary condition at infinity and a source condition is unique, these two representations, (3.48) and (3.63), are equivalent, resulting in the following identity:

$$H_0^{(2)}(k\rho) = \frac{1}{j\pi^2} \int\limits_{-\infty}^{\infty}\int\limits_{-\infty}^{\infty} dk_x\, dk_y\, \frac{1}{\left(k^2 - k_x^2 - k_y^2\right)} e^{-jk_x x - jk_y y} \tag{3.64}$$

where $\rho = \sqrt{x^2 + y^2}$ and k is the wave number of the medium. Note that this representation of Green's function is nothing but the plane wave expansion of an otherwise cylindrical wave. One can argue that the nature of the waves generated by an infinite line source has to be cylindrically symmetrical and expanding cylindrically outward from the location of the infinite extend source, with a uniform amplitude along the axis of the source. However, for some problems, although it might not be the natural mode of propagation for such a source, it can be beneficial to expand a cylindrical wave in terms of plane waves. In such circumstances, this is the representation of Green's function one can employ with ease.

Hybrid Method

The third alternative for the solution of the scalar wave equation is to use both the Fourier transform method (in one dimension only) and the direct solution of a differential equation. The approach starts with assuming that a one-dimensional Fourier transform of the scalar Green's function exists; that is, one of the space variables, say y, is fixed, while the Fourier transformation is performed over the other variable. So, defining the Fourier-transform pair as

$$\tilde{G}(k_x, y) = \int\limits_{-\infty}^{\infty} dx\, g(x,y) e^{jk_x x} \tag{3.65}$$

$$g(x,y) = \frac{1}{2\pi} \int\limits_{-\infty}^{\infty} dk_x\, \tilde{G}(k_x, y) e^{-jk_x x} \tag{3.66}$$

and the impulse function as

$$\delta(x) = \frac{1}{2\pi} \int\limits_{-\infty}^{\infty} dk_x\, e^{-jk_x x} \tag{3.67}$$

facilitates the use of the Fourier-transform technique in one dimension only. Substituting (3.66) and (3.67) into the governing equation (3.58) and performing an interchange of the order of differentiation and integration result in

$$\int\limits_{-\infty}^{\infty} dk_x \left(\frac{\partial^2}{\partial y^2} + k^2 - k_x^2\right)\tilde{G}(k_x, y) e^{-jk_x x} = -\delta(y)\int\limits_{-\infty}^{\infty} dk_x\, e^{-jk_x x}$$

As this equality must be true for every x, the integrands of both sides must be equal, resulting in the following differential equation:

$$\left(\frac{\partial^2}{\partial y^2} + k^2 - k_x^2\right)\tilde{G}(k_x, y) = -\delta(y) \tag{3.68}$$

where $k^2 - k_x^2 = k_y^2$ is used hereafter. Since the goal is to find the spectral representation of the scalar Green's function, and in turn Green's function itself via the inverse Fourier transformation, this second-order, ordinary, inhomogeneous differential equation needs to be solved, of course with a boundary condition $\tilde{G}(k_x, y \to \pm\infty) = 0$ and the source condition at $y = 0$. As is often the approach for the solution of an inhomogeneous differential equation, the homogeneous part is first solved with the boundary conditions, resulting in

$$\tilde{G}(k_x, y) = \begin{cases} A e^{-jk_y y} & y > 0 \\ B e^{jk_y y} & y < 0 \end{cases} \tag{3.69}$$

where the imaginary part of k_y must be negative to satisfy the boundary condition at infinity. Investigating this solution with the differential equation (3.68), one can observe that the second derivative with respect to y can only result in the impulse function on the right-hand side. Consequently, the first derivative of $\tilde{G}(k_x, y)$ with respect to y needs to be discontinuous at $y = 0$ by the amount of the constant before the impulse function, and $\tilde{G}(k_x, y)$ itself must be continuous at $y = 0$. Putting these into mathematical terms as

$$\frac{\partial \tilde{G}(k_x, y = 0^+)}{\partial y} - \frac{\partial \tilde{G}(k_x, y = 0^-)}{\partial y} = -1 \tag{3.70}$$

$$\tilde{G}(k_x, y = 0^+) = \tilde{G}(k_x, y = 0^-) \tag{3.71}$$

and implementing the boundary conditions on the homogeneous solution result in

$$\tilde{G}(k_x, y) = \frac{1}{j2k_y} e^{-jk_y|y|} \tag{3.72}$$

Hence, the scalar Green's function can be obtained from the inverse Fourier transform of its spectral representation as

$$g(x, y) = \frac{1}{2\pi} \int_{-\infty}^{\infty} dk_x \frac{e^{-jk_x x - jk_y|y|}}{j2k_y} \tag{3.73}$$

Using again the argument of uniqueness for the solution of the scalar wave equation with boundary conditions, this representation must be equivalent to the previous two representations; that is, (3.73) must be equal to (3.63) and (3.48), which results in the following integral identity of the Hankel function:

$$H_0^{(2)}(k\rho) = \frac{1}{\pi} \int_{-\infty}^{\infty} dk_x \frac{e^{-jk_x x - jk_y|y|}}{k_y} \tag{3.74}$$

where $k^2 = k_x^2 + k_y^2$. Through the comparison of two identities of the Hankel function, namely (3.64) and (3.74), it can be deduced that (3.74) can be obtained by integrating (3.64) over k_y, using the residue integral method.

Existence of Fourier Transform

Before completing this section, it may be worth questioning one of the central assumptions of the Fourier-transform technique and the hybrid method, the existence of the Fourier transform of the scalar Green's function [12]. To answer the question of the existence of the Fourier transform for a given function $f(x)$, one might use the following simple and not quite complete, but successful, lemma [13]: Suppose f is Riemann integrable on every interval $[a, b]$ and absolutely integrable, that is,

$$\int_{-\infty}^{\infty} |f(x)| dx < \infty \qquad (3.75)$$

Then,

$$F(k_x) = \int_{-\infty}^{\infty} f(x) e^{jk_x x} dx \qquad (3.76)$$

is well defined for all $k_x \in \mathbf{R}$. Since the integrand in (3.76) is less than or equal to the absolute value of the function, $|f(x)|$, if the integral exists for the absolute value of the function, that is, if $f(x)$ is absolutely integrable, then the Fourier transform integral (3.76) is guaranteed to converge. However, we should note that these conditions stated in the above lemma do not guarantee the absolute convergence of $F(k_x)$, which is the reason for the noncompleteness of the above conditions. To give an example, taken from [13], consider $f(x) = 1$ for $x \in [-1, 1]$ and $f(x) = 0$ elsewhere. It is obvious that the conditions of the lemma are satisfied, but $F(k_x) = 2 \sin k_x / k_x$ is not absolutely convergent. Once the existence of the Fourier transform is established, we can go back to the original question, the existence of the Fourier transform of the scalar Green's function, which is the Hankel function of zero order, second kind (3.48). To see if the Hankel function is absolutely convergent, let us remember and rewrite its limiting form for the large argument, (3.47),

$$H_0^{(2)}(k\rho) \approx \sqrt{\frac{2}{j\pi k \rho}} e^{-jk\rho}, \quad k\rho \to \infty$$

where the propagation constant k is real for a lossless medium and complex for a lossy medium. The absolute value of (3.47) would have an algebraic decay for lossless media, of $1/\sqrt{\rho}$ and an exponential decay for lossy ones. Therefore, the Hankel function of zero order, second kind is not absolutely integrable for lossless media, as the algebraic behavior is not integrable over an infinite domain. However, it is integrable if the medium of propagation is lossy; that is the exponential decay in addition to the algebraic behavior makes it integrable over an infinite domain of integration. Hence, the Fourier transform of the scalar Green's function $\tilde{G}(k_x, k_y)$ due to a line source (i.e., the Fourier transform of the Hankel function) can be rigorously defined only if there is a loss in the medium. Actually, this is not a restriction at all, as all realistic media incorporate some amount of loss in nature. In other words, lossless material is just a mathematical approximation to low-loss material to ease the computations.

For the calculation of the scalar Green's function from the inversion integral of the Fourier transform, (3.73), it can be observed that the inversion integral cannot be performed for lossless media, and hence is not defined. From the dispersion relation, it is obvious that $k_y = \sqrt{k^2 - k_x^2}$ is a multivalued function of k_x, and hence the integrand of the inverse Fourier transform, (3.73), has two branch-point singularities ($k_x = \pm k$, where k is real for lossless media) on the integration path, which is the real axis on the k_x-plane. So adding a small loss into the medium of propagation makes k complex with a small negative imaginary part as $k = k' - jk''$, and pushes the branch points out of the way of integration, from the real k_x into the complex plane of k_x. As for the discussion of the existence of the Fourier transform of the Hankel function, a small loss renders the inversion integral convergent and well defined. Perhaps a little note is in order for the sign of the imaginary part of k: the sign must be negative because it makes the wave decay since the Hankel function of second kind corresponds to the outgoing wave, as discussed earlier, with $\exp(-jk\rho)$ space dependence for large distances from the line source.

3.3.3 Scalar Green's Function for a Point Source

As compared to a line source, which is a point source in two-dimensional space, a point source in three-dimensional space has much use in real-life problems, as objects in nature are always three-dimensional, and two-dimensional problems are just simplified models of three-dimensional problems. Therefore, the fields and potentials due to such a source play a crucial role in the analysis of an arbitrary source distribution in a material medium, as discussed in Section 3.2. In this section, Green's function of a scalar wave equation will be derived by three different approaches in addition to the direct solution in the spherical coordinate system as detailed in Section 3.2.1. Moreover, some useful integral identities are deduced from these derivations, and salient features of Green's functions, as well as the integral identities, are discussed with the necessary mathematics and accompanying physical intuitions. For the sake of illustration, let us consider an electric point source located at the origin in an unbounded medium and polarized in z-direction, as shown in Figure 3.4.

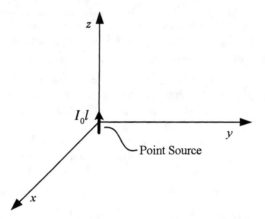

Figure 3.4 A point electrical current source polarized in z-direction in unbounded medium.

Remembering that the mathematical description of a point source located at an arbitrary point of (x', y', z') has been given in Cartesian coordinates as in (3.34), a point current source at the origin, with its polarization direction, can be written as

$$\mathbf{J} = \hat{z} I_0 l \, \delta(\mathbf{r}) = \hat{z} I_0 l \, \delta(x)\delta(y)\delta(z) \tag{3.77}$$

As discussed in the previous section, the vector nature of the source can be eliminated by considering the Cartesian components of the fields and vector potentials. Therefore, it would be sufficient to solve for the scalar wave equation

$$\left[\frac{\partial^2}{\partial x^2} + \frac{\partial^2}{\partial y^2} + \frac{\partial^2}{\partial z^2} + k_0^2 \right] \Phi(x, y, z) = -\delta(x)\delta(y)\delta(z) \tag{3.78}$$

with a scalar point source in a homogenous material medium characterized by $(\varepsilon_r, \mu_r = 1)$ and $k = \omega\sqrt{\mu_0 \varepsilon_0}\sqrt{\varepsilon_r}$. Although the solution of this partial differential equation in spherical coordinates, subject to a radiation boundary condition at infinity, has already been obtained in Section 3.2.1 as

$$g(r) = \frac{e^{-jkr}}{4\pi r} \tag{3.79}$$

its procedure is generally not suitable for bounded geometries, especially for those with planar layered structures with interfaces perpendicular to z. Therefore, it would be instructive to study the solution of the scalar wave equation (3.78) in Cartesian coordinates and cylindrical coordinates.

Fourier Transform Technique in 3D

Let us start with the solution of the scalar wave equation (3.78) in Cartesian coordinates, assuming that the 3D Fourier transform of $g(x, y, z)$ exists and is denoted by $\tilde{G}(k_x, k_y, k_z)$, with the following definition of the Fourier transformation pair:

$$\tilde{G}(k_x, k_y, k_z) = \int_{-\infty}^{\infty} \int_{-\infty}^{\infty} \int_{-\infty}^{\infty} dx\, dy\, dz\, g(x, y, z) e^{j(k_x x + k_y y + k_z z)} \tag{3.80}$$

$$g(x, y, z) = \frac{1}{(2\pi)^3} \int_{-\infty}^{\infty} \int_{-\infty}^{\infty} \int_{-\infty}^{\infty} dk_x\, dk_y\, dk_z\, \tilde{G}(k_x, k_y, k_z) e^{-j(k_x x + k_y y + k_z z)} \tag{3.81}$$

It is easily recognized that differentiations with respect to x, y, and z in the scalar wave equation (3.78) can be performed analytically if this representation of the solution is substituted. Of course, this can be done with the justification of the interchange of the order of differentiation and integration by the assumption that the triple integral in (3.81) converges. Hence, substituting (3.81) into (3.78), together with the spectral representation of the impulse function (3.67), results in

$$\int_{-\infty}^{\infty} \int_{-\infty}^{\infty} \int_{-\infty}^{\infty} dk_x\, dk_y\, dk_z \left[k^2 - k_x^2 - k_y^2 - k_z^2 \right] \tilde{G}(k_x, k_y, k_z) e^{-j(k_x x + k_y y + k_z z)} =$$

$$- \int_{-\infty}^{\infty} \int_{-\infty}^{\infty} \int_{-\infty}^{\infty} dk_x\, dk_y\, dk_z\, e^{-j(k_x x + k_y y + k_z z)}$$

Since the equality must hold for every (x, y, z), the integrands on both sides need to be equal, resulting in the following representation of the spectral-domain Green's function:

$$\tilde{G}(k_x, k_y, k_z) = \frac{-1}{k^2 - k_x^2 - k_y^2 - k_z^2} \tag{3.82}$$

Hence, taking the inverse Fourier transform provides the spatial-domain counterpart as a triple integral:

$$g(x, y, z) = \frac{-1}{(2\pi)^3} \int_{-\infty}^{\infty} \int_{-\infty}^{\infty} \int_{-\infty}^{\infty} dk_x\, dk_y\, dk_z\, \frac{e^{-j(k_x x + k_y y + k_z z)}}{k^2 - k_x^2 - k_y^2 - k_z^2} \tag{3.83}$$

Note that although this is not an efficient way to calculate Green's function, it provides the plane wave spectrum of the solution due to a point source, as the exponential kernel denotes the plane wave in three-dimensional space and the rest is the corresponding amplitude.

Fourier Transform Technique in 2D

From the study of the previous approach for the solution of the scalar wave equation due to a point source, it can easily be inferred that one can use the spectral representation in two dimensions while solving the second-order differential equation with respect to the third dimension analytically. Therefore, choosing any two of the three coordinate variables as the Fourier transform variables, say x and y, the scalar Green's function can be written as

$$g(x,y,z) = \frac{1}{(2\pi)^2} \int\limits_{-\infty}^{\infty}\int\limits_{-\infty}^{\infty} dk_x dk_y \, \tilde{G}(k_x,k_y,z) e^{-j(k_x x + k_y y)} \tag{3.84}$$

Substituting this into the scalar wave equation (3.78) with the spectral representation of the impulse functions $\delta(x)$ and $\delta(y)$, the following second-order, ordinary differential equation is obtained:

$$\left(\frac{d^2}{dz^2} + k^2 - k_x^2 - k_y^2 \right) \tilde{G}(k_x,k_y,z) = -\delta(z) \tag{3.85}$$

where $k^2 - k_x^2 - k_y^2 = k_z^2$ is used. Since the solution procedure of this differential equation is exactly the same as the one in the hybrid method, (3.68), the solution can be written directly by comparing the differential equations, (3.68) and (3.85), and the solution (3.72):

$$\tilde{G}(k_x,k_y,z) = \frac{1}{j2k_z} e^{-jk_z|z|}, \quad \text{Im}[k_z] < 0 \tag{3.86}$$

Hence, using the inverse Fourier transform definition (3.84), the scalar Green's function can be written as

$$g(x,y,z) = \frac{1}{(2\pi)^2} \int\limits_{-\infty}^{\infty}\int\limits_{-\infty}^{\infty} dk_x dk_y \, e^{-j(k_x x + k_y y)} \frac{e^{-jk_z|z|}}{j2k_z} \tag{3.87}$$

By comparing this representation and the one obtained from the 3D Fourier transformation, (3.83), it can be inferred that (3.87) can be obtained by integrating (3.83) over k_z. As this integration requires some background in complex calculus, it is given without a detailed introduction of such topics in Example 3.4 for interested readers.

Example 3.4

Derive the 2D inverse Fourier transform representation of the scalar Green's function (3.87) from its 3D inverse Fourier transform representation (3.83) by performing k_z integration analytically.

To perform the inversion over one of the Fourier transform variables, say k_z, (3.83) is rewritten as

$$g(x, y, z) = \frac{-1}{(2\pi)^3} \int\limits_{-\infty}^{\infty}\int\limits_{-\infty}^{\infty} dk_x dk_y \, e^{-j(k_x x + k_y y)} \int\limits_{-\infty}^{\infty} dk_z \, \frac{e^{-jk_z z}}{k^2 - k_x^2 - k_y^2 - k_z^2} \qquad (3.88)$$

where the inner integrand has two pole singularities at $k_z = \pm\sqrt{k^2 - k_x^2 - k_y^2}$, lying on the real axis of k_z for lossless media (k being real). As the integration path is defined over real k_z, these singularities will make the Fourier inversion not defined unless a slight loss is introduced into the medium, by which the wave number of the medium becomes complex ($k = k_r - jk_i$), and in turn, the singularities are pushed out of the integration path, as shown in Figure 3.5.

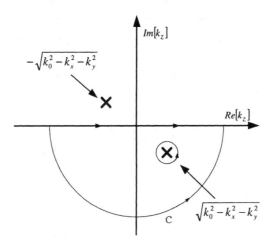

Figure 3.5 Application of contour integration to perform the Fourier inversion integral (3.88) analytically.

Note that for lossless media, the scalar Green's function in the spatial domain (3.79) is not absolutely integrable, as detailed for the case of the 2D scalar Green's function. Therefore, to have its spectral-domain representation and to define the Fourier inversion integral of the assumed spectral-domain representation (3.88), it is necessary to introduce small loss into the medium.

Once it is assumed that the pole singularities are out of the path of integration, the inner integral in (3.88) is written as

$$\int\limits_{FIP} \frac{dk_z \, e^{-jk_z z}}{k^2 - k_x^2 - k_y^2 - k_z^2} = -\int\limits_{FIP} \frac{dk_z \, e^{-jk_z z}}{\left(\underbrace{k^2 - \sqrt{k^2 - k_x^2 - k_y^2}}_{k_z'} \right)\left(k_z + \sqrt{k^2 - k_x^2 - k_y^2} \right)} \qquad (3.89)$$

where *FIP* denotes the Fourier inversion path, and it is along the real axis of k_z-plane. Assuming that readers have some knowledge of complex calculus, especially integration in the complex plane, Cauchy's theorem and the residue theorem are employed to cast (3.89) into

$$\underbrace{\int_C dk_z \frac{e^{-jk_z z}}{(k_z - k_z')(k_z + k_z')}}_{\rightarrow 0 \text{ due to Jordan's lemma}} - j2\pi \, \text{Res}(k_z = k_z') - \int_{FIP} dk_z \frac{e^{-jk_z z}}{(k_z - k_z')(k_z + k_z')} = 0 \qquad (3.90)$$

where Res( ) stands for the residue at the pole enclosed by the paths of integration C and *FIP*, as shown in Figure 3.5. The first integral goes to zero due to Jordan's lemma, as the exponential term vanishes over the path C for $z > 0$. Hence, the integral over *FIP* becomes equal to $-j2\pi$ times the residue at the pole k_z', which is $\exp(-jk_z'z)/2k_z'$. Following the same line of argument for $z < 0$, now with the semicircle enclosed from the upper half-space ($\text{Im}[z] < 0$) to ensure the applicability of Jordan's lemma, and in turn, accounting for the contribution of the pole at $-k_z'$ in the calculations, the Fourier inversion integral can be written as

$$-\int_{FIP} dk_z \frac{e^{-jk_z z}}{(k_z - k_z')(k_z + k_z')} = \begin{cases} j\pi \dfrac{e^{-jk_z'z}}{k_z'} & z > 0 \\[2mm] j\pi \dfrac{e^{jk_z'z}}{k_z'} & z < 0 \end{cases} = j\pi \frac{e^{-jk_z'|z|}}{k_z'} \qquad (3.91)$$

Hence, the spatial-domain Green's function is obtained as

$$g(x, y, z) = \frac{1}{(2\pi)^2} \int_{-\infty}^{\infty} \int_{-\infty}^{\infty} dk_x dk_y \frac{e^{-j(k_x x + k_y y + k_z'|z|)}}{j2k_z'}$$

which is the same as (3.87).

It is now clear that the scalar Green's functions can be represented in terms of their plane wave constituents, as is made evident by the representations in 3D and 2D, (3.83) and (3.87), respectively. Considering the original problem, investigating the solution to a wave equation due to a point source in space, it may be visualized that the waves generated from such a source would have to be cylindrically symmetrical. In other words, if the xy-plane is chosen to be the plane of the source with no boundary, any disturbance caused by the source will propagate as circular wavefronts on this plane. Therefore, it seems to be possible to further simplify the 2D inverse Fourier representation of the scalar Green's function in Cartesian coordinates (3.87) by representing it in cylindrical coordinates. To do so, let us define the propagation vector $\mathbf{k} = k_x \hat{x} + k_y \hat{y} = k_\rho \cos k_\phi \hat{x} + k_\rho \sin k_\phi \hat{y}$ and the position vector $\boldsymbol{\rho} = x\hat{x} + y\hat{y} = \rho \cos\phi \, \hat{x} + \rho \sin\phi \, \hat{y}$ in terms of cylindrical coordinates. Then, the scalar Green's function can be written as

$$g(\rho,z) = \frac{1}{(2\pi)^2} \int\limits_0^\infty dk_\rho k_\rho \int\limits_0^{2\pi} dk_\varphi \frac{e^{-jk_\rho\rho\cos(k_\varphi-\varphi)-jk_z|z|}}{j2k_z} \tag{3.92}$$

where $\mathbf{k}\cdot\boldsymbol{\rho} = k_x x + k_y y = k_\rho\rho\cos(k_\phi - \phi)$ is used. In addition, recognizing the integral definition of Bessel's functions of the first kind as

$$J_0(k_\rho\rho) = \frac{1}{2\pi} \int\limits_0^{2\pi} dk_\phi \, e^{-jk_\rho\rho\cos(k_\phi-\phi)} \tag{3.93}$$

results in the simplified representation of the scalar Green's function,

$$g(\rho,z) = \frac{1}{2\pi} \int\limits_0^\infty dk_\rho k_\rho J_0(k_\rho\rho) \frac{e^{-jk_z|z|}}{j2k_z} \tag{3.94}$$

Another form of this expression, which is also known as the Hankel transform, can be obtained by substituting the following identities into (3.94):

$$J_0(k_\rho\rho) = \frac{1}{2}\left[H_0^{(1)}(k_\rho\rho) + H_0^{(2)}(k_\rho\rho)\right] \tag{3.95}$$

$$H_0^{(2)}(e^{-j\pi}k_\rho\rho) = -H_0^{(1)}(k_\rho\rho) \tag{3.96}$$

Then, the scalar Green's function is written as

$$g(\rho,z) = \frac{1}{4\pi} \int\limits_{SIP} dk_\rho k_\rho H_0^{(2)}(k_\rho\rho) \frac{e^{-jk_z|z|}}{k_z} \tag{3.97}$$

Let us note some facts about this integral representation and give the reasoning behind changing the integration path from the real axis to a path named SIP: (1) $H_0^{(2)}(k_\rho\rho)$ has a logarithmic branch-point singularity at $k_\rho = 0$; (2) $k_z = \sqrt{k^2 - k_\rho^2}$ has algebraic branch-point singularities at $k_\rho = \pm k$; and (3) the integration is over the real axis of k_ρ-plane. These facts would be enough to make the integral undefined, let alone if the medium is lossless, unless the path of integration is deformed legitimately, as shown in Figure 3.6 and known as the Sommerfeld integration path (SIP). Due to the branch-point singularity at $k_\rho = 0$ and the associated branch cut along the negative real axis of the k_ρ-plane, the integration path needs to be displaced off the axis towards the opposite direction where the branch points will move with additional loss. The detailed discussion on this issue and related topics can be found in the literature [11, 12].

Since the solution of the scalar wave equation with a boundary condition at infinity and a source condition is unique, combining these representations — (3.79), (3.87), (3.94), and (3.97) — provides the following two well-known integral identities: the Weyl identity

$$\frac{e^{-jkr}}{r} = \frac{1}{\pi} \int_{-\infty}^{\infty} \int_{-\infty}^{\infty} dk_x dk_y \, e^{-j(k_x x + k_y y)} \frac{e^{-jk_z|z|}}{j2k_z} \tag{3.98}$$

and the Sommerfeld identity

$$\frac{e^{-jkr}}{r} = \int_{0}^{\infty} dk_\rho k_\rho J_0(k_\rho \rho) \frac{e^{-jk_z|z|}}{jk_z} = \int_{SIP} dk_\rho k_\rho H_0^{(2)}(k_\rho \rho) \frac{e^{-jk_z|z|}}{j2k_z} \tag{3.99}$$

The physical interpretations of these identities are simple and yet very important to understanding the derivation of Green's functions in planar multilayer media: the Weyl identity is nothing but the plane wave expansion of spherical waves, which plays an important role in defining the reflection and transmission of spherical waves at the planar dielectric interfaces. The Sommerfeld identity is the cylindrical wave expansion of the spherical wave in the ρ direction and a plane wave in the z direction. Note that both identities include evanescent waves as a part of the plane wave spectrum of a spherical wave.

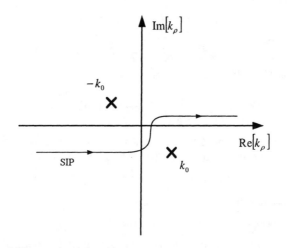

Figure 3.6 Sommerfeld integration path.

3.4 ANALYSIS OF PLANAR MULTILAYER MEDIA

So far in this chapter, the following topics have been covered to be prepared for the derivation of Green's function in a planar multilayer environment: (1) the concept of Green's functions; (2) derivation of the spatial-domain Green's function for the scalar wave equation in an unbounded and homogeneous medium as a spherical wave; (3) derivation of the dyadic Green's function for the electric field in terms of Green's function of the scalar wave equation; (4) study of the point sources in different coordinate systems; and (5) the derivation of the plane wave spectrum of a spherical wave. Now, it is time to find Green's functions of fields and potentials in more a realistic and practical environment used for modern microwave circuits and antennas, like a planar multilayer medium. A typical planar multilayer structure is demonstrated in Figure 3.7, where the electromagnetic properties of isotropic layers, μ and ε, vary in one direction only, the z-direction in Figure 3.7 and throughout the discussions on planar media in this book. In addition, the source layer is denoted by the subscript i; the origin of the coordinate system is defined at the source; and the thickness, permittivity, and permeability of any layer, layer-i for example, are denoted by d_i, ε_i, and μ_i, respectively, where ε_i could be a complex number to incorporate the losses in the layers.

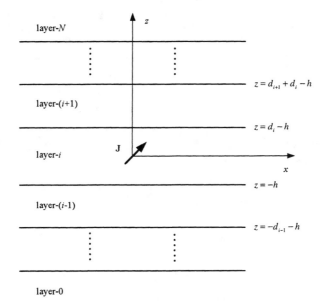

Figure 3.7 An electric current embedded in a multilayer planar medium.

With the introduction of the layered media and the relevant parameters, Green's function of the vector wave equation for the electric field can now be obtained with the dipole source embedded in layer-i. To find Green's function for the electric field, we first need to write the governing differential equation for the electric field, with the source term on the right-hand side. As is often the case for the solution of inhomogeneous differential equations, one needs to solve for the homogeneous part of the differential equation first. So, using Maxwell's equations in the isotropic and source-free region, the following vector wave equations for the electric and magnetic fields are obtained and must be solved:

$$\mu \nabla \times \frac{1}{\mu} \nabla \times \mathbf{E}(\mathbf{r}) - \omega^2 \mu \varepsilon \, \mathbf{E}(\mathbf{r}) = 0 \qquad (3.100)$$

$$\varepsilon \nabla \times \frac{1}{\varepsilon} \nabla \times \mathbf{H}(\mathbf{r}) - \omega^2 \mu \varepsilon \mathbf{H}(\mathbf{r}) = 0 \qquad (3.101)$$

where $\mu(z)$ and $\varepsilon(z)$ are assumed. For source-free cases, the vector wave equations can be reduced to decoupled scalar wave equations, which are characterized as the TE and TM waves in the z-direction. Let us first assume that TE-z and TM-z waves exist, and then the scalar wave equations for both cases are derived to demonstrate the existence of such waves. In addition, the solutions of these wave equations are shown to be independent of each other; that is, these two forms of the waves are decoupled. For the TE-z formulation, for which the electric field is linearly polarized in the xy-plane, the electric field is assumed in the direction of the y-coordinate with no loss of generality. This is because the coordinate system can always be rotated about the z-axis to align the polarization of the electric field in the y-direction. Likewise, for the TM-z formulation, only the y component of the magnetic field is assumed to exist. Hence,

$$\mathbf{E}(\mathbf{r}) = \hat{a}_y E_y(\mathbf{r}) \qquad (3.102)$$

$$\mathbf{H}(\mathbf{r}) = \hat{a}_y H_y(\mathbf{r}) \qquad (3.103)$$

and with the substitution of (3.102) and (3.103) into the vector wave equations, (3.100) and (3.101), respectively, the following scalar wave equations for the electric and magnetic fields are obtained:

$$\left[\frac{\partial^2}{\partial x^2} + \mu(z) \frac{\partial}{\partial z} \mu^{-1}(z) \frac{\partial}{\partial z} + \omega^2 \mu \varepsilon \right] E_y = 0 \qquad (3.104)$$

$$\left[\frac{\partial^2}{\partial x^2} + \varepsilon(z) \frac{\partial}{\partial z} \varepsilon^{-1}(z) \frac{\partial}{\partial z} + \omega^2 \mu \varepsilon \right] H_y = 0 \qquad (3.105)$$

The derivations of the above equations are demonstrated in the next example.

Example 3.5

Derive the governing equation (3.104) for TE-z waves in planar media with stratification in the z-direction, as in Figure 3.7, starting with the vector wave equation (3.100).

For TE-z waves, the electric field may have x- and y-vector components. Since the choice of the coordinate axes is arbitrary, it is convenient to choose the direction of the electric field in the transverse domain as the y-axis. Therefore,

$$\mathbf{E}(\mathbf{r}) = \hat{a}_y E_y(\mathbf{r})$$

Substituting this electric field into the vector wave equation in (3.100),

$$\mu \left[\left(\nabla \frac{1}{\mu} \right) \times \left(\nabla \times \hat{a}_y E_y \right) + \frac{1}{\mu} \nabla \times \nabla \times \hat{a}_y E_y \right] - \omega^2 \mu \varepsilon \hat{a}_y E_y = 0$$

and using the fact that the permeability varies only in the z-direction, the following expression can be written:

$$\mu\left[\left(\frac{\partial}{\partial z}\mu^{-1}\right)\hat{a}_z\times\left(\nabla\times\hat{a}_yE_y\right)+\frac{1}{\mu}\underbrace{\nabla\times\nabla\times\hat{a}_yE_y}_{\nabla\nabla\cdot\hat{a}_yE_y-\nabla^2\hat{a}_yE_y}\right]-\omega^2\mu\varepsilon\hat{a}_yE_y=0$$

where the first triple vector product and the divergence of the electric field can be written as

$$\hat{a}_z\times\left(\nabla\times\hat{a}_yE_y\right)=\nabla\underbrace{\left(\hat{a}_z\cdot\hat{a}_yE_y\right)}_{=0}-\left(\hat{a}_z\cdot\nabla\right)\hat{a}_yE_y=-\frac{\partial}{\partial z}\hat{a}_yE_y$$

$$\nabla\cdot\mathbf{D}=0\Rightarrow\nabla\cdot\varepsilon(z)\hat{a}_yE_y=\underbrace{\nabla\varepsilon(z)\cdot\hat{a}_yE_y}_{=0}+\varepsilon(z)\nabla\cdot\hat{a}_yE_y=0\Rightarrow\nabla\cdot\hat{a}_yE_y=0$$

Note that the choice of the coordinate system has led us to the fields with no y variation. Hence, the vector wave equation in (3.100) can be reduced to

$$\left[-\mu\left(\frac{\partial}{\partial z}\mu^{-1}\right)\frac{\partial}{\partial z}-\nabla^2-\omega^2\mu\varepsilon\right]\hat{a}_yE_y=0$$

and, in addition, with the use of the product rule of the derivative

$$\left(\frac{\partial}{\partial z}\mu^{-1}\right)\frac{\partial}{\partial z}\equiv\frac{\partial}{\partial z}\mu^{-1}\frac{\partial}{\partial z}-\mu^{-1}\frac{\partial^2}{\partial z^2}$$

it is further simplified to

$$\left(-\mu\frac{\partial}{\partial z}\mu^{-1}\frac{\partial}{\partial z}+\frac{\partial^2}{\partial z^2}-\nabla^2-\omega^2\mu\varepsilon\right)\hat{a}_yE_y=0$$

Since $\nabla\cdot\hat{a}_yE_y=\partial E_y/\partial y=0$, the governing equation is finally obtained as

$$\left(\frac{\partial^2}{\partial x^2}+\mu\frac{\partial}{\partial z}\mu^{-1}\frac{\partial}{\partial z}+\omega^2\mu\varepsilon\right)E_y=0$$

where μ and ε are functions of z only.

Now, it has been shown that for a source-free and planar medium, the vector wave equations for electric and magnetic fields can be reduced down to two scalar equations, as given in (3.104) and (3.105). However, nothing has been done so far to show that the solutions of these two scalar differential equations are independent, that is, decoupled. This can be demonstrated by finding the field components available for TE-z and TM-z formulations and by showing that they form mutually exclusive vector components. For TE-z, because E_y is the only electric field component and because it is not a function of y, as demonstrated in

Example 3.5, the following field components can be shown to exist from Maxwell's equations:

$$\nabla \times \hat{a}_y E_y = -j\omega\mu \mathbf{H}(x,z) \Rightarrow \mathbf{H}(x,z) = \hat{a}_x H_x(x,z) + \hat{a}_z H_z(x,z)$$

$$\nabla \times (\hat{a}_x H_x + \hat{a}_z H_z) = j\omega\varepsilon\, \mathbf{E}(x,z) \Rightarrow \mathbf{E}(x,z) = \hat{a}_y E_y$$

For TM-z, doing the same thing with H_y, the available field components are obtained as

$$\nabla \times \hat{a}_y H_y = j\omega\varepsilon E_y(x,z) \Rightarrow \mathbf{E}(x,z) = \hat{a}_x E_x(x,z) + \hat{a}_z E_z(x,z)$$

$$\nabla \times (\hat{a}_x E_x + \hat{a}_z E_z) = -j\omega\mu \mathbf{H}(x,z) \Rightarrow \mathbf{H}(x,z) = \hat{a}_y H_y$$

To summarize, the field components for TE-z are (E_y, H_x, H_z), while those for TM-z are (H_y, E_x, E_z); they are mutually exclusive, and hence decoupled.

To help find the solutions of the scalar wave equations, (3.104) and (3.105), which are two-dimensional partial differential equations, they can be further simplified, or rather reduced, by considering the salient feature of the planar geometry. Because the geometry (Figure 3.7) is unbounded in the transverse-to-z plane (xy-plane) and because there is no point on this plane that could be distinguished from the other points on the same plane, the fields are translationally invariant, and the magnitudes of the fields must be uniform on the plane. In other words, since there is no variation of the boundary in the transverse-to-z plane, which would force the magnitudes of the fields to vary, the magnitudes must be constant (no variation with respect to the x variable), while there should be phase progress (propagation) along the x-direction. Hence, the fields in such environments can be represented mathematically as follows:

$$E_y(x,z) = E_y(z)e^{\pm jk_x x} \tag{3.106}$$

$$H_y(x,z) = H_y(z)e^{\pm jk_x x} \tag{3.107}$$

Note that the above argument is true in every layer in a multilayer geometry, and the fields in all layers will have the same form as (3.106) and (3.107). In addition, imposing the continuity of the tangential fields at the interfaces between the layers (E_y and H_y are tangential at the interfaces) dictates that the fields in all layers must have the same propagation constant in the x-direction; that is, the solution for all z values must have the same phase variation in the x-direction, $\exp(\pm jk_x x)$, which is called the phase-matching condition. Consequently, the two-dimensional scalar wave equations are reduced down to one-dimensional scalar wave equations:

$$\left[\mu(z)\frac{\partial}{\partial z}\mu^{-1}(z)\frac{\partial}{\partial z} + \underbrace{\omega^2\mu\varepsilon - k_x^2}_{k_z^2} \right] E_y = 0 \quad \text{for TE-}z \tag{3.108}$$

$$\left[\varepsilon(z)\frac{\partial}{\partial z}\varepsilon^{-1}(z)\frac{\partial}{\partial z}+\underbrace{\omega^2\mu\varepsilon-k_x^2}_{k_z^2}\right]H_y=0 \quad \text{for TM-}z \qquad (3.109)$$

Instead of starting to solve these equations in the whole geometry at once, they can first be solved in each piecewise-constant region, and then the solution accounting for the complete geometry is obtained by matching the boundary conditions at the interfaces between the layers. So, in the homogeneous medium of layer-i, these one-dimensional scalar wave equations, (3.108) and (3.109), can be written as

$$\left[\frac{\partial^2}{\partial z^2}+\underbrace{\omega^2\mu_i\varepsilon_i-k_x^2}_{k_{zi}^2}\right]E_{yi}(z)=0 \quad \text{for TE-}z \qquad (3.110)$$

$$\left[\frac{\partial^2}{\partial z^2}+\underbrace{\omega^2\mu_i\varepsilon_i-k_x^2}_{k_{zi}^2}\right]H_{yi}(z)=0 \quad \text{for TM-}z \qquad (3.111)$$

which resemble the governing equations of voltage and current in TEM-supporting transmission lines. In other words, in a stratified medium, where z is the direction of stratification and the transverse-to-z domain is unbounded, as in Figure 3.7, the transverse components of the electric (magnetic) field can be split into two transverse electric (magnetic) fields, TE to z and TM to z, both governed by the transmission line equation. This interpretation of the transverse fields in a multilayer planar environment provides a means to perform the calculation; that is, the transverse fields in any layer in a stratified medium can be obtained by treating each layer as a transmission line with a given characteristic impedance and propagation constant [11].

Example 3.6

In a stratified medium, where stratification is in the z-direction and transverse domain is unbounded, show that the transverse-to-z fields can be split into two decoupled components, TE-z and TM-z, and that they satisfy the transmission line equations with corresponding characteristic impedances and propagation constants.

Note that the procedure provided in this example has been taken from the classic book by L. B. Felsen and N. Marcuvitz [11]. As was discussed in Section 1.3.1, fields of guided waves in a source-free region have $\exp(-jk_z z)$ dependence. In addition, the fields of such waves in transversely unbounded and homogenous media can be represented in terms of the fields of the form

$$\mathbf{E}(\mathbf{r})=\mathbf{E}_0(\mathbf{k}_\rho)e^{-j\mathbf{k}_\rho\cdot\mathbf{\rho}}e^{-jk_z z} \qquad (3.112)$$

$$\mathbf{H}(\mathbf{r})=\mathbf{H}_0(\mathbf{k}_\rho)e^{-j\mathbf{k}_\rho\cdot\mathbf{\rho}}e^{-jk_z z} \qquad (3.113)$$

where the propagation vector is defined as $\mathbf{k} = \mathbf{k}_\rho + k_z \hat{z}$, and the amplitude vectors are uniform in the transverse domain. Note that even though the amplitudes are uniform in the spatial transverse domain, they are functions of the transverse propagation constant $\mathbf{k}_\rho$ because the plane waves propagating in different transverse directions may have different amplitudes. Once the building blocks of the fields in each layer, homogeneous and unbounded in the transverse domain, have been introduced as the plane waves, decomposing the gradient operator and fields in Maxwell's equations as

$$\nabla = \nabla_t + \hat{z}\frac{\partial}{\partial z} = -j\mathbf{k}_\rho + \hat{z}\frac{\partial}{\partial z}; \quad \mathbf{E} = \mathbf{E}_t + \hat{z}E; \quad \mathbf{H} = \mathbf{H}_t + \hat{z}H_z \quad (3.114)$$

and grouping the transverse and longitudinal components results in

$$-j\mathbf{k}_\rho \times \hat{z}E_z + \frac{\partial}{\partial z}\hat{z}\times\mathbf{E}_t = -j\omega\mu\,\mathbf{H}_t \quad (3.115)$$

$$-j\mathbf{k}_\rho \times \hat{z}H_z + \frac{\partial}{\partial z}\hat{z}\times\mathbf{H}_t = j\omega\varepsilon\,\mathbf{E}_t \quad (3.116)$$

for transverse components and

$$-j\mathbf{k}_\rho \times \mathbf{E}_t = -j\omega\mu H_z\hat{z} \quad (3.117)$$

$$-j\mathbf{k}_\rho \times \mathbf{H}_t = j\omega\varepsilon E_z\hat{z} \quad (3.118)$$

for longitudinal components. Note that $\nabla_t \equiv -j\mathbf{k}_\rho$ is used, and $\partial/\partial z$ is retained as it is (not replaced by $-jk_z z$), because our goal is to get the transmission line equations that involve derivatives in the longitudinal direction. The longitudinal components of the fields can be written in terms of the transverse components, from (3.117) and (3.118), as

$$\omega\mu H_z = \left(\mathbf{k}_\rho \times \mathbf{E}_t\right)\cdot\hat{z} = -\mathbf{k}_\rho \cdot\left(\hat{z}\times\mathbf{E}_t\right) \quad (3.119)$$

$$\omega\varepsilon E_z = -\left(\mathbf{k}_\rho \times \mathbf{H}_t\right)\cdot\hat{z} = -\mathbf{k}_\rho \cdot\left(\mathbf{H}_t \times\hat{z}\right) \quad (3.120)$$

where the vector identity $\mathbf{a}\cdot(\mathbf{b}\times\mathbf{c}) = \mathbf{b}\cdot(\mathbf{c}\times\mathbf{a}) = \mathbf{c}\cdot(\mathbf{a}\times\mathbf{b})$ is employed. With the substitution of these longitudinal field components into (3.115) and (3.116), differential equations involving only the transverse field components are obtained. For the sake of illustration, let us substitute (3.120) into (3.115) and take the cross-product with $\hat{z}$ from the right:

$$\frac{j}{\omega\varepsilon}\big[\mathbf{k}_\rho \cdot(\mathbf{H}_t \times\hat{z})\big]\underbrace{(\mathbf{k}_\rho \times\hat{z})\times\hat{z}}_{-\mathbf{k}_\rho} + \frac{\partial}{\partial z}\underbrace{(\hat{z}\times\mathbf{E}_t)\times\hat{z}}_{\mathbf{E}_t} = -j\omega\mu\,\mathbf{H}_t \times\hat{z}$$

where $\mathbf{a}\times(\mathbf{b}\times\mathbf{c}) = \mathbf{b}(\mathbf{a}\cdot\mathbf{c}) - \mathbf{c}(\mathbf{a}\cdot\mathbf{b})$ is used. Then, with a little reorganization, the above equation can be cast into

$$\frac{\partial}{\partial z}\mathbf{E}_t = \left[-j\omega\mu\,\overline{\mathbf{I}} - \frac{\mathbf{k}_\rho\mathbf{k}_\rho}{j\omega\varepsilon}\right]\cdot\left(\mathbf{H}_t\times\hat{z}\right) \tag{3.121}$$

and similarly, substituting (3.119) into (3.116) and performing similar operations results in

$$\frac{\partial}{\partial z}\mathbf{H}_t = \left[-j\omega\varepsilon\,\overline{\mathbf{I}} - \frac{\mathbf{k}_\rho\mathbf{k}_\rho}{j\omega\mu}\right]\cdot\left(\hat{z}\times\mathbf{E}_t\right) \tag{3.122}$$

Now, let us try to figure out how we can decompose the transverse field components by investigating the terms in the square brackets in (3.121) and (3.122). We first notice that the second terms in the square brackets result in transverse fields only in the k_ρ-direction,

$$\mathbf{k}_\rho\mathbf{k}_\rho\cdot\left\{\begin{matrix}\left(\mathbf{H}_t\times\hat{z}\right)\\\left(\hat{z}\times\mathbf{E}_t\right)\end{matrix}\right\} = \hat{k}_\rho\,k_\rho^2\left[\underbrace{\hat{k}_\rho\cdot\left\{\begin{matrix}\left(\mathbf{H}_t\times\hat{z}\right)\\\left(\hat{z}\times\mathbf{E}_t\right)\end{matrix}\right\}}_{\text{scalar}}\right]$$

Hence, we can state that $\mathbf{E}_t$ and $\mathbf{H}_t$ have nonzero components in the k_ρ-direction. Then, when these components of $\mathbf{E}_t$ and $\mathbf{H}_t$ are substituted into (3.122) and (3.121), respectively, it is observed that the k_ρ component of $\mathbf{E}_t$ produces $\left(\hat{z}\times\hat{k}_\rho\right)$ component of $\mathbf{H}_t$, while the k_ρ component of $\mathbf{H}_t$ produces $\left(\hat{k}_\rho\times\hat{z}\right)$ component of $\mathbf{E}_t$. Therefore, the transverse field components can be decomposed as

$$\mathbf{E}_t(z) = \mathbf{E}_t' + \mathbf{E}_t'' = E_t'(z)\hat{k}_\rho + E_t''(z)\hat{k}_\rho\times\hat{z} \tag{3.123}$$

$$\mathbf{H}_t(z) = \mathbf{H}_t' + \mathbf{H}_t'' = H_t'(z)\hat{z}\times\hat{k}_\rho + H_t''(z)\hat{k}_\rho \tag{3.124}$$

With the substitution of these transverse field representations into the differential equations (3.121) and (3.122), the following first-order, scalar differential equations are obtained:

$$\frac{\partial E_t'}{\partial z} = -j\omega\left(\mu - \frac{k_\rho^2}{\omega^2\varepsilon}\right)H_t';\quad \frac{\partial H_t'}{\partial z} = -j\omega\varepsilon\,E_t' \tag{3.125}$$

$$\frac{\partial E_t''}{\partial z} = -j\omega\mu\,H_t'';\quad \frac{\partial H_t''}{\partial z} = -j\omega\left(\varepsilon - \frac{k_\rho^2}{\omega^2\mu}\right)E_t'' \tag{3.126}$$

where the differential equations relating E_t' and H_t', (3.125), are uncoupled from those relating E_t'' and H_t'', (3.126). Since E_t' and H_t' are the magnitudes of the vector fields in the $\hat{k}_\rho$- and $\hat{z}\times\hat{k}_\rho$-directions, respectively, substituting them into (3.119) and (3.120) results in $H_z = 0$ and $E_z \neq 0$. Similarly, it can be shown that

E_t'' and H_t'' produce the longitudinal fields of $E_z = 0$ and $H_z \neq 0$. Therefore, the fields resulting from (E_t', H_t') are independent of the fields resulting from (E_t'', H_t'') and are referred to as the TM-z (E-mode) and TE-z (H-mode) modes, respectively.

Once it is established that the fields in a stratified medium, where the stratification is in z-direction and the layers are unbounded in the transverse-to-z direction, can be decomposed into two independent modes, TM-z (E_t', H_t', E_z) and TE-z (E_t'', H_t'', H_z), now it is time to solve for the field components in each mode. To do so, we need to solve for (E_t', H_t') from (3.125) and (E_t'', H_t'') from (3.126), where both sets of first-order, coupled differential equations resemble the transmission line equations relating voltage and current. Hence, remembering the transmission line equations in a lossless transmission line

$$\frac{\partial V}{\partial z} = -j\omega L I; \quad \frac{\partial I}{\partial z} = -j\omega C V; \quad Z_0 = \sqrt{\frac{L}{C}}$$

and making the analogy with (3.125) and (3.126) leads to

$$L' = \mu - \frac{k_\rho^2}{\omega^2 \varepsilon}; \quad C' = \varepsilon; \quad Z_0' = \frac{k_z'}{\omega \varepsilon}$$

for TM-z and to

$$L'' = \mu; \quad C'' = \varepsilon - \frac{k_\rho^2}{\omega^2 \mu}; \quad Y_0'' = \frac{k_z''}{\omega \mu}$$

for TE-z, where $k_z' = k_z'' = \sqrt{\omega^2 \mu \varepsilon - k_\rho^2}$. Reorganizing (3.125) and (3.126) with the introduced characteristic impedance, admittance, and propagation constants results in the following two sets of coupled differential equations:

$$\frac{\partial E_t'}{\partial z} = -jk_z' Z_0' H_t'; \quad \frac{\partial H_t'}{\partial z} = -jk_z' Y_0' E_t' \qquad (3.127)$$

$$\frac{\partial E_t''}{\partial z} = -jk_z'' Z_0'' H_t''; \quad \frac{\partial H_t''}{\partial z} = -jk_z'' Y_0'' E_t'' \qquad (3.128)$$

which have the same form as the transmission line equations we discussed in Chapter 1. Once these first-order, coupled differential equations are combined, second-order differential equations for E_t'' and H_t' are obtained for TE-z and TM-z waves, respectively, as

$$\left[\frac{\partial^2}{\partial z^2} + k_z^2 \right] E_t''(z) = 0 \quad \text{for TE-}z$$

$$\left[\frac{\partial^2}{\partial z^2} + k_z^2 \right] H_t'(z) = 0 \quad \text{for TM-}z$$

Since we have only the transverse components of the E-field (E_t'') and H-field (H_t') for TE-z and TM-z waves, respectively, the choices of E_t'' for TE-z waves and H_t' for TM-z waves cover all the components of the corresponding fields. In addition, as we can arbitrarily align the transverse coordinate axis, choosing the y-direction as the direction of these transverse components results in exactly the same equations as the reduced scalar wave equations given in (3.110) and (3.111).

So far, we have demonstrated that the fields in layered media, whose stratification is in the z-direction and unbounded in the transverse domain, can be decomposed into independent TE-z and TM-z wave modes and that the transverse field components of these waves are governed by the simple transmission line equations, (3.110) and (3.111). To build a general solution in a multilayer medium, it would be instructive to start with the simplest form of a layered medium, two semi-infinite dielectric materials, where we can define the reflection and transmission coefficients at the interface.

3.4.1 Fresnel's Reflection and Transmission Coefficients

For the sake of demonstration, the solution of the scalar wave equation for TE-z waves is investigated in one of the simplest layered geometries: a two-medium one-interface structure, as shown in Figure 3.8. For layer-i, the scalar wave equation for TE-z waves can be written as

$$\left[\frac{\partial^2}{\partial z^2} + \underbrace{\omega^2 \mu_i \varepsilon_i - k_x^2}_{k_{zi}^2} \right] E_{yi}(z) = 0 \qquad (3.129)$$

where $E_{yi}(x,z) = E_{yi}(z)e^{\pm jk_x x}$ is used as discussed before, with the understanding that all the field components in every layer will have the same x dependence due to the phase-matching condition. Since the incident wave is assumed to propagate toward the positive x-direction in Figure 3.8, the negative sign is chosen for the exponent of the x variation to represent the positive x-propagating wave, remembering the $e^{j\omega t}$ time dependence assumed throughout this book. The electric fields in both layers, as the solution of (3.129) in layer-i and the solution of a similar equation for layer-$(i-1)$, can be written as

$$E_{yi}(z) = E_i^+ e^{-jk_{zi}z} + E_i^- e^{+jk_{zi}z} \qquad (3.130)$$

$$E_{y(i-1)}(z) = E_{i-1}^+ e^{-jk_{z(i-1)}z} + E_{i-1}^- e^{+jk_{z(i-1)}z} \qquad (3.131)$$

where $k_{zi} = \sqrt{k_i^2 - k_x^2}$, $k_i = \omega\sqrt{\mu_i \varepsilon_i}$, and there are four unknown coefficients to be determined. Let us first consider layer-i, where there is an incident wave propagating in the negative z-direction and a reflected wave propagating in the positive z-direction. Therefore, $E_i^- = E_0$ is the amplitude of the incident wave that can be determined by the source, and E_i^+ is the amplitude of the reflected wave from the interface, determined as the reflection coefficient R times the amplitude of the incident wave. In layer-$(i-1)$, since there is only a transmitted wave propagating in the negative z-direction, the coefficient of the positive propagating wave, E_{i-1}^+, must be equal to zero. As a result, the electric fields in both layers, given in (3.130) and (3.131), can be rewritten as

$$E_{yi}(z) = R_{i,i-1}^{\text{TE}} E_0 e^{-jk_{zi}z} + E_0 e^{+jk_{zi}z} \tag{3.132}$$

$$E_{y(i-1)}(z) = T_{i,i-1}^{\text{TE}} E_0 e^{+jk_{z(i-1)}z} \tag{3.133}$$

where $R_{i,i-1}^{\text{TE}}$ and $T_{i,i-1}^{\text{TE}}$ are the reflection and transmission coefficients for TE-z waves defined at the interface between layer-i and layer-$(i-1)$.

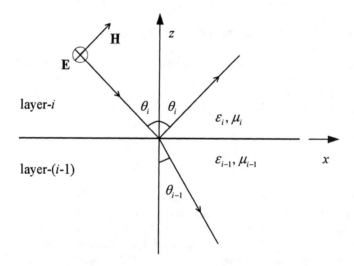

Figure 3.8 A typical two-layer medium showing incident, reflected, and transmitted waves.

Note that the unknown coefficients in (3.130) and (3.131) have been replaced by the unknown reflection and transmission coefficients. To determine these two unknowns, an additional boundary condition to the continuity of the tangential electric fields at the interface is required and readily obtained from the continuity of the tangential magnetic fields at the interface. Hence, by substituting the already obtained electric fields in both layers, (3.132) and (3.133), into Maxwell's equation

$$\nabla \times \mathbf{E} = -j\omega\mu\mathbf{H} \Rightarrow \begin{cases} H_x = \dfrac{1}{j\omega\mu}\dfrac{\partial}{\partial z}E_y \\[2mm] H_z = -\dfrac{1}{j\omega\mu}\dfrac{\partial}{\partial x}E_y \end{cases}$$

the tangential components of the magnetic fields are obtained as

$$H_{xi}(z) = \frac{jk_{zi}}{j\omega\mu_i}E_0\left[e^{jk_{zi}z} - R_{i,i-1}^{\text{TE}}\,e^{-jk_{zi}z}\right] \tag{3.134}$$

$$H_{x(i-1)}(z) = \frac{jk_{z(i-1)}}{j\omega\mu_{i-1}}T_{i,i-1}^{\text{TE}}E_0\,e^{+jk_{z(i-1)}z} \tag{3.135}$$

Since both E_y and H_x are the tangential field components at the interface (i.e., $z = 0$), implementing the boundary conditions on the continuity of the tangential electric and magnetic fields results in two equations for two unknowns:

$$1 + R_{i,i-1}^{\text{TE}} = T_{i,i-1}^{\text{TE}} \tag{3.136}$$

$$\frac{k_{zi}}{\mu_i}\left[1 - R_{i,i-1}^{\text{TE}}\right] = \frac{k_{z(i-1)}}{\mu_{i-1}}T_{i,i-1}^{\text{TE}} \tag{3.137}$$

Simultaneous solution of these equations gives the following reflection and transmission coefficients for TE-z waves in terms of known quantities of the layers; these solutions are called Fresnel's reflection and transmission coefficients:

$$R_{i,i-1}^{\text{TE}} = \frac{\mu_{i-1}k_{zi} - \mu_i k_{z(i-1)}}{\mu_{i-1}k_{zi} + \mu_i k_{z(i-1)}} \tag{3.138}$$

$$T_{i,i-1}^{\text{TE}} = \frac{2\mu_{i-1}k_{zi}}{\mu_{i-1}k_{zi} + \mu_i k_{z(i-1)}} \tag{3.139}$$

Note that the reflection and transmission coefficients for TE-z waves are defined for the electric field; see (3.132) and (3.133). Since most of the materials used in microwave and antenna circuits are nonmagnetic, assuming the relative permeability of the layers are equal and unity (i.e., $\mu_{i-1} = \mu_i = \mu_0$ in Figure 3.8) would not impose much of a restriction.

For TM-z waves, since the governing differential equation is a scalar wave equation for H_y (3.111), following a similar procedure results in

$$H_{yi}(z) = R_{i,i-1}^{\text{TM}}H_0 e^{-jk_{zi}z} + H_0 e^{+jk_{zi}z} \tag{3.140}$$

$$H_{y(i-1)}(z) = T_{i,i-1}^{\text{TM}}H_0 e^{+jk_{z(i-1)}z} \tag{3.141}$$

where the reflection and transmission coefficients are defined for the magnetic field. Using the generalized Ampere's circuital law, the associated electric field is obtained as

$$\nabla \times \mathbf{H} = j\omega\varepsilon\mathbf{E} \Rightarrow \begin{cases} E_x = -\dfrac{1}{j\omega\varepsilon}\dfrac{\partial}{\partial z}H_y \\[2mm] E_z = \dfrac{1}{j\omega\varepsilon}\dfrac{\partial}{\partial x}H_y \end{cases}$$

from which the tangential component of the electric field in both media can be explicitly written as

$$E_{xi}(z) = \frac{-jk_{zi}}{j\omega\varepsilon_i}H_0\left[e^{jk_{zi}z} - R_{i,i-1}^{\mathrm{TM}}\,e^{-jk_{zi}z}\right] \tag{3.142}$$

$$E_{x(i-1)}(z) = \frac{-jk_{z(i-1)}}{j\omega\varepsilon_{i-1}}T_{i,i-1}^{\mathrm{TM}}H_0\,e^{+jk_{z(i-1)}z} \tag{3.143}$$

Implementing the boundary conditions on the continuity of the tangential components of magnetic and electric fields results in the reflection and transmission coefficients for TM-z waves as

$$R_{i,i-1}^{\mathrm{TM}} = \frac{\varepsilon_{i-1}k_{zi} - \varepsilon_i k_{z(i-1)}}{\varepsilon_{i-1}k_{zi} + \varepsilon_i k_{z(i-1)}} \tag{3.144}$$

$$T_{i,i-1}^{\mathrm{TM}} = \frac{2\varepsilon_{i-1}k_{zi}}{\varepsilon_{i-1}k_{zi} + \varepsilon_i k_{z(i-1)}} \tag{3.145}$$

Note that $T_{i,i-1} = 1 + R_{i,i-1}$ and $R_{i,i-1} = -R_{i-1,i}$ for both TE-z and TM-z waves. After having obtained the general expressions for the Fresnel's reflection and transmission coefficients for two different polarizations, TE-z and TM-z, it would be instructive to discuss some of their salient features.

Reflection and Transmission

Note that the reflection and transmission coefficients given above describe what happens when a plane wave is incident (oblique or normal) upon an interface separating two electrically different media. From the derivation presented above, it can be concluded that a plane wave impinging on an interface is partially reflected and partially transmitted to satisfy the boundary conditions at the interface, and their amplitudes depend on the polarization of the incident wave.

TE and TM Waves (H- and E-Waves or s- and p-Waves)

There are two independent polarizations that can make up any wave in a stratified medium, and they are characterized by the polarization of the fields with respect to the plane of incidence. The plane of incidence is defined by the plane formed by the normal of the surface and the propagation direction of the incident plane wave, that is, z-direction and $\mathbf{k}_i = k_{xi}\hat{x} + k_{yi}\hat{y} + k_{zi}\hat{z}$ in Figure 3.8, respectively. If the electric field is in the same plane as the plane of incidence, the polarization is named transverse magnetic, as the magnetic field in a plane wave would be normal to the direction of propagation and the electric field, hence transverse to the plane of incidence. If the electric field is perpendicular to the plane of incidence, the polarization is named transverse electric. Since the use of the terms TE and TM might mean different things in different topics or might be referred to differently in different fields, it would be better to clarify their use or their alternative referrals. In the topic of waveguides, transverse means normal to the direction of propagation, and TE and TM waves are sometimes referred to as *H*- and *E*-waves, as the magnetic and electric fields have longitudinal components, and the governing equations are the wave equations for these components (refer to Chapter 1). In optics, TE and TM waves are referred to as s- and p-waves, respectively, as the letters "s" and "p" stand for perpendicular (in German) and parallel to the plane of incidence, respectively.

Brewster Angle

Note that the reflection coefficients, (3.138) and (3.144), can be written in terms of the wave numbers of the media and the angles of incidence, reflection, and transmission. To do so, it should be recognized that these waves are plane waves and their propagation constants in both regions can be written, from Figure 3.8, as

$$k_{xi} = k_i \sin\theta_i; \quad k_{zi} = k_i \cos\theta_i \tag{3.146}$$

$$k_{x(i-1)} = k_{i-1} \sin\theta_{i-1}; \quad k_{z(i-1)} = k_{i-1} \cos\theta_{i-1} \tag{3.147}$$

where $k_i = k_0\sqrt{\varepsilon_{ri}}$. Remember that the phase-matching condition at the interface [i.e., $k_{xi} = k_{x(i-1)}$] results in Snell's law

$$\frac{\sin\theta_i}{\sin\theta_{i-1}} = \frac{\sqrt{\varepsilon_{r(i-1)}}}{\sqrt{\varepsilon_{ri}}} \tag{3.148}$$

which connects the angle of transmission to the angle of incidence, as well as to the electrical parameters of the media involved. Substituting the propagation constants into the reflection coefficient expressions and utilizing Snell's law result in the following expressions:

$$R_{i,i-1}^{TE} = \frac{\cos\theta_i - \dfrac{\sqrt{\varepsilon_{r(i-1)}}}{\sqrt{\varepsilon_{ri}}}\cos\theta_{i-1}}{\cos\theta_i + \dfrac{\sqrt{\varepsilon_{r(i-1)}}}{\sqrt{\varepsilon_{ri}}}\cos\theta_{i-1}} = -\frac{\sin(\theta_i - \theta_{i-1})}{\sin(\theta_i + \theta_{i-1})} \tag{3.149}$$

$$R_{i,i-1}^{TM} = \frac{\cos\theta_i - \dfrac{\sqrt{\varepsilon_{ri}}}{\sqrt{\varepsilon_{r(i-1)}}}\cos\theta_{i-1}}{\cos\theta_i + \dfrac{\sqrt{\varepsilon_{ri}}}{\sqrt{\varepsilon_{r(i-1)}}}\cos\theta_{i-1}} = \frac{\tan(\theta_i - \theta_{i-1})}{\tan(\theta_i + \theta_{i-1})} \tag{3.150}$$

With a little investigation, it is observed that the reflection coefficient for TE waves cannot vanish for any angle of incidence, unless both media have the same electrical properties. However, for TM waves, the reflection coefficient becomes zero when $\theta_i + \theta_{i-1} = \pi/2$, as the denominator goes to infinity. The angle of incidence that results in this equality is known as the Brewster angle and is defined as follows:

$$\theta_B = \frac{\pi}{2} - \theta_{i-1} \Rightarrow \cos\theta_B = \sin\theta_{i-1} \Rightarrow \theta_B = \tan^{-1}\left(\sqrt{\frac{\varepsilon_{r(i-1)}}{\varepsilon_{ri}}}\right) \qquad (3.151)$$

where Snell's law is employed.

Critical Angle and Total Reflection

From Snell's law, (3.148), the transmission angle is written as

$$\sin\theta_{i-1} = \frac{\sqrt{\varepsilon_{ri}}}{\sqrt{\varepsilon_{r(i-1)}}}\sin\theta_i \qquad (3.152)$$

In cases of $\varepsilon_{ri} > \varepsilon_{r(i-1)}$, there must be an angle of incidence that makes the transmission angle $\pi/2$ [i.e., no transmission into medium-$(i-1)$], and this angle of incidence is called the critical angle. Using this definition, $\theta_{i-1} = \pi/2$ for $\theta_i = \theta_c$, we can find the expression of the critical angle as

$$\theta_c = \sin^{-1}\left(\sqrt{\frac{\varepsilon_{r(i-1)}}{\varepsilon_{ri}}}\right) \qquad (3.153)$$

If the angle of incidence is larger than the critical angle, that is $\theta_i > \theta_c$, then

$$\sin\theta_i > \sin\theta_c = \sqrt{\frac{\varepsilon_{r(i-1)}}{\varepsilon_{ri}}} \quad \text{where } \theta_i > 0, \quad \theta_c < \pi/2$$

and, from (3.152), the transmission angle needs to satisfy

$$\sin\theta_{i-1} > 1$$

which can assume only complex solutions for θ_{i-1}. Since the angles have been introduced in defining the propagation constants, (3.146) and (3.147), the interpretation of complex angles can be better understood in terms of these propagation constants. If the propagation vector of the transmitted wave is denoted by $\mathbf{k}_{i-1}$, then it is written as follows according to Figure 3.8:

$$\mathbf{k}_{i-1} = \hat{x}k_{x(i-1)} - \hat{z}k_{z(i-1)}$$

where

$$k_{x(i-1)} = k_{i-1}\sin\theta_{i-1} = k_i\sin\theta_i \quad \text{(phase matching)}$$

$$k_{z(i-1)} = k_{i-1}\cos\theta_{i-1} = k_0\sqrt{\varepsilon_{r(i-1)}}\sqrt{1 - \frac{\varepsilon_{ri}}{\varepsilon_{r(i-1)}}\sin^2\theta_i}$$

$$= k_0\sqrt{\varepsilon_{ri}}\sqrt{\frac{\varepsilon_{r(i-1)}}{\varepsilon_{ri}} - \sin^2\theta_i} = jk_i\sqrt{\sin^2\theta_i - \frac{\varepsilon_{r(i-1)}}{\varepsilon_{ri}}}$$

Upon substituting these propagation constants into the definition of Fresnel's reflection coefficients, (3.138) and (3.144), the reflection coefficients for the incident angles larger than the critical angle can be obtained as

$$R_{i,i-1}^{TE} = \frac{\cos\theta_i - j\sqrt{\sin^2\theta_i - \dfrac{\varepsilon_{r(i-1)}}{\varepsilon_{ri}}}}{\cos\theta_i + j\sqrt{\sin^2\theta_i - \dfrac{\varepsilon_{r(i-1)}}{\varepsilon_{ri}}}} = e^{j2\phi^{TE}} \tag{3.154}$$

$$R_{i,i-1}^{TM} = \frac{\varepsilon_{r(i-1)}\cos\theta_i - j\varepsilon_{ri}\sqrt{\sin^2\theta_i - \dfrac{\varepsilon_{r(i-1)}}{\varepsilon_{ri}}}}{\varepsilon_{r(i-1)}\cos\theta_i + j\varepsilon_{ri}\sqrt{\sin^2\theta_i - \dfrac{\varepsilon_{r(i-1)}}{\varepsilon_{ri}}}} = e^{j2\phi^{TM}} \tag{3.155}$$

Note that the magnitudes of both reflection coefficients are unity with different phase terms, which implies the equality of the amplitudes of incident and reflected waves. In other words, the total reflection occurs if the incident wave impinges on the interface between the two media with an angle larger than the critical angle, provided the condition for the existence of the critical angle is satisfied: the medium of incidence must be denser than the transmission medium.

Reflection from a Perfectly Conducting Plane

Assume that layer-$(i-1)$ in Figure 3.8 is a perfect conductor, where no field can exist. Since there will be no transmitted wave, the fields of incident and reflected waves in layer-i need to satisfy the boundary condition at the conducting interface, where the tangential electric fields must be zero. For TE waves, since E_y is the only tangential electric field component at the conducting interface, (3.132), implementing the boundary condition at $z = 0$ results in

$$E_{yi}(z = 0) = R_{i,i-1}^{TE}E_0 + E_0 = 0 \Rightarrow R_{i,i-1}^{TE} = -1$$

Likewise, for TM waves, E_x is the only tangential electric field component at the conducting surface, (3.142), and setting it to zero at $z = 0$ results in

$$E_{xi}(z = 0) = \frac{-jk_{zi}}{j\omega\varepsilon_i}H_0\left[1 - R_{i,i-1}^{TM}\right] = 0 \Rightarrow R_{i,i-1}^{TM} = 1$$

Remember that the reflection and transmission coefficients for TE-z and TM-z waves are for the electric and magnetic fields, respectively. Therefore, for TE-z waves, the reflection coefficient must be −1 to cancel the effect of the incident electric field at the conducting surface. However, for TM-z waves, it should be +1 to sustain the tangential magnetic field at the conducting surface, which, in turn, induces surface current density on the surface of the conductor, $\mathbf{J}_s = \hat{n} \times \mathbf{H}_i$.

3.4.2 Generalized Reflection and Transmission Coefficients

Once Fresnel's reflection and transmission coefficients have been introduced at the interface between two semi-infinite layers, where there are no multiple reflections involved, the concept of reflection and transmission coefficients must be generalized to take multiple reflections into account, as multiple reflections inevitably occur in geometries with more than one interface. Therefore, before jumping into the analysis of electromagnetic waves in the most general multilayered geometry, it would be easier to understand the mechanisms of multiple reflections and their incorporations into the solutions of wave equations in a three-layer geometry as shown in Figure 3.9.

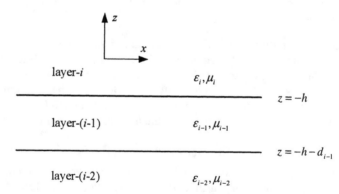

Figure 3.9 A typical three-layer planar geometry showing the coordinate system.

As in the study of two-layer media, we assume source-free, homogeneous, and isotropic layers in Figure 3.9; then, the solutions of the governing equations for TE-z and TM-z waves have similar forms and natures, consisting of positive-z and negative-z propagating waves with unknown amplitudes in each layer. Therefore, for TE-z waves, the electric field in layer-i is written, by inspection of (3.130), as

$$E_{yi}(z) = A_i' e^{+jk_{zi}(z+h)} + B_i' e^{-jk_{zi}(z+h)} \tag{3.156}$$

where $(z+h)$ is used on the exponents to define the unknown coefficients, A_i' and B_i', at the interface between layer-i and layer-$(i-1)$. In other words, the amplitudes of the down-going and up-going waves at the interface ($z = -h$) are explicitly seen in the expression as A_i' and B_i', respectively, and they are due to multiple reflections from the interfaces at the top and bottom of the layer.

Therefore, their ratio, B_i'/A_i', can be defined as the generalized reflection coefficient at the interface and denoted by $\tilde{R}_{i,i-1}^{TE}$. Using the definition of the generalized reflection coefficient, the electric fields for TE-z waves can be written in the layers as follows:

$$E_{yi}(z) = A_i \left[e^{jk_{zi}z} + \tilde{R}_{i,i-1}^{TE} \, e^{-jk_{zi}z} \, e^{-jk_{zi}2h} \right] \tag{3.157}$$

$$E_{y(i-1)}(z) = A_{i-1} \left[e^{jk_{z(i-1)}z} + R_{i-1,i-2}^{TE} \, e^{-jk_{z(i-1)}z} \, e^{jk_{z(i-1)}2(-h-d_{i-1})} \right] \tag{3.158}$$

$$E_{y(i-2)}(z) = A_{i-2} \, e^{jk_{z(i-2)}z} \tag{3.159}$$

where the reflection coefficient defined at the interface between layer-$(i-1)$ and layer-$(i-2)$ is a simple Fresnell reflection coefficient, as it is the last interface and no multiple reflections to account for. However, at the interface between layer-i and layer-$(i-1)$, the generalized reflection coefficient has to be used to take multiple reflections into account from the layers below. Although there seems to be three parameters to be determined, namely A_{i-1}, A_{i-2}, and $\tilde{R}_{i,i-1}^{TE}$, excluding the incident wave amplitude A_i, it would be sufficient to find an expression for the amplitude transfer between the layers and an expression for the generalized reflection coefficient. To do so, one needs to relate the down-going wave amplitude at $z = -h$ in layer-$(i-1)$ to that in layer-i as

$$\underbrace{A_{i-1} \, e^{-jk_{z(i-1)}h}}_{\substack{\text{Down-going wave} \\ \text{in layer-}(i-1)\text{ at }z=-h}} = \underbrace{A_i \, e^{-jk_{zi}h} \, T_{i,i-1}^{TE}}_{\substack{\text{Transmission of down-going} \\ \text{wave in layer-}i}} + \underbrace{A_{i-1} R_{i-1,i-2}^{TE} \, e^{jk_{z(i-1)}h} \, e^{jk_{z(i-1)}2(-h-d_{i-1})} R_{i-1,i}^{TE}}_{\substack{\text{Reflection of up-going} \\ \text{wave in layer-}(i-1)\text{ at }z=-h}}$$

Hence, the amplitude of the down-going wave in layer-$(i-1)$ is related to the amplitude of the incident wave in layer-i as

$$A_{i-1} \, e^{-jk_{z(i-1)}h} = A_i \, \frac{T_{i,i-1}^{TE} \, e^{-jk_{zi}h}}{1 - R_{i-1,i-2}^{TE} R_{i-1,i}^{TE} \, e^{-jk_{z(i-1)}2d_{i-1}}} \tag{3.160}$$

To find the other unknown, the generalized reflection coefficient, the up-going wave amplitude in layer-i can be written as the sum of the amplitude of the reflected direct wave and the amplitude of the transmitted waves into layer-i from the up-going waves in layer-$(i-1)$:

$$\underbrace{A_i \tilde{R}_{i,i-1}^{TE} \, e^{-jk_{zi}h}}_{\substack{\text{Up-going wave} \\ \text{in layer-}i\text{ at }z=-h}} = \underbrace{A_i R_{i,i-1}^{TE} \, e^{-jk_{zi}h}}_{\substack{\text{Reflection of down-going} \\ \text{wave in layer-}i\text{ at }z=-h}} + \underbrace{A_{i-1} R_{i-1,i-2}^{TE} \, e^{jk_{z(i-1)}h} \, e^{jk_{z(i-1)}2(-h-d_{i-1})} T_{i-1,i}^{TE}}_{\substack{\text{Transmission of up-going} \\ \text{wave in layer-}(i-1)\text{ at }z=-h}}$$

Hence, the generalized reflection coefficient can be obtained, with the substitution of the amplitude transfer relation (3.160), as

$$\tilde{R}_{i,i-1}^{\text{TE}} = R_{i,i-1}^{\text{TE}} + \frac{T_{i,i-1}^{\text{TE}} T_{i-1,i}^{\text{TE}} R_{i-1,i-2}^{\text{TE}} e^{-jk_{z(i-1)}2d_{i-1}}}{1 - R_{i-1,i-2}^{\text{TE}} R_{i-1,i}^{\text{TE}} e^{-jk_{z(i-1)}2d_{i-1}}} \tag{3.161}$$

Consequently, with the knowledge of the amplitude transfer (3.160) and the generalized reflection coefficient (3.161), the field components given in (3.157) to (3.159) can be uniquely defined, apart from the amplitude of the incident wave A_i.

Note that the amplitude of the incident plane wave can be defined uniquely by the source or set by the user as a known amplitude of the incident plane wave. Incidentally, the main motivation for finding reflection and transmission coefficients at an interface between two electrically different media for an incident plane wave is that fields generated by any source, even by a point source, can be written in terms of plane waves, known as plane wave expansion and discussed in this chapter in detail. Therefore, knowing how a plane wave is reflected and transmitted at an interface will help us to analyze problems of arbitrary source in a multilayer medium.

As was mentioned in the introduction to generalized reflection coefficients, they are supposed to account for multiple reflections in a multilayer medium. Therefore, the same expression for the generalized reflection coefficient, (3.161), should be obtained by counting the multiple reflections. For example, let us derive (3.161) by counting the multiple reflections in a three-layer medium, as shown in Figure 3.10:

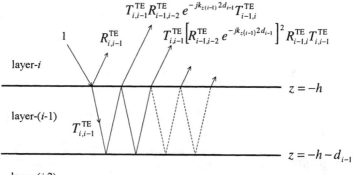

Figure 3.10 Multiple reflections in a three-layer geometry.

Since the generalized reflection coefficient is the sum of the multiple reflections due to a plane wave with a unity amplitude, it is written as an infinite summation,

$$\tilde{R}_{i,i-1}^{TE} = \underbrace{R_{i,i-1}^{TE}}_{\substack{\text{Direct} \\ \text{reflected}}} + T_{i,i-1}^{TE} \underbrace{R_{i-1,i-2}^{TE} \, e^{-jk_{z(i-1)}2d_{i-1}}}_{\substack{\text{One round-trip} \\ \text{in layer-}(i-1)}} T_{i-1,i}^{TE}$$

$$+ T_{i,i-1}^{TE} \underbrace{\left[R_{i-1,i-2}^{TE} \, e^{-jk_{z(i-1)}2d_{i-1}} \right]^2}_{\substack{\text{Two round-trips} \\ \text{in layer-}(i-1)}} R_{i-1,i}^{TE} \, T_{i-1,i}^{TE} + \dots \tag{3.162}$$

Note that the multiple reflection is due to the transmission of waves propagating back and forth in the middle layer; that is, after every round-trip in layer-$(i-1)$, a part of the wave is transmitted back into the topmost layer and the rest goes through one more round-trip, and this process continues indefinitely. As a result, there is a common factor in each term, due to the same round-trip, and once this is factored out, we get the following series:

$$\tilde{R}_{i,i-1}^{TE} = R_{i,i-1}^{TE} + T_{i,i-1}^{TE} R_{i-1,i-2}^{TE} \, e^{-jk_{z(i-1)}2d_{i-1}} T_{i-1,i}^{TE} \underbrace{\left[1 + R_{i-1,i-2}^{TE} R_{i-1,i}^{TE} \, e^{-jk_{z(i-1)}2d_{i-1}} + \dots \right]}_{\frac{1}{1 - R_{i-1,i-2}^{TE} R_{i-1,i}^{TE} \, e^{-jk_{z(i-1)}2d_{i-1}}}}$$

where it is recognized that the expression in the square bracket is a power series with each term less than unity. Hence, using the following closed-form expression for this infinite summation,

$$\lim_{\substack{N \to \infty \\ |x|<1}} \left(1 + x^2 + x^3 + \dots + x^{N-1} \right) = \lim_{\substack{N \to \infty \\ |x|<1}} \frac{1 - x^N}{1 - x} = \frac{1}{1 - x} \tag{3.163}$$

the same expression for the generalized reflection coefficient as (3.161) is obtained.

Extension to Multilayer Geometries

It should be noted that the amplitude transfer relation (3.160) and the generalized reflection coefficient (3.161) are derived specifically for the three-layer geometry. If the geometry is generalized one step further to a geometry with multiple layers down below layer-$(i-2)$ in Figure 3.9, then the Fresnell reflection coefficient $R_{i-1,i-2}^{TE}$ in (3.158), defined at the interface between layer-$(i-1)$ and layer-$(i-2)$, should be replaced by the generalized reflection coefficient to account for the multiple reflection caused by the layers below. Therefore, the only difference in the field expression in layer-$(i-1)$, (3.158), is the use of the generalized reflection coefficient instead of the Fresnel reflection coefficient. Hence, replacing $R_{i-1,i-2}^{TE}$ with $\tilde{R}_{i-1,i-2}^{TE}$ in (3.160) and (3.161) results in the following general definition of the amplitude transfer and the generalized reflection coefficient, defined at the interface between layer-i and layer-$(i-1)$:

$$A_{i-1} \, e^{-jk_{z(i-1)}h} = A_i \, \frac{T_{i,i-1}^{\text{TE}} \, e^{-jk_{zi}h}}{1 - \tilde{R}_{i-1,i-2}^{\text{TE}} R_{i-1,i}^{\text{TE}} \, e^{-jk_{z(i-1)}2d_{i-1}}} \tag{3.164}$$

$$\tilde{R}_{i,i-1}^{\text{TE}} = R_{i,i-1}^{\text{TE}} + \frac{T_{i,i-1}^{\text{TE}} T_{i-1,i}^{\text{TE}} \tilde{R}_{i-1,i-2}^{\text{TE}} \, e^{-jk_{z(i-1)}2d_{i-1}}}{1 - \tilde{R}_{i-1,i-2}^{\text{TE}} R_{i-1,i}^{\text{TE}} \, e^{-jk_{z(i-1)}2d_{i-1}}} \tag{3.165}$$

Note that both the amplitude transfer relation (3.164) and the generalized reflection coefficient expression (3.165), defined between layer-i and layer-$(i-1)$, are dependent on the generalized reflection coefficient defined at the interface between layer-$(i-1)$ and layer-$(i-2)$. Hence, to find the amplitude of the field in a layer-j, where $j < i$, it is necessary first to find the generalized reflection coefficients at each interface below layer-i, recursively from (3.165), starting from the bottom of the multilayer structure, where the reflection coefficient is known, and moving up to the layer where the field amplitude is known, say layer-i. Then, the amplitude in layer-j can be determined by calculating the amplitudes in each layer, between layer-i and layer-j, recursively using (3.164).

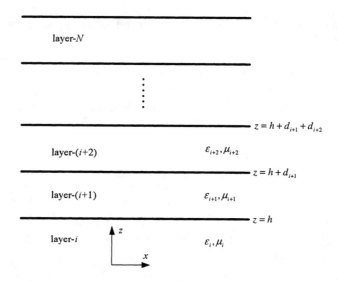

Figure 3.11 Multilayer planar geometry showing the coordinate system.

So far, the source layer or the origin of the coordinate system is assumed to be somewhere in the top layer of the structure. Consequently, the above definitions of the amplitude transfer and the generalized reflection coefficient are derived to transfer the amplitude of the wave downward and to find the generalized reflection coefficient looking downward, respectively. Although the expressions for the amplitude transfer and the generalized reflection, when moving upward, can be deduced from the ones derived when moving downward, it would be a good exercise to go through the important steps of the derivation, without going into detail. For the sake of illustration, let us consider the geometry in Figure 3.11 as a reference, where there are layers above layer-i, denoted by the vertical dotted line. First, the field expressions in layer-i and layer-$(i + 1)$ are written as follows:

$$E_{yi} = A_i \left[e^{-jk_{zi}z} + \widetilde{R}_{i,i+1}^{\text{TE}} e^{jk_{zi}(z-2h)} \right]$$

$$E_{y(i+1)} = A_{i+1} \left[e^{-jk_{z(i+1)}z} + \widetilde{R}_{i+1,i+2}^{\text{TE}} e^{jk_{z(i+1)}z} e^{-jk_{z(i+1)}2(h+d_{i+1})} \right]$$

where the down-going wave in layer-$(i + 1)$ is multiplied by the generalized reflection coefficient defined at the interface between layer-$(i + 1)$ and layer-$(i + 2)$, because of the multiple layers above layer-$(i + 1)$. Then, matching the up-going wave in layer-$(i + 1)$ to the sum of the up-going wave in layer-i and the reflected wave of the down-going wave in layer-$(i + 1)$, all at $z = h$, results in the following amplitude transfer expression:

$$A_{i+1} e^{-jk_{z(i+1)}h} = A_i e^{-jk_{zi}h} \frac{T_{i,i+1}}{1 - \widetilde{R}_{i+1,i+2}^{\text{TE}} R_{i+1,i}^{\text{TE}} e^{-jk_{z(i+1)}2d_{i+1}}} \tag{3.166}$$

Likewise, matching the down-going wave in layer-i to the sum of the reflected wave of the incident wave in layer-i and the transmitted wave of the down-going wave in layer-$(i + 1)$ results in the following generalized reflection coefficient seen looking from layer-i into layer-$(i + 1)$:

$$\widetilde{R}_{i,i+1}^{\text{TE}} = R_{i,i+1}^{\text{TE}} + \frac{T_{i,i+1}^{\text{TE}} T_{i+1,i}^{\text{TE}} \widetilde{R}_{i+1,i+2}^{\text{TE}} e^{-jk_{z(i+1)}2d_{i+1}}}{1 - \widetilde{R}_{i+1,i+2}^{\text{TE}} R_{i+1,i}^{\text{TE}} e^{-jk_{z(i+1)}2d_{i+1}}} \tag{3.167}$$

Throughout this section, only TE-z waves are considered, and therefore, all the reflection and transmission coefficients are of TE types. Since the same analysis holds true for TM-z waves, the amplitude transfer relations and the generalized reflection coefficients for TM-z waves would be the same as those for TE-z waves, (3.164) to (3.167), provided that the reflection and transmission coefficients are replaced by TM types.

3.4.3 Green's Functions in Planar Multilayer Media

Before getting started, it should be remembered that all the derivations so far in this section have assumed planar layered media with no source; that is, the homogenous parts of the governing equations, (3.100) and (3.101), were solved or simplified. Therefore, the solutions so obtained are the natural modes or eigen-solutions of the wave equations and have turned out to be plane waves, for which reflection and transmission coefficients have been defined analytically in the previous sections. Now, in this section, fields and/or potentials due to point current sources, such as horizontal electric, horizontal magnetic, vertical electric, and vertical magnetic dipoles (HED, HMD, VED, and VMD, respectively), embedded in a layered structure will be derived in terms of reflection and transmission coefficients defined for the plane waves.

Since the electric and magnetic fields can be written in terms of vector and scalar potentials, as detailed in Section 3.2.2, Green's functions for these fields can also be obtained using the same relations from Green's function of potentials. Therefore, for the sake of the coherence of the discussion, the expressions of the electric field, (3.168), magnetic field, (3.169), and the Lorentz gauge, (3.170), are rewritten here as

$$\mathbf{E}(\mathbf{r}) = -j\omega \mathbf{A}(\mathbf{r}) - \nabla \phi(\mathbf{r}) \qquad (3.168)$$

$$\mu \mathbf{H}(\mathbf{r}) = \nabla \times \mathbf{A}(\mathbf{r}) \qquad (3.169)$$

$$\nabla \cdot \mathbf{A}(\mathbf{r}) = -j\omega\varepsilon\mu\,\phi(\mathbf{r}) \qquad (3.170)$$

Substituting the scalar potential expression in terms of vector potential from the Lorentz gauge into the electric field expression (3.168), the electric field can be written as

$$\mathbf{E}(\mathbf{r}) = -j\omega \left[\bar{\mathbf{I}} + \frac{\nabla\nabla}{k^2} \right] \cdot \mathbf{A}(\mathbf{r}) \qquad (3.171)$$

Since the vector potential satisfies the wave equation with a source term $-\mu \mathbf{J}$ [see (3.24)], each vector component satisfies the scalar wave equation with the corresponding component of the source. Therefore, for an electric current dipole oriented in an arbitrary direction $\hat{\alpha}$, represented as $\mathbf{J} = \hat{\alpha}Il\,\delta(\mathbf{r} - \mathbf{r}')$, the vector potential in a homogenous and unbounded medium can be written as

$$\mathbf{A}(\mathbf{r},\mathbf{r}') = \hat{\alpha}\mu\,Il \frac{e^{-jk|\mathbf{r}-\mathbf{r}'|}}{4\pi\,|\mathbf{r} - \mathbf{r}'|} \qquad (3.172)$$

where the scalar Green's function is convolved with the source term. Hence, substituting (3.172) into (3.171) yields the electric field in a homogenous and unbounded medium as

$$\mathbf{E}(\mathbf{r}) = -j\omega \left[\bar{\mathbf{I}} + \frac{\nabla\nabla}{k^2} \right] \cdot \hat{\alpha}\mu\,Il \frac{e^{-jk|\mathbf{r}-\mathbf{r}'|}}{4\pi\,|\mathbf{r} - \mathbf{r}'|} \qquad (3.173)$$

It should be stressed here that the vector potential (3.172) and the electric field (3.173) are not for layered media yet, but they are now ready to be written in terms of plane waves by using Weyl's identity, (3.98), or Sommerfeld's identity, (3.99), for the spherical wave terms in the expressions. As Weyl's identity is in a more familiar form for those who are accustomed to the Fourier transformation, the spherical wave term is written as

$$\frac{e^{-jk|\mathbf{r}-\mathbf{r}'|}}{|\mathbf{r} - \mathbf{r}'|} = \frac{1}{\pi} \int\limits_{-\infty}^{\infty} \int\limits_{-\infty}^{\infty} dk_x dk_y\, e^{-jk_x(x-x')-jk_y(y-y')} \frac{e^{-jk_z|z-z'|}}{j2k_z} \qquad (3.174)$$

where the translational invariance of the spherical wave due to a point source in a homogenous and unbounded medium is used. Once the fields are written in terms of plane waves, the field expressions can be obtained in any layer using the definitions of reflection and transmission coefficients at the interfaces between the adjacent layers. Note that writing the spherical wave in terms of plane waves opens the way to using the reflection and transmission coefficients derived for plane waves, and hence to finding the transmission and reflection of spherical waves as a sum of their plane wave constituents. As was noted and demonstrated earlier, the fields in a planar medium (layered in the z-direction and unbounded in the transverse plane) can be decomposed into TE-z (E_y, H_x, H_z) and TM-z (H_y, E_x, E_z) modes, whose field components are mutually exclusive. Note that the TE-z and TM-z modes may be characterized by their longitudinal components H_z and E_z, respectively, and the rest of the fields can be obtained from these components. Therefore, it would be sufficient to write the longitudinal components of the fields, E_z and H_z, in terms of plane waves. To perform the derivation of Green's functions, the source needs to be defined, together with the physical and electrical properties of the layers.

Green's Functions of Horizontal Electric Dipole
For a layered medium, Figure 3.12 is considered the geometrical setting, where the origin is at the bottom of layer-i, and source (HED) is in layer-i and is positioned above the interface between layer-$(i-1)$ and layer-i by a distance of z'. Here are some of the parameters (electrical and physical) of layer-i: dielectric constant of the medium including the loss tangent $\varepsilon_i = \varepsilon_0 \varepsilon_{ri}(1 - j \tan \delta_i)$; permeability of the medium $\mu_i = \mu_0 \mu_{ri}$; and the thickness of the layer d_i.

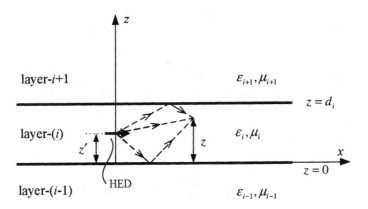

Figure 3.12 Multilayer planar geometry showing the source point z' and observation point z, both in layer-i. Direct wave and reflected waves at the observation point are also demonstrated.

   Note that the origin of the coordinate system is assumed to be at the bottom of
the source layer, as opposed to the location of the source, which was chosen in the
derivations of the generalized reflection and transmission coefficients in Section
3.4.2. So, in cases of using the field expressions and the amplitude transfer
expressions, one needs to use a simple coordinate transformation, like z' replacing
h and $z - z'$ replacing z in the expressions.

   The derivation starts with assuming that the source HED is in a homogenous
and unbounded medium with the electrical properties of layer-i. Hence, the
longitudinal field components can be written from (3.173) and (3.19) as

$$E_{zi} = -\frac{jIl}{4\pi\omega\varepsilon_i} \frac{\partial^2}{\partial z \partial x} \frac{e^{-jk_i|\mathbf{r}-\mathbf{r}'|}}{|\mathbf{r}-\mathbf{r}'|} \tag{3.175}$$

$$H_{zi} = -\frac{Il}{4\pi} \frac{\partial}{\partial y} \frac{e^{-jk_i|\mathbf{r}-\mathbf{r}'|}}{|\mathbf{r}-\mathbf{r}'|} \tag{3.176}$$

Then, using Weyl's identity (3.174) and interchanging the order of differentiation
and the integration result in the following longitudinal field components:

$$E_{zi} = \frac{\pm Il}{8\pi^2\omega\varepsilon_i} \int\limits_{-\infty}^{\infty}\int\limits_{-\infty}^{\infty} dk_x dk_y k_x e^{-jk_x(x-x')-jk_y(y-y')} e^{-jk_{zi}|z-z'|} \tag{3.177}$$

$$H_{zi} = \frac{Il}{8\pi^2} \int\limits_{-\infty}^{\infty}\int\limits_{-\infty}^{\infty} dk_x dk_y k_y e^{-jk_x(x-x')-jk_y(y-y')} \frac{e^{-jk_{zi}|z-z'|}}{k_{zi}} \tag{3.178}$$

where $k_{zi} = \sqrt{k_i^2 - k_x^2 - k_y^2}$, and the $\pm$ sign is due to the derivative of
$\exp(-jk_{zi}|z - z'|)$ with respect to z variable: $+$ for $z > z'$, $-$ for $z > z'$. Note that
although the source seems to be positioned at the point $\mathbf{r}' = (0, y', z')$, as shown in
Figure 3.12, the field expressions in (3.177) and (3.178) have been written for a
more general source location. To extend the field representations, (3.177) and
(3.178), for a layered structure, we should remember that the propagation
constants along the x- and y-directions, k_x and k_y, respectively, must be equal in all
layers due to the phase matching conditions at the interfaces. Therefore, adding
layers above and below the source layer i will only modify the z-dependent
exponential terms in (3.177) and (3.178). For the waves in the source layer; that is,
when both source and observation points are in layer-i, the longitudinal field
components can be written as

$$E_{zi} = \frac{Il}{8\pi^2\omega\varepsilon_i} \int\limits_{-\infty}^{\infty}\int\limits_{-\infty}^{\infty} dk_x dk_y k_x e^{-jk_x(x-x')-jk_y(y-y')} F_{TM}(z,z') \tag{3.179}$$

$$H_{zi} = \frac{Il}{8\pi^2} \int\limits_{-\infty}^{\infty}\int\limits_{-\infty}^{\infty} dk_x dk_y k_y \frac{e^{-jk_x(x-x')-jk_y(y-y')}}{k_{zi}} F_{TE}(z,z') \tag{3.180}$$

where

$$F_{TE}(z,z') = e^{-jk_{zi}|z-z'|} + A_h^e\, e^{jk_{zi}(z-z')} + C_h^e\, e^{-jk_{zi}(z-z')} \qquad (3.181)$$

$$F_{TM}(z,z') = \pm e^{-jk_{zi}|z-z'|} + B_h^e\, e^{jk_{zi}(z-z')} + D_h^e\, e^{-jk_{zi}(z-z')} \qquad (3.182)$$

and A_h^e, B_h^e, C_h^e, and D_h^e are the coefficients of down-going and up-going waves in layer-i. The subscript h and superscript e in the coefficients represent the orientation of the source as horizontal and the source type as electrical, respectively. As E_{zi} and H_{zi} are the only longitudinal components in the TM and TE modes, respectively, the terms in the corresponding integrands are referred to as F_{TM} and F_{TE}. Once the terminology has been settled, let us try to understand these new functions in the integrands, (3.181) and (3.182), as their first terms account for the direct wave and the other two terms for the up-going and down-going waves in layer-i. The down-going waves, the second terms in (3.181) and (3.182), are the consequence of the reflection of the up-going waves from the top of the layer, $z = d_i$ while the up-going waves, the third terms, are the result of the down-going waves. Therefore, the down-going wave for the TM case at the upper interface, $z = d_i$, can be written as the product of the up-going wave at $z = d_i$ and the generalized reflection coefficient defined at the interface between layer-i and layer-$(i + 1)$ as

$$B_h^e e^{jk_{zi}(d_i-z')} = \tilde{R}_{TM}^{i,i+1}\left[e^{-jk_{zi}(d_i-z')} + D_h^e e^{-jk_{zi}(d_i-z')} \right] \qquad (3.183)$$

and likewise, the up-going wave at the lower interface, $z = 0$, is written as the product of the down-going wave at $z = 0$ and the generalized reflection coefficient defined between layer-i and layer-$(i - 1)$ as

$$D_h^e e^{jk_{zi}z'} = \tilde{R}_{TM}^{i,i-1}\left(- e^{-jk_{zi}z'} + B_h^e e^{-jk_{zi}z'} \right) \qquad (3.184)$$

Note that the negative sign in front of the direct wave in (3.184) is due to evaluating (3.182) at $z = 0$, where $z < z'$. The simultaneous solution of (3.183) and (3.184) results in

$$B_h^e = \frac{\tilde{R}_{TM}^{i,i+1} e^{-j2k_{zi}(d_i-z')} - \tilde{R}_{TM}^{i,i+1}\tilde{R}_{TM}^{i,i-1} e^{-j2k_{zi}d_i}}{1 - \tilde{R}_{TM}^{i,i+1}\tilde{R}_{TM}^{i,i-1} e^{-j2k_{zi}d_i}} \qquad (3.185)$$

$$D_h^e = \frac{- \tilde{R}_{TM}^{i,i-1} e^{-j2k_{zi}z'} + \tilde{R}_{TM}^{i,i-1}\tilde{R}_{TM}^{i,i+1} e^{-j2k_{zi}d_i}}{1 - \tilde{R}_{TM}^{i,i+1}\tilde{R}_{TM}^{i,i-1} e^{-j2k_{zi}d_i}} \qquad (3.186)$$

Following the same procedure for TE waves using (3.181), the coefficients A_h^e and C_h^e can be obtained as

$$A_h^e = \frac{\tilde{R}_{TE}^{i,i+1} e^{-j2k_{zi}(d_i-z')} + \tilde{R}_{TE}^{i,i+1}\tilde{R}_{TE}^{i,i-1} e^{-j2k_{zi}d_i}}{1 - \tilde{R}_{TE}^{i,i+1}\tilde{R}_{TE}^{i,i-1} e^{-j2k_{zi}d_i}} \qquad (3.187)$$

$$C_h^e = \frac{\tilde{R}_{TE}^{i,i-1} e^{-j2k_{zi}z'} + \tilde{R}_{TE}^{i,i-1} \tilde{R}_{TE}^{i,i+1} e^{-j2k_{zi}d_i}}{1 - \tilde{R}_{TE}^{i,i+1} \tilde{R}_{TE}^{i,i-1} e^{-j2k_{zi}d_i}} \tag{3.188}$$

Thus, as we know all the coefficients in the representations of the longitudinal electric and magnetic fields in terms of known quantities of the geometry, when the source and observation points are in the same layer, the spectral-domain representations of Green's functions for these field components can be written from (3.179) and (3.180) as

$$\tilde{G}_{zx}^E = \frac{1}{2\omega\varepsilon_i} k_x \left[\pm e^{-jk_{zi}|z-z'|} + B_h^e e^{jk_{zi}(z-z')} + D_h^e e^{-jk_{zi}(z-z')} \right] \tag{3.189}$$

$$\tilde{G}_{zx}^H = \frac{k_y}{2k_{zi}} \left[e^{-jk_{zi}|z-z'|} + A_h^e e^{jk_{zi}(z-z')} + C_h^e e^{-jk_{zi}(z-z')} \right] \tag{3.190}$$

where $\sim$ denotes the spectral-domain representation, except for the reflection coefficient for which $\sim$ refers to the generalized reflection coefficient; superscripts E and H stand for the electric and magnetic fields; and subscript zx denotes the z component of the electric or magnetic field Green's function due to the x component of the source. Once the spectral-domain Green's functions are known, their spatial-domain counterparts can be obtained by the two-dimensional inverse Fourier transform, (3.179) and (3.180), or equivalently by the inverse Hankel transform.

With the knowledge of the longitudinal field components as the integral sum of their spectral representations, (3.179) and (3.180), the transverse components of the electric and magnetic fields can be obtained, as in the case of the general cylindrical waveguide in Section 1.3. For the sake of completeness, let us demonstrate some of the key steps of the derivation for the expressions relating the transverse components to the longitudinal components. First of all, it is assumed that field components have their unique spectral-domain representations, and hence the fields can be written as the inverse Fourier transform

$$\left\{ \begin{matrix} \mathbf{E(r)} \\ \mathbf{H(r)} \end{matrix} \right\} = \frac{1}{4\pi^2} \int_{-\infty}^{\infty} \int_{-\infty}^{\infty} dk_x dk_y \left\{ \begin{matrix} \tilde{\mathbf{E}}(k_x,k_y;\mathbf{r}) \\ \tilde{\mathbf{H}}(k_x,k_y;\mathbf{r}) \end{matrix} \right\} \tag{3.191}$$

or Hankel transform

$$\left\{ \begin{matrix} \mathbf{E(r)} \\ \mathbf{H(r)} \end{matrix} \right\} = \frac{1}{4\pi} \int_{SIP} dk_\rho \left\{ \begin{matrix} \tilde{\mathbf{E}}(k_\rho;\mathbf{r}) \\ \tilde{\mathbf{H}}(k_\rho;\mathbf{r}) \end{matrix} \right\} \tag{3.192}$$

of these spectral-domain representations. As each spectral component satisfies Maxwell's equations, Maxwell's curl equations operating on the spectral components are split into longitudinal and transverse components, with the fields and gradient operator split likewise. Therefore, by rearranging the vector components and applying some vector identities, the following spectral-domain representations of the transverse components can be obtained in terms of the longitudinal components in the same domain:

$$\tilde{\mathbf{H}}_t = \frac{1}{k^2 - k_z^2}\left[\frac{\partial}{\partial z}\nabla_t \tilde{H}_z + j\omega\varepsilon\,\nabla_t \times \hat{z}\tilde{E}_z\right] \tag{3.193}$$

$$\tilde{\mathbf{E}}_t = \frac{1}{k^2 - k_z^2}\left[\frac{\partial}{\partial z}\nabla_t \tilde{E}_z - j\omega\mu\,\nabla_t \times \hat{z}\tilde{H}_z\right] \tag{3.194}$$

which are the same equations as the ones derived in Section 1.3.1. Note that all the field quantities in these expressions are in the spectral domain; that is, they are the integrands of the integral representations (3.191) or (3.192), not the integrals themselves.

With the use of (3.193) and (3.194) for the spectral components of the longitudinal fields that have already been obtained in the source layer, (3.189) and (3.190), the following components of Green's function for electric and magnetic fields are obtained in the source layer i:

1. x-polarized electric field

$$\tilde{G}_{xx}^E = \frac{-j\omega\mu}{2\left(k_i^2 - k_{zi}^2\right)}\left[\frac{k_x^2}{k_i^2}\frac{\partial}{\partial z}F_{TM}(z,z') + \frac{k_y^2}{jk_{zi}}F_{TE}(z,z')\right] \tag{3.195}$$

2. y-polarized electric field

$$\tilde{G}_{yx}^E = \frac{-j\omega\mu k_x k_y}{2\left(k_i^2 - k_{zi}^2\right)}\left[\frac{1}{k_i^2}\frac{\partial}{\partial z}F_{TM}(z,z') + \frac{j}{k_{zi}}F_{TE}(z,z')\right] \tag{3.196}$$

3. x-polarized magnetic field

$$\tilde{G}_{xx}^H = \frac{k_x k_y}{2\left(k_i^2 - k_{zi}^2\right)}\left[F_{TM}(z,z') - \frac{j}{k_{zi}}\frac{\partial}{\partial z}F_{TE}(z,z')\right] \tag{3.197}$$

4. y-polarized magnetic field

$$\tilde{G}_{yx}^H = \frac{-1}{2\left(k_i^2 - k_{zi}^2\right)}\left[k_x^2 F_{TM}(z,z') + \frac{jk_y^2}{k_{zi}}\frac{\partial}{\partial z}F_{TE}(z,z')\right] \tag{3.198}$$

due to x-polarized electric current dipole. Note that although these Green's functions have been obtained specifically for an x-directed HED, Green's functions for a y-directed HED can be easily deduced from them by interchanging x and y in the subscripts and k_x and k_y in the expressions. Hence, the spectral-domain Green's functions for the electric and magnetic fields due to an HED in a multilayer environment have been obtained in closed forms in the source layer.

Since vector and scalar potentials play important roles in the analysis of planar multilayer structures, this section would not be complete without giving the derivations of their Green's functions. As used in (3.172) and the preceding discussion, the vector potential due to a dipole in an unbounded medium has a single vector component in the direction of the dipole. However, it is a well-known fact that this is not the case if there is a plane boundary, for which a single vector component of a vector potential cannot describe the electromagnetic field everywhere [14, 15]. The phrase "cannot describe" means the boundary conditions for the electric and magnetic fields to be satisfied at the interface cannot be satisfied with a single vector potential. As originated by Sommerfeld, vector potential due to a horizontal dipole over a layered medium, whose planes are on the xy-plane, is generally assumed to have two vector components: one in the direction of the dipole and the other in the direction perpendicular to the xy-plane (i.e., longitudinal). For a vertical dipole, a single component of the vector potential in the direction of the dipole becomes sufficient to satisfy all the boundary conditions of the field quantities. As a result, vector components of Green's functions for vector potential due to a general dipole can be written in dyadic form as

$$\overline{\mathbf{G}}^A = \left(\hat{x}\hat{x} + \hat{y}\hat{y} \right) G_{xx}^A + \hat{z}\hat{x} G_{zx}^A + \hat{z}\hat{y} G_{zy}^A + \hat{z}\hat{z} G_{zz}^A \tag{3.199}$$

which is called the traditional form. However, it has been demonstrated that this set of components of the vector potential is not unique, as one may choose the y component of the vector potential to go along with the x component for an x-directed electric dipole [16, 17]. This choice of the vector components leads to an alternative form of the dyadic Green's function for the vector potential:

$$\overline{\mathbf{G}}^A = \hat{x}\hat{x} G_{xx}^A + \hat{y}\hat{y} G_{yy}^A + \left(\hat{x}\hat{y} + \hat{y}\hat{x} \right) G_{xy}^A + \hat{z}\hat{z} G_{zz}^A \tag{3.200}$$

Although there are other possible combinations of vector components for the vector potential in layered media, the traditional form is the one commonly used in the analysis of printed structures in layered media. Therefore, the vector components in (3.199) are studied and derived in this section.

Once the components of the vector potentials are chosen for a given dipole orientation, the scalar potential can be obtained via the Lorentz gauge (3.170). Because the Lorentz gauge requires the divergence of the vector potential and because the vector potentials associated with HED and VED have different components, the scalar potential may not be unique for these sources. With this background in mind, an expression that relates Green's function of the scalar potential to those of the vector potential is derived, starting from the Lorentz gauge as

$$\nabla \cdot \mathbf{A}(\mathbf{r}) = -j\omega\varepsilon\mu\,\phi(\mathbf{r})$$

Then, the vector and scalar potentials are defined as follows and are substituted into the above equation:

$$\mathbf{A}(\mathbf{r}) = \int_V d\mathbf{r}' \overline{\mathbf{G}}^A(\mathbf{r}, \mathbf{r}') \cdot \mathbf{J}(\mathbf{r}') \tag{3.201}$$

$$\varphi(\mathbf{r}) = \int_V d\mathbf{r}' G^q(\mathbf{r}, \mathbf{r}') \rho(\mathbf{r}') \tag{3.202}$$

where the Green's function of the scalar potential G^q is for a single, time-harmonic point charge, not for a double charge associated with a dipole. Hence, the Lorentz gauge can be written in terms of Green's functions as

$$\nabla \cdot \int_V dr' \overline{\mathbf{G}}^A(\mathbf{r},\mathbf{r}') \cdot \mathbf{J}(\mathbf{r}') = -j\omega\mu_i\varepsilon_i \int_V dr' G^q(\mathbf{r},\mathbf{r}')\rho(\mathbf{r}') \qquad (3.203)$$

As the divergence on the left-hand side is on the observation coordinate $\mathbf{r}$, it can be brought under the integral sign operating on the dyadic Green's function of the vector potential. Moreover, using the continuity equation for the charge density on the right-hand side, (3.203) can be cast into the following form:

$$\int_V dr' \left[\nabla \cdot \overline{\mathbf{G}}^A(\mathbf{r},\mathbf{r}')\right] \cdot \mathbf{J}(\mathbf{r}') = -j\omega\mu_i\varepsilon_i \int_V dr' G^q(\mathbf{r},\mathbf{r}')\frac{\nabla' \cdot \mathbf{J}(\mathbf{r}')}{-j\omega} \qquad (3.204)$$

To relate the integrand on the left to the one on the right, it is necessary to transfer the gradient on the right, operating on the current density, onto Green's function of the scalar potential. To do so, the following vector identity is used for the integral on the right-hand side,

$$\nabla' \cdot \left[G^q(\mathbf{r},\mathbf{r}')\mathbf{J}(\mathbf{r}')\right] = G^q(\mathbf{r},\mathbf{r}')\left[\nabla' \cdot \mathbf{J}(\mathbf{r}')\right] + \nabla' G^q(\mathbf{r},\mathbf{r}') \cdot \mathbf{J}(\mathbf{r}') \qquad (3.205)$$

leading to

$$\int_V dr' \left[\nabla \cdot \overline{\mathbf{G}}^A(\mathbf{r},\mathbf{r}')\right] \cdot \mathbf{J}(\mathbf{r}') = \mu_i\varepsilon_i \int_V dr' \nabla' \cdot \left[G^q(\mathbf{r},\mathbf{r}')\mathbf{J}(\mathbf{r}')\right]$$
$$- \mu_i\varepsilon_i \int_V dr' \nabla' G^q(\mathbf{r},\mathbf{r}') \cdot \mathbf{J}(\mathbf{r}') \qquad (3.206)$$

Since the current densities are defined over conducting bodies for the problems of printed geometries in layered media, as a physical constraint, the normal components of the current density at the surface of the conducting body must be zero. Therefore, applying the divergence theorem for the first integral on the right-hand side, and using this fact, the following relation between the vector and scalar potential Green's functions is obtained:

$$\frac{j\omega}{k_i^2}\nabla \cdot \overline{\mathbf{G}}^A = \frac{1}{j\omega}\nabla' G^q \qquad (3.207)$$

So, with the knowledge of the components of the vector potential (3.199) and the relation between Green's functions of the scalar and vector potentials (3.207), one may find the spectral-domain representations of these potentials from those of the field quantities.

Remembering that $\mu_i\mathbf{H} = \nabla \times \mathbf{A}$, and $\mathbf{A} = \hat{x}A_x + \hat{z}A_z$ for an x-directed electric dipole, and that such relations hold for the spectral representations of these quantities, one can simply write

$$\tilde{A}_x = \frac{\mu_i}{jk_y}\tilde{H}_z; \quad \tilde{A}_z = -\frac{\mu_i}{jk_y}\tilde{H}_x \qquad (3.208)$$

Hence, the components of Green's function of the vector potential for an x-directed dipole are obtained as

$$\tilde{G}^A_{xx} = \frac{\mu_i}{j2k_{zi}}\left[e^{-jk_{zi}|z-z'|} + A^e_h e^{jk_{zi}(z-z')} + C^e_h e^{-jk_{zi}(z-z')}\right] \qquad (3.209)$$

$$\tilde{G}^A_{zx} = \frac{j\mu_i k_x}{2(k_i^2 - k_{zi}^2)}\left[(A^e_h + B^e_h)e^{jk_{zi}(z-z')} + (D^e_h - C^e_h)e^{-jk_{zi}(z-z')}\right] \qquad (3.210)$$

For the scalar potential Green's function, using (3.207) together with the available components of the vector potential, the following relation is obtained in the spatial domain:

$$\frac{\partial}{\partial x'}G^q_x = \frac{-1}{\mu_i \varepsilon_i}\left[\frac{\partial}{\partial x}G^A_{xx} + \frac{\partial}{\partial z}G^A_{zx}\right] \qquad (3.211)$$

Considering the spectral-domain representations of Green's functions, with $\partial/\partial x' = -\partial/\partial x$ and $\partial/\partial x \to -jk_x$ substitutions, (3.211) can be written in the spectral domain as follows:

$$\tilde{G}^q_x = \frac{-1}{\mu_i \varepsilon_i}\left[-\tilde{G}^A_{xx} + \frac{1}{jk_x}\frac{\partial}{\partial z}\tilde{G}^A_{zx}\right] \qquad (3.212)$$

Substituting (3.209) and (3.210) into (3.212) and rearranging the terms would result in the following spectral-domain Green's function of scalar potential due to an x-directed electric dipole:

$$\tilde{G}^q_x = \frac{1}{j2\varepsilon_i k_\rho^2}\left[\frac{k_\rho^2}{k_{zi}}e^{-jk_{zi}|z-z'|} + \frac{k_i^2 A^e_h + k_{zi}^2 B^e_h}{k_{zi}}e^{jk_{zi}(z-z')}\right.$$
$$\left. + \frac{k_i^2 C^e_h - k_{zi}^2 D^e_h}{k_{zi}}e^{-jk_{zi}(z-z')}\right] \qquad (3.213)$$

As the above derivations are for the dipole oriented in the x-direction, Green's functions for a dipole in y-direction (still a HED) can simply be obtained by setting $\tilde{G}^A_{yy} = \tilde{G}^A_{xx}$, $\tilde{G}^A_{zy}/k_y = \tilde{G}^A_{zx}/k_x$, and $\tilde{G}^q_y = \tilde{G}^q_x$.

Let us note one more time that all of the Green's functions obtained in this section are due to an HED, are defined in the source layer, and are given in the spectral domain. As it must have been noticed, the spectral-domain Green's functions are in closed forms (i.e., analytically represented), and their spatial-domain counterparts can only be obtained using either Fourier or Hankel inversion integrals. As these inversion integrals are computationally expensive due to their oscillatory kernels and the slow converging nature of the spectral-domain Green's functions for some cases, the use of the spatial-domain Green's functions was not popular until the introduction of an efficient approach to evaluate these integrals was introduced [18, 19]. Since such numeric algorithms are out of the scope of this book, they are not discussed here any further, except to state that the spatial-domain Green's functions are to be obtained from the given spectral-domain representations.

Green's Functions of Vertical Electric Dipole

The procedure to derive the spectral-domain Green's function for a VED is almost the same as the one for an HED. So, without repeating all the details, the longitudinal components of the electric and magnetic fields in a homogenous and unbounded medium can be written for a VED as

$$E_{zi} = -\frac{jIl\omega\mu_i}{4\pi k_i^2}\left(k_i^2 + \frac{\partial^2}{\partial z^2}\right)\frac{e^{-jk_i|\mathbf{r}-\mathbf{r}'|}}{|\mathbf{r}-\mathbf{r}'|} \tag{3.214}$$

$$H_{zi} = 0 \tag{3.215}$$

where the vector potential (3.172) with $\hat{\alpha} = \hat{z}$ is used in (3.173) and (3.19). Note that having a zero longitudinal magnetic field implies the absence of TE waves for this source orientation. Substituting Weyl's identity (3.174) into (3.214) and interchanging the order of differentiation and integration result in

$$E_{zi} = \frac{-Il}{8\pi^2\omega\varepsilon_i}\int_{-\infty}^{\infty}\int_{-\infty}^{\infty}dk_x dk_y\, k_\rho^2\, e^{-jk_x(x-x')-jk_y(y-y')}\frac{e^{-jk_{zi}|z-z'|}}{k_{zi}} \tag{3.216}$$

for a homogenous and unbounded medium. Note that since the second derivative with respect to z operating on the direct term produces an impulse-type singularity when $z = z'$, this field representation is not valid at the plane of the source. As one can find a detailed treatment of this singularity in [12], it is sufficient to state here that this singularity is the result of interchanging the order of the gradient and the integral in the derivation of the dyadic Green's function of the electric field (3.30). To extend this to a multilayer medium, as was done in the case of HED, only the z-dependent term needs to be modified to account for the reflected waves as

$$E_{zi} = \frac{-Il}{8\pi^2\omega\varepsilon_i}\int_{-\infty}^{\infty}\int_{-\infty}^{\infty}dk_x dk_y\,\frac{k_\rho^2}{k_{zi}}e^{-jk_x(x-x')-jk_y(y-y')}F_{TM}(z,z') \tag{3.217}$$

where

$$F_{TM}(z,z') = e^{-jk_{zi}|z-z'|} + A_v^e\, e^{jk_{zi}(z-z')} + B_v^e\, e^{-jk_{zi}(z-z')} \tag{3.218}$$

and A_v^e and B_v^e are the coefficients of the down-going and up-going waves due to the reflections from layers above and below the source layer, respectively. Again matching the down-going wave at the upper interface to the reflection of the up-going waves, and the up-going wave at the lower interface to the reflection of the down-going waves, these coefficients can be found as

$$A_v^e = \frac{\tilde{R}_{TM}^{i,i-1} e^{-j2k_{zi}z'} + \tilde{R}_{TM}^{i,i-1} \tilde{R}_{TM}^{i,i+1} e^{-j2k_{zi}d_i}}{1 - \tilde{R}_{TM}^{i,i+1} \tilde{R}_{TM}^{i,i-1} e^{-j2k_{zi}d_i}} \tag{3.219}$$

$$B_v^e = \frac{\tilde{R}_{TM}^{i,i+1} e^{-j2k_{zi}(d_i-z')} + \tilde{R}_{TM}^{i,i+1} \tilde{R}_{TM}^{i,i-1} e^{-j2k_{zi}d_i}}{1 - \tilde{R}_{TM}^{i,i+1} \tilde{R}_{TM}^{i,i-1} e^{-j2k_{zi}d_i}} \tag{3.220}$$

Hence, Green's function of the longitudinal electric field in the source layer can be written as

$$\tilde{G}_{zz}^E = \frac{-1}{2\omega\varepsilon_i} \frac{k_\rho^2}{k_{zi}} F_{TM}(z,z') \tag{3.221}$$

and Green's functions of the other field components can be obtained by substituting the integrands of (3.217) into (3.193) and (3.194).

For Green's function of the vector potential, since there is only the z component of the vector potential, according to the traditional choice (3.199), using (3.19) provides a simple expression relating the x component or y component of the magnetic field to the z component of the vector potential. Then, it is a simple matter to find Green's function of the vector potential as

$$\tilde{G}_{zz}^A = \frac{\mu_i}{j2k_{zi}} F_{TM}(z,z') \tag{3.222}$$

However, finding Green's function for the scalar potential due to a VED is simple methodically but requires a bit of algebra and manipulation. It starts with using the definition (3.207), which results in

$$\frac{\partial}{\partial z'} \tilde{G}_z^q = -\frac{1}{\mu_i \varepsilon_i} \frac{\partial}{\partial z} \tilde{G}_{zz}^A \tag{3.223}$$

Then, the z-derivative on the right-hand side is changed to the z'-derivative after F_{TM} has been written as an explicit function of z and z' and the signs in the interchange have been properly chosen. Hence, Green's function for the scalar potential is obtained as

$$\tilde{G}_z^q = \frac{1}{j2\varepsilon_i k_{zi}} \left[e^{-jk_{zi}|z-z'|} + C_v^e e^{jk_{zi}(z-z')} + D_v^e e^{-jk_{zi}(z-z')} \right] \tag{3.224}$$

where

$$C_v^e = \frac{-\tilde{R}_{TM}^{i,i-1} e^{-j2k_{zi}z'} + \tilde{R}_{TM}^{i,i-1} \tilde{R}_{TM}^{i,i+1} e^{-j2k_{zi}d_i}}{1 - \tilde{R}_{TM}^{i,i+1} \tilde{R}_{TM}^{i,i-1} e^{-j2k_{zi}d_i}} \tag{3.225}$$

$$D_v^e = \frac{-\tilde{R}_{TM}^{i,i+1} e^{-j2k_{zi}(d_i-z')} + \tilde{R}_{TM}^{i,i+1} \tilde{R}_{TM}^{i,i-1} e^{-j2k_{zi}d_i}}{1 - \tilde{R}_{TM}^{i,i+1} \tilde{R}_{TM}^{i,i-1} e^{-j2k_{zi}d_i}} \tag{3.226}$$

Hence, all the necessary Green's functions in the source layer, due to the sources of HED and VED, have been obtained. It should be noted here that Green's functions of the scalar potential for the HED and VED are different, as it was pointed out in the discussion of the choice of the traditional form for the components of the vector potential in a layered medium. However, it has been demonstrated in [17] that the choice of the alternative representation of Green's function of the vector potential (3.200) results in an identical Green's function for the scalar potential associated with the HED and VED in layered media.

Different Source and Observation Layers

This issue has already been discussed in Section 3.4.2 for TE and TM waves with no explicit direct term. That is, the field expressions in layer-j were assumed to be in the form of

$$A_j \left[e^{jk_{zj}(z-z')} + \tilde{R}_{j,j-1} e^{-jk_{zj}(z-z')} e^{-jk_{zj}2z_{jl}} \right] \quad \text{for } j < i \qquad (3.227)$$

$$A_j \left[e^{-jk_{zj}(z-z')} + \tilde{R}_{j,j+1} e^{jk_{zj}(z-z')} e^{-jk_{zj}2z_{ju}} \right] \quad \text{for } j > i \qquad (3.228)$$

where the reflection coefficients are defined at the reference planes $z = -z_{jl}$ (lower interface of layer-j) and $z = z_{ju}$ (upper interface of layer-j) for $j < i$ and $j > i$, respectively, and layer-i is the source layer. In addition, the unknown coefficient A_j is related to the known coefficient of the same form as derived in (3.164) and (3.166). However, in Green's function problems, the only known amplitude must be at the source layer, as obtained in (3.179) and (3.180), and they are given in the spectral-domain as follows

$$\varepsilon_i \tilde{E}_{zi} \propto \left[\pm e^{-jk_{zi}|z-z'|} + B_h^e e^{jk_{zi}(z-z')} + D_h^e e^{-jk_{zi}(z-z')} \right] \qquad (3.229)$$

$$\tilde{H}_{zi} \propto \left[e^{-jk_{zi}|z-z'|} + A_h^e e^{jk_{zi}(z-z')} + C_h^e e^{-jk_{zi}(z-z')} \right] \qquad (3.230)$$

apart from a multiplicative constant for each. Therefore, one needs to start to transfer the amplitudes from the source layer to the observation layer, downward or upward, depending upon the location of the observation point. To do so, the field representations in the source layer need to be cast into the form of (3.227) or (3.228), which requires the substitution of the coefficients and rearranging the resulting expressions into one of the required forms to be able to apply the amplitude transfer expressions (3.164) and (3.166). For example, the magnetic field expression (3.230) can be written as

$$\tilde{H}_{zi} \propto \underbrace{\frac{1 + e^{-jk_{zi}2z'} \tilde{R}_{i,i-1}^{TE}}{1 - \tilde{R}_{i,i+1}^{TE} \tilde{R}_{i,i-1}^{TE} e^{-jk_{zi}2d_i}}}_{A_i^+} \left[e^{-jk_{zi}(z-z')} + \tilde{R}_{i,i+1}^{TE} e^{jk_{zi}(z-z')} e^{-jk_{zi}2(d_i-z')} \right]$$

for $z > z'$, and

$$\tilde{H}_{zi} \propto \underbrace{\frac{1 + e^{-jk_{zi}2(d_i-z')} \tilde{R}_{i,i+1}^{TE}}{1 - \tilde{R}_{i,i+1}^{TE} \tilde{R}_{i,i-1}^{TE} e^{-jk_{zi}2d_i}}}_{A_i^-} \left[e^{jk_{zi}(z-z')} + \tilde{R}_{i,i-1}^{TE} e^{-jk_{zi}(z-z')} e^{-jk_{zi}2z'} \right]$$

for $z < z'$. Then, the field representations in any layer can be obtained by just implementing the amplitude transfer expressions onto A_i^+ or A_i^-, recursively up to or down to the observation layer. Once the longitudinal fields are obtained in the observation layer, the rest of the calculations follow the same procedure and equations as those for the same source and observation layers.

3.5 APPLICATION OF MOM TO PRINTED CIRCUITS

As discussed in the introduction of this chapter, the MoM is one of the most common numerical techniques for the solution of integral equations in the analysis of printed geometries [20–32]. Although the method is very suitable for such structures, it requires additional information as compared to other rigorous techniques: the knowledge of Green's functions, relevant for the integral equation used for the analysis. Since Green's functions of different field and potential parameters have already been derived as the first step, a governing equation in the form of an integral equation needs to be written as the second step for electromagnetic fields in planar multilayer environments. Integral equations are the equations with unknown functions under the integral operator, and generally they do not have closed-form solutions. As there may be a number of integral equations that govern the same problem for different fields or potentials, the class of integral equations may be restricted for the geometries of interest in this book. Let us first state the common geometrical features of the problems that are of interest: geometries are (1) built on a planar multilayered dielectric medium; (2) printed on a plane; and/or (3) to use vertical thin metallic strips or pins between the layers. Based on the first feature, Green's functions have been obtained to account for the planar multilayer structure; hence, there will be no more implementation of boundary conditions between the layers. From the second and third features, one may deduce that defining surface current densities on the conductors will be sufficient for such structures, resulting in surface integrals in the definition of the field quantities rather than volume integrals. As a result, integral equations will be derived using the superposition integral of Green's functions and the surface current densities, based on the linearity of Maxwell's equations in linear media with linear boundaries.

3.5.1 Integral Equations

In view of the above discussion, integral equations in EM are nothing but the integral representations of the field quantities in terms of Green's functions and the current densities. For example, let us write the electric field integral equation (EFIE) for a typical printed geometry on the xy-plane, as shown in Figure 3.13, where the excitation is an incident electric field $\mathbf{E}^i$. Implementing the boundary condition on the tangential electric field at the conductor yields the following equation:

$$\hat{n} \times \left[\mathbf{E}^i(\mathbf{r}) + \mathbf{E}^s(\mathbf{r}) \right] = 0 \qquad \mathbf{r} \in \mathbf{r}' \qquad (3.231)$$

where $\hat{n} = \hat{z}$ is the unit normal vector at the surface of the conductor. As the incident field is known and the scattered field is due to the induced current density over the conductor, $\mathbf{J}_s$, the boundary condition (3.231) can be written more explicitly as

$$\hat{n} \times \iint_{D(\mathbf{J}_s)} \overline{\mathbf{G}}^E(\mathbf{r},\mathbf{r}') \cdot \mathbf{J}_s(\mathbf{r}') = -\hat{n} \times \mathbf{E}^i(\mathbf{r}) \qquad \mathbf{r} \in \mathbf{r}' \qquad (3.232)$$

where $D(\mathbf{J}_s)$ denotes the domain of integration over the domain of the current density $\mathbf{J}_s$, and $\overline{\mathbf{G}}^E$ is the dyadic Green's function of the electric field. The only unknown in (3.232) is a function, which is the surface current density, and under the integral operator, this equation is referred to as electric field integral equation (EFIE), as it employs Green's function of the electric field. Although this equation has been written for an assumed source of incident plane wave, it is also valid for an impressed current source connected directly to the printed conductor. In the case of an impressed current source, one can assume that the current under the integral operator has two parts: one is induced and unknown, and the other is impressed and known. Hence, if there is no incident wave, then the right-hand side of (3.232) becomes zero, but then the part of the integral due to the known part of the current density can be taken to the right-hand side, making the integral equation inhomogeneous.

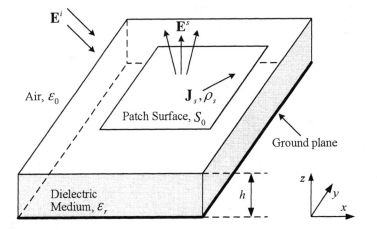

Figure 3.13 A typical printed circuit on a planar layered medium.

Once the governing integral equation has been derived, the only thing left to do is to solve for the unknown function under the integral sign via a numerical technique. It should be stated that using the boundary condition for the magnetic field, instead of the electric field as in the case of (3.232), results in an integral equation known as magnetic field integral equation (MFIE), which requires the knowledge of Green's functions of the magnetic field. The excitation for general microstrip circuits can be modeled using an impressed current source(s) at the excitation location. Then, each component of the electric field tangential to the plane of conductor ($z = z' = h$) can be written as follows:

$$E_x(x, y) = \iint\limits_{D(J_x)} dx'dy' G_{xx}^E(x - x', y - y') J_x(x', y') +$$
$$\iint\limits_{D(J_y)} dx'dy' G_{xy}^E(x - x', y - y') J_y(x', y') \tag{3.233}$$

$$E_y(x, y) = \iint\limits_{D(J_x)} dx'dy' G_{yx}^E(x - x', y - y') J_x(x', y') +$$
$$\iint\limits_{D(J_y)} dx'dy' G_{yy}^E(x - x', y - y') J_y(x', y') \tag{3.234}$$

where the current density and spatial-domain Green's functions are evaluated at the planes of the conductors, and $D(J_{x,y})$ denotes the domain of $J_{x,y}$ as the domain of integration. Since these field expressions are valid over the whole plane, not only on the conductors, the field components on the left-hand sides of (3.233) and (3.234) are unknowns, as are the current density components on the conductors. With the application of the boundary conditions, either pointwise or in the integral sense, it may be possible to eliminate the unknown field components, resulting in an EFIE formulated in the spatial domain with unknown surface current densities. However, if one uses the Fourier-transformation property of the convolution operator, (3.233) and (3.234) can be written in the spectral domain as

$$\tilde{E}_x(k_x, k_y) = \tilde{G}_{xx}^E(k_x, k_y)\tilde{J}_x(k_x, k_y) + \tilde{G}_{xy}^E(k_x, k_y)\tilde{J}_y(k_x, k_y) \tag{3.235}$$

$$\tilde{E}_y(k_x, k_y) = \tilde{G}_{yx}^E(k_x, k_y)\tilde{J}_x(k_x, k_y) + \tilde{G}_{yy}^E(k_x, k_y)\tilde{J}_y(k_x, k_y) \tag{3.236}$$

where all quantities are in the spectral domain. Since no integrals are involved in this representation of the field quantities, they are just linear equations with more unknown functions than equations. To eliminate the unknown field quantities on the left-hand sides of (3.235) and (3.236), boundary conditions need to be imposed over the conductor, but this is not as straightforward as it is for the spatial-domain implementation. This is because the spectral-domain tangential field components at the conductors are not equal to zero. Because the details of the spectral-domain MoM are out of the scope of this book, and because the spatial-domain formulation of the integral equation and its solution in planar multilayer media have been proved more computationally efficient [24–30], interested readers may refer to [33, 34] for the introduction to the spectral-domain approach and to the wealth of papers in *IEEE Transactions on Microwave Theory and Techniques* and *IEEE Transactions on Antennas and Propagation* for the application of the spectral-domain method of moments.

As a result of the above discussions, it can be concluded that spatial-domain approaches have become more efficient after the introduction of the closed-form Green's functions. In addition, remembering that the electric-field Green's functions are more singular when compared to those of the vector and scalar potentials, as discussed in Section 3.2.2, the electric field can be written in terms of vector and scalar potentials in (3.231). Consequently, a new form of integral equation is obtained, which involves Green's functions of both scalar and vector potentials, and it is referred to as a mixed potential integral equation (MPIE):

$$\hat{n} \times \left[\underbrace{- j\omega \int_{V'} d\mathbf{r}' \overline{\mathbf{G}}^A (\mathbf{r},\mathbf{r}') \cdot \mathbf{J}(\mathbf{r}')}_{\mathbf{A}(\mathbf{r})} - \underbrace{\nabla \int_{V'} d\mathbf{r}' G^q(\mathbf{r},\mathbf{r}') \frac{-\nabla' \cdot \mathbf{J}(\mathbf{r}')}{j\omega}}_{\varphi(\mathbf{r})} \right] = 0 \quad \mathbf{r} \in \mathbf{r}' \qquad (3.237)$$

Hence, a more manageable (from a computational point of view) integral equation has been derived with the assumption that Green's functions are known a priori. Instead of writing the MPIE in a general form, one may prefer to write the field components explicitly as

$$E_x = -j\omega G_{xx}^A * J_x + \frac{1}{j\omega} \frac{\partial}{\partial x} \left(G^q * \nabla \cdot \mathbf{J} \right) \qquad (3.238)$$

$$E_y = -j\omega G_{yy}^A * J_y + \frac{1}{j\omega} \frac{\partial}{\partial y} \left(G^q * \nabla \cdot \mathbf{J} \right) \qquad (3.239)$$

$$E_z = -j\omega G_{zx}^A * J_x - j\omega G_{zy}^A * J_y - j\omega G_{zz}^A * J_z + \frac{1}{j\omega} \frac{\partial}{\partial z} \left(G^q * \nabla \cdot \mathbf{J} \right) \qquad (3.240)$$

and then to apply the necessary boundary conditions. Note that the convolution operator $*$ denotes the superposition integral in the cases of z and/or z' integrations, as the associated Green's functions are not translational invariant in this direction. Since the traditional choice of the vector potential components has been used in this formulation, (3.199), the scalar potentials for the horizontal and vertical dipoles are not equal to each other. Therefore, it would be instructive to explicitly write the terms of (3.238) to (3.240) involving Green's function of the scalar potential:

$$G^q * \nabla \cdot \mathbf{J} = G_x^q * \frac{\partial J_x}{\partial x} + G_y^q * \frac{\partial J_y}{\partial y} + G_z^q * \frac{\partial J_z}{\partial z} \qquad (3.241)$$

where $G_x^q (=G_y^q)$ is different from G_z^q. Once the integral equation has been derived to govern physical mechanisms in the structure, the goal becomes to find the current density over the conductors, and in turn, to characterize the structure. To do so, a numerical method for the solution of an integral equation needs to be employed, and in this book, the choice is the MoM. In the following section, the spatial-domain MoM will be introduced briefly for the solution of MPIE.

3.5.2 Method of Moments

MoM is a general and rigorous approach to finding the solution of an operator equation, such as differential, integral, or integrodifferential equations. Therefore, to keep the introduction general, the method will be introduced in a general setting as follows:

$$\mathbf{L} f = g \qquad (3.242)$$

where $\mathbf{L}$ is any linear operator, f is the unknown function to be determined, and g is the known excitation or source function. Because the unknown is a function and because the closed-form solutions of such an operator equation are not generally available, the solution is approximated in general. As the MoM is one of those techniques that approximate the solution of an operator equation, it is necessary to provide its main steps for a general operator equation (3.242) to understand the method:

1. Expand the unknown function f in terms of known basis functions B_n with unknown coefficients

$$f = \sum_{n=1}^{N} a_n B_n \qquad (3.243)$$

    Hence, using the linearity of the operator, the operator equation (3.242) can be written as

$$\sum_{n=1}^{N} a_n \mathbf{L} B_n = g \qquad (3.244)$$

2. Choose a set of testing functions T_m (also known as weighting functions), and take the inner product of both sides of (3.244) for each T_m,

$$\sum_{n=1}^{N} a_n < T_m, \mathbf{L} B_n > = < T_m, g > \qquad m = 1,2,...,N \qquad (3.245)$$

    where the inner product is defined as

$$< u, v > = \int_{\Omega} u\, v^* d\Omega \qquad (3.246)$$

Note that $*$ and Ω denote the complex conjugate and the domain of integration, respectively. Then, the resulting set of linear equations (3.245) can be cast into a matrix equation form as

$$\overline{L}x = b$$

where $L_{ij} = <T_i, LB_j>$, $b_i = <T_i, g>$ and $x_j = a_j$.

3. Solve the matrix equation for the unknown coefficient vector x. Substituting the coefficients into the expansion of the unknown function, (3.243), results in the approximate solution to the original operator equation.

As could be expected, each step of the method outlined above can be further discussed and detailed from a rigorous mathematical stand point. Although such a discussion would be illuminating for advanced readers, it could be discouraging for those who are merely interested in the application of the method. As this book is more about the applications of such methods, those who are interested in studying the MoM more thoroughly are referred to [5, 6, 35–38].

Perhaps one point that may be worth mentioning with a theoretical aspect is the reasoning behind the testing process (step 2 in the above explanation). Other than to generate an equal number of equations as the number of unknowns, testing process can be thought as a minimization of the residual. To understand this, let us introduce the residual (or error) in the approximation as

$$R = \underbrace{Lf}_{\text{Exact}} - \underbrace{\sum_{n=1}^{N} a_n LB_n}_{\text{Approximate}} = g - \sum_{n=1}^{N} a_n LB_n \qquad (3.247)$$

In an attempt to reduce the residual over the domain of the problem, its projection over the range space of the operator can be set to zero. This is accomplished by taking the inner product of the residual with a set of testing functions spanning the range space of the operator:

$$<T_m, R> = <T_m, g> - \sum_{n=1}^{N} a_n <T_m, LB_n> = 0 \qquad m = 1, \ldots, N \qquad (3.248)$$

leading to a set of simultaneous linear equations for the unknown coefficients. Because of this, the MoM is sometimes referred to as the method of weighted residual.

Another issue is the selection of basis and testing functions. If the operator L is positive, then one can define a Hilbert space with an energy inner product and vector norm. Now, if the expansion and testing functions are *complete* in the Hilbert space, Lf converges weakly to g. If the operator L is a positive definite, then Lf converges to g [15]. There are numerous functional forms of the basis functions, such as rooftop functions, piecewise sinusoidal functions, RWG basis functions, and so forth, and some are very flexible to use for arbitrary geometrical features, like RWG basis functions. However, due to their simplicity and ease of use, rooftop functions have been preferred for the demonstration of the method in this book. In addition, testing functions have been chosen to be the same as the basis functions, leading to the commonly known Galerkin MoM.

While the basis functions are the "known" functions with "unknown" amplitudes to represent the current density on the region where the current is induced, testing functions are the weighting functions defined over the regions where the boundary conditions for the electric and/or magnetic fields are applied in an integral sense. Considering only the electric current density in the geometry (i.e., the geometries with only conducting traces), the total electric field on a region specified by the testing function is the sum of the contributions of the basis functions, each of which acts like a separate source. If the basis functions are chosen in such a way that the current density on the conductors of the geometry can be accurately represented by a linear combination of these basis functions, then the accuracy of the results and the computational cost of the algorithm are improved significantly. In other words, if the functional form of the basis functions cannot match the physical form of the actual current density on the conductors of the geometry, then the MoM may not converge (or may converge very slowly) with the increased number of basis functions, resulting in a very inefficient algorithm and inaccurate results [39, 40].

For the sake of demonstrating the steps, let us consider a narrow microstrip line over a substrate backed by a ground plane. The very first step is to get the spectral-domain Green's functions of vector and scalar potentials for the specific geometry at hand. Then, these Green's functions are somehow transformed into the spatial-domain, either using the discrete complex image method (DCIM) or numerically integrating the Sommerfeld integrals. Because the longitudinal direction of the line is chosen to be in the x-direction, and because the width of the line is rather narrow as compared to the wavelength of operation, the current is assumed to only flow in the x-direction. Therefore, Green's functions due to an x-directed dipole need to be calculated, which are namely G_{xx}^A, G_{zx}^A, and G_x^q. In addition, since there is no vertical conductor involved in the geometry, there is no need to write the z component of the electric field to apply the boundary condition, and in turn, no need for G_{zx}^A. Hence, the governing MPIE can be written as the x-directed electric field from (3.238) as

$$E_x = -j\omega G_{xx}^A * J_x + \frac{1}{j\omega}\frac{\partial}{\partial x}\left(G_x^q * \frac{\partial J_x}{\partial x}\right) \qquad (3.249)$$

where the unknown to be determined is the x-directed current density. As the main step of the MoM, the unknown function can be expanded in terms of known basis functions as follows:

$$J_x(x,y) = \sum_{n=0}^{N} I_{nx} B_{nx}(x,y) \qquad (3.250)$$

where B_{nx} is a known basis function with the unknown coefficient I_{nx}, except $I_{0x} = 1$ amp denotes the source amplitude with the corresponding half-rooftop function B_{0x}, as shown in Figure 3.14. With the choice of the rooftop function, the basis function can be written mathematically as

$$B_{nx}(x, y) = \begin{cases} \dfrac{1}{h_y}\left(\dfrac{|x - x_n|}{h_x} - 1\right) & \begin{aligned} x_n - h_x < x < x_n + h_x \\ 0 < y < h_y \end{aligned} \\ 0 & \text{elsewhere} \end{cases}$$ (3.251)

where h_x and h_y are the spans of the basis function in the x- and y-directions, respectively. Substituting (3.250) into (3.249) and testing with a set of testing functions yield the following matrix entries:

$$Z_{xx}^{mn} = -j\omega\left\langle T_{mx}, G_{xx}^A * B_{nx}\right\rangle + \frac{1}{j\omega}\left\langle T_{mx}, \frac{\partial}{\partial x}\left[G_x^q * \frac{\partial B_{nx}}{\partial x}\right]\right\rangle$$ (3.252)

for $m, n = 1, 2, \ldots, N$.

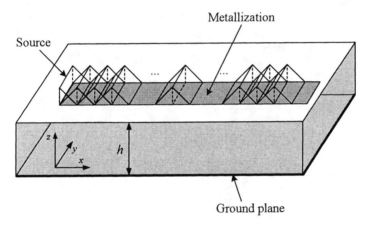

Figure 3.14 Basis and testing functions along a microstrip line.

When the source term is taken to the right-hand side, it yields the following excitation vector:

$$V_x^{m0} = j\omega\left\langle T_{mx}, G_{xx}^A * B_{0x}\right\rangle - \frac{1}{j\omega}\left\langle T_{mx}, \frac{\partial}{\partial x}\left[G_x^q * \frac{\partial B_{0x}}{\partial x}\right]\right\rangle$$ (3.253)

for $m = 1, 2, \ldots, N$. Hence, the matrix equation is formed as

$$\mathbf{Z}\mathbf{x} = \mathbf{V}$$

where $\mathbf{Z}$ is the MoM matrix, $\mathbf{x} = (I_{1x}, I_{2x}, \ldots, I_{Nx})^T$ is the unknown coefficient vector, and $\mathbf{V}$ is the excitation vector. Assuming that all the inner-product terms are calculated, the only thing left is to solve the matrix equation for the unknown coefficients. The solution of a matrix equation is straightforward, like using LU decomposition, for small- to medium-size problems (in terms of wavelength, and, in turn, in terms of number of unknowns) but out of the scope of this book for large problems. After filling in the matrix entries and solving the matrix equation for the unknown coefficients, current distribution on the conductors can be approximated by (3.250), and then, any circuit parameter required for the characterization of the circuit can simply be extracted from the current distribution. The computational efficiency of the MoM, as far as the filling in the MoM matrix is concerned, can be improved by using the closed-form Green's functions in the spatial domain [27–30]. The improvement is in the computation of the inner-product terms and is quite significant, which is mainly due to the analytical evaluation of the MoM matrix entries when the closed-form Green's functions are used in conjunction with the MoM [30].

Before completing this short summary on the MoM, a brief discussion of a seemingly problematic point in the evaluation of the inner-product terms is necessary. Since the derivative of the discontinuous source basis function is involved in the evaluation of the excitation vector, (3.253), the question of singularity as a result of the derivative needs to be answered. Note that such singularities are inevitable at the source, sink, and intersections of horizontal and vertical conductors in the geometry.

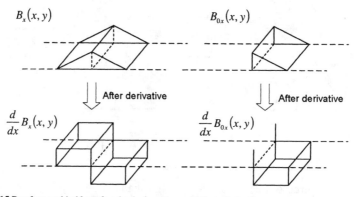

Figure 3.15 Rooftop and half-rooftop basis functions and their derivatives.

For the sake of illustration, a rooftop function and a half-rooftop function are shown in Figure 3.15, together with their derivatives. Since the basis functions are written to approximate the unknown current density, their derivatives represent the charge density due to the continuity equation. Therefore, the chosen basis functions should ensure the continuity of the current, as well as the conservation of charges on the conductor. Note that the total charges corresponding to a basis function, representing a small part of the induced current along a conductor, equal zero, as can be observed from Figure 3.15. This fact seems to be the result of the choice of the rooftop functions as the basis functions for the current density, but, on the contrary, it is the requirement for the current and charge distributions over a conductor. Remembering that rooftop basis functions are defined over rectangular cells, the current entering into and exiting from a cell is equal to zero, as the amplitudes at these edges become zero. Hence, there is no loss or gain of charges in this rectangular region (i.e., no sink or source in this particular cell). Therefore, the total charges in the circuit must be equal to zero, because charges supplied by the source basis functions can be taken out of the circuit by the sink basis functions. In other words, the rest of the basis functions defined over the conductor can neither introduce nor lose any charge. Only source and sink basis functions, which are chosen to be half-rooftop functions in this example, result in finite total charges on the domain where they are defined, provided that the impulse introduced by the differentiation of a half-rooftop function is ignored. Considering a real physical connection of the source to a microstrip line, the half-rooftop function representing the source current on the microstrip line must be due to the current flowing into the line. Therefore, since there is no discontinuity on the current flow in the real picture, it is only in the mathematical model, which can be corrected by ignoring the singularity in the charge distribution.

References

[1]   Roland Schinzinger and Patricio A. A. Laura, *Conformal Mapping: Methods and Applications*, New York: Elsevier, 1991.

[2]   P. Silvester and P. Benedek, "Equivalent Capacitance of Microstrip Open Circuits," *IEEE Trans. on Microwave Theory Tech.*, vol. 20, pp. 511–516, Aug. 1972.

[3]   P. Silvester and P. Benedek, "Equivalent Capacitance of Microstrip Gaps and Steps," *IEEE Trans. on Microwave Theory Tech.*, vol. 20, pp. 729–733, Nov. 1972.

[4]   P. Silvester and P. Benedek, "Equivalent Discontinuities Capacitances for Right-Angle Bends, T-Junctions and Crossings," *IEEE Trans. on Microwave Theory Tech.*, vol. 21, pp. 341–346, May 1973.

[5]   R. F. Harrington, *Field Computations by Moment Methods*, New York: Krieger Publishing, 1983.

[6]   R. F. Harrington, "Matrix Methods for Field Problems," *IEEE Proceedings*, vol. 55, pp. 136–149, Feb. 1967.

[7]   K. Kunz and R. Luebber, *The Finite Difference Time Domain Method for Electromagnetics*, Boca Raton, FL: CRC Press, 1993.

[8]   Allen Taflove and Susan C. Hagness, *Computational Electrodynamics: The Finite-Difference Time-Domain Method*, 2nd ed., Norwood, MA: Artech House, 2000.

[9]   J. Jin, *The Finite Element Method in Electromagnetics*, New York: John Wiley and Sons, 1993.

[10] M. I. Aksun and R. Mittra, "Choices of Expansion and Testing Functions for the Method of Moment Applied to a Class of Electromagnetic Problems," *IEEE Trans. on Microwave Theory Tech.*, vol. 44, pp. 503–509, Mar. 1993.

[11] L. B. Felsen and N. Marcuvitz, *Radiation and Scattering of Waves*, Oxford: Oxford University Press, 1996.

[12] W. C. Chew, *Waves and Fields in Inhomogeneous Media*, New York: Van Nostrand Reinhold, 1990.

[13] T. W. Korner, *Fourier Analysis*, Cambridge: Cambridge University Press, 1990.

[14] A. Sommerfeld, *Partial Differential Equations in Physics*, New York: Academic Press, 1949.

[15] Donald G. Dudley, *Mathematical Foundations for Electromagnetic Theory*, New York: IEEE Press, 1994.

[16] A. Erteza and B. K. Park, "Nonuniqueness of Resolution of Hertz Vector in Presence of a Boundary, and the Horizontal Dipole Problem," *IEEE Trans. Antennas Propagat.*, vol. AP-17, pp. 376–378, May 1969.

[17] K. A. Michalski, "On the Scalar Potential of a Point Charge Associated with a Time-Harmonic Dipole in a Layered Medium," *IEEE Trans. Antennas Propagat.*, vol. AP-35, pp. 1299–1301, Nov. 1987.

[18] D. C. Fang, J. J. Yang, and G. Y. Delisle, "Discrete Image Theory for Horizontal Electric Dipoles in a Multilayered Medium," *Proc. Inst. Elect. Eng.*, pt. H, vol. 135, pp. 297–303, Oct. 1988.

[19] L. Chow et al., "A Closed-Form Spatial Green's Function for the Thick Microstrip Substrate," *IEEE Trans. on Microwave Theory Tech.*, vol. 39, pp. 588–592, Mar. 1991.

[20] J. R. Mosig, "Arbitrarily Shaped Microstrip Structures and Their Analysis with a Mixed Potential Integral Equation," *IEEE Trans. on Microwave Theory Tech.*, vol. MTT-36, pp. 314–323, Feb. 1988.

[21] David C. Chang and Jian X. Zheng, "Electromagnetic Modeling of Passive Circuit Elements in MMIC," *IEEE Trans. on Microwave Theory Tech.*, vol. 40, pp. 1741–1747, Sept. 1992.

[22] I. Park, R. Mittra, and M. I. Aksun, "Numerically Efficient Analysis of Planar Microstrip Configurations Using Closed-Form Green's Functions," *IEEE Trans. on Microwave Theory Tech.*, vol. 43, pp. 394–400, Feb. 1995.

[23] K. Naishadham and T. W. Nuteson, "Efficient Analysis of Passive Microstrip Elements in MMICs," *Int. J. MIMICAE*, vol. 4, pp. 219–229, July 1994.

[24] N. Kinayman and M. I. Aksun, *EMPLAN: Electromagnetic Analysis of Printed Structures in Planarly Layered Media*, Norwood, MA: Artech House, 2000.

[25] N. Kinayman and M. I. Aksun, "Efficient and Accurate EM Simulation Technique for Analysis and Design of MMICs," *Int J. MIMICAE*, vol. 7, pp. 344–357, Sept. 1997.

[26] N. Kinayman and M. I. Aksun, "Efficient Use of Closed-Form Green's Functions for the Analysis of Planar Geometries with Vertical Connections," *IEEE Trans. on Microwave Theory Tech.*, vol. 45, pp. 593–603, May 1997.

[27] N. Kinayman and M. I. Aksun, "Efficient Evaluation of Spatial-Domain MoM Matrix Entries," *IEEE Trans. on Microwave Theory Tech.*, vol. 48, pp. 309–312, Feb. 2000.

[28] M. I. Aksun and R. Mittra, "Derivation of Closed-Form Green's Functions for a General Microstrip Geometry," *IEEE Trans. on Microwave Theory Tech*, vol. 40, pp. 2055–2062, Nov. 1992.

[29] M. I. Aksun and R. Mittra, "Spurious Radiation from Microstrip Interconnects," *IEEE Trans. Electromagnetic Compat.*, vol. EMC-35, pp. 148–158, May 1993.

[30] L. Alatan et al., "Analytical Evaluation of the MoM Matrix Elements," *IEEE Trans. on Microwave Theory Tech.*, vol. 44, pp. 519–525, Apr. 1996.

[31] G. Dural and M. I. Aksun, "Closed-Form Green's Functions for General Sources and Stratified Media," *IEEE Trans. on Microwave Theory Tech.*, vol. 43, pp. 1545–1552, July 1995.

[32] M. I. Aksun, "A Robust Approach for the Derivation of Closed-Form Green's Functions," *IEEE Trans. on Microwave Theory Tech.*, vol. 44, pp. 651–658, May 1996.

[33] T. Itoh and R. Mittra, "Spectral-Domain Approach for Calculating the Dispersion Characteristics of Microstrip Line," *IEEE Trans. on Microwave Theory Tech.*, vol. 21, pp. 496–499, July 1973.

[34] T. Itoh and R. Mittra, "A Technique for Computing Dispersion Characteristics of Shielded Microstrip Lines," *IEEE Trans. on Microwave Theory Tech.*, vol. 22, pp. 896–898, Oct. 1974.

[35] Johnson J. H. Wang, *Generalized Moment Methods in Electromagnetics,* New York: John Wiley and Sons, 1991.

[36] Alvin Wexler, "Computation of Electromagnetic Fields," *IEEE Trans. on Microwave Theory Tech.*, vol. 17, pp. 416–439, Aug. 1969.

[37] Michel M. Ney, "Method of Moments as Applied to Electromagnetic Problems," *IEEE Trans. on Microwave Theory Tech.*, vol. 33, pp. 972–980, Oct. 1985.

[38] S. G. Mikhlin, *Variational Methods in Mathematical Physics,* New York: The Macmillan Company, 1964.

[39] T. K. Sarkar, "A Note on the Choice of Weighting Functions in the Method of Moments," *IEEE Trans. Antennas Propagat.*, vol. AP-33, pp. 436–441, Apr. 1985.

[40] T. K. Sarkar, "On the Choice of Expansion and Weighting Functions in the Numerical Solution of Operator Equations," *IEEE Trans. Antennas Propagat.*, vol. AP-33, pp. 988–996, Sept. 1985.

Chapter 4

Microstrip Patch Antennas

The microstrip patch antenna is one of the simplest radiating structures that can be built using printed circuits. Single patch antennas and patch antenna arrays are widely used in communication systems and airborne applications because of their light weight, precise reproduction through photolithographic techniques, conformal properties, suitability to integrate with active circuits, and low cost. Although the microstrip patch antenna is not the best in terms of electrical properties, it is the preferred structure used for radiation in the vast majority of low-cost applications because of its unique properties.

The first reported study on microstrip antennas is apparently due to Deschamps, Cutton, and Baissinot in the early 1950s [1]. Later, Munson and his colleagues reported their work on conformal microstrip antennas and arrays in 1974 [2]. That work described a wrap-around strip element fed from multiple points by means of a corporate power divider, which is suitable to apply to the surface of missiles. The antenna had an omnidirectional pattern in the plane perpendicular to the missile axis. This was clearly an important step in simplifying the radiation elements, where weight and aerodynamical properties are important. During the same period, Howell published the experimental results on rectangular and circular patch antennas [3]. Then in 1979, Lo and his coworkers published on the cavity model for many canonical microstrip antenna shapes [4]. At that point, there had been enough work accumulated to set up the stage for a workshop devoted to microstrip antennas at New Mexico State University in November 1979. This was followed by a special issue of *IEEE Transactions on Antenna and Propagation* on microstrip antennas in January 1981. This issue contained two classic review papers on microstrip antenna [5] and array technology [6], as well as other publications covering both theoretical and practical aspects [7–12]. Some other good papers [13–19] and books [20–25] have also been published both during that period and recently.

In the ensuing years, the fast pace of the development of microstrip circuits in general fueled microstrip antenna studies, and microstrip antennas became the natural choice as a radiation structure in most microstrip circuits. Due to the aforementioned advantages, microstrip antennas are now being used in a vast array of applications, including, but not limited to, wireless communications (e.g., local multipoint distribution systems), automotive applications (e.g., autonomous cruise control radar and high-resolution radar), space communications, military applications (e.g., GPS systems), and marine applications (e.g., marine radars) [22].

4.1 DESIGN OF MICROSTRIP PATCH ANTENNAS

Many geometrical shapes for microstrip patch antennas exist in the literature, albeit the most commonly used shapes are rectangular and circular patches. Circular patches consume somewhat less area than rectangular patches and might be more suitable in array applications in terms of interelement spacing and substrate coverage. In spite of the shape differences, the radiation characteristics of patch antennas are similar. A typical patch antenna has a directive gain of 5 to 7 dB, and exhibits a 3-dB beamwidth between 70° and 90° [23]. Figure 4.1 shows a microstrip-fed rectangular patch antenna printed on a dielectric layer.

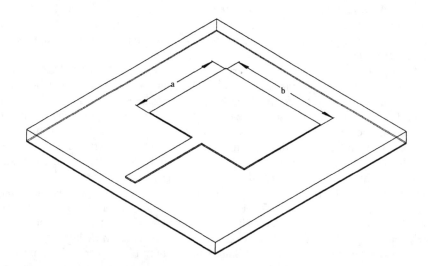

Figure 4.1 A rectangular microstrip patch antenna printed on a substrate. The patch is fed by a microstrip line from left. Bottom of the substrate is covered with ground plane.

The patch antenna can be fed using a microstrip line, coaxial probe, or slot on the ground plane. Usually, slot- or probe-fed antenna configuration results in more compact designs. Besides, these latter techniques hide the feed transmission lines behind the ground plane, hence reducing their adverse effect on the radiation pattern. Probe-fed configuration is the natural choice when a coaxial cable is used to feed the antenna because the center conductor of the cable itself can be utilized as the probe. When active circuits and antenna are placed on opposite sides of a printed circuit, one can also employ probe-feeding through a via-hole between the antenna and circuit layers. Note that in the probe-fed configuration, the input impedance becomes more inductive at the actual antenna resonance due to the length of the probe/via conductor. This point should be taken into account in designing antenna matching networks. The microstrip-fed configuration is usually used when active circuits are mounted on the same substrate surface with the antenna patch. It is also used in microstrip arrays when more than one patch is printed on the same substrate and shares the feed lines.

The typical dimensions of a rectangular patch antenna are given by the following simple expressions [5]:

$$a = \frac{c}{2 f_0 \sqrt{\varepsilon_r \mu_r}} - 2\Delta l \tag{4.1}$$

$$b = \frac{c}{2 f_0} \tag{4.2}$$

where a and b are the antenna dimensions (see Figure 4.1), ε_r is the relative dielectric constant, μ_r is the relative magnetic constant, f_0 is the resonance frequency, and c is the speed of light in free space. The factor Δl takes into account the fringing fields at the edges and can be determined using the standard microstrip formula:

$$\Delta l = 0.412 \frac{\varepsilon_{eff} + 0.300}{\varepsilon_{eff} - 0.258} \frac{b/h + 0.262}{b/h + 0.813} \tag{4.3}$$

where

$$\varepsilon_{eff} = \frac{\varepsilon_r + 1}{2} + \frac{\varepsilon_r - 1}{2} \left(1 + \frac{10h}{b} \right)^{-1/2} \tag{4.4}$$

More accurate and general expressions for the resonance frequency can be found in the literature [23, 26–28]. Note that the fundamental resonance frequency of the antenna is determined by the dimension a. As will be demonstrated later, the dimension b then controls radiation resistance at the resonance. The resonance resistance can be decreased by increasing the width b as long as the b/a ratio is kept to less than approximately two to prevent a decrease in efficiency [5]. This is an important feature because, for example, one can control the radiated power of individual elements by adjusting the respective b dimensions in a series antenna array, thus providing the desired amplitude tapering.

4.1.1 Quality Factor

For a microstrip patch antenna, there are four main loss mechanisms that need to be considered: radiation loss (Q_{rad}), surface-wave loss (Q_{sw}), dielectric loss (Q_d), and metallization loss (Q_c). Radiation loss, as the name implies, represents the loss due to radiation into space waves. The surface-wave loss, on the other hand, represents the amount of power coupled into surface waves, which needs to be minimized for a typical design. Surface-wave effects reduce the overall efficiency and are one of the disadvantages of patch antennas. The last two loss mechanisms have the usual definitions used in general microstrip circuits. The total quality factor of the antenna can be given by [5, 22]:

$$\frac{1}{Q} = \frac{1}{Q_{rad}} + \frac{1}{Q_{sw}} + \frac{1}{Q_d} + \frac{1}{Q_c} \tag{4.5}$$

Determination of the Q-factors in the above equation analytically for a patch with arbitrary shape can be difficult. However, for rectangular patches, the following simple relations have been derived [22, 29]:

$$Q_d = \frac{1}{\tan \delta} \tag{4.6}$$

$$Q_c = h\sqrt{\mu\pi f\sigma} \tag{4.7}$$

$$Q_{sw} = Q_{rad}\left(\frac{\eta_r^0}{1-\eta_r^0}\right) \tag{4.8}$$

$$Q_{rad} = \frac{3}{16}\frac{\varepsilon_r}{p}\frac{a_e}{b_e}\frac{\lambda_0}{h}\frac{1}{1-\frac{1}{\varepsilon_r\mu_r}+\frac{2}{5\varepsilon_r^2\mu_r^2}} \tag{4.9}$$

where η_r^0 is the radiation efficiency (assuming no dielectric and conductor losses), a_e is the effective length of the antenna, b_e is the effective width of the antenna, h is the substrate thickness, and p is the ratio of the power radiated into space by the patch to the power radiated by an equivalent dipole with a moment m. Other parameters have the usual definitions. The effective width and length of the antenna can be calculated using equations available in the literature for microstrip lines. Although the definitions of (4.6) and (4.7) are complete, we still need to define η_r^0 and p to be able to calculate (4.8) and (4.9), which can be done as follows [22, 29]:

$$\eta_r^0 \cong \frac{P_{sp}^{HED}}{P_{sp}^{HED} + P_{sw}^{HED}} \tag{4.10}$$

where P_{sp}^{HED} and P_{sw}^{HED} refer to radiated power into space and surface waves for a horizontal electric dipole, respectively, and are given by [22, 29]:

$$P_{sp}^{HED} = \frac{1}{\lambda_0^2}(k_0 h)^2 80\pi^2\mu_r^2\left(1-\frac{1}{\varepsilon_r\mu_r}+\frac{2}{5\varepsilon_r^2\mu_r^2}\right) \tag{4.11}$$

$$P_{sw}^{HED} = \frac{1}{\lambda_0^2}(k_0 h)^3 60\pi^3 \mu_r^3 \left(1 - \frac{1}{\varepsilon_r \mu_r}\right)^3 \qquad (4.12)$$

Finally, the *p*-factor is defined as follows [22, 29]:

$$p = \frac{3}{\pi} \int_0^{\pi/2} \int_0^{\pi/2} \left(\sin^2\phi\cos^2\theta + \cos^2\phi\right) T_1^2(\theta,\phi) T_2^2(\theta,\phi)\sin\theta\, d\theta\, d\phi$$

where

$$T_1(\theta,\phi) = \frac{\sin(k_y b_e/2)}{k_y b_e/2}$$

$$T_2(\theta,\phi) = \frac{\cos(k_x a_e/2)}{1 - \frac{4}{\pi^2}(k_x a_e/2)^2}$$

$$k_x = k_0 \sin\theta\cos\phi$$

$$k_y = k_0 \sin\theta\sin\phi$$

Evaluation of the integrals in the *p*-factor can easily be carried out numerically, so no closed-form approximations will be presented here.

At this point, it would be beneficial to discuss the above equations in a little bit more detail. First, the radiation efficiency of the patch antenna in (4.10) is approximated with the efficiency of a horizontal electric dipole with an appropriate moment. Note that this approximation does not take into account the array factor due to the distribution of current across the patch surface. However, the calculation of efficiencies for various antennas with different feeding techniques has led to the conclusion that the radiation efficiency, η_r^0, mainly depends on the substrate thickness and permittivity; it is not affected strongly by either the antenna shape or its feed [30]. This suggests that the efficiency can be approximated reasonably well by an assumed current distribution, or an infinitesimal current source, as demonstrated here. Second, it can be seen from (4.7) that increasing the substrate thickness, *h*, increases *Q* because conductor loss is inversely proportional to substrate thickness in patch antennas. However, there is no such dependency in the *Q*-factor due to dielectric loss; see (4.6). Therefore, for most of the modern microwave substrate materials, the conductor losses dominate over the dielectric losses. Third, the radiated power into space waves by a horizontal electric dipole is proportional to the square of the substrate height as shown in (4.11), whereas the power radiated into surface waves by the same dipole varies with the third power of substrate height as shown in (4.12). The consequence of this behavior is that the overall antenna efficiency drops after a threshold value in substrate thickness due to increased surface wave excitation. Thus, increasing the substrate thickness after the threshold value does not increase the radiated power into space waves. To study this behavior further, we can define the overall antenna efficiency (when the dielectric and conductor losses are present) as

$$\eta_r = \frac{Q}{Q_{rad}} \tag{4.13}$$

One can plot (4.13) versus normalized substrate height for different dielectric constants for a rectangular patch antenna as shown in Figure 4.2. In the figure, the effects of conductor and surface-wave losses are clearly visible. For small substrate heights, the conductor losses dominate, reducing the antenna efficiency considerably. As the substrate thickness is increased, radiated power starts to increase, thus increasing efficiency until it reaches a maximum. After this point, the surface waves start to dominate, and efficiency drops again. Therefore, it is important to select the proper dielectric properties and patch geometry for maximum efficiency at a given resonance frequency.

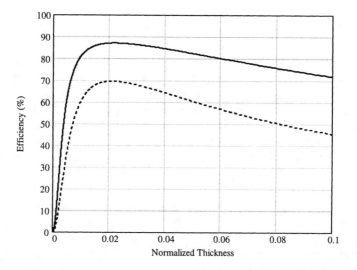

Figure 4.2 Radiation efficiency of a microstrip patch antenna versus normalized substrate thickness (h/λ_0) for two different dielectric constants. Solid and dotted lines correspond to $\varepsilon_r = 2.2$ and $\varepsilon_r = 9.6$, respectively ($f_0 = 5$ GHz, $\tan\delta = 0.002$, $\sigma = 4.1 \times 10^7$, $b = 1.5a$, $\mu_r = 1$).

4.1.2 Resonance Frequency

Resonance frequency for a patch antenna is defined as the frequency where input impedance has no reactive part. At this frequency, the input impedance of the antenna is approximately equal to the radiation resistance provided that all other loss mechanisms (e.g., conductor, dielectric) are relatively small. Note that this definition assumes that the antenna can be represented by a simple parallel *RLC* tank-circuit near the resonance; thus, the reference plane of the antenna input impedance is important. Besides, as we will see in Section 4.2, the effect of higher order modes can also be approximated using an inductive shift near resonance affecting the actual resonance frequency.

The resonance frequency of a rectangular patch antenna on a single dielectric substrate can be determined using the simple equation given in (4.1). Note that in the literature, sometimes ε_{eff} is used instead of ε_r in (4.1). As long as the fringe effects are taken into account at the open ends, both of these approaches are acceptable since the patch actually forms a very wide microstrip transmission line (i.e., $\varepsilon_{eff} \to \varepsilon_r$). A study comparing different models and providing expressions for higher-order resonances can be found in the literature [26–28].

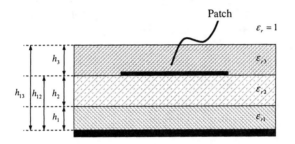

Figure 4.3 A patch antenna in a multilayer dielectric. (*After:* [28].)

For the multilayer dielectric structure shown in Figure 4.3, (4.1) can still be used provided that the proper effective permittivity is determined. For this purpose, one can use the following quasi-static equation [28]:

$$\varepsilon_{eff} = \varepsilon_{r1}\varepsilon_{r2}\frac{(q_1+q_2)^2}{\varepsilon_{r1}q_2+\varepsilon_{r2}q_1} + \varepsilon_{r3}\frac{(1-q_1-q_2)^2}{\varepsilon_{r3}(1-q_1-q_2-q_3)+q_3} \tag{4.14}$$

where q_1, q_2, and q_3 are the filling factors defined as follows:

$$q_1 = \frac{h_1}{2h_{12}}\left\{1+\frac{\pi}{4}-\frac{h_{12}}{w_e}\ln\left[\frac{2w_e}{h_1}\sin\left(\frac{\pi h_1}{2h_{12}}\right)+\cos\left(\frac{\pi h_1}{2h_{12}}\right)\right]\right\}$$

$$q_2 = 1-q_1-\frac{h_{12}}{2w_e}\ln\left(\frac{\pi w_e}{h_{12}}-1\right)$$

$$q_3 = 1-q_1-q_2-\frac{h_{12}-v_e}{2w_e}\ln\left[\frac{2w_e}{2h_{13}-h_{12}+v_e}\cos\left(\frac{\pi v_e}{2h_{12}}\right)+\sin\left(\frac{\pi v_e}{2h_{12}}\right)\right]$$

with

$$w_e = w+\frac{2h_{12}}{\pi}\ln\left[17.08\left(\frac{w}{2h_{12}}+0.92\right)\right]$$

$$v_e = \frac{2h_{12}}{\pi} \tan^{-1}\left[\frac{2\pi}{\pi w_e - 4h_{12}}(h_{13} - h_{12}) \right]$$

Note here that $w = b$ (see Figure 4.1). After determining ε_{eff} using (4.14), it is plugged to (4.1) instead of ε_r, and the resonance frequency of a patch embedded in a multilayer structure as shown in Figure 4.3 can be calculated.

4.1.3 Radiation Resistance

The radiation resistance of an antenna is defined as the resistance that, if inserted in place of an antenna, would consume the same amount of power that is radiated by the antenna. The radiation resistance, therefore, can be calculated by first expressing the total radiated power and then evaluating the resulting integrals. The input resistance of a rectangular patch antenna obtained through this approach can be given as follows at resonance frequency [31]:

$$R_r \cong \begin{cases} \varepsilon_{eff} Z_0^2 \dfrac{1}{120 I_1}, & \varepsilon_r \le 5 \\[3mm] \varepsilon_{eff} Z_0^2 \dfrac{I_2}{120 I_3}, & 5 < \varepsilon_r < 10 \end{cases} \tag{4.15}$$

where

$$I_1 = (k_0 h)^2 \left[0.53 - 0.03795 \cdot \left(k_0 \frac{w}{2} \right)^2 - \frac{0.03553}{\varepsilon_{eff}} \right]$$

$$I_2 = \left(1.29 - 3.57h \frac{\sqrt{\varepsilon_r}}{\lambda_0} \right) \cdot \frac{\varepsilon_r - 1}{9}$$

$$I_3 = (k_0 h)^2 \cdot \left\{ \begin{array}{l} \left[1.3 - \dfrac{4}{3\varepsilon_r} + \dfrac{0.53}{\varepsilon_r^2} - \left[0.08856 - \dfrac{0.08856}{\varepsilon_r} + \dfrac{0.03795}{\varepsilon_r^2} \right] (0.5 k_0 w)^2 \right. \\[3mm] \left. - \left[0.248714 - \dfrac{0.373071}{\varepsilon_r} + \dfrac{0.159887}{\varepsilon_r^2} \right] \left(\dfrac{1}{\varepsilon_{eff}} \right)^2 \right\} \end{array} \right.$$

In the above equations, Z_0 is the characteristic impedance of the microstrip line, which has width w; ε_r is the relative permittivity; and ε_{eff} is the relative effective permittivity that can be calculated using (4.4). Note here that $w = b$ (see Figure 4.1).

The above approach can be used to obtain the real part of the input impedance at resonance frequency, provided that all other loss mechanisms are small. A more general expression for the input impedance of a patch antenna can be found using the cavity method for various patch shapes, as we will demonstrate in Section 4.2.

Interested readers can refer to literature for more discussion on the input impedance of patch antennas [32–35].

4.1.4 Bandwidth

Bandwidth is the other important design parameter that needs to be introduced for microstrip patch antennas. There are basically three definitions for bandwidth of a microstrip patch antenna or an array: (1) impedance bandwidth, (2) pattern bandwidth, and (3) polarization bandwidth [5]. The limiting factor for a single antenna element is the impedance bandwidth; the pattern and polarization bandwidths vary relatively slowly with frequency. Note that impedance bandwidth of a simple patch antenna is around 1% to 3% without any compensation.

One way to increase the impedance bandwidth is to increase the thickness of the dielectric substrate. However, as the thickness of the substrate increases, the impedance locus of the antenna becomes inductive, which makes matching the antenna difficult, and surface wave excitation becomes higher, which causes spurious radiation [5]. Note that the input impedance is dominantly inductive on thicker substrates because of the higher-order modes [13]. The effect of the higher-order modes can be so dominant that they can prevent the patch antenna from resonating (i.e., impedance locus does not intersect with the real axis on the Smith chart) for substrate thicknesses greater than approximately 0.1λ [13]. For probe-fed configurations, an additional inductance term appears due to the length of the probe. Note that it is important to distinguish the inductive shifts due to higher-order modes and probe feed. Therefore, increasing the substrate thickness alone to control the input impedance is not usually employed.

The alternative techniques that are widely used to increase the impedance bandwidth of a patch antenna can be listed as follows: (1) using a matching network to match the feed to the antenna over a broadband [36–38], (2) using multiple resonators that are tuned to slightly different frequencies [39–42], (3) using stacked patches [43–52], and (4) modifying feed configuration of the antenna [53–60]. In the following sections, we will provide examples of each of these techniques. At this point, it would be worthwhile to provide brief explanations of each technique.

In the first method, an impedance matching network, such as a microstrip stub, is used to match the antenna impedance. Although it is theoretically possible to match input impedance through this technique, it is not always a good solution because it needs a relatively large substrate area and may deteriorate the far-field pattern if it is placed on the same side as the antenna because of the spurious radiation from microstrip discontinuities [61]. In the second method, additional passive (not directly excited) patch resonators that are slightly tuned to different frequencies are placed next to the nonradiating edges of the antenna. However, using multiple resonators in this manner also has the disadvantage of consuming a large substrate area. Stacking resonators in a multilayer fashion, as indicated in the third method, also increases the impedance bandwidth. In this configuration, one patch becomes the driven patch, whereas other patches act as parasitic elements. The disadvantage of this approach is the requirement of relatively thick element spacing in the vertical direction to control the impedance. Hence, it may not be the preferred choice in printed circuit technology where laminating and procurement of thick dielectrics may be problematic. Therefore, perhaps the most efficient way of increasing the impedance bandwidth of a microstrip patch antenna is through modification of the feed configuration as proposed in the last method. By *modification of the feed configuration*, we mean using one of the following approaches: proximity coupling, aperture coupling, or changing the feed location. In proximity coupling, the patch and microstrip feed are placed on different dielectric layers, resulting in capacitive coupling. Then, the input impedance is tuned by adjusting the amount of overlap between the patch and microstrip. Aperture coupling is accomplished by adding a ground layer with a slot opening between the patch and microstrip of the proximity-coupled structure, hence making a three-layer construction. This provides additional flexibility for controlling the input impedance by adjusting the slot dimensions, as well as microstrip and slot overlap. Lastly, changing the physical feed location of the antenna on the patch surface also controls the input impedance. In probe-fed antennas, this is done relatively easily by inserting the probe at the desired location, which is typically on the symmetry axis and far from the radiating edges. In microstrip-fed antennas, the same effect can be obtained to some extent by the technique called inset-fed, which takes the effective feeding location inside of the antenna with the help of an inset.

It is also important to note that the usual 3-dB definition of bandwidth is not used for the impedance bandwidth of patch antennas. Instead, the following definition has been adopted, which defines the bandwidth for a given voltage standing wave ratio (VSWR) at the lower and upper band-edge frequencies [36]:

$$BW = \frac{1}{Q} \frac{VSWR - 1}{\sqrt{VSWR}} \qquad (4.16)$$

where Q is the quality factor of the antenna. The bandwidth of patch antennas is typically given for a VSWR of 1:2.

4.1.5 Radiation Pattern

There are fundamentally two ways of obtaining the radiation pattern of a microstrip patch antenna: the electric current model and the magnetic current model [62]. In the first method, the radiation pattern is directly obtained from the currents flowing on the patch surface, which are calculated using Green's function of the medium as explained in detail in Chapter 3. In the second method, the equivalence principle is applied to a surface surrounding the patch and substrate below the patch by assuming the patch cavity has perfect magnetic walls. Then, the radiation is calculated from the resulting magnetic currents on the walls. The electric current method yields the most rigorous calculation because there are fewer approximations involved. However, the magnetic current model (or slot model) is relatively simple and computationally more efficient because closed-form analytical expressions can be obtained.

In the simplest form of the magnetic current model, the radiation pattern of a rectangular patch antenna is calculated by modeling the radiator as two parallel, uniform, magnetic line sources of length b separated by a distance a [5]. A more general approach would be to model all four edges of the rectangular patch as radiating slots (see Figure 4.4) so that field patterns of the higher-order modes could also be calculated [63]. By using this latter approach, it can be shown that the far-field pattern of the mnth mode of a rectangular patch antenna can be written as follows [63]:

$$E_\theta = j \cdot k \cdot \frac{e^{-jkR}}{2\pi R} \left[E_x \cos(\phi) + E_y \sin(\phi)\right] \tag{4.17}$$

$$E_\phi = j \cdot k \frac{e^{-jkR}}{2\pi R} \left[-E_x \sin(\phi)\cos(\theta) + E_y \cos(\phi)\cos(\theta)\right] \tag{4.18}$$

where

$$E_x = \left[\left(-1-(-1)^m\right) \cdot j \cdot \sin(\xi B) + \left(1-(-1)^m\right) \cdot \cos(\xi B)\right] \cdot h \cdot E_0 \cdot A$$
$$\cdot \operatorname{sinc}(\xi H) \cdot j^n \cdot \left[\operatorname{sinc}\left(\eta A + \frac{n\pi}{2}\right) + (-1)^n \operatorname{sinc}\left(\eta A - \frac{n\pi}{2}\right)\right] \tag{4.19}$$

$$E_y = \left[\left(-1-(-1)^n\right) \cdot j \cdot \sin(\eta A) + \left(1-(-1)^n\right) \cdot \cos(\eta A)\right] \cdot h \cdot E_0 \cdot B$$
$$\cdot \operatorname{sinc}(\eta H) \cdot j^m \cdot \left[\operatorname{sinc}\left(\xi B + \frac{m\pi}{2}\right) + (-1)^m \operatorname{sinc}\left(\xi B - \frac{m\pi}{2}\right)\right] \tag{4.20}$$

with

$$\xi = k \cdot \sin\theta \cdot \cos\phi \tag{4.21}$$

$$\eta = k \cdot \sin\theta \cdot \sin\phi \tag{4.22}$$

$$k = \frac{2\pi}{\lambda_0} \tag{4.23}$$

Coordinates of the far-field pattern are shown in Figure 4.5 (dimension $2A$ is the radiating edge of the fundamental mode). Note that the resonance frequency of each mode should be calculated using a separate method. In the above expressions, E_0 is the maximum amplitude of the E_z field, which can be determined by normalizing the input voltage at the feed point (x_0', y_0') to 1V:

$$h \cdot E_z(x_0', y_0') = 1 \implies E_0 = \left[h \cdot \cos\left(\frac{m\pi x_0'}{2B}\right) \cdot \cos\left(\frac{n\pi y_0'}{2A}\right)\right]^{-1}$$

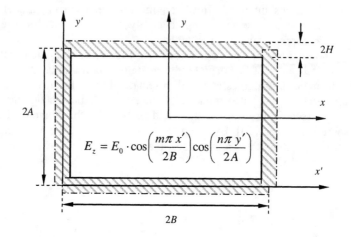

Figure 4.4 A generic patch surface and coordinate system of the radiating slots for calculating the far field. The slot widths $2H$ are assumed to be equal to the substrate thickness, h. Note that the dimension $2A$ is the radiating edge for the fundamental mode to comply with the coordinate system given in Figure 4.5. (*After:* [63].)

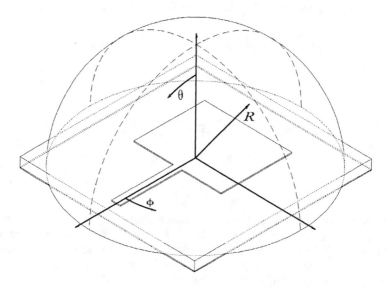

Figure 4.5 Coordinate system for the far-field pattern of the microstrip patch antenna.

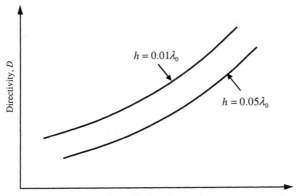

Figure 4.6 Qualitative plot of the directivity for a rectangular microstrip antenna over a large ground plane.

Knowing the far-field expression, it is then straightforward to determine the radiation resistance of the antenna using

$$R_r = \frac{1}{2P_r} \qquad (4.24)$$

where P_r is the radiated power that can be determined by numerical integration:

$$P_r = \lim_{R \to \infty} \frac{1}{Z_0} \int_0^{\pi/2} \int_0^{2\pi} \left\{ \left| E_\phi \right|^2 + \left| E_\theta \right|^2 \right\} \sin\theta \cdot R^2 \cdot d\phi\, d\theta \qquad (4.25)$$

with Z_0 being the wave impedance of free space.

The directivity of a single element can also be determined by using numerical integration of the field pattern [5, 64]:

$$D = \frac{4\pi \cdot U_m}{P_r} \qquad (4.26)$$

where U_m is the maximum radiation intensity [64]. A qualitative plot of the directivity versus the length of the radiating edge, b, is shown in Figure 4.6. As can be seen from the figure, the directivity increases as the length of the radiating edge is increased (as long as only the fundamental mode is excited). Another observation is that thicker substrates cause a decrease in directivity because of the destructive interference between the patch and image currents [5]. It should also be noted that a single patch mounted on a small ground plane will have less directivity than one mounted on large plane because of spillover into the region behind the ground.

4.1.6 Surface Waves

It can be shown that the air-dielectric interface of microstrip circuits supports guided modes called surface waves [30, 65–68]. Both TE and TM surface waves can be excited on a grounded dielectric. Since they decay by $\rho^{-1/2}$, in contrast to the ρ^{-2} decay of space waves, they can travel relatively far, thus increasing crosstalk between the antenna elements. Furthermore, they are diffracted by the edges of finite-sized substrates, deteriorating the antenna pattern. Note that as the substrate thickness is increased, the number of supported surface-wave modes also increases.

The cutoff frequencies of the TM_0 surface-wave modes is given by the following equation:

$$f_c = \frac{nc}{4h\sqrt{\varepsilon_r - 1}} \qquad (4.27)$$

where n is the mode index, h is the substrate thickness, and c is the speed of light. Note that the cutoff frequency of the TM_0 surface wave is zero, which means that at least one mode of surface wave exists at all frequencies and substrate thicknesses.

4.2 ANALYSIS TECHNIQUES FOR PATCH ANTENNAS

In the literature, various techniques exist that can be used in the analysis of microstrip patch antennas. Carver and Mink give a very good review of the basic analysis techniques for patch antennas [5]. An extensive review of both analytical and numerical techniques can be found in Garg et al. [23].

The most direct and precise means of modeling the patch antennas is to employ numerical techniques, as introduced in Chapter 3. However, approximate analytical methods have also been used frequently because they can provide more insight into the underlying physics, and they are fast compared to numerical techniques. The basic analytical model for a rectangular microstrip patch antenna is the transmission line model, which considers the antenna as a transmission line that connects two parallel, radiating-slot impedances. In its simplest form, the transmission line model has several disadvantages: (1) it is only useful for rectangular-shaped patches, (2) it ignores the field variations along the radiating edges, and (3) corrections for fringe fields must be empirically determined. In another analysis method, where some of these disadvantages are eliminated, the antenna is modeled as a thin TM_z mode cavity. In that method, fields between the patch and ground plane are expanded in terms of the cavity-resonant modes. The effect of radiation is accounted for by artificially increasing dielectric loss or by placing an impedance boundary condition on the cavity walls [5]. In the following sections, these two analysis techniques are explained in more detail.

4.2.1 Transmission Line Model

The transmission line model is the simplest model used to obtain the input impedance of a rectangular microstrip patch antenna [5, 69–73]. The transmission line model for a rectangular patch antenna is shown in Figure 4.7. As can be seen from the figure, the patch is modeled by two slots (i.e., radiating edges) connected through a transmission line (i.e., patch metallization that forms a very low impedance line). By assuming that the slots are radiating into a half-space, their admittances can be approximately given as [5]:

$$G_1 + jB_1 \cong \frac{\pi b}{\lambda_0 \eta}\left[1 + j\left(1 - 0.636 \ln k_0 w\right)\right] \tag{4.28}$$

where λ_0 is the free-space wavelength, $\eta = \sqrt{\mu_0/\varepsilon_0}$, $k_0 = 2\pi/\lambda_0$, and w is the slot width, which is equal to the substrate thickness, h. In the transmission line model, it is assumed that the patch acts as a low-impedance transmission line that is terminated by the above slot admittances at both ends. The characteristic admittance of this low-impedance transmission line due to a rectangular patch is given by

$$Y_0 = \frac{b\sqrt{\varepsilon_r}}{h\eta} \tag{4.29}$$

where h is the substrate thickness. Then, input impedance is found by using simple transmission line theory as shown in Figure 4.7. Note that the rectangular patch forms a very-low-impedance transmission line compared to the slot impedances. This places the slot impedances near the horizontal axis and close to the open circuit point on the Smith chart during impedance transformation. Since the length of the patch is selected to be slightly less than one-half of guided wavelength to excite the slots 180° out of phase, imaginary parts of slot admittances nearly cancel each other, and real parts get doubled after impedance transformation. Thus, the total input admittance of the antenna seen at the input terminal is obtained approximately as follows:

$$\begin{aligned}
Y_{in} &= \left(G_1 + jB_1\right) + \left(\tilde{G}_1 + j\tilde{B}_1\right) \\
&\cong \left(G_1 + jB_1\right) + \left(G_1 - jB_1\right) \\
&= 2G_1
\end{aligned} \tag{4.30}$$

In a typical design, $b = \lambda_0/2$ so that $G_1 = 0.00417$ mhos (i.e., $R_{in} = 120$ ohms). Note that (4.28) readily suggests that increasing the b dimension would reduce the radiation resistance of the antenna.

There are improved versions of transmission line models in the literature, which overcome some of the restrictions of the simplest approach described above. For example, the model developed by Pues and Capelle adds the effects of mutual coupling between the radiating slots and improves the equations for slot self-admittance [71]. In another extended version of the transmission line model, which is called the generalized transmission line model, it is possible to analyze nonrectangular but symmetrical (e.g., circular) patches as well [23, 70, 72].

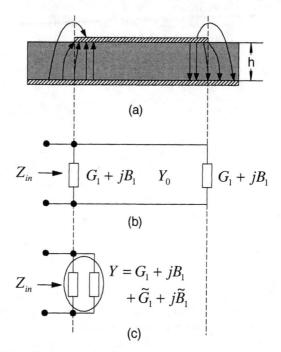

Figure 4.7 Explanation of the transmission line model used to analyze rectangular patch antennas: (a) E-field distribution and radiating slots for a rectangular patch antenna; (b) radiating slot admittances connected through a low-impedance transmission line, which is the patch metallization; and (c) final model for the input impedance.

4.2.2 Cavity Model

Although the transmission line model for microstrip patch antennas is easy to apply, in its simplest form, it fails to address some important problems, such as analysis of general shape patches and impedance variations with respect to feed location [7]. Extensions to the method might address these difficulties, but then it becomes more complicated to apply. To address the shortcomings of the transmission line model in a more rigorous way, the cavity model has been developed, where the substrate beneath the patch antenna is viewed as a cavity with magnetic walls on its sides [4, 5, 7, 32] [see Figure 4.8(a, b)]. This approximation yields good results for electrically thin substrates where the tangential magnetic fields can be assumed to be zero around the periphery of the patch (hence, the magnetic wall approximation). It is also assumed in the cavity model that the electric field is normal to the patch surface.

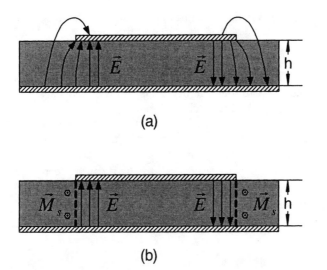

(a)

(b)

Figure 4.8 Cavity model to analyze patch antennas: (a) E-field distribution for a rectangular patch antenna; and (b) equivalent approximate model where the patch is viewed as a TM_z-mode cavity with magnetic walls.

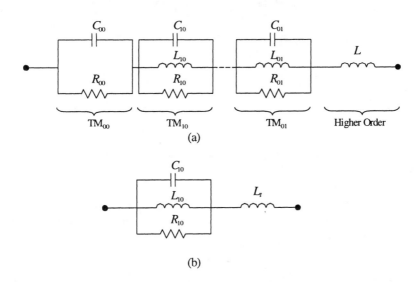

Figure 4.9 Equivalent lumped model representing the patch antenna according to the cavity model: (a) general network; and (b) simplified network for the isolated TM_{10} mode. (*After:* [5].)

According to the cavity model, the input impedance of a patch antenna can be approximated by series connection of parallel resonators, each responsible for the resonance phenomenon of a particular cavity mode, as shown in Figure 4.9(a).

Note that TM_{10} mode is the fundamental mode where the rectangular patches are operated. It can also be shown that around the vicinity of TM_{10} resonance, the model can be simplified, resulting in only a parallel resonator in series with an inductor [Figure 4.9(b)]. In this simple, lumped model, all losses, including the radiation, are combined into the resistive component R_{10}. The series inductance L_t accounts for the effect of higher-order modes. This equivalent model is applicable to microstrip-fed or probe-fed patch configurations.

The resonant resistance of a patch antenna, R_{10}, shown in the equivalent model is a function of substrate parameters and patch dimensions. The resonant resistance of a typical rectangular patch antenna varies between 100 and 250 ohms, depending on the patch aspect ratio a/b. It can be decreased by increasing the width b provided that the length a is kept approximately equal to one-half of the substrate wavelength.

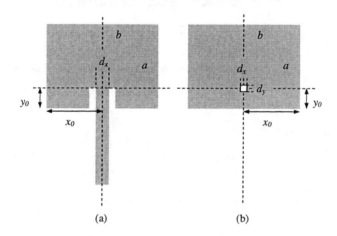

(a)   (b)

Figure 4.10 Coordinate system used in cavity model formulation: (a) microstrip inset-fed antenna; and (b) probe-fed antenna.

Figure 4.10(a, b) shows the coordinate system that will be used in cavity-model formulation both for microstrip- and probe-fed configurations. It can be shown that the input impedance of a probe-fed microstrip patch antenna can be written using the cavity formulation as follows [5]:

$$Z_{in} = -jZ_0 kt \sum_{m=0}^{\infty} \sum_{n=0}^{\infty} \frac{\psi_{mn}^2(x_0, y_0)}{k^2 - k_{mn}^2} G_{mn} \qquad (4.31)$$

where k and h are the wave number and the thickness of the substrate, respectively, and Z_0 is the wave impedance in free space. The coordinates of the feed-line position are represented by x_0 and y_0 (see Figure 4.10). The parameters ψ_{mn}, G_{mn}, k_{mn}, and X_{mn} are given by [5]

$$\psi_{mn}(x, y) = \frac{\chi_{mn}}{\sqrt{ab}} \cos\left(\frac{n\pi x}{b}\right) \sin\left(\frac{m\pi y}{a}\right) \tag{4.32}$$

$$G_{mn} = \frac{\sin(n\pi d_x / 2b)}{n\pi d_x / 2b} \cdot \frac{\sin(m\pi d_y / 2a)}{m\pi d_y / 2a} \tag{4.33}$$

$$k_{mn} = \sqrt{\left(\frac{n\pi}{b}\right)^2 + \left(\frac{m\pi}{a}\right)^2} \tag{4.34}$$

$$\chi_{mn} = \begin{cases} 1 & m = n \text{ and } n = 0 \\ \sqrt{2} & m = 0 \text{ or } n = 0 \\ 2 & m \neq 0 \text{ and } n \neq 0 \end{cases} \tag{4.35}$$

In the above expressions, the factor G_{mn} accounts for the width of the feed line. For a coaxial feed, $d_x = d_y$ and the cross-section $d_x d_y$ is set equal to the effective cross-section of the feed. For a microstrip feed, one sets $y_0 = 0$ and $d_y = 0$, and uses d_x as the feed-line width as a zero-order approximation, ignoring the transition effects. Note that the above formulation is correct for a nonradiating cavity. For a radiating cavity, the eigenvalues given in (4.34) are no longer valid, and complex eigenvalues must be found to solve the input impedance of the antenna through the above formulation. Fortunately, except the near TM_{10}-mode resonant frequency, the losses in the system can be neglected, and eigenvalues can be still approximated by (4.34) for nonresonant modes. For the resonant mode, however, the losses cannot be neglected, and corresponding complex eigenvalues, k_{10}, are found either by approximating all the losses through a complex permittivity or by assuming impedance-type boundary conditions on the radiating slot walls, which leads to a complex transcendental equation given as follows [5]:

$$\tan(k_{10}a) = \frac{2k_{10}\alpha_{10}}{k_{10} - \alpha_{10}^2} \tag{4.36}$$

$$\alpha_{10} = j\frac{2\pi Z_0}{\lambda_0} \frac{h}{b} Y_w \tag{4.37}$$

where Y_w and h are the admittance of the radiating walls and substrate thickness, respectively. The admittance of the radiating walls can be found from [5]

$$Y_w = G_w + jB_w \tag{4.38}$$

$$G_w = \frac{\pi}{376} \frac{b}{\lambda_0} \tag{4.39}$$

$$B_w = 0.01668 \frac{\Delta l}{h} \frac{b}{\lambda_0} \varepsilon_{eff} \tag{4.40}$$

$$\frac{\Delta l}{h} = 0.412 \cdot \left[\frac{\varepsilon_{eff} + 0.300}{\varepsilon_{eff} - 0.258} \right] \cdot \left[\frac{\dfrac{b}{h} + 0.262}{\dfrac{b}{h} + 0.813} \right] \tag{4.41}$$

$$\varepsilon_{eff} = \frac{\varepsilon_r + 1}{2} + \frac{\varepsilon_r + 1}{2} \left(1 + \frac{10h}{b} \right)^{-1/2} \tag{4.42}$$

At this point, it is also important to note that the wall admittance given by the above expressions is also not exact because the wall impedance is a function of both frequency and the angle of incidence. To correct this problem, it was proposed that an empirical correction factor be multiplied with the wall admittance as follows [5]:

$$F_y(b/a) = 0.7747 + 0.5977 \left(\frac{b}{a} - 1 \right) - 0.1638 \left(\frac{b}{a} - 1 \right)^2 \tag{4.43}$$

Then, the modified radiating wall admittance is expressed as

$$Y_w' = F_y(b/a) \cdot (G_w + jB_w) \tag{4.44}$$

The above formulation can be used in synthesis and/or analysis of microstrip patch antennas. It can be programmed into a digital computer very easily.

Despite the simplicity of the cavity model and other approximate analytical models in general, rigorous full-wave analysis of microstrip patch antennas is still needed to account for radiation, surface waves, electromagnetic coupling, and fringing fields accurately. We will discuss these full-wave methods briefly in the next section.

4.2.3 Full-Wave Analysis Methods

So far, we have introduced two analytical methods that are used in microstrip antenna design: the transmission line method and the cavity method. They are simple and easy to use. However, they have some serious disadvantages, which can be listed as follows [13]: (1) they can be applied only to electrically thin substrates, (2) characterization of mutual coupling between the elements is not possible (due to space and surface waves), and (3) rigorous analysis of arbitrary metallization shapes is not possible. The first point is crucial because, as the substrate thickness is increased, the simple cavity assumption starts to fail due to the fact that the z-derivative of the electric field cannot be ignored, and fringing fields around the edges of the patch become significant. Second, it is important to note that surface-wave excitation and mutual coupling start to be more pronounced on thick substrates. As the substrate thickness is increased, more surface-wave modes can exist, and power can be coupled to those modes, reducing the efficiency and/or increasing the coupling between other elements. Besides, the excited surface waves can be diffracted from substrate edges, causing increases in sidelobe levels in antenna patterns. Mutual coupling between the elements due to reactive near fields is also an important factor in array design and should be taken into account. Finally, the analytical methods are restrictive to simple patch shapes and feed techniques. Thus, although the simple analytical models are very convenient in analyzing and understanding the general operation principles of simple isolated patch shapes on thin dielectrics, they are insufficient to address the wide class of problems due to the aforementioned difficulties.

Full-wave analysis methods can provide the rigorous characterization necessary to address all shortcomings of the analytical models [23, 24]. Full-wave analysis is a term used to describe any numerical technique that is based on rigorous solution of Maxwell's equations. Since the governing differential or integral equations are solved numerically and few approximations are made during the formulation, the results are typically an order of magnitude more accurate than one can obtain from the approximate analytical models. For single elements or small arrays, essentially any full-wave method that is designed for the analysis of open microstrip structures (i.e., not in a metallic box) can be used to analyze patch antennas as well. From the theoretical point of view, there is no difference in analyzing, let's say, a power divider or a microstrip patch antenna, provided that the numerical technique is designed for open structures.

One must be cautious, however, in analyzing large antenna arrays because brute-force application of a numerical technique to electrically large structures may require prohibitively large computer resources. Note that the matrix solution time is usually the dominant factor for large problems. One can speed up the solution time by using an iterative solver in such cases. The fast multipole method (FMM) and the conjugate-gradient fast Fourier transform method (CG-FFT) are examples of methods that can be used for that purpose [74–77].

The most commonly used full-wave analysis techniques for patch antennas are the spectral-domain method, the mixed-potential integral equation (MPIE) method, the finite-difference time-domain (FDTD) technique, and the finite-element method (FEM). We have already introduced MPIE in Chapter 3. It would be beneficial to briefly review the salient features of these methods here. In the spectral-domain method, the problem is formulated using spectral-domain Green's functions. One of the advantages of this approach is that it eliminates the source singularity of Green's functions in the spatial domain. However, the resulting integrals become oscillatory and slowly converging. In the MPIE, on the other hand, the problem is formulated in the spatial domain. But now one has to deal with the singularity of Green's functions when the source and testing domains overlap. Fortunately, this can be addressed relatively easily, and MPIE can be very a efficient method for the analysis of microstrip patch antennas. The important point that one has to keep in mind is that both of these methods are Green's function approaches. Therefore they are suitable only to planar geometries with the possibility of vertical metallizations (which is sometimes called 2.5-D geometries). The FDTD or FEM, on the other hand, are true 3D methods and can be applied to cases where arbitrary metallization shapes exist (including finite-size ground planes) or the substrate is not planar. The paid price is that they are slower than Green's function approaches on 2.5-D structures.

Example 4.1

In this example, we will calculate the input impedance of the square patch antenna shown in Figure 4.11 by using both the cavity model and full-wave electromagnetic simulation. Surface current density and far-field plots will also be provided.

The approximate resonance frequency of this patch antenna can be determined using (4.1) as follows:

$$f_0 = \frac{3 \times 10^{10}}{2 \cdot (4.02 + 2 \cdot 0.0502) \cdot \sqrt{2.55}}$$
$$\cong 2.279 \text{ GHz}$$

Figures 4.12 and 4.13 show real and imaginary parts, respectively, of the input impedance of the antenna given in Figure 4.11, calculated using the cavity formulation described in Section 4.2.

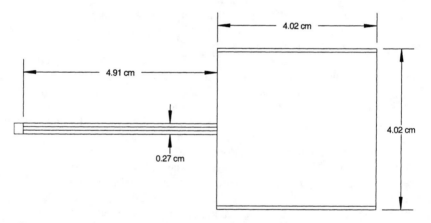

Figure 4.11 Full-wave electromagnetic simulation model of a microstrip patch antenna ($h = 1.58$ mm, $\varepsilon_r = 2.55$).

By inspecting Figure 4.13, one can see that the resonance frequency of the antenna is approximately $f = 2.27$ GHz. Radiation resistance can be found as 400 ohms from Figure 4.12. Note that this radiation resistance is somewhat higher than the radiation resistance of a typical patch antenna because the antenna studied here has a square shape (i.e., the radiating edges are relatively shorter).

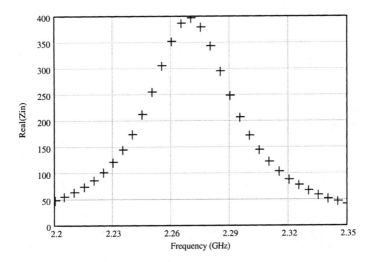

Figure 4.12 Real part of the input impedance of the patch antenna shown in Figure 4.11 calculated using the cavity model.

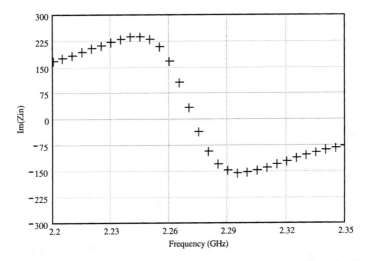

Figure 4.13 Imaginary part of the input impedance of the patch antenna shown in Figure 4.11 calculated using the cavity model.

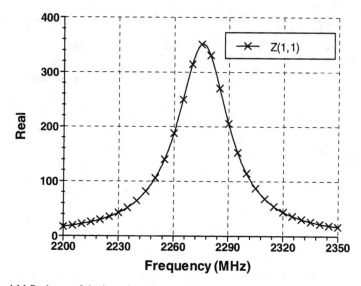

Figure 4.14 Real part of the input impedance of the patch antenna shown in Figure 4.11 calculated using MPIE.

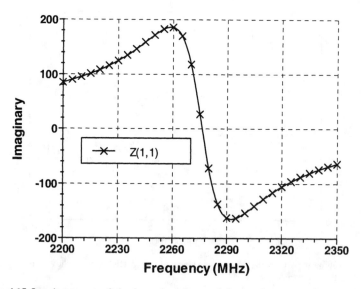

Figure 4.15 Imaginary part of the input impedance of the patch antenna shown in Figure 4.11 calculated using MPIE.

Next, we will compare the results of the cavity model with full-wave electromagnetic simulations. For this purpose, the same antenna is also analyzed by MPIE formulation, giving the results shown in Figures 4.14 and 4.15. As can be seen from comparison of the plots, the cavity model could only approximately predict the resonance frequency and the input impedance of the antenna. The resonance frequency predicted by the full-wave simulation is $f = 2.275$ GHz, which is 5 MHz higher than the cavity-model result. Radiation resistance is predicted to be approximately 15% higher by the cavity model (350 ohms versus 400 ohms). The cavity model predicts low Q because it overestimates the loss due to radiation. Even though there are inaccuracies in the cavity model when compared with full-wave simulations, it still provides a good starting point for patch antenna designs. Comparing the resonance frequency given by (4.1) with the full-wave results would also be interesting. The approximate closed-form equation predicts the resonance frequency will be 2.279 GHz, which is 0.2% higher than the frequency given by the full-wave solution.

As a next step, we will investigate the change in the input impedance of the antenna at resonance with respect to change in the patch width, b. As stated before, input impedance will be reduced as the ratio of b/a is increased. To demonstrate this, the patch antenna is simulated for different widths, and results are shown in Figure 4.16. As can be observed from the figure, the input impedance is approximately 175 ohms for $b = 1.5a$, which is the typical value.

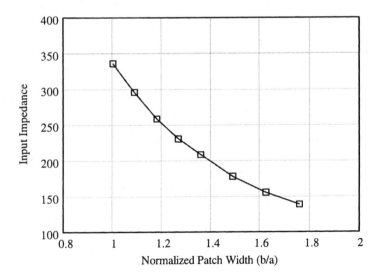

Figure 4.16 Change in resonance input impedance of the antenna shown in Figure 4.11 versus different antenna widths.

To study the patch antenna further, return loss is calculated for different substrate dielectric constants using full-wave simulation, giving the results shown in Figures 4.17 and 4.18. As expected, the resonance frequency increases as the dielectric constant of the substrate is decreased. This is because the guided wavelength in the substrate is inversely proportional with the square root of dielectric constant. For this particular antenna, the change in resonance frequency is $\pm 1\%$ when there is a $\mp 2\%$ change in the dielectric constant. This point might be important in designing for substrate parameter variations [78].

Far-field pattern plots are another important tool in studying microstrip patch antennas. Normalized far-field E-field patterns of the patch antenna under study are given in Figures 4.19 to 4.22. From the first two plots, it can be seen that the single patch antenna has a wide 3-dB E-plane beamwidth (approximately 120°). However, the beam width is smaller in the H-plane. The reason for this is that the E-field component in the H-plane, E_ϕ, is tangential to ground plane (which is assumed to be infinitely large in the transverse plane). Therefore, as the angle θ starts to approach 90° or −90°, this component of the E-field starts to get shorted by the ground plane, reducing the beam width faster on this plane. Next, two field plots provide the field patterns for constant θ. By observing variation in the field magnitude as ϕ varies, one can get information about the polarization of the antenna. In an ideal circular polarization, one would see that field patterns form a perfect circle, an indication that field magnitude does not change for different ϕ. In this case, however, the field patterns show a cardioid shape because the antenna simulated here is linearly polarized. Note that when $\theta = 0°$, the field pattern of E_ϕ equals to the field pattern of E_ϕ rotated by 90° and vice versa.

Spectral plots of current densities on the patch antenna are given in Figures 4.23 and 4.24 near resonance to illustrate current distribution on a patch antenna. Note that at the edges of the patch parallel to the x-axis, parallel components of the electric current density exhibit peaks due to the phenomenon called edge singularity. The reason for this behavior is that the tangential electric field on the patch is almost zero, and this necessity requires accumulation of a strong charge density at the edges to cancel the field generated by the charges deposited in the interior regions of the patch [79]. Note that this singularity is not dictated by the choice of singular basis functions that are located at the edges through the formulation of a full-wave solution; it automatically results from the solution of method-of-moment matrix system, which employs subdomain basis functions. Another point worth mentioning is that the y-directed current is not excited as much as the x-directed current, which is an indication of a linearly polarized far-field pattern. However, there are techniques that can excite the y-directed current as well as the x-directed current to produce circular polarization. These methods will be demonstrated later in this chapter.

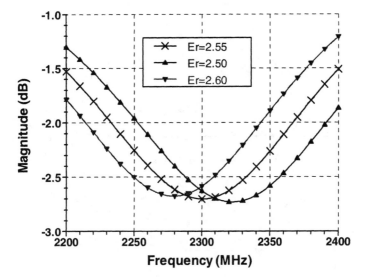

Figure 4.17 Magnitude of S_{11} of the patch antenna shown in Figure 4.11 for different substrate dielectric constants.

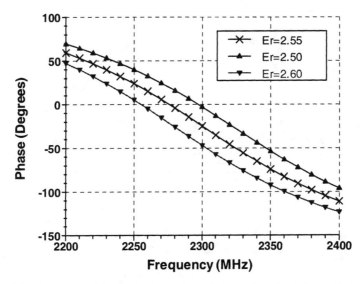

Figure 4.18 Phase of S_{11} of the patch antenna shown in Figure 4.11 for different substrate dielectric constants. Note that there is a shift in resonance frequency.

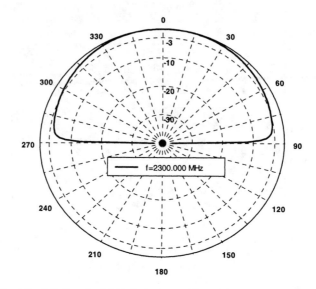

Figure 4.19 Normalized *E*-plane far-field radiation pattern (E_θ) of the microstrip patch shown in Figure 4.11 at $f = 2,300$ MHz ($\phi = 0°$, $-90° < \theta < 90°$).

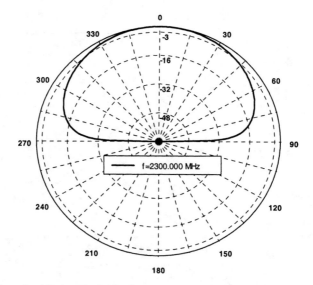

Figure 4.20 Normalized *H*-plane far-field radiation pattern (E_ϕ) of the microstrip patch shown in Figure 4.11 at $f = 2,300$ MHz ($\phi = 0°$, $-90° < \theta < 90°$).

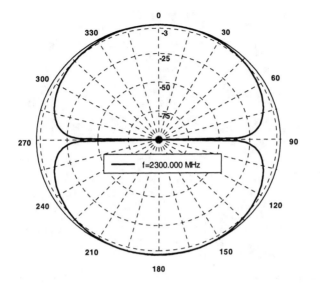

Figure 4.21 Normalized far-field radiation pattern (E_θ) of the microstrip patch shown in Figure 4.11 at $f = 2{,}300$ MHz ($\theta = 0°$, $-180° < \phi < 180°$).

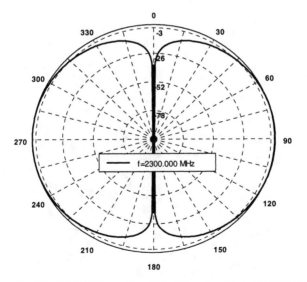

Figure 4.22 Normalized far-field radiation pattern (E_ϕ) of the microstrip patch shown in Figure 4.11 at $f = 2{,}300$ MHz ($\theta = 90°$, $-180° < \phi < 180°$).

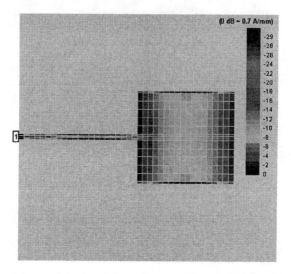

Figure 4.23 *x*-directed current density of the patch antenna shown in Figure 4.11 at *f* = 2,300 MHz (ε_r = 2.55). Note singularity effect at the edges.

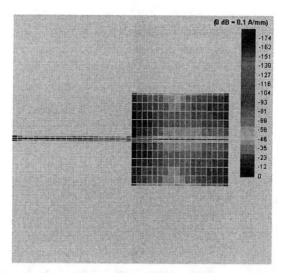

Figure 4.24 *y*-directed current density of the patch antenna shown in Figure 4.11 at *f* = 2,300 MHz (ε_r = 2.55). Note that the *y*-directed current is not excited as much as the *x*-directed current.

4.3 PROXIMITY-COUPLED MICROSTRIP PATCH ANTENNAS

A proximity-coupled patch antenna is constructed by placing the feed line and patch on separate dielectric layers, as shown in Figure 4.25. Then, the antenna is tuned by adjusting the coupling between the line and patch, which can be done by changing the amount of overlap. The maximum coupling of the feed line to the patch occurs approximately at $s = a/2$ where s is the amount of overlap. Note that since the feed line is capacitively coupled to the patch, the inductive nature of the input impedance due to higher-order modes can be compensated for. Proximity coupling was first proposed by Pozar and is one of the methods commonly used to increase the impedance bandwidth of microstrip patch antennas [53].

One can further improve the impedance bandwidth in proximity coupling further by placing a shunt tuning stub at the feed line [53]. In this way, it is possible to achieve an impedance bandwidth of more than 10%, which is quite an improvement considering the bandwidth of an uncompensated patch. If it is decided to use stub matching, it would be appropriate to place stubs on both sides of the feed line to keep the structure symmetric.

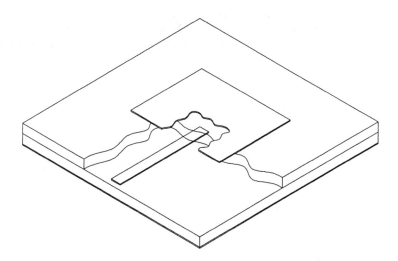

Figure 4.25 Proximity-coupled microstrip patch antenna. Note how the top patch and feed line are positioned on separate layers. The top layer and patch are cut open for easy visualization.

Note that proximity coupling has the disadvantage of requiring two dielectric layers. However, with the advance in PCB technology, producing multiple-layer RF circuits can easily be done even with microwave circuit boards. The designer must be aware that there might be some registration errors between the top layer and bottom layer because each layer in a multilayer PCB is separately prepared and then bonded together. Any shift in the layers during this process will affect the position of the feed line with respect to the patch.

Example 4.2

In this example, the proximity-coupled patch antenna shown in Figure 4.26 is analyzed through full-wave electromagnetic simulations, and the effect of overlap between the feed line and the patch is investigated for different coupling lengths.

From the simulation results, we observe that the impedance locus of the antenna moves from the capacitive reactance region to the inductive reactance region as the coupling length increases; see Figures 4.27 and 4.28. This is an expected result because we already know that the input impedance locus of a patch antenna around resonance on thick substrate is dominantly inductive. Thus, increasing the series capacitance (i.e., increasing the overlap), would shift the input impedance to the inductive region. The resonance frequency of this antenna is approximately 3,600 MHz. Note that since the impedance locus is on the left side of the Smith chart, resonance frequency is determined by the point where the phase of reflection coefficient jumps from 180° to −180°. It is important to state that the coupling length of the microstrip line also changes the apparent resonance frequency of the antenna slightly. At $s = 1.0$ cm, the impedance bandwidth of the antenna is approximately 4% for a return loss better than 10 dB. As indicated before, this figure could be further improved by adding a small stub on the feed line.

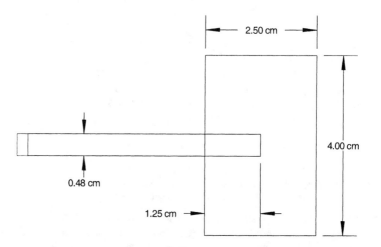

Figure 4.26 Full-wave electromagnetic simulation model of a proximity-coupled microstrip patch antenna ($h_1 = 1.58$ mm, $\varepsilon_{r1} = 2.2$, $h_2 = 1.58$ mm, $\varepsilon_{r2} = 2.2$). (*After:* [53].)

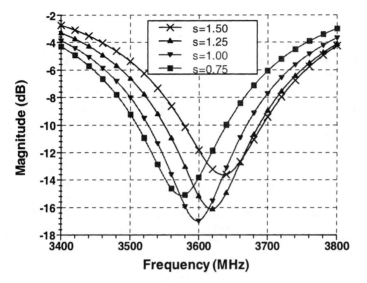

Figure 4.27 Magnitude of S_{11} of the proximity-coupled patch antenna shown in Figure 4.26 for different coupling lengths, s (cm).

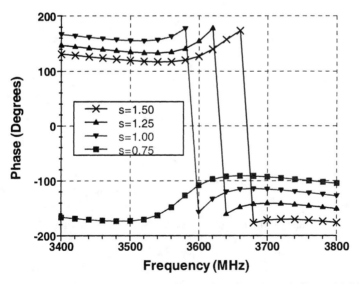

Figure 4.28 Phase of S_{11} of the proximity-coupled patch antenna shown in Figure 4.26 for different coupling lengths, s (cm).

4.4 APERTURE-COUPLED MICROSTRIP PATCH ANTENNAS

Aperture coupling is another frequently employed configuration for patch antennas where multilayer PCB construction is possible. A typical aperture-coupled rectangular patch antenna is shown in Figure 4.29. In this configuration, the radiating patch is placed over a ground plane which has an opening. Then, the feed line is placed on the other side of the ground plane. Thus, the energy propagated through the feed line is coupled to the patch through the opening. Note that the feed line continues slightly after the aperture, effectively forming an open circuited stub for the input impedance seen at the aperture plane. By controlling the aperture length, stub length, patch width and substrate thicknesses, one can achieve a wide impedance bandwidth. Another advantage of aperture coupling is that since the feed lines are hidden by the ground plane, their adverse effect on the radiation pattern is minimized in array applications. It is also possible to select aperture shapes other than a simple rectangle to optimize the coupling further. Aperture coupling in microstrip patch antennas was first proposed by Pozar [80]. There is an extensive literature on aperture-coupled patch antennas and some of them are given in the references [50, 51, 80–83].

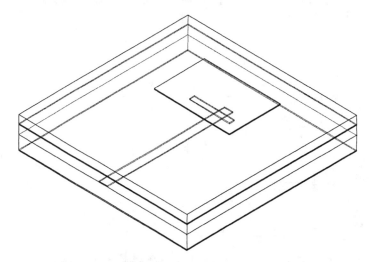

Figure 4.29 Aperture-coupled microstrip patch antenna. Note that a ground layer with a slot opening is placed between the patch antenna and microstrip feed line.

It is possible to design the aperture coupling using a stripline feed (i.e., the feed line has two ground planes on both sides) or microstrip. In the stripline configuration, the coupling from the feed line to the slot can be lower than the microstrip case. Therefore, it may be beneficial to utilize an inhomogeneous stripline to maximize the field density on the slot side of the feed. In the microstrip configuration, on the other hand, one must be careful not to resonate the slot because a resonating slot would also radiate, which would increase the radiation to the back side of the circuit. A qualitative plot of the input impedance of an aperture-coupled patch antenna is given in Figure 4.30.

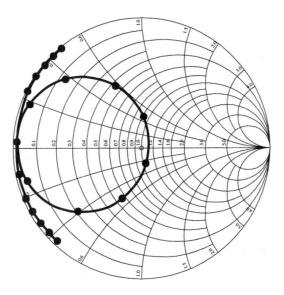

Figure 4.30 Qualitative plot of the input impedance for an aperture-coupled microstrip antenna.

Due to their complexity, aperture-coupled antennas are best analyzed through numerical methods. However, the transmission line and cavity methods have been applied to aperture-coupled antennas as well [56, 57]. For the sake of completeness, we will now present the transmission line analysis for the aperture-coupled rectangular patch antenna.

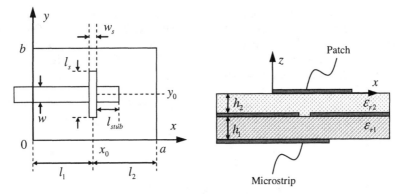

Figure 4.31 Top and side views of an aperture-coupled antenna showing the important dimensions.

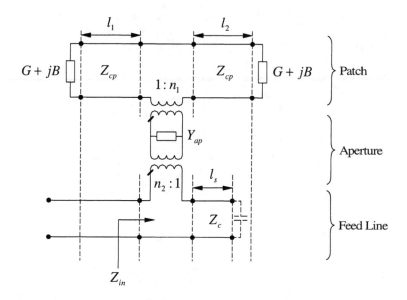

Figure 4.32 Equivalent model of the aperture-coupled rectangular patch antenna.

Let's first consider the coordinate system and equivalent model of the structure as shown in Figures 4.31 and 4.32, respectively. Here, we will assume a microstrip feed line as depicted in Figure 4.31. As a rule of thumb, it is recommended to select the slot width and length according to [60]

$$l_s \approx 10 \cdot w_s \qquad (4.45)$$

Moreover, the slot length should be long enough to avoid interactions between the feed line and slot edges [59]:

$$l_s > w + 10 \cdot h_1 \qquad (4.46)$$

Perhaps the most important parameters of the equivalent model shown in Figure 4.32 are the transformer ratios n_1, n_2 and the aperture admittance Y_{ap}. Once these parameters are calculated accurately, determining the input impedance Z_{in} is relatively straightforward. By following the circuit model given in Figure 4.32, the input impedance is written as follows:

$$Z_{in} = \frac{n_2^2}{\left(n_1^2 Y_{patch} + Y_{ap}\right)} - jZ_c \cot\left(\beta l_{stub}\right) \qquad (4.47)$$

where

$$Y_{patch} = Y_1 + Y_2 \qquad (4.48)$$

$$Y_{ap} = -2jY_{cs} \cot\left(\beta_s l_s / 2\right) \qquad (4.49)$$

$$Y_1 = Y_{cp} \frac{(G + jB) + jY_{cp} \tan(\beta_p \, l_1)}{Y_{cp} + j(G + jB)\tan(\beta_p \, l_1)}$$

$$Y_2 = Y_{cp} \frac{(G + jB) + jY_{cp} \tan(\beta_p \, l_2)}{Y_{cp} + j(G + jB)\tan(\beta_p \, l_2)}$$

$$n_1 \cong \frac{l_s}{b} \tag{4.50}$$

$$n_2 = n_0 \left[\frac{\beta^2 k_2 \varepsilon_r}{k_2 \varepsilon_r \cos(k_1 h) - k_1 \sin(k_1 h)} + \frac{\beta_s^2 k_1}{k_1 \cos(k_1 h) + k_2 \sin(k_1 h)} \right] \tag{4.51}$$

with

$$n_0 = \frac{J_0(\beta_s \, w/2) J_0(\beta \, w_s/2)}{\beta_s^2 + \beta^2}$$

$$k_1 = k_0 \sqrt{|\varepsilon_r - \varepsilon_{em} - \varepsilon_{es}|} \quad k_2 = k_0 \sqrt{|\varepsilon_{em} + \varepsilon_{es} - 1|}$$

$$\beta_s = k_0 \sqrt{\varepsilon_{es}} \quad \beta = k_0 \sqrt{\varepsilon_{em}}$$

In the above equations, β, β_p, and β_s are the propagation constants on the microstrip feed line, rectangular patch, and slotline, respectively. Similarly, Z_c, Z_{cp}, and Z_{cs} are the characteristic impedances of the microstrip feed line, rectangular patch, and slotline, respectively. The effective relative dielectric constants of the slotline and microstrip are denoted by ε_{es} and ε_{em}, respectively. The microstrip line to slotline coupling has been extensively addressed and readers are referred to the literature for a more detailed discussion on this subject [84–92].

The transformer ratio n_2 is found according to the theory of Das [84]. An improved approximation is given by Kim [87]. An alternative, simpler expression for the transformer ratio n_2 is also given as follows [56]:

$$n_2 \cong \frac{l_s}{\sqrt{w_{eff} h_1}} \tag{4.52}$$

where w_{eff} is the effective width of the feed line.

Note that characterization of the slotline (i.e., determination of the characteristic impedance, Z_{cs} and the effective dielectric constant, ε_{es}) is an important step in the analysis of slot-coupled microstrip patch antenna. Characteristic impedance of the slotline, Z_{cs}, can be approximately given by the following expression [88, 89, 92, 93]:

$$Z_{cs} = \frac{60 \cdot \pi}{\sqrt{\varepsilon_{es}}} \frac{K(k_0)}{K(k_0')} \tag{4.53}$$

where

$$k_0 = \sqrt{2\frac{\alpha_0}{1+\alpha_0}} \quad k_0' = \sqrt{1-k_0^2}$$

$$\alpha_0 = \tanh\left(\frac{\pi w_s}{2h_0}\right)$$

where $K(\ )$ is the complete elliptic integral of the first kind. In the above equation, h_0 is the virtual distance below and above the slot where the fields are assumed to be concentrated. Note that because of the non-TEM nature of the propagation mode, the characteristic impedance and phase velocity of the slotline are more dispersive than a microstrip line. On the other hand, slotline differs from waveguides in the sense that it has no cutoff frequency.

4.5 STACKED MICROSTRIP PATCH ANTENNAS

A typical stacked-patch configuration is shown in Figure 4.33. In this configuration, there is one driven (bottom) and one parasitic (top) antenna separated by one or multiple dielectric layers, although more than one parasitic antenna is also possible. The stacked-patch configuration can be used to improve impedance bandwidth of microstrip patch antennas [45–49]. When it is implemented with aperture coupling and multiple resonators, an impedance bandwidth of nearly 70% can be achieved for VSWR < 2 [44, 52]. To improve the impedance bandwidth further, one can also utilize the inset-fed technique on the driven patch if microstrip feed is being used. If the driven patch is fed by a probe, then the probe location can be optimized in that case.

In the figure, each of the antennas has its own substrate separated by a third substrate to provide the necessary spacing. This configuration will provide the flexibility of easily adjusting the separation between the patches. By changing the thickness of this third layer, coupling between the antennas is controlled. Since the antenna substrates are separate in this configuration, one can use a variety of materials as the middle layer, such as air (patches must be supported by spacers in that case), foam, PTFE, or regular printed circuit board materials so that various experiments can be performed. The main design parameters of a stacked antenna are the dimensions of the parasitic element and spacing between the parasitic and driven elements. Typical separation between the parasitic and driven elements is in the order of 0.1 to 0.15λ. It should be noted that due to relatively thick spacing requirements, design of stacked antennas in a laminate process might be difficult because a single substrate with the required thickness may not be found easily. In that case, many layers can be laminated to achieve the required thickness by using a low-loss and low-thickness attachment material (i.e., prepreg). Multilayer PCBs are discussed in more detail in Chapter 2.

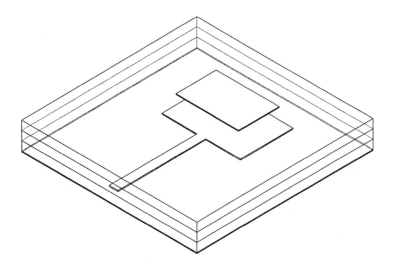

Figure 4.33 Stacked patch antennas. Note that the driven and parasitic patches are separated using a dielectric layer. Depending on the separation between two patches, spacers might be needed to provide enough clearance.

It has been shown that for a stacked configuration similar to that given in Figure 4.33, there would be three fundamental regions of operation [48, 49]. In the first region, where the separation between the parasitic and drive patches is less than approximately 0.15λ, the antenna configuration provides wide impedance bandwidth (up to 10%) provided that the inset-fed technique is also used on the driven patch. The antenna gain in this region is around 7 dB, which is typical. Then, in the second region, where the separation falls between 0.15λ and 0.3λ, the radiation pattern becomes anomalous. Thus, operation in this region should be avoided. When the separation is greater than 0.3λ, the impedance bandwidth drops to around 2% to 3%, but antenna gain increases to approximately 10 dB. One can think of this last region as analogous to a Yagi-Uda antenna [64]. Thus, the third region can be useful when high directive gain is desirable.

Example 4.3

In this example, we will study a stacked patch antenna configuration for different resonator element dimensions using full-wave electromagnetic simulation. As indicated previously, the main design parameters are the separation between the driven and parasitic elements and dimensions of the parasitic element. Here, we will keep the separation constant and vary the parasitic patch dimensions until we obtain a good input return loss. The two patches are separated by a PTFE material with a thickness of 2.54 mm. The simulated model is shown in Figure 4.34.

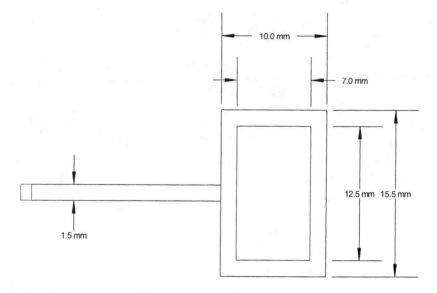

Figure 4.34 Full-wave electromagnetic simulation model of stacked microstrip patch antennas ($h_1 = 0.254$ mm, $\varepsilon_{r1} = 2.1$, $h_2 = 2.54$ mm, $\varepsilon_{r2} = 2.17$, $h_3 = 0.254$ mm, $\varepsilon_{r3} = 2.2$). (*After:* [49].)

Note that the resonance frequency of a single patch antenna for the given substrate and dimensions should be around 10 GHz. Typically, when the size of the parasitic patch is reduced, it is reduced in both coordinates (i.e., x and y) by keeping the symmetry as shown in the figure. Although asymmetrical changes in the parasitic patch with respect to the driven patch are also possible, this may have an adverse effect on the radiation pattern.

Simulation results are provided in Figures 4.35 and 4.36 for different d values for which the parasitic patch dimensions are given by $(10 - 2 \cdot d) \times (15.5 - 2 \cdot d)$. The plot for $d = 0.0$ in Figure 4.35 corresponds to when the parasitic and given patches have the same sizes. In that case, the input return loss is almost equivalent to that of a single rectangular patch on a thin substrate fed by a microstrip line. As we reduce the size of the parasitic element, it is seen that the input return loss starts to improve. And when the parasitic element is shorter than the driven element by 3 mm in each dimension, one can obtain reasonably good input return loss. Note that the resonance frequency is also affected when we change the size of the parasitic element. The resonance frequency of the stacked configuration given in Figure 4.34 is around 9.75 GHz when optimum return loss is obtained. As one might anticipate, there is an optimum size for the parasitic patch in terms of the return loss. For this particular example, reducing the parasitic patch size more would not improve the return loss further.

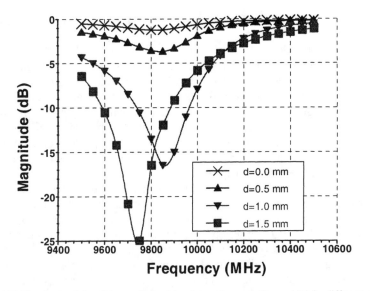

Figure 4.35 Magnitude of S_{11} of the stacked patch antenna shown in Figure 4.34 for different parasitic patch dimensions $(10 - 2 \cdot d) \times (15.5 - 2 \cdot d)$.

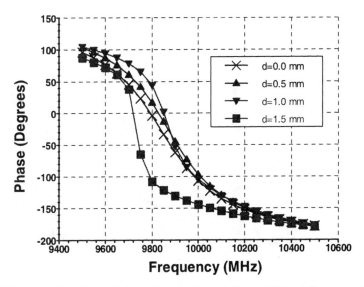

Figure 4.36 Phase of S_{11} of the stacked patch antenna shown in Figure 4.34 for different parasitic patch dimensions $(10 - 2 \cdot d) \times (15.5 - 2 \cdot d)$.

4.6 MICROSTRIP PATCH ANTENNAS WITH PARASITIC ELEMENTS

The use of multiple parasitic elements (resonators) tuned to slightly different frequencies along the nonradiating and/or radiating edges of an antenna is another method to increase the impedance bandwidth of microstrip patch antennas [39–42]. Figure 4.37 shows a microstrip patch antenna with resonators placed along the nonradiating edges. By changing the resonator lengths and spacing between the patch and the resonators, input impedance of the antenna can be controlled. Note that one can also place the resonators along the radiating edges or even along all four edges. In those cases, the probe feed must be employed because there will be no clearance to place a microstrip feed.

Although this method can be convenient for single-patch antennas, it might not be appropriate for patch antenna arrays because of the increased antenna area and interelement coupling (especially if wider parasitics are employed). Another disadvantage, which is more serious, is the fast deterioration of the antenna pattern as the frequency changes due to additional elements. To minimize the distortion on the pattern, the parasitic elements should be as narrow as possible. Perhaps the biggest advantage of the method is obtained when two narrow, different length resonators are used along the nonradiating edges. In that case, antenna input impedance can be matched at two different frequencies, resulting in dual-frequency operation. This point will be demonstrated in the next example.

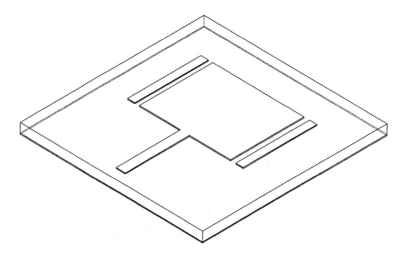

Figure 4.37 Microstrip patch antenna with resonators next to nonradiating sides.

Example 4.4

In this example, we will demonstrate the effects of parasitic elements along the nonradiating edges of the microstrip patch antenna shown in Figure 4.38, using full-wave electromagnetic simulations. The input reflection coefficient of the patch antenna is obtained for different lengths of the parasitic elements (s_1 and s_2) using MPIE, and the results are given in Figures 4.39 and 4.40. Note that during the experiment, the width of the parasitic elements and the gap between the patch and parasitic elements are kept constant.

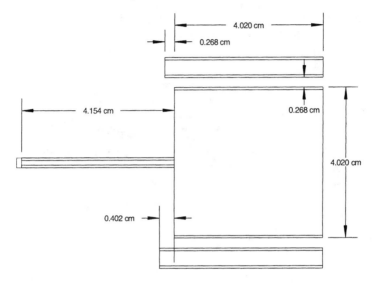

Figure 4.38 Full-wave electromagnetic simulation model of a microstrip patch antenna with parasitic elements (h = 1.58 mm, ε_r = 2.55). (*After:* [42].)

It is interesting to compare the input reflection coefficient of the patch without parasitic elements with the case when parasitic elements are included. As expected, the simple antenna without parasitic elements has very poor return loss at and around the resonance. Placing the parasitic elements next to nonradiating edges significantly improves the return loss as shown in the figure. When the resonator lengths are equal to each other, the input reflection coefficient has one minimum in the frequency range of interest and the antenna can only be operated around that point. On the other hand, when different resonator lengths are selected, the input reflection coefficient has two minimums resulting in dual-frequency operation of the antenna, which is the main benefit of the technique.

Spectral plots of the current density on the patch antenna are also given in Figures 4.41 and 4.42, at 2,225 and 2,295 MHz, respectively, to illustrate the current distribution on the resonators. In the first figure, the bottom resonator is active, whereas in the second one, the top resonator is active. It is clearly seen that each resonator resonates at its designated frequency, thus providing dual-frequency operation.

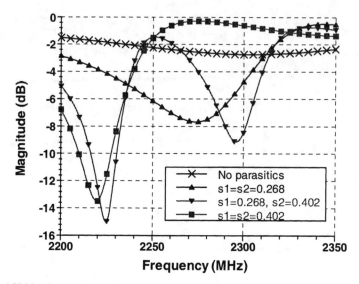

Figure 4.39 Magnitudes of S_{11} of the microstrip patch antenna with parasitic elements shown in Figure 4.38 for different resonator lengths s_1 and s_2. (cm).

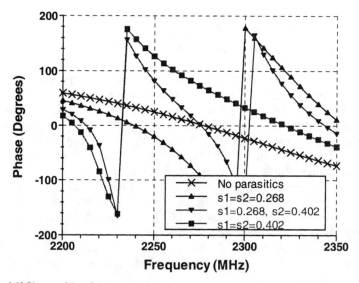

Figure 4.40 Phases of S_{11} of the microstrip patch antenna with parasitic elements shown in Figure 4.38 for different resonator lengths s_1 and s_2 (cm).

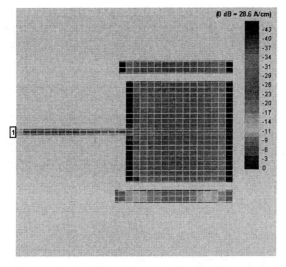

Figure 4.41 *x*-directed current density of the patch antenna shown in Figure 4.38 at $f = 2,225$ MHz for $s_1 = 0.268$ cm and $s_2 = 0.402$ cm.

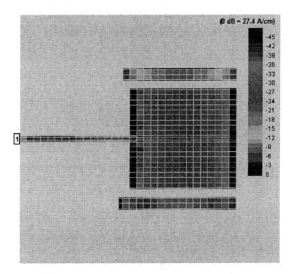

Figure 4.42 *x*-directed current density of the patch antenna shown in Figure 4.38 at $f = 2,295$ MHz for $s_1 = 0.268$ cm and $s_2 = 0.402$ cm.

4.7 INSET-FED MICROSTRIP PATCH ANTENNAS

An inset-fed configuration is perhaps the most common method to improve the impedance bandwidth of microstrip patch antennas because of its simplicity [25, 78]. In this scheme, an inset is created at the point where the microstrip feed line is connected to the patch as shown in Figure 4.43. By changing depth of the inset, d, the input impedance of the antenna can be controlled. Note that the inset-fed configuration is analogous to the probe-fed configuration, where the probe position is moved along the center line of the antenna. Since the vertical component of the electric field seen at the input is changing as we move the feed point along that line, the input impedance is altered.

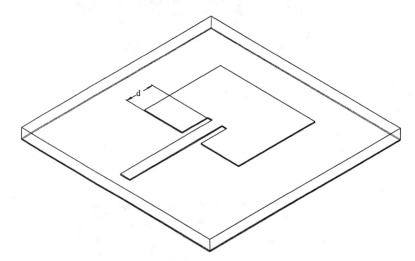

Figure 4.43 Inset-fed microstrip patch antenna.

At this point, it would be useful to provide an approximate expression for the input impedance of the inset-fed patch antenna based on the cavity model. According to the cavity model, the resonance input impedance of a patch antenna is given by the following expression:

$$R_{in} = R_0 \cdot \cos^2\left(2\pi \frac{d}{\lambda}\right) \qquad (4.54)$$

where R_0 is the input resistance when the patch is fed at the edge, d is the feed distance (or inset depth) from the edge, and λ is the guided wavelength at resonance frequency. This equation tells us that it is possible to modify the input impedance of a patch antenna by changing the feed position. Therefore, one can match the antenna to the driving circuitry by properly selecting the feed distance from the edge. Note that the above equation is derived for the probe-fed configuration. However, it can also be used for the microstrip inset-fed configuration. One must be careful in the microstrip case, however, that the reference plane for the impedance value given by (4.54) is at the edge of the inset (i.e., at the actual feeding point). Therefore, the impedance transformation due to the narrow transmission line segment of length d should be taken into account in practice where the reference plane is usually taken at the edge of the patch. This might be important because the feed line section in the inset usually forms a high-impedance transmission line.

Note that to disturb the field distribution in the patch cavity in a least possible way, width of the microstrip feed line is typically selected less than one-tenth of the length of the radiating edge. The gap widths of the inset should be kept as small as possible for the same reason.

Example 4.5

In this example, we will investigate the effect of inset depth on the input impedance of the patch antenna shown in Figure 4.44 using full-wave electromagnetic simulations.

The results of comparisons of the simple theory (4.54) and full-wave simulation are given in Figure 4.45. In order to carry out the comparison, the antenna input impedance is first found using the full-wave simulation when the antenna is fed at the edge (i.e., $d = 0$). Then, using this found impedance in place of R_0, input impedances for different feed positions are determined both by using the approximate expression given in (4.54) and separate full-wave simulations for each case, which gives the plot in Figure 4.45.

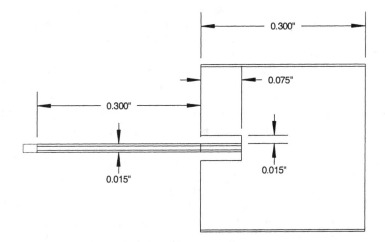

Figure 4.44 Full-wave electromagnetic simulation model of an inset-fed microstrip patch antenna ($h = 1.58$ mm, $\varepsilon_r = 2.33$).

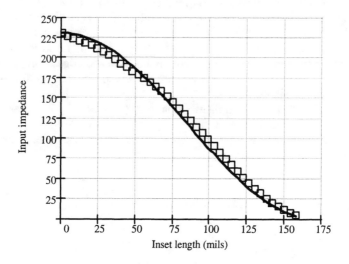

Figure 4.45 Input impedance of microstrip patch shown in Figure 4.44 versus inset length (line: approximate model, square: full-wave simulation). The resonance frequency of the antenna is $f = 11.3$ GHz.

The first observation that can be made is that once R_0 is determined accurately (in this case, using full-wave simulations), then (4.54) can provide input impedance quite well for most cases. By inspecting the plot, it is also seen that input impedance equals approximately 50 ohms when the inset depth equals 125 mils. Note that the reference plane for the impedance values shown in the figure is at the edge of the inset as indicated before (full-wave simulations are also de-embedded appropriately). But this is not practical for microstrip-fed applications. Although we have the flexibility of de-embedding results in the simulation domain, in practice, the reference plane is determined at the point where the antenna is connected to the rest of the circuitry where the microstrip line is 50 ohms (typically). Therefore, the input impedance values given in Figure 4.45 must be transferred to the edge of the patch antenna for 50-ohm normalization impedance because the original feed line has very high characteristic impedance (unless the rest of the circuit does not have the same impedance as normalization impedance).

Figures 4.46 and 4.47 show the results for different inset depths d, when the reference plane is set at the patch edge to demonstrate the above point. In this case, optimum performance is obtained when the inset depth is around $d = 70$ mils. Note that this value is different from the value obtained in Figure 4.45 because impedance transformation occurs due to the high characteristic impedance microstrip feed line, as indicated above.

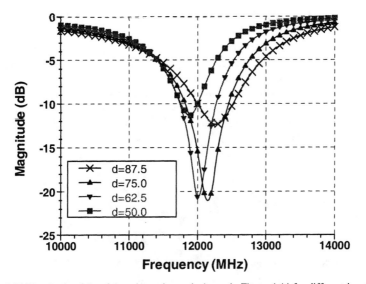

Figure 4.46 Magnitude of S_{11} of the microstrip patch shown in Figure 4.44 for different inset depths d (mils).

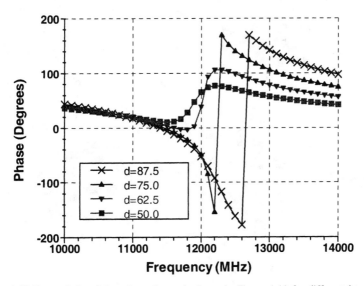

Figure 4.47 Phase of S_{11} of the microstrip patch shown in Figure 4.44 for different inset depths d (mils).

4.8 CIRCULARLY POLARIZED MICROSTRIP PATCH ANTENNAS

In some applications, it is necessary to have a circularly polarized antenna because the receiving antenna orientation changes with respect to the transmitter (e.g., GPS receivers) [9]. Circular polarization provides nearly constant receiving power levels in such cases. The basic microstrip patch antenna dominantly excites currents only in one direction as demonstrated before. In order to create circularly polarized radiation, it is required to excite two orthogonal patch modes on the antenna with 90° phase difference.

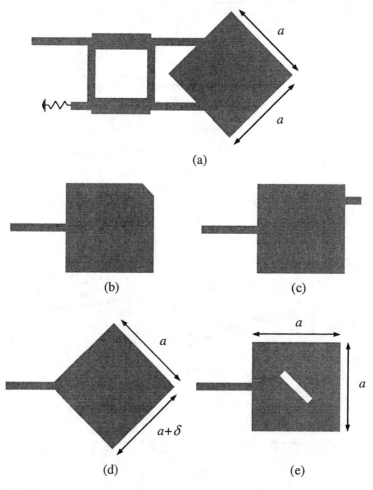

Figure 4.48 Different methods to create circular polarization using a square patch antenna: (a) using 90° 3-dB hybrid; (b) mitered corner; (c) stub on the radiating edge; (d) corner fed; and (e) slot center.

Polarization of antennas is usually quantified using a figure of merit called axial ratio [64]:

$$AR = \frac{OA}{OB} \qquad (4.55)$$

where OA and OB are the semimajor and semiminor axes of the polarization ellipse. For circular polarization, AR becomes unity.

There are different techniques to obtain circular polarization, and a good summary of these methods is given in the literature [9, 52, 94–97]. The two orthogonal modes can be excited on a square patch by using separate feeds on two adjacent sides and a 90° 3-dB hybrid, or by creating a slight irregularity on a patch so that two degenerate modes are excited, as shown in Figure 4.48(a–e). Note that the irregularity on the patch surface will shift the resonance frequencies of the degenerate modes in opposite directions. With proper design, a 90° phase difference is achieved between the mode voltages, thus producing circular polarization. For instance, by adjusting the amount of mitering in Figure 4.48(b) or the length and position of the stub in Figure 4.48(c), the desired circular polarization is achieved. It is usually possible to obtain a more compact structure by using the latter techniques because the 90° hybrid is eliminated. However, the axial ratio of the patch will degenerate more rapidly compared to the separate orthogonal feeds.

Example 4.6

In this example, we will analyze a patch antenna with a stub as shown in Figure 4.49 using full-wave simulations to demonstrate circularly polarized radiation. Here, the stub length is fixed, but the stub position will be changed to see its effect on the polarization.

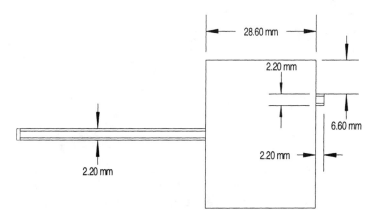

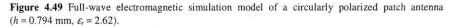

Figure 4.49 Full-wave electromagnetic simulation model of a circularly polarized patch antenna ($h = 0.794$ mm, $\varepsilon_r = 2.62$).

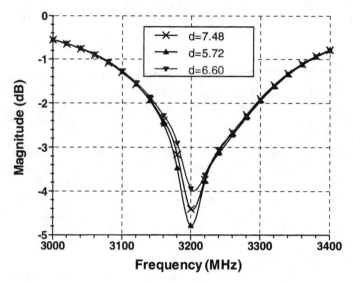

Figure 4.50 Magnitude of S_{11} of the microstrip patch shown in Figure 4.49 for different stub positions d (mm).

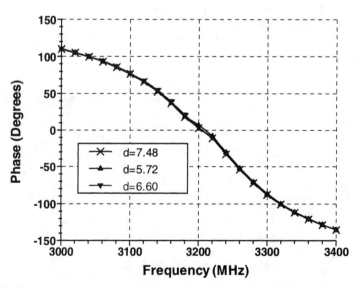

Figure 4.51 Phase of S_{11} of the microstrip patch shown in Figure 4.49 for different stub positions d (mm).

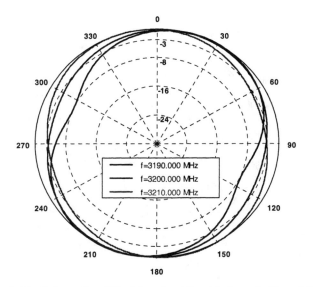

Figure 4.52 Far-field radiation pattern (E_θ) of the microstrip patch shown in Figure 4.49 for $d = 7.48$ mm ($\theta = 0°$, $0° < \phi < 360°$).

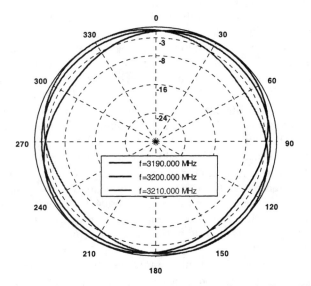

Figure 4.53 Far-field radiation pattern (E_θ) of the microstrip patch shown in Figure 4.49 for $d = 5.72$ mm ($\theta = 0°$, $0° < \phi < 360°$).

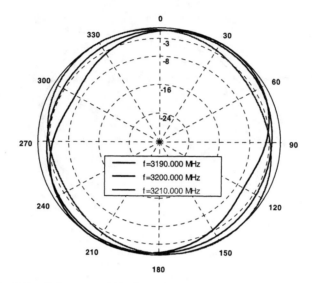

Figure 4.54 Far-field radiation pattern (E_θ) of the microstrip patch shown in Figure 4.49 for $d = 6.60$ mm ($\theta = 0°$, $0° < \phi < 360°$).

Figures 4.50 and 4.51 show the input impedance of the patch antenna with respect to different stub positions. From the plots, we can see that the input impedance and resonance frequency of the patch antenna are relatively insensitive to the stub position, d. Note that the return loss is not very good, but our aim here is to demonstrate circular polarization.

The next step is to investigate the effect of the stub position on the far-field pattern by plotting the field plots. As can be expected, good circular polarization is only generated at a specific stub position as shown in Figures 4.52 through 4.54. The frequency bandwidth of circular polarization is also provided in each field plot by plotting E_θ at different frequencies. The best axial ratio is obtained when the stub position equals 5.72 mm as shown in Figure 4.53. Note that for a perfect circular polarization, the E_θ pattern should be a circle for $\theta = 0°$ and $0° < \phi < 360°$ at the resonant frequency. This would indicate that the received power by the antenna will be constant regardless of the value of ϕ. The field plot for $f = 3,200$ MHz given in Figure 4.53 is the closest one to being a circle compared with the other field plots (compare Figure 4.53 with Figure 4.21 of the linear patch antenna as well). Besides, the same stub position also provides the best frequency insensitivity for circular polarization (i.e., wide frequency bandwidth for circular polarization).

Current distribution of the antenna near resonance is also given in Figures 4.55 and 4.56 for $d = 5.72$ mm. Note that both x- and y-directed currents are excited with almost same magnitude (while comparing the figures, pay attention to the change in the magnitude scale), another indication of possible circular polarization.

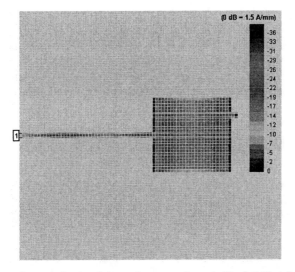

Figure 4.55 *x*-directed current density of the patch antenna shown in Figure 4.49 at f = 3,200 MHz for d = 5.72 mm.

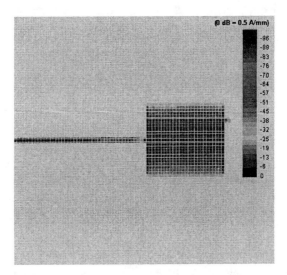

Figure 4.56 *y*-directed current density of the patch antenna shown in Figure 4.49 at f = 3,200 MHz for d = 5.72 mm.

4.9 COUPLING BETWEEN MICROSTRIP PATCH ANTENNAS

Electromagnetic coupling between patch antennas is an important figure for array applications [98–100]. Although for separations greater than approximately 0.2λ, where λ is the free-space wavelength, the input impedance of each element is practically insensitive to the patch separation, the same cannot be said for the far-field array pattern. It has been indicated that coupling levels greater than roughly 20 to 30 dB may have an adverse effect on array performance [13]. Therefore, coupling between the elements needs to be studied, especially for arrays with tight side-lobe requirements.

Two patch antennas placed on the same substrate can couple either in the E- or H-plane through space waves and surface waves. It should be noted that couplings at moderate to high distances are mainly due to surface waves because surface waves attenuate more slowly than the space waves. Space-wave attenuation is proportional to ρ^{-2}, whereas surface waves attenuate at a rate of $\rho^{-1/2}$ and $\rho^{-3/2}$ in the E- and H-planes, respectively [100]. Thus, the surface-wave attenuation rate in the H-plane is faster than in the E-plane. Another interesting feature of mutual coupling between patch antennas is the nonmonotonic behavior observed primarily in the H-plane [100]. This is because the space and surface waves add up constructively or destructively, depending on the separation between antennas. Therefore, it is difficult to establish a simple rule for coupling because of this complex behavior. The best way to investigate coupling between the patch antennas is to employ a full-wave computer simulation. Unlike the transmission line model and the cavity model, rigorous full-wave formulation includes the surface-wave and coupling effects, and therefore, it can be used to predict the mutual coupling between the microstrip antennas very accurately.

Example 4.7

In this example, coupling between two microstrip patch antennas will be investigated using full-wave electromagnetic simulations. Figure 4.57 shows two microstrip patch antennas separated by distance s. This configuration was simulated, and the coupling shown in Figure 4.58 was obtained for different patch separations. As expected, the coupling increases as the separation between patches is decreased. It also reaches its maximum value at resonance frequency for a given separation. The input reflection coefficients of the antennas are also affected if the distance between the patches is too small, as shown in Figure 4.59. However, after a certain value, change in the input reflection coefficient is practically insignificant.

To study coupling effects better, Figures 4.60 and 4.61 are generated showing H- and E-plane couplings versus coupling distance between two patch antennas at the resonance frequency. These figures are obtained by sweeping the distance between two patch antennas and performing full-wave simulation at each separation. It is important to perform such a study in patch array designs to understand the amount of coupling between the elements. For this particular structure, both couplings reduce below −25 dB after one-quarter of the free-space wavelength separation. Note that excitation of surface waves can increase the *E*-plane coupling between the patch antennas, especially on thicker dielectrics (this is because *H*-plane attenuation of surface waves is faster than the *E*-plane attenuation as indicated before).

Current distributions on the coupled antennas are also given in Figures 4.62 and 4.63. In both figures, the top antenna is the excited antenna. Current excitation on the passive antenna is due to electromagnetic coupling.

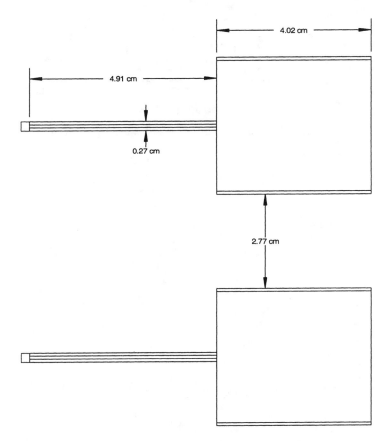

Figure 4.57 Full-wave electromagnetic simulation model of two rectangular microstrip patch antennas placed in close proximity (*h* = 1.58 mm, ε_r = 2.55).

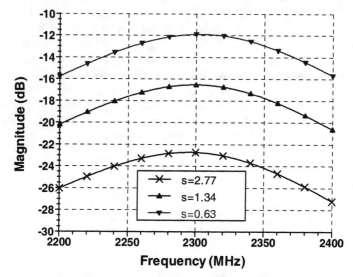

Figure 4.58 Coupling (S_{21}) between the two patch antennas shown in Figure 4.57 for different patch spacings s (cm).

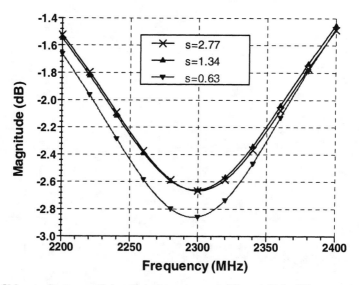

Figure 4.59 Input reflection coefficient (S_{11}) of the antennas in Figure 4.57 for different patch spacings s (cm).

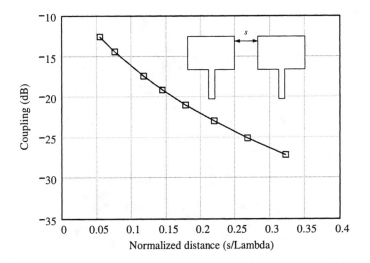

Figure 4.60 *H*-plane coupling (S_{21}) between the two patch antennas shown in Figure 4.57 ($h = 1.58$ mm, $\varepsilon_r = 2.55$, $a = b = 4.02$ cm).

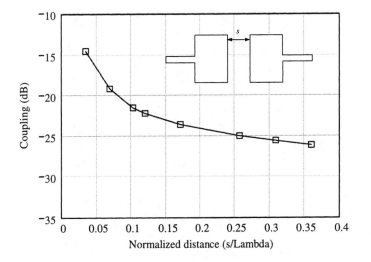

Figure 4.61 *E*-plane coupling (S_{21}) between the two patch antennas shown in Figure 4.57 ($h = 1.58$ mm, $\varepsilon_r = 2.55$, $a = b = 4.02$ cm).

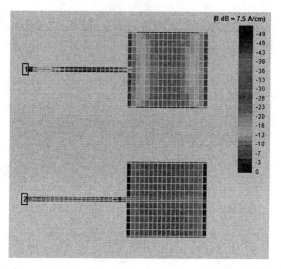

Figure 4.62 x-directed current density of the patch antennas shown in Figure 4.57 at f = 2,300 MHz for s = 2.77 cm. The top antenna is the excited one.

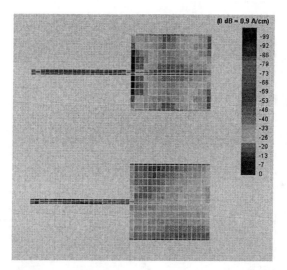

Figure 4.63 y-directed current density of the patch antennas shown in Figure 4.57 at f = 2,300 MHz for s = 2.77 cm. The top antenna is the excited one.

References

[1] David M. Pozar, "An Update on Microstrip Antenna Theory and Design Including Some Novel Feeding Techniques," *IEEE Antennas and Propagat. Society Newsletter*, pp. 5–9, Oct. 1986.

[2] Robert E. Munson, "Conformal Microstrip Antennas and Microstrip Phased Arrays," *IEEE Trans. Antennas Propagat.*, vol. 22, pp. 74–78, Jan. 1974.

[3] John Q. Howell, "Microstrip Antennas," *IEEE Trans. Antennas Propagat.*, vol. 29, pp. 90–93, Jan. 1975.

[4] Y. T. Lo, D. Solomon, and W. F. Richards, "Theory and Experiment on Microstrip Antennas," *IEEE Trans. Antennas Propagat.*, vol. 27, pp. 137–145, Mar. 1979.

[5] K. R. Carver and J. W. Mink, "Microstrip Antenna Technology," *IEEE Trans. on Antennas and Propagat.*, vol. 29, pp. 2–24, Jan. 1981.

[6] R. J. Mailloux, J. F. McIlvenna, and N. P. Kernweis, "Microstrip Array Technology," *IEEE Trans. on Antennas and Propagat.*, vol. AP-29, pp. 25–37, 1981.

[7] William F. Richards, Yuen T. Lo, and Daniel D. Harrison, "An Improved Theory for Microstrip Antennas and Applications," *IEEE Trans. Antennas Propagat.*, vol. 29, pp. 38–46, Jan. 1981.

[8] James R. James et al., "Some Recent Developments in Microstrip Antenna Design," *IEEE Trans. Antennas Propagat.*, vol. 29, pp. 124–128, Jan. 1981.

[9] D. H. Schaubert et al., "Microstrip Antennas with Frequency Agility and Polarization Diversity," *IEEE Trans. on Antennas and Propagat.*, vol. 29, pp. 118–123, Jan. 1981.

[10] W. F. Richards, Y. T. Lo, and J. Brewer, "A Simple Experimental Method for Separating Loss Parameters of a Microstrip Antenna," *IEEE Trans. Antennas Propagat.*, vol. 29, pp. 150–151, Jan. 1981.

[11] David C. Chang, "Analytical Theory of an Unloaded Rectangular Microstrip Patch," *IEEE Trans. Antennas Propagat.*, vol. 29, pp. 54–62, Jan. 1981.

[12] M. A. Weiss, "Microstrip Antennas for Millimeter Waves," *IEEE Trans. Antennas Propagat.*, vol. 29, pp. 171–174, Jan. 1981.

[13] David M. Pozar, "Considerations for Millimeter Wave Printed Antennas," *IEEE Trans. Antennas Propagat.*, vol. 31, pp. 740–747, Sept. 1983.

[14] J. R. James and G. J. Wilson, "Microstrip Antennas and Arrays, Pt. I — Fundamental Action and Limitations," *IEE J. Microwaves, Optics, and Acoustics*, pp. 165–174, 1977.

[15] J. R. James and P. S. Hall, "Microstrip Antennas and Arrays, Pt. II — New Array-Design Technique," *IEE J. Microwaves, Optics, and Acoustics*, pp. 175–181, 1977.

[16] Anders G. Derneryd, "Linearly Polarized Microstrip Antennas," *IEEE Trans. on Antennas and Propagat.*, pp. 846–851, Nov. 1976.

[17] Anders G. Derneryd and Anders G. Lind, "Extended Analysis of Rectangular Microstrip Resonator Antennas," *IEEE Trans. on Antennas and Propagat.*, vol. 27, pp. 846–849, Nov. 1979.

[18] Anders G. Derneryd, "A Theoretical Investigation of the Rectangular Microstrip Antenna Element," *IEEE Trans. on Antennas and Propagat.*, vol. 26, pp. 532–535, July 1978.

[19] J. R. James, "Printed Antennas," in *Proc. SBMO Int. Microwave Symp.*, Rio de Janeiro, Brazil, pp. 597–606, 1987.

[20]   I. J. Bahl and P. Bhartia, *Microstrip Antennas*, Dedham, MA: Artech House, 1982.

[21]   Y. T. Lo and S. W. Lee (Eds.), *Antenna Handbook*, New York: Van Nostrand Reinhold, 1993.

[22]   Kai Fong Lee and Wei Chen (Eds.), *Advances in Microstrip and Printed Antennas*, New York: John Wiley and Sons, 1997.

[23]   Ramesh Garg et al., *Microstrip Antenna Design Handbook*, Norwood, MA: Artech House, 2001.

[24]   D. M. Pozar and J. R. James, "A Review of CAD for Microstrip Antenna and Arrays," in *Microstrip Antennas*, New York: IEEE Press, 1995.

[25]   Daniel H. Schaubert, "A Review of Some Microstrip Antenna Characteristics," in *Microstrip Antennas*, New York: IEEE Press, 1995.

[26]   Russell W. Dearnley and Alain R. F. Barel, "A Comparison of Models to Determine the Resonant Frequencies of a Rectangular Microstrip Antenna," *IEEE Trans. Antennas Propagat.*, vol. 37, pp. 114–118, Jan. 1989.

[27]   Xu Gang, "On the Resonant Frequencies of Microstrip Antennas," *IEEE Trans. Antennas Propagat.*, vol. 37, pp. 245–247, Feb. 1989.

[28]   Shun-Shi Zhong, Gnag Liu, and Ghulam Qasim, "Closed-Form Expression for Resonant Frequency of Rectangular Patch Antenna with Multidielectric Layers," *EEE Trans. Antennas Propagat.*, vol. 42, pp. 1360–1363, Sept. 1994.

[29]   David R. Jackson and Nicolaous G. Alexopoulos, "Simple Approximate Formulas for Input Resistance, Bandwidth, and Efficiency of a Resonant Rectangular Patch," *IEEE Trans. Antennas Propagat.*, vol. 39, pp. 407–410, Mar. 1991.

[30]   D. M. Pozar, "Rigorous Closed-Form Expressions for the Surface Wave Loss of Printed Antennas," *Electron. Letters*, vol. 26, pp. 954–955, June 1990.

[31]   Amos E. Gera, "The Radiation Resistance of a Microstrip Element," *IEEE Trans. Antennas Propagat.*, vol. 38, pp. 568–570, Apr. 1990.

[32]   D. Thouroude, M. Himdi, and J. P. Daniel, "CAD-Oriented Cavity Model for Rectangular Patches," *Electron. Lett.*, vol. 26, pp. 842–844, June 1990.

[33]   P. Pichon, J. Mosig, and A. Papiernik, "Input Impedance of Arbitrarily Shaped Microstrip Antennas," *Electron. Lett.*, vol. 24, pp. 1214–1215, Sept. 1988.

[34]   Z. Sipus, J. Bartolic, and B. Stipetic, "Input Impedance of Rectangular Patch Antenna Fed by Microstrip Line," *Electron. Lett.*, vol. 28, pp. 1886–1888, Sept. 1992.

[35]   M. D. Deshpande and M. C. Bailey, "Input Impedance of Microstrip Antennas," *IEEE Trans. on Antennas and Propagat.*, vol. AP-30, pp. 645–650, July 1982.

[36]   H. F. Pues and A. R. Van de Capelle, "An Impedance-Matching Technique for Increasing the Bandwidth of Microstrip Antennas," *IEEE Trans. on Antennas and Propagat.*, vol. AP-37, pp. 1345–1354, Nov. 1989.

[37]   H. F. Pues and A. R. Van de Capelle, "Wide-Band Impedance-Matched Microstrip Resonator Antennas," *IEE Second Inter. Conf. on Antennas And Propagat.*, Pt. 1, pp. 402–405, 1981.

[38]   H. An, Bart K. J. C. Nauwelaers, and A. R. Van de Capelle, "Broadband Microstrip Antenna Design With the Simplified Real Frequency Design," *IEEE Trans. on Antennas and Propagat.*, vol. 42, pp. 129–136, Feb. 1994.

[39]   C. Wood, "Improved Bandwidth of Microstrip Antennas Using Parasitic Elements," *IEE Proceedings*, vol. 127, Pt. H., pp. 231–234, 1980.

[40]   G. Kumar and K. C. Gupta, "Non-Radiating Edges and Four-Edges Gap-Coupled with Multiple Resonator, Broadband Microstrip Antennas," *IEEE Trans. on Antennas and Propagat.*, vol. AP-33, pp. 173–178, 1985.

[41]   Tsien Ming Au and Kwai Man Luk, "Effect of Parasitic Element on the Characteristic of Microstrip Patch Antenna," *IEEE Trans. Antennas Propagat.*, vol. 39, pp. 1247–1251, Aug. 1991.

[42]   D. H. Schaubert and F. G. Farrar, "Some Conformal Printed Circuit Antenna Designs," *in Proc. Workshop Printed Circuit Antenna Tech.*, New Mexico State Univ., Las Cruces, NM, pp. 1–21, Oct. 1979.

[43]   Frederic Croq and D. M. Pozar, "Millimeter-Wave Design of Wide-Band Aperture-Coupled Stacked Microstrip Antennas," *IEEE Trans. Antennas Propagat.*, vol. 39, pp. 1770–1776, Dec. 1991.

[44]   S. D. Targonski, R. B. Waterhouse, and D. M. Pozar, "Design of Wide-Band Aperture-Stacked Patch Microstrip Antennas," *IEEE Trans. Antennas Propagat.*, vol. 46, pp. 1245–1251, Sept. 1998.

[45]   Georg Splitt and Marat Davidovitz, "Guidelines for Design of Electromagnetically Coupled Microstrip Patch Antennas on Two-Layer Substrates," *IEEE Trans. Antennas Propagat.*, vol. 38, pp. 1136–1140, July 1990.

[46]   G. Dubost et al., "Patch Antenna Bandwidth Increase by Means of a Director," *Electron. Letters*, vol. 22, pp. 1345–1347, 1986.

[47]   H. K. Smith and Paul E. Mayes, "Stacking Resonators to Increase the Bandwidth of Low-Profile Antennas," *IEEE Trans. Antennas Propagat.*, vol. 35, pp. 1473–1476, Dec. 1987.

[48]   R. Q. Lee, K. F. Lee, and J. Bobinchak, "Characteristics of a Two-Layer Electromagnetically Coupled Rectangular Patch Antenna," *Electron. Letters*, vol. 23, pp. 1070–1072, Sept. 1987.

[49]   Richard Q. Lee and Kai-Fong Lee, "Experimental Study of the Two-Layer Electromagnetically Coupled Rectangular Patch Antenna," *IEEE Trans. Antennas Propagat.*, vol. 38, pp. 1298–1302, Aug. 1990.

[50]   Nirod K. Das and David M. Pozar, "Multiport Scattering Analysis of General Multilayered Printed Antennas Fed by Multiple Feed Ports: Part I — Theory," *IEEE Trans. Antennas Propagat.*, vol. 40, pp. 469–481, May 1992.

[51]   Nirod K. Das and David M. Pozar, "Multiport Scattering Analysis of General Multilayered Printed Antennas Fed by Multiple Feed Ports: Part II — Applications," *IEEE Trans. Antennas Propagat.*, vol. 40, pp. 482–491, May 1992.

[52]   S. D. Targonski and D. M. Pozar, "Design of Wideband Circularly Polarized Aperture-Coupled Microstrip Antennas," *IEEE Trans. Antennas Propagat.*, vol. 41, pp. 214–220, Feb. 1993.

[53]   D. M. Pozar and B. Kaufman, "Increasing the Bandwidth of a Microstrip Antenna by Proximity Coupling," *Electron. Letters*, vol. 23, pp. 368–369, Apr. 1987.

[54]   D. M. Pozar and S. D. Tagonski, "Improved Coupling for Aperture Coupled Microstrip Antennas," *Electron. Lett.*, vol. 27, pp. 1129–1131, June 1991.

[55]   K. M. Luk et al., "Broadband Microstrip Patch Antenna," *Electron. Lett.*, vol. 34, pp. 1442–1443, July 1998.

[56]   M. Himdi, J. P. Daniel, and C. Terret, "Analysis of Aperture Coupled Microstrip Antenna Using Cavity Method," *Electron. Lett.*, vol. 25, pp. 391–392, Mar. 1989.

[57]   M. Himdi, J. P. Daniel, and C. Terret, "Transmission Line Analysis of Aperture-Coupled Microstrip Antenna," *Electron. Lett.*, vol. 25, pp. 1229–1230, Aug. 1989.

[58]   Peter L. Sullivan and Daniel H. Schaubert, "Analysis of an Aperture Coupled Microstrip Antenna," *IEEE Trans. Antennas Propagat.*, vol. 34, pp. 977–984, Aug. 1986.

[59]   N. Herscovici and D. M. Pozar, "Full-Wave Analysis of Aperture Coupled Microstrip Lines," *IEEE Trans. Microwave Theory Tech.*, vol. 39, pp. 1108–1114, July 1991.

[60]   D. M. Pozar, *A Review of Aperture Coupled Microstrip Antennas: History, Operation, Developments and Applications*, ECE Department Report, University of Massachusetts, Amherst, May 1996.

[61]   Shih-Chang Wu, Nicolas G. Alexopoulos, and Owen Fordham, "Feeding Structure Contribution to Radiation by Patch Antennas with Rectangular Boundaries," *IEEE Trans. on Antennas and Propagat.*, vol. 40, pp. 1245–1249, Oct. 1992.

[62]   David R. Jackson and Jeffery T. Williams, "A Comparison of CAD Models for Radiation from Rectangular Microstrip Patches," *International Journal Microwave and Millimeter Wave Computer-Aided Engineering*, pp. 236–248, Apr. 1991.

[63]   P. Hammer et al., "A Model for Calculating the Radiation Field of Microstrip Antennas," *IEEE Trans. on Antennas and Propagat.*, vol. 27, pp. 267–270, Mar. 1979.

[64]   John D. Kraus, *Antennas*, New York: McGraw-Hill, 1988.

[65]   Robert E. Collin, *Field Theory of Guided Waves*, New York: IEEE Press, 1991.

[66]   Arun K. Bhattacharyya, "Characteristic of Space and Surface Waves in a Multilayered Structures," *IEEE Trans. Antennas Propagat.*, vol. 38, pp. 1231–1238, Aug. 1990.

[67]   B. Nauwelaers and A Van de Capelle, "Surface Wave Losses of Rectangular Microstrip Antennas," *Electron. Lett.*, vol. 25, pp. 696–697, May 1989.

[68]   Miguel A. Marin, Sina Barkeshli, and Prabhakar H. Pathak, "On the Location of Proper and Improper Surface Wave Poles for the Grounded Dielectric Slab," *IEEE Trans. Antennas Propagat.*, vol. 38, pp. 570–573, Apr. 1990.

[69]   A. Benalla and K. C. Gupta, "Transmission-Line Model for Two-Port Rectangular Microstrip Antenna," *Electron. Lett.*, vol. 23, pp. 882–884, 1987.

[70]   A. K. Bhattacharyya and R. Garg, "Generalized Transmission Line Model for Microstrip Patches," *IEE Proceedings*, 132, pt. H, pp. 93–98, 1985.

[71]   H. Pues and A. Van de Capelle, "Accurate Transmission-Line Model for the Rectangular Microstrip Antenna," *IEE Proceedings*, 131, pt. H, pp. 334–340, Dec. 1984.

[72]   G. Dubost and A. Zerguerras, "Transmission Line Model Analysis of Arbitrary Shape Symmetrical Patch Antenna Coupled with a Director," *Electron. Letters*, vol. 26, pp. 952–954, 1990.

[73]   Russel W. Dearnley and Alain R. F. Barel, "A Broad-Band Transmission Line Model for a Rectangular Microstrip Antenna," *IEEE Trans. on Antennas and Propagat.*, vol. 37, pp. 6–14, Jan. 1989.

[74]   Ning Yuan et al., "A Fast Analysis of Scattering and Radiation of Large Microstrip Antenna Arrays," *IEEE Trans. on Antennas and Propagat.*, vol. 31, pp. 2218–2226, Sept. 2003.

[75]   Chao-Fu Wang, Feng Ling, and Jian-Ming Jin, "A Fast Full-Wave Analysis of Scattering and Radiation from Large Finite Arrays of Microstrip Antennas," *IEEE Trans. on Antennas and Propagat.*, vol. 46, pp. 1467–1474, Oct. 1998.

[76]   Feng Ling, Jiming Song, and Jian-Ming Jin, "Multilevel Fast Multipole Algorithm for Analysis of Large-Scale Microstrip Structures," *IEEE Microwave and Wireless Comp. Lett.*, vol. 9, pp. 508–510, Dec. 1999.

[77]   Jun-Sheng Zhao et al., "Thin-Stratified Medium Fast-Multipole Algorithm for Solving Microstrip Structures," *IEEE Trans. Microwave Theory Tech.*, vol. 46, pp. 395–403, Apr. 1998.

[78]   Daniel H. Schaubert, David M. Pozar, and Andres Adrian, "Effect of Microstrip Antenna Substrate Thickness and Permittivity: Comparison of Theories with Experiment," *IEEE Trans. Antennas Propagat.*, vol. 37, pp. 677–682, June 1989.

[79]   J. Van Bladel, *Singular Electromagnetic Fields and Sources*, Oxford Science Publications, 1991.

[80]   D. M. Pozar, "Microstrip Antenna Aperture-Coupled to a Microstrip Line," *Electron. Lett.*, vol. 21, pp. 49–50, Jan. 1985.

[81]   Mikael Dich, Allan Ostergaard, and Ulrich Gothelf, "A Network Model for the Aperture Coupled Microstrip Patch," *International Journal Microwave and Millimeter Wave Computer-Aided Engineering*, pp. 326–329, Oct. 1993.

[82]   A. Ittipiboon, R. Oostlander, and Y. M. M. Antar, "Modal Expansion Method of Analysis for Slot-Coupled Microstrip Antenna," *Electron. Letters*, vol. 25, pp. 1338–1339, Sept. 1989.

[83]   Filip De Meulenaere and Jean Van Bladel, "Polarizability of Some Small Apertures," *IEEE Trans. on Antennas and Propagat.*, vol. 25, pp. 198–205, Mar. 1977.

[84]   N. K. Das, "Generalized Multiport Reciprocity Analysis of Surface-to-Surface Transitions Between Multiple Printed Transmission Lines," *IEEE Trans. Microwave Theory Tech.*, vol. 41, pp. 1164–1177, June/July 1993.

[85]   Jeong Phill Kim and Wee Sang Park, "Analysis and Network Modeling of an Aperture-Coupled Microstrip Patch Antenna," *IEEE Trans. on Antennas and Propagat.*, vol. 49, pp. 649–654, June 2001.

[86]   R. Janaswamy and D. H. Schaubert, "Characteristic Impedance of a Wide Slotline on Low-Permittivity Substrates," *IEEE Trans. Microwave Theory Tech.*, vol. 34, pp. 900–902, Aug. 1986.

[87]   J. P. Kim and W. S. Park, "An Improved Network Modeling of Slot-Coupled Microstrip Lines," *IEEE Trans. Microwave Theory Tech.*, vol. 46, pp. 1484–1491, Oct. 1998.

[88]   S. B. Cohn, "Slotline on a Dielectric Substrate," *IEEE Trans. Microwave Theory Tech.*, vol. 17, pp. 768–778, Oct. 1969.

[89]   Jiri Svacina, "Dispersion Characteristics of Multilayered Slotlines – A Simple Approach," *IEEE Trans. Microwave Theory Tech.*, vol. 47, pp. 1826–1829, Sept. 1999.

[90]   J. S. Rao, K. K. Joshi, and B. N. Das, "Analysis of Small Aperture Coupling between Rectangular Waveuide and Microstrip Line," *IEEE Trans. Microwave Theory Tech.*, vol. 29, pp. 150–154, Feb. 1981.

[91]   N. B. Das and K. K. Koshi, "Impedance of a Radiating Slot in the Ground Planes of a Microstrip Line," *IEEE Trans. on Antennas and Propagat.*, vol. 30, pp. 922–926, 1982.

[92]   K. C. Gupta, *Microstrip Lines and Slotlines*, Norwood, MA: Artech House, 1996.

[93]   M. A. R. Gunston, *Microwave Transmission Line Impedance Data*, Noble Publishing, 1997.

[94]   P. S. Hall, "Review of Techniques for Dual and Circularly Polarized Microstrip Antennas," in *Microstrip Antennas*, New York: IEEE Press, 1995.

[95]   M. D. Deshpande and N. K. Das, "Rectangular Microstrip Antenna for Circular Polarization," *IEEE Trans. on Antennas and Propagat.*, vol. AP–34, pp. 744–746, May 1986.

[96]   Young-Ho Suh, Chunlei Wang, and Kai Chang, "Circularly Polarized Truncated-Corner Square Patch Microstrip Antenna for Wireless Power Transmission," *Electron. Lett.*, vol. 36, pp. 600–602, Mar. 2000.

[97]   Kin-Lu Wong and Yi-Fang Lin, "Circularly Polarized Microstrip Antenna with Tuning Stub," *Electron. Lett.*, vol. 34, pp. 831–832, Apr. 1998.

[98]   David M. Pozar, "Input Impedance and Mutual Coupling of Rectangular Microstrip Antennas," *IEEE Trans. on Antennas and Propagat.*, vol. AP-30, pp. 1191–1196, Nov. 1982.

[99]   E. H. Newman, J. H. Richmond, and B. W. Kwan, "Mutual Impedance Computation Between Microstrip Antennas," *IEEE Trans. Microwave Theory Tech.*, vol. 31, pp. 941–945, Nov. 1983.

[100] P. R. Haddad and D. M. Pozar, "Anomalous Mutual Coupling Between Microstrip Antennas," *IEEE Trans. Antennas Propagat.*, vol. 42, pp. 1545–1549, Nov. 1994.

Chapter 5

Microstrip Coupled Lines

The study of coupled TEM or quasi-TEM transmission lines is important in microwave and digital electrical engineering. Traditionally, coupled transmission lines have been extensively used in RF circuits to design directional couplers and electric filters [1–3]. Directional couplers implemented using rectangular waveguides were an extremely useful tool for microwave engineers because of their unique property of separating forward- and backward-traveling waves. Perhaps the most important device made using directional couplers was the reflectometer, which enabled the invention of many important devices such as power detectors and network analyzers.

The coupling theory developed by physicist Hans Albrecht Bethe during his brief appointment in the radiation laboratory at the Massachusetts Institute of Technology during World War II, explaining the coupling mechanisms through small apertures between two waveguides, was the first rigorous way of designing waveguide directional couplers. Although rectangular waveguides were extensively used in designing directional couplers, TEM structures, such as coaxial lines, were also employed [1]. Then, as the development of printed circuits advanced, printed TEM or quasi-TEM structures started to be used extensively in designing directional couplers, although for very high-power applications hollow waveguides were still employed. The fast development of printed circuit technology eventually enabled compact and cost-effective designs of filters in microstrip or stripline structures. Microwave filter design using coupled lines also has a very well-established theory in RF and microwave engineering [4–7]. The book by Matthaei, Young, and Jones has been a classic reference for electrical engineers in this area for many years [3]. As will be shown in Chapter 6, one can methodically design coupled-line filters using canonical coupled-line elements starting from low-pass prototype values. Thus, understanding coupled-line theory is quite important to have a good background in microwave filter design. Coupled-line theory is also applied to the calibration of differential circuits, because the differential circuits are best characterized by considering both of the propagating modes [8–10].

The application of coupled-line theory was not limited to microwave circuits; it had a wide impact on digital and telecommunications circuits too. The historic example of the application of coupled-line theory to telecommunications is the analysis of telephone lines where signal propagation in a tightly packed bundle of wires is analyzed [11]. The extensive use of PCB technology in modern electronics is the other example. Due to growing demand for smaller and faster circuits, PCB signal traces have been placed in close proximity, demanding that the lines be treated as coupled systems instead of isolated conductors. Examples of such modern technologies, where the coupled transmission line theory is being used heavily are untwisted pair (UTP) lines for Ethernet connections, low-voltage differential signaling (LVDS) used in high-speed digital communications, digital bus architectures for communication between CPUs and other components (such as Direct Rambus), and flexible printed circuits used in hard-disk trace suspension assemblies (TSA) and liquid crystal display (LCD) connections in portable computers [12]. To elaborate further, it can also be indicated that the digital subscriber line (DSL) connections that bring high-speed Internet to consumers over standard telephony lines enjoy the benefits of the controlled impedance structure of twisted-pair cables. In the Direct Rambus technology, where a high-speed data bus is established between the CPU and RAM, the impedance of the transmission lines must be strictly controlled as well. As can be seen from all these examples, the coupled lines are not just limited to high-end microwave electronic circuits, as one might think, but they are very much being used in consumer electronics as well, thus increasing the importance and practicality of the subject.

In summary, it can be said that there are two chief application areas for coupled-line theory in electrical engineering: The first is in microwave engineering, where the coupling between the line elements is enhanced in a such a way that coupled lines can be used to design directional couplers, electrical filters, and so forth. The second is in digital and telecommunications engineering, where the coupling between the lines is controlled so that the signals on each line propagate with minimum distortion on the pulse shape.

In this chapter, we will focus on TEM or quasi-TEM coupled lines. The general analysis of TEM coupled lines is performed through multiconductor transmission line (MTL) equations, where matrix theory is used, resulting in a very compact analysis [13–19]. An interesting result of matrix analysis of coupled lines is the possibility of decoupling the governing differential equations of the structure using similarity transformations, which will be demonstrated later.

5.1 ANALYSIS OF COUPLED TEM LINES

In this section, we will focus on uniform coupled lines in a homogeneous medium consisting of only two signal conductors because this is the fundamental structure used in most microwave applications. Besides, the simple analysis of two-conductor lines will provide us with the necessary background to understand the coupled lines in general.

The fundamental parameters of TEM coupled lines are the per-unit-length inductance, **L**, and capacitance, **C**, matrixes. These are full matrixes completely describing the magnetic and electric coupling between the lines. Computer programs can be used to extract these parameters for a given cross-section of a structure (assuming the lines are uniform). By uniform, we mean that the widths of the lines are constant in the direction of propagation. Uniform coupled transmission lines can be further divided into subgroups as asymmetrical lines and symmetrical lines. By symmetrical, we mean that the signal lines are identical, and there is a symmetry in the geometry (hence, the capacitance and inductance matrixes are symmetrical). Although analysis of the asymmetrical case certainly includes the symmetrical case, assuming symmetry at the beginning considerably simplifies the formulation. Here, we will cover these two cases separately for the sake of discussion.

5.1.1 Analysis of Symmetrical Coupled TEM Lines

Analysis of symmetrical, uniformly coupled TEM lines is the most common form encountered in the microwave literature to demonstrate the coupling phenomena. Figure 5.1 shows two uniform symmetrical TEM coupled lines terminated with resistive loads. This structure supports two modes, which are historically named even and odd modes based on the electric field patterns of each mode in the cross-sectional plane.

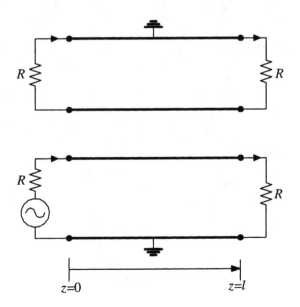

Figure 5.1 Symmetrical and uniform TEM coupled lines terminated with the same impedance at all ports. This structure supports two linearly independent modes of propagation.

Note that the transmission lines present different characteristic impedances to each propagation mode. This is the result of placing two transmission lines in close proximity. If we had an isolated TEM transmission line, then it would have a single characteristic impedance as usual.

The analysis hinges on the expansion of terminal voltages and currents using two different propagating modes. Thus, unlike the coupled-line formulation of general asymmetrical lines, which will be reviewed in the next section, the capacitance and inductance matrixes of the coupled line are implicit in the characteristic impedances and propagation constant of each mode; therefore they are not referenced directly. Our aim is to find voltage values at the terminals of the coupled lines in the presence of two different modes of propagation. For this purpose, we first write the voltage and current phasors on both lines as follows:

$$V_1(z) = A_1 e^{-\gamma z} + A_2 e^{\gamma z} + A_3 e^{-\gamma z} + A_4 e^{\gamma z}$$
$$V_2(z) = A_1 e^{-\gamma z} + A_2 e^{\gamma z} - A_3 e^{-\gamma z} - A_4 e^{\gamma z}$$
(5.1)

$$I_1(z) = \frac{A_1}{Z_{c1}} e^{-\gamma z} - \frac{A_2}{Z_{c1}} e^{\gamma z} + \frac{A_3}{Z_{c2}} e^{-\gamma z} - \frac{A_4}{Z_{c2}} e^{\gamma z}$$
$$I_2(z) = \frac{A_1}{Z_{c1}} e^{-\gamma z} - \frac{A_2}{Z_{c1}} e^{\gamma z} - \frac{A_3}{Z_{c2}} e^{-\gamma z} + \frac{A_4}{Z_{c2}} e^{\gamma z}$$
(5.2)

where Z_{c1} and Z_{c2} are the characteristic impedances corresponding to two TEM waves propagating along the lines. For symmetrical coupled lines, these two modes are the even and odd modes, respectively, and the impedances of the two modes are traditionally represented by $Z_{c1} = Z_{0e}$ and $Z_{c2} = Z_{0o}$ in that case. The coefficients A_n are unknown amplitudes of the waves corresponding to each mode. Note that since it is assumed the mode of propagation is TEM, both of the modes have the same propagation constant. For quasi-TEM cases, this assumption does not strictly hold, which has some implications that will be discussed later.

To determine voltage values on the terminals, we must find the unknown wave coefficients, A_n, in the above equations. However, before attempting this, we first find a relationship between the characteristic impedance of the two modes (Z_{c1} and Z_{c2}) and the termination resistances (R) in order to match the coupled lines to the termination impedances. We will then use this relationship in determining the unknown wave coefficients. To do this, we write input impedances seen by the generator for the two modes as follows:

$$Z_{in}^{(1)} = Z_{c1} \frac{Z_L + jZ_{c1} \tan(\theta_1)}{Z_{c1} + jZ_L \tan(\theta_1)}$$
(5.3)

$$Z_{in}^{(2)} = Z_{c2} \frac{Z_L + jZ_{c2} \tan(\theta_2)}{Z_{c2} + jZ_L \tan(\theta_2)}$$
(5.4)

where $Z_L = R$ (see Figure 5.1). Note that for a homogenous medium, phase velocities of both modes equal (i.e., $\theta_1 = \theta_2$). Then, the total input impedance seen from the first port can be written as

$$Z_{in} = \frac{V^{(1)} + V^{(2)}}{I^{(1)} + I^{(2)}}$$

$$Z_{in} = \frac{\dfrac{Z_{in}^{(1)}}{R + Z_{in}^{(1)}} + \dfrac{Z_{in}^{(2)}}{R + Z_{in}^{(2)}}}{\dfrac{1}{R + Z_{in}^{(1)}} + \dfrac{1}{R + Z_{in}^{(2)}}} = \frac{\left(R + Z_{in}^{(1)}\right) \cdot Z_{in}^{(2)} + \left(R + Z_{in}^{(2)}\right) \cdot Z_{in}^{(1)}}{2R + Z_{in}^{(1)} + Z_{in}^{(2)}}$$

where superscript denotes the modes. In order to have the coupled lines matched to the rest of the circuit, we must have $Z_{in} = R$; that is,

$$\frac{\left(R + Z_{in}^{(1)}\right) \cdot Z_{in}^{(2)} + \left(R + Z_{in}^{(2)}\right) \cdot Z_{in}^{(1)}}{2R + Z_{in}^{(1)} + Z_{in}^{(2)}} = R$$

After some mathematical operations, we find the following equation:

$$R = \sqrt{Z_{in}^{(1)} Z_{in}^{(2)}} \tag{5.5}$$

Now, let's replace the input impedances of the modes in the above equation:

$$R^2 = Z_{c1} \frac{R + jZ_{c1}\tan(\theta)}{Z_{c1} + jR\tan(\theta)} Z_{c2} \frac{R + jZ_{c2}\tan(\theta)}{Z_{c2} + jR\tan(\theta)}$$

$$= Z_{c1} Z_{c2} \frac{R + jRZ_{c2}\tan(\theta) + jRZ_{c1}\tan(\theta) - Z_{c1}Z_{c2}\tan^2(\theta)}{Z_{c1}Z_{c2} + jZ_{c1}R\tan(\theta) + jZ_{c2}R\tan(\theta) - R^2\tan^2(\theta)}$$

After some mathematical operations, the following expression is obtained:

$$j\tan(\theta)R^3\left(Z_{c1} + Z_{c2}\right) - R^4\tan^2(\theta) = j\tan(\theta)RZ_{c1}Z_{c2}\left(Z_{c1} + Z_{c2}\right) - Z_{c1}^2 Z_{c2}^2 \tan^2(\theta)$$

In order to have the above equation hold, one should have

$$R = \sqrt{Z_{c1} Z_{c2}} \tag{5.6}$$

Equation (5.6) provides the relationship between the characteristic impedances of the modes and the termination resistance in order to have the system matched. It is one of the classical equations of TEM directional-coupler theory. It is important to realize that (5.6) merely establishes a relation between the characteristic impedances of each mode and termination resistances so that the coupled lines are matched to the rest of the RF circuitry. That is, no reflections occur at the ports of the directional coupler if (5.6) holds as far as the external circuit is concerned. It does not, however, ensure that the propagating modes on the coupled lines are not reflected as they are terminated by single-mode transmission lines. In fact, the modes reflect in the coupled structure itself in that case and this is the mechanism of how directional coupler works. We will elaborate on this later when introducing the matrix theory of multiconductor transmission lines.

The next step is to determine voltages on the terminals. For this purpose, we use boundary conditions (i.e., voltage and current relationships on the terminals) on the lines as follows:

$$R\left(\frac{A_1}{Z_{c1}} - \frac{A_2}{Z_{c1}} + \frac{A_3}{Z_{c2}} - \frac{A_4}{Z_{c2}}\right) - 1 = A_1 + A_2 + A_3 + A_4 \tag{5.7}$$

$$R\left(\frac{A_1}{Z_{c1}} - \frac{A_2}{Z_{c1}} - \frac{A_3}{Z_{c2}} + \frac{A_4}{Z_{c2}}\right) = -A_1 - A_2 + A_3 + A_4 \tag{5.8}$$

$$R\left(\frac{A_1}{Z_{c1}}e^{-\gamma l} - \frac{A_2}{Z_{c1}}e^{\gamma l} + \frac{A_3}{Z_{c2}}e^{-\gamma l} - \frac{A_4}{Z_{c2}}e^{\gamma l}\right) = A_1 e^{-\gamma l} + A_2 e^{\gamma l} + A_3 e^{-\gamma l} + A_4 e^{\gamma l} \tag{5.9}$$

$$R\left(\frac{A_1}{Z_{c1}}e^{-\gamma l} - \frac{A_2}{Z_{c1}}e^{\gamma l} - \frac{A_3}{Z_{c2}}e^{-\gamma l} + \frac{A_4}{Z_{c2}}e^{\gamma l}\right) = A_1 e^{-\gamma l} + A_2 e^{\gamma l} - A_3 e^{-\gamma l} - A_4 e^{\gamma l} \tag{5.10}$$

Note that the generator is assumed to have unity amplitude. After some mathematical operations, we find the wave amplitudes:

$$A_1 = \frac{Z_{c1}}{2}\frac{1}{R - Z_{c1}}\frac{e^{2\gamma l}}{\dfrac{R + Z_{c1}}{R - Z_{c1}}e^{2\gamma l} - \dfrac{R - Z_{c1}}{R + Z_{c1}}} \tag{5.11}$$

$$A_2 = \frac{Z_{c1}}{2}\frac{1}{R + Z_{c1}}\frac{1}{\dfrac{R + Z_{c1}}{R - Z_{c1}}e^{2\gamma l} - \dfrac{R - Z_{c1}}{R + Z_{c1}}} \tag{5.12}$$

$$A_3 = \frac{Z_{c2}}{2}\frac{1}{R - Z_{c2}}\frac{e^{2\gamma l}}{\dfrac{R + Z_{c2}}{R - Z_{c2}}e^{2\gamma l} - \dfrac{R - Z_{c2}}{R + Z_{c2}}} \tag{5.13}$$

$$A_1 = \frac{Z_{c2}}{2}\frac{1}{R - Z_{c2}}\frac{1}{\dfrac{R + Z_{c2}}{R - Z_{c2}}e^{2\gamma l} - \dfrac{R - Z_{c2}}{R + Z_{c2}}} \tag{5.14}$$

where

$$\frac{R + Z_{c1}}{R - Z_{c1}} = -\frac{R + Z_{c2}}{R - Z_{c2}} \tag{5.15}$$

which is the consequence of (5.6). After determining the wave coefficients, one can find voltages at each terminal of the structure as follows:

$$V_1(0) = A_1 + A_2 + A_3 + A_4$$

$$= \frac{1}{2} \frac{Z_{c1}(R+Z_{c2})e^{2\gamma l} - Z_{c2}(R-Z_{c1}) + Z_{c2}(R+Z_{c1})e^{2\gamma l} - Z_{c1}(R-Z_{c2})}{(R+Z_{c1})(R+Z_{c2})e^{2\gamma l} + (R-Z_{c2})(R-Z_{c1})}$$

$$= \frac{1}{2}$$

$$(5.16)$$

$$V_2(0) = A_1 + A_2 - A_3 - A_4$$

$$= \frac{1}{2} \frac{Z_{c1}(R+Z_{c2})e^{2\gamma l} + Z_{c2}(R-Z_{c1}) - Z_{c2}(R+Z_{c1})e^{2\gamma l} - Z_{c1}(R-Z_{c2})}{(R+Z_{c1})(R+Z_{c2})e^{2\gamma l} + (R-Z_{c2})(R-Z_{c1})}$$

$$= \frac{1}{2} \frac{(Z_{c1} - Z_{c2})\sinh(\gamma l)}{2R\cosh(\gamma l) + (Z_{c1} + Z_{c2})\sinh(\gamma l)}$$

$$(5.17)$$

$$V_1(l) = A_1 e^{-\gamma l} + A_2 e^{\gamma l} + A_3 e^{-\gamma l} + A_4 e^{\gamma l}$$

$$= \frac{1}{2} \frac{Z_{c1}(R+Z_{c2})e^{\gamma l} - Z_{c2}(R-Z_{c1})e^{\gamma l} + Z_{c2}(R+Z_{c1})e^{\gamma l} - Z_{c1}(R-Z_{c2})e^{\gamma l}}{(R+Z_{c1})(R+Z_{c2})e^{2\gamma l} + (R-Z_{c2})(R-Z_{c1})}$$

$$= \frac{1}{2} \frac{2R}{2R\cosh(\gamma l) + (Z_{c1} + Z_{c2})\sinh(\gamma l)}$$

$$(5.18)$$

$$V_2(l) = A_1 e^{-\gamma l} + A_2 e^{\gamma l} - A_3 e^{-\gamma l} - A_4 e^{\gamma l}$$

$$= \frac{1}{2} \frac{Z_{c1}(R+Z_{c2})e^{\gamma l} + Z_{c2}(R-Z_{c1})e^{\gamma l} - Z_{c2}(R+Z_{c1})e^{\gamma l} - Z_{c1}(R-Z_{c2})e^{\gamma l}}{(R+Z_{c1})(R+Z_{c2})e^{2\gamma l} + (R-Z_{c2})(R-Z_{c1})}$$

$$= 0$$

$$(5.19)$$

The above equations can be simplified further by defining a coupling parameter C such that

$$C = \frac{Z_{c1} - Z_{c2}}{Z_{c1} + Z_{c2}} \tag{5.20}$$

Then,

$$V_1(0) = \frac{1}{2} \tag{5.21}$$

$$V_2(0) = \frac{1}{2} \frac{C\sinh(\gamma l)}{\sqrt{1-C^2}\,\cosh(\gamma l) + \sinh(\gamma l)} \tag{5.22}$$

$$V_1(l) = \frac{1}{2} \frac{\sqrt{1-C^2}\,\sinh(\gamma l)}{\sqrt{1-C^2}\,\cosh(\gamma l) + \sinh(\gamma l)} \tag{5.23}$$

$$V_2(l) = 0 \tag{5.24}$$

This completes the analysis of two uniform, symmetrical, coupled TEM lines. By inspecting the above equations, we first observe that the coupled port, $V_2(0)$, is the adjacent port to the input. This configuration is called contradirectional coupling. Also, coupling is a periodic function of length of the coupled section. It is seen that maximum coupling occurs when the electrical length of the coupled section equals $n/4 + n\lambda/2$, where n is an integer. In addition, $V_2(l)$ is zero because the even- and odd-mode phase velocities are equal. This result is valid for coupled lines embedded in homogeneous medium. For microstrip lines, this assumption will not strictly hold, and $V_2(l)$ will be nonzero deteriorating the isolation.

5.1.2 Analysis of Asymmetrical Coupled TEM Lines

The derivation presented in the previous section assumes a pair of coupled symmetric lines in a homogeneous medium. This assumption results in contradirectional coupling (or backward coupling) between the lines, which is also called the normal mode of coupling. In this section, we will generalize this derivation using coupled-mode formulation, which assumes an asymmetric structure [20–25]. The formulation that will be followed here is based on the telegrapher's equation utilizing per-unit-length (PUL) line parameters. This formulation is more general and provides more physical insight than the previous approach.

For this purpose, let's consider the equivalent circuit of coupled TEM transmission lines shown in Figure 5.2. In this equivalent circuit, the coupling between the lines is modeled by mutual capacitance and inductance. The telegrapher's equations in the presence of mutual capacitance and inductance can then be written as [20, 25]

$$\frac{\partial V_1}{\partial z} = -L_1 \frac{\partial I_1}{\partial t} - L_m \frac{\partial I_2}{\partial t}$$

$$\frac{\partial V_2}{\partial z} = -L_2 \frac{\partial I_2}{\partial t} - L_m \frac{\partial I_1}{\partial t}$$
(5.25)

$$\frac{\partial I_1}{\partial z} = -C_1 \frac{\partial V_1}{\partial t} + C_m \frac{\partial V_2}{\partial t}$$

$$\frac{\partial I_2}{\partial z} = -C_2 \frac{\partial V_2}{\partial t} + C_m \frac{\partial V_1}{\partial t}$$
(5.26)

which can be cast in matrix form as follows:

$$\frac{\partial \mathbf{V}}{\partial z} = -j\omega \begin{bmatrix} L_1 & L_m \\ L_m & L_2 \end{bmatrix} \times \mathbf{I}$$
(5.27)

$$\frac{\partial \mathbf{I}}{\partial z} = -j\omega \begin{bmatrix} C_1 & -C_m \\ -C_m & C_2 \end{bmatrix} \times \mathbf{V}$$
(5.28)

where $e^{j\omega t}$ time dependency is assumed. The vectors $\mathbf{V}$ and $\mathbf{I}$ are the column vectors representing voltage and current along the lines, respectively. We also define the forward- and backward-traveling waves as

$$a_1 = \frac{V_1 + Z_1 I_1}{\sqrt{2Z_1}} \quad b_1 = \frac{V_1 - Z_1 I_1}{\sqrt{2Z_1}} \qquad (5.29)$$

$$a_2 = \frac{V_2 + Z_2 I_2}{\sqrt{2Z_2}} \quad b_2 = \frac{V_2 - Z_2 I_2}{\sqrt{2Z_2}} \qquad (5.30)$$

where $Z_i = \sqrt{L_i / C_i}$. The traveling waves on the coupled lines are pictorially shown in Figure 5.3.

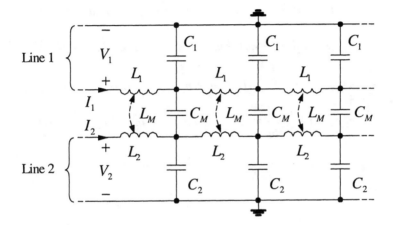

Figure 5.2 Equivalent circuit of coupled TEM transmission lines. Note that the coupling between the lines is modeled by mutual capacitances and inductances.

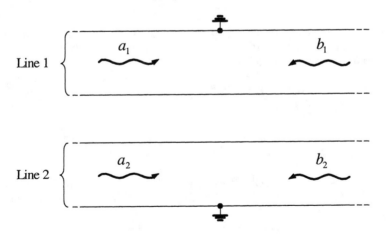

Figure 5.3 Forward- and backward-traveling waves on the coupled TEM lines.

In the above expressions, the waves a and b represent the forward- and backward-traveling waves on each line, respectively. Note that they are normalized in the usual way so that the power traveling in each direction is given by the magnitude square of the wave quantity.

The next step is to link the traveling waves to the line parameters (L, C, L_m, and C_m) using (5.25), (5.26), (5.29), and (5.30). By substituting the definition of the traveling waves into telegraphers' equations, one can obtain the following sets of equations:

$$\frac{\partial}{\partial z}\left(m_{11}a_1 + m_{12}b_1\right) = -j\omega L_1\left(m_{21}a_1 + m_{22}b_1\right) - j\omega L_m\left(n_{21}a_2 + n_{22}b_2\right)$$

$$\frac{\partial}{\partial z}\left(m_{21}a_1 + m_{22}b_1\right) = -j\omega C_1\left(m_{11}a_1 + m_{12}b_1\right) + j\omega C_m\left(n_{11}a_2 + n_{12}b_2\right)$$

$$\frac{\partial}{\partial z}\left(n_{11}a_2 + n_{12}b_2\right) = -j\omega L_2\left(n_{21}a_2 + n_{22}b_2\right) - j\omega L_m\left(m_{21}a_1 + m_{22}b_1\right)$$

$$\frac{\partial}{\partial z}\left(n_{21}a_2 + n_{22}b_2\right) = -j\omega C_2\left(n_{11}a_2 + n_{12}b_2\right) + j\omega C_m\left(m_{11}a_1 + m_{12}b_1\right)$$

where

$$m_{11} = m_{12} = \sqrt{Z_1/2}$$
$$m_{21} = -m_{22} = \sqrt{1/2Z_1}$$

and

$$n_{11} = n_{12} = \sqrt{Z_2/2}$$
$$n_{21} = -n_{22} = \sqrt{1/2Z_2}$$

The above equations can be solved to give a matrix equation for the traveling waves. For example, the forward-traveling waves (a_1 and a_2) can be related to both types of waves on the lines as follows:

$$\frac{\partial}{\partial z}a_1\left(\frac{m_{11}m_{22} - m_{12}m_{21}}{m_{22}}\right) = -j\omega\left[-\frac{m_{11}m_{12}}{m_{22}}C_1 + m_{21}L_1 \quad \frac{n_{11}m_{12}}{m_{22}}C_m + n_{21}L_m\right] \times \begin{bmatrix} a_1 \\ a_2 \end{bmatrix}$$

$$-j\omega\left[-\frac{m_{12}m_{12}}{m_{22}}C_1 + m_{22}L_1 \quad \frac{n_{12}m_{12}}{m_{22}}C_m + n_{22}L_m\right] \times \begin{bmatrix} b_1 \\ b_2 \end{bmatrix}$$

$$\frac{\partial}{\partial z}a_2\left(\frac{n_{11}n_{22} - n_{12}n_{21}}{n_{22}}\right) = -j\omega\left[\frac{m_{11}n_{12}}{n_{22}}C_m + m_{21}L_m \quad -\frac{n_{11}n_{12}}{n_{22}}C_2 + n_{21}L_2\right] \times \begin{bmatrix} a_1 \\ a_2 \end{bmatrix}$$

$$-j\omega\left[\frac{n_{12}m_{12}}{n_{22}}C_m + m_{22}L_m \quad -\frac{n_{12}n_{12}}{n_{22}}C_2 + n_{22}L_2\right] \times \begin{bmatrix} b_1 \\ b_2 \end{bmatrix}$$

which can be put in matrix form as:

$$\frac{\partial}{\partial z}\begin{bmatrix} a_1 \\ a_2 \end{bmatrix} = -j\begin{bmatrix} \beta_1 & K \\ K & \beta_2 \end{bmatrix} \times \begin{bmatrix} a_1 \\ a_2 \end{bmatrix} + j\begin{bmatrix} 0 & L \\ L & 0 \end{bmatrix} \times \begin{bmatrix} b_1 \\ b_2 \end{bmatrix} \tag{5.31}$$

where

$$K = \frac{1}{2}\sqrt{\beta_1\beta_2}(k_L - k_C) \qquad L = \frac{1}{2}\sqrt{\beta_1\beta_2}(k_L + k_C) \tag{5.32}$$

$$k_L = \frac{L_m}{\sqrt{L_1 L_2}} \qquad k_C = \frac{C_m}{\sqrt{C_1 C_2}} \tag{5.33}$$

$$\beta_i = \omega\sqrt{L_i C_i} \tag{5.34}$$

The same derivation can also be carried out for backward-traveling waves (b_1 and b_2) resulting in the following matrix equation:

$$\frac{\partial}{\partial z}\begin{bmatrix} b_1 \\ b_2 \end{bmatrix} = -j\begin{bmatrix} 0 & L \\ L & 0 \end{bmatrix} \times \begin{bmatrix} a_1 \\ a_2 \end{bmatrix} + j\begin{bmatrix} \beta_1 & K \\ K & \beta_2 \end{bmatrix} \times \begin{bmatrix} b_1 \\ b_2 \end{bmatrix} \tag{5.35}$$

Now, we have obtained a set of matrix equations relating the forward- and backward-traveling waves. By inspecting (5.31) and (5.35), one can observe that a forward-traveling wave on one of the lines, in general, can couple to both forward- and backward-traveling waves on the other line. Thus, contradirectional coupling is just a special case of the coupling mechanism between two TEM lines. By selecting the line parameters appropriately, one can also achieve codirectional coupling where forward-traveling waves on one line couple to forward-traveling waves on the other line and vice versa. To elaborate this behavior further, we will now combine (5.31) and (5.35) to give the following matrix equation:

$$\frac{\partial}{\partial z}\begin{bmatrix} a_1 \\ a_2 \\ b_1 \\ b_2 \end{bmatrix} = -j\begin{bmatrix} \beta_1 & K & 0 & -L \\ K & \beta_2 & -L & 0 \\ 0 & L & -\beta_1 & -K \\ L & 0 & -K & -\beta_2 \end{bmatrix} \times \begin{bmatrix} a_1 \\ a_2 \\ b_1 \\ b_2 \end{bmatrix} \tag{5.36}$$

By assuming a z-variation of the form $a_i = a_{0i}e^{\gamma z}$ and $b_i = b_{0i}e^{\gamma z}$, the above equation can be cast into

$$\begin{bmatrix} \beta_1 & K & 0 & -L \\ K & \beta_2 & -L & 0 \\ 0 & L & -\beta_1 & -K \\ L & 0 & -K & -\beta_2 \end{bmatrix} \times \begin{bmatrix} a_1 \\ a_2 \\ b_1 \\ b_2 \end{bmatrix} = j\gamma\begin{bmatrix} a_1 \\ a_2 \\ b_1 \\ b_2 \end{bmatrix} \tag{5.37}$$

which can immediately be recognized as an eigenvalue equation, where the propagation constant, γ, and the traveling wave amplitudes are the eigenvalues and eigenvectors, respectively. Closed-form expressions for the eigenvalues of full 4×4 matrixes are very complicated. Fortunately, the special symmetry of the matrix given in (5.37) allows closed-form expressions of the eigenvalues to be obtained as follows:

$$\gamma = \pm j \frac{1}{2} \left[\pm 2 \left(\beta_2^4 - 2\beta_1^2 \beta_2^2 - 4\beta_2^3 \beta_1 k_L k_C + \beta_1^4 - 4\beta_1^3 \beta_2 k_L k_C + 4\beta_1^2 \beta_2^2 k_C^2 + 4\beta_1^2 \beta_2^2 k_L^2 \right)^{1/2} \right. $$
$$\left. + 2\beta_2^2 + 2\beta_1^2 - 4\beta_1 \beta_2 k_L k_C \right]^{1/2}$$

Adopting the notation in [20], one can show that

$$\gamma = \pm j\beta_0 \sqrt{1 \pm \delta} \tag{5.38}$$

where

$$\beta_0 = \sqrt{\frac{\beta_1^2 + \beta_2^2}{2} - \beta_1 \beta_2 k_L k_C} \tag{5.39}$$

and

$$\delta = \sqrt{1 - \left(\frac{\beta_1^2 \beta_2^2}{\beta_0^4} \right) \left(1 - k_L^2 \right)\left(1 - k_C^2 \right)} \tag{5.40}$$

Examination of (5.38) reveals some interesting features of general coupled TEM lines. First of all, in a homogenous medium, the constant δ must be zero since only one phase velocity is permitted in a homogenous medium. In that case, β_0 becomes ω/c where c is the speed of light in the medium. This can easily be verified using (5.39) and noting that $\mathbf{LC} = \mu\varepsilon\mathbf{U}$ in a homogeneous medium, where $\mathbf{U}$ is the identity matrix. For symmetrical lines, having $\delta = 0$ further implies that $k_L = k_C$ or $k_L = -k_C$. When the medium is inhomogeneous, two phase velocities will result and in this case, $k_L \neq k_C$ in general. The two cases ($k_L = k_C$ and $k_L = -k_C$) in a homogenous medium are of special interest, and we will investigate them next. For $k_L = k_C$, (5.36) becomes

$$\frac{\partial}{\partial z} \begin{bmatrix} a_1 \\ a_2 \\ b_1 \\ b_2 \end{bmatrix} = -j \begin{bmatrix} \beta & 0 & 0 & -L \\ 0 & \beta & -L & 0 \\ 0 & L & -\beta & 0 \\ L & 0 & 0 & -\beta \end{bmatrix} \times \begin{bmatrix} a_1 \\ a_2 \\ b_1 \\ b_2 \end{bmatrix} \tag{5.41}$$

and for $k_L = -k_C$,

$$\frac{\partial}{\partial z} \begin{bmatrix} a_1 \\ a_2 \\ b_1 \\ b_2 \end{bmatrix} = -j \begin{bmatrix} \beta & K & 0 & 0 \\ K & \beta & 0 & 0 \\ 0 & 0 & -\beta & -K \\ 0 & 0 & -K & -\beta \end{bmatrix} \times \begin{bmatrix} a_1 \\ a_2 \\ b_1 \\ b_2 \end{bmatrix} \tag{5.42}$$

Equation (5.41) shows that the coupling mechanism in a pair of coupled TEM lines is contradirectional when $k_L = k_C$ (also see Figure 5.3). This is the same result obtained in previous section. On the other hand, when $k_L = -k_C$, (5.42) shows that codirectional coupling results. The condition $k_L = -k_C$ can be obtained by adding an excess capacitance between the lines so that the capacitive coupling changes its sign [26]. The same effect can be obtained by reversing the direction of the coupled inductors (in case of lumped couplers) in Figure 5.2.

5.2 THE MTL FORMULATION

The general solution of n-coupled TEM lines can be achieved through the multiconductor transmission line (MTL) formulation [14–17]. The MTL formulation is important because it allows us to develop SPICE circuit models of coupled lines for computer simulation, which is very valuable in signal integrity studies. Note that it would be computationally prohibitive to carry out 3D full-wave electromagnetic analysis of very long, coupled transmission lines (e.g., high-speed flex cables connecting two circuit assemblies). Instead, SPICE models based on PUL parameters that are extracted using numerical techniques are employed to characterize such structures (assuming the cross-section remains uniform). We will show that under certain conditions, the differential equations of the coupled system can be decoupled, yielding a simple representation of the original coupled system.

The MTL formulation can be carried out in either frequency or time domains [14, 15]. In this section, we will concentrate on the time-domain formulation only. To demonstrate the method, let's consider the general representation of n-coupled TEM lines shown in Figure 5.4.

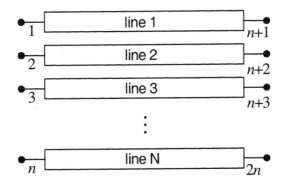

Figure 5.4 General representation of n-coupled TEM lines.

For the sake of simplicity, we will assume lossless lines. The time-domain MTL equations for the coupled system can be written as follows:

$$\frac{\partial}{\partial z}\mathbf{V}(z,t) = -\mathbf{L}\frac{\partial}{\partial t}\mathbf{I}(z,t) \qquad (5.43)$$

$$\frac{\partial}{\partial z}\mathbf{I}(z,t) = -\mathbf{C}\frac{\partial}{\partial t}\mathbf{V}(z,t) \qquad (5.44)$$

where $\mathbf{V}$ and $\mathbf{I}$ are the voltage and current vectors along the line, and $\mathbf{L}$ and $\mathbf{C}$ are the PUL inductances and capacitances, respectively:

$$\mathbf{V}(z,t) = \begin{bmatrix} V_1(z,t) & V_2(z,t) & \cdots & V_n(z,t) \end{bmatrix}^T \qquad (5.45)$$

$$\mathbf{I}(z,t) = \begin{bmatrix} I_1(z,t) & I_2(z,t) & \cdots & I_n(z,t) \end{bmatrix}^T \qquad (5.46)$$

$$\mathbf{L} = \begin{bmatrix} L_{11} & L_{12} & \cdots & L_{1n} \\ L_{21} & L_{22} & \ddots & L_{2n} \\ \vdots & \ddots & \ddots & \vdots \\ L_{n1} & L_{n2} & \cdots & L_{nn} \end{bmatrix} \qquad (5.47)$$

$$\mathbf{C} = \begin{bmatrix} C_{11} & C_{12} & \cdots & C_{1n} \\ C_{21} & C_{22} & \ddots & C_{2n} \\ \vdots & \ddots & \ddots & \vdots \\ C_{n1} & C_{n2} & \cdots & C_{nn} \end{bmatrix} \qquad (5.48)$$

Note that the PUL matrixes $\mathbf{L}$ and $\mathbf{C}$ are, in general, full matrixes representing the coupling between the lines. To decouple (5.43) and (5.44), we will utilize similarity transformations. For this purpose, we define the following transformations between the line quantities to modal quantities:

$$\mathbf{V}(z,t) = \mathbf{T}_V \mathbf{V}_m(z,t) \qquad (5.49)$$

$$\mathbf{I}(z,t) = \mathbf{T}_I \mathbf{I}_m(z,t) \qquad (5.50)$$

where $\mathbf{T}_V$ and $\mathbf{T}_I$ are the transformation matrixes yet to be determined. The subscript m denotes the modal quantities. Substituting these transformations into (5.43) and (5.44) yields

$$\frac{\partial}{\partial z}\mathbf{V}_m(z,t) = -\mathbf{L}_m\frac{\partial}{\partial t}\mathbf{I}_m(z,t) \qquad (5.51)$$

$$\frac{\partial}{\partial z}\mathbf{I}_m(z,t) = -\mathbf{C}_m\frac{\partial}{\partial t}\mathbf{V}_m(z,t) \qquad (5.52)$$

where

$$\mathbf{L}_m = \mathbf{T}_V^{-1}\mathbf{L}\mathbf{T}_I \qquad (5.53)$$

$$\mathbf{C}_m = \mathbf{T}_I^{-1}\mathbf{C}\mathbf{T}_V \qquad (5.54)$$

Now it is possible to find suitable $\mathbf{T}_V$ and $\mathbf{T}_I$ so that $\mathbf{L}_m$ and $\mathbf{C}_m$ become diagonal matrixes; that is,

$$\mathbf{L}_m = \begin{bmatrix} L_{11}^{(m)} & 0 & \cdots & 0 \\ 0 & L_{22}^{(m)} & \ddots & 0 \\ \vdots & \ddots & \ddots & \vdots \\ 0 & 0 & \cdots & L_{nn}^{(m)} \end{bmatrix} \tag{5.55}$$

$$\mathbf{C}_m = \begin{bmatrix} C_{11}^{(m)} & 0 & \cdots & 0 \\ 0 & C_{22}^{(m)} & \ddots & 0 \\ \vdots & \ddots & \ddots & \vdots \\ 0 & 0 & \cdots & C_{nn}^{(m)} \end{bmatrix} \tag{5.56}$$

where the superscript (m) denotes that the quantity is a modal quantity. Note that $\mathbf{T}_V$ and $\mathbf{T}_I$ are related such that

$$\mathbf{T}_V^t \mathbf{T}_I = \mathbf{D} \tag{5.57}$$

where $\mathbf{D}$ is a diagonal matrix if all eigenvalues are distinct; in general, it is a block diagonal matrix. The transformation matrixes can also be redefined so that $\mathbf{D}$ becomes the identity matrix [19]. The crucial requirement of this process is that the redefined transformation matrixes retain the ability to simultaneously diagonalize (5.43) and (5.44). One way of achieving this is to redefine the transformation matrixes in the following form [19]:

$$\mathbf{T}_V' = \mathbf{T}_V (\mathbf{D}^t)^{-1} \tag{5.58}$$

$$\mathbf{T}_I' = \mathbf{T}_I \tag{5.59}$$

Hence, it can be assumed that the transformation matrixes are selected so that $\mathbf{D}$ is the diagonal matrix.

Note that (5.51) and (5.52) are the same equations for isolated n two-conductor lines with characteristic impedances and propagation constants of

$$Z_{Ci}^{(m)} = \sqrt{\frac{L_{ii}^{(m)}}{C_{ii}^{(m)}}} \tag{5.60}$$

$$\beta_{Ci}^{(m)} = \omega \sqrt{L_{ii}^{(m)} C_{ii}^{(m)}} \tag{5.61}$$

Thus, the coupled system is reduced to an isolated system of transmission lines each of them carrying only one mode. The remaining task is to solve the modal unknown parameters on each isolated line using the terminal conditions. This can be done because the general solution of a single TEM transmission line characterized by (5.60) and (5.61) is known. Perhaps the simplest way of doing this is to employ a circuit-simulation tool. For this purpose, we rewrite the transformations between the modal and terminal quantities as follows:

$$
\begin{bmatrix} V_1(z,t) \\ V_2(z,t) \\ \vdots \\ V_n(z,t) \end{bmatrix} = \begin{bmatrix} T_{V11} & T_{V12} & \cdots & T_{V1n} \\ T_{V21} & T_{V22} & \ddots & T_{V2n} \\ \vdots & \ddots & \ddots & \vdots \\ T_{Vn1} & T_{Vn2} & \cdots & T_{Vnn} \end{bmatrix} \begin{bmatrix} V_1^{(m)}(z,t) \\ V_2^{(m)}(z,t) \\ \vdots \\ V_n^{(m)}(z,t) \end{bmatrix} \tag{5.62}
$$

$$
\begin{bmatrix} I_1^{(m)}(z,t) \\ I_2^{(m)}(z,t) \\ \vdots \\ I_n^{(m)}(z,t) \end{bmatrix} = \begin{bmatrix} T_{I11} & T_{I12} & \cdots & T_{I1n} \\ T_{I21} & T_{I22} & \ddots & T_{I2n} \\ \vdots & \ddots & \ddots & \vdots \\ T_{In1} & T_{In2} & \cdots & T_{Inn} \end{bmatrix}^{-1} \begin{bmatrix} I_1(z,t) \\ I_2(z,t) \\ \vdots \\ I_n(z,t) \end{bmatrix} \tag{5.63}
$$

The equivalent circuit model of the above equations is given in Figure 5.5, where transmission of each mode is accomplished by a two-conductor transmission line, whose parameters are given by (5.60) and (5.61), and the interactions between the nodal and modal quantities [i.e., (5.62) and (5.63)] are taken into account by controlled sources [14, 22, 27, 28].

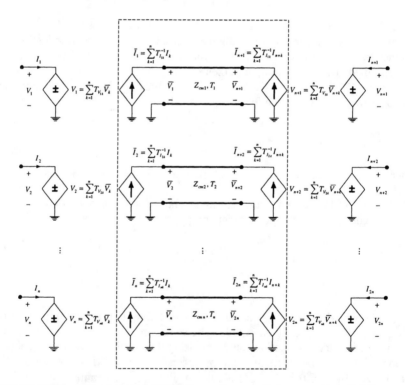

Figure 5.5 SPICE model of n-coupled TEM lines using controlled sources. Note that the coupled system is decomposed into propagating modes such that each mode is transmitted on a fictitious TEM line with proper characteristic impedance and electrical length.

Note that this model has the flexibility of being simulated with other linear and/or nonlinear circuit components, thus it is very suitable for computer simulations. For instance, once the PUL line parameters are determined and differential equations are decoupled, long PCB traces can be simulated with this model. One can also design generic subcircuits for a different number of coupled sections, increasing the flexibility of the approach. Note that one could also determine eigenvectors and modal characteristic impedances required in the decoupling process through full-wave electromagnetic simulations as well [27].

Another important parameter of coupled lines that is found in the literature is the characteristic impedance matrix, $\mathbf{Z}_C$ [17, 19, 29–35]. The characteristic impedance matrix is a full-matrix linking the forward- and backward-traveling waves on the coupled-line system. It is important to distinguish this matrix from the modal characteristic impedance matrix defined previously in (5.60). Although many different forms exist in the literature, two of them are given here [19, 29]:

$$\begin{aligned}\mathbf{Z}_C &= \mathbf{L}\mathbf{T}_I\gamma^{-1}\mathbf{T}_I^{-1} \\ &= \mathbf{T}_V\gamma\,\mathbf{T}_V^{-1}\mathbf{C}^{-1}\end{aligned} \tag{5.64}$$

where γ is a diagonal matrix whose elements are given by the square root of the eigenvalues of the $\mathbf{L}\times\mathbf{C}$ matrix. An alternative form of the characteristic impedance matrix can be written in terms of the modal characteristic impedance as follows [19, 29]:

$$\mathbf{Z}_C = \left(\mathbf{T}_I^t\right)^{-1}\mathbf{Z}_C^{(m)}\mathbf{T}_I^{-1} \tag{5.65}$$

The physical significance of the characteristic impedance matrix is that no reflection occurs only if the coupled system is terminated with a network whose impedance matrix is equal to the characteristic impedance matrix. In general, to satisfy this, it is not sufficient to only place impedance from each line to the reference conductor; impedance connections between the lines are also required. A pictorial description of this termination is depicted in Figure 5.6. In the figure, the termination block represents the resistive network consisting of $n(n+1)/2$ resistors whose values are given as [30, 36, 37]

$$R_{ii} = \left(\sum_{k=1}^{n}Y_{Cik}\right)^{-1}, \quad i = 1,2,\ldots,n \tag{5.66}$$

$$R_{ij} = \left(-Y_{Cij}\right)^{-1}, \quad 1 \le i \le j \le n \tag{5.67}$$

where

$$\mathbf{Y}_C = \mathbf{Z}_C^{-1} \tag{5.68}$$

Obviously, implementation of such a network for a large number of coupled conductors is quite difficult because it requires a connection from each line to every other line. Implementation on planar circuits is even more difficult because it requires a 3D connection scheme. Therefore, an alternative scheme has been proposed in [30], where each line is only connected to the reference conductor (i.e., ground) and the nearest other conductor, enabling a planar solution; that is,

$$R_{ij} = 0, \quad \text{if } |i - j| \geq 2$$

It must be noted, however, that this solution is an approximate one. Nevertheless, it provides a good compromise between termination of lines and ease of implementation.

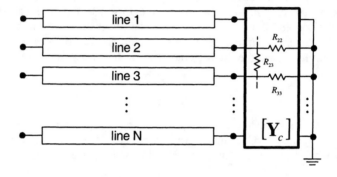

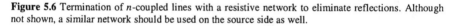

Figure 5.6 Termination of n-coupled lines with a resistive network to eliminate reflections. Although not shown, a similar network should be used on the source side as well.

As an example, for a pair of symmetrical coupled TEM lines in a homogeneous medium, the characteristic impedance and admittance matrixes can be found using (5.65) as follows:

$$\mathbf{Z}_C = \begin{bmatrix} \dfrac{1}{\sqrt{2}} & -\dfrac{1}{\sqrt{2}} \\ \dfrac{1}{\sqrt{2}} & \dfrac{1}{\sqrt{2}} \end{bmatrix} \times \begin{bmatrix} Z_{0e} & 0 \\ 0 & Z_{0o} \end{bmatrix} \times \begin{bmatrix} \dfrac{1}{\sqrt{2}} & \dfrac{1}{\sqrt{2}} \\ -\dfrac{1}{\sqrt{2}} & \dfrac{1}{\sqrt{2}} \end{bmatrix}$$

$$= \frac{1}{2} \begin{bmatrix} Z_{0e} + Z_{0o} & Z_{0e} - Z_{0o} \\ Z_{0e} - Z_{0o} & Z_{0e} + Z_{0o} \end{bmatrix}$$

$$\mathbf{Y}_C = \mathbf{Z}_C^{-1}$$

$$= \frac{1}{2 Z_{0e} Z_{0o}} \begin{bmatrix} Z_{0e} + Z_{0o} & Z_{0o} - Z_{0e} \\ Z_{0o} - Z_{0e} & Z_{0e} + Z_{0o} \end{bmatrix}$$

where Z_{0e} and Z_{0o} are the even- and odd-mode characteristic impedances of the symmetrical coupled lines, respectively. Then, using (5.66) and (5.67), the termination resistor values are found as

$$R_{11} = R_{22} = Z_{0e}$$

$$R_{12} = 2 \frac{Z_{0e} Z_{0o}}{Z_{0e} - Z_{0o}}$$

The resistive network, whose elements are given with the above equations, terminating the coupled lines, is shown in Figure 5.7.

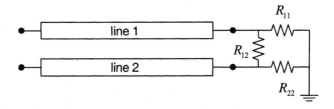

Figure 5.7 Termination of coupled lines with a resistive network to eliminate reflections.

Despite the elegance of the diagonalization method described above, it is not always possible to find the transformation matrixes to decouple the MTL equations in the time domain [14]. For example, if the lines are lossless, one can always find the matrixes $\mathbf{T}_V$ and $\mathbf{T}_I$ that simultaneously diagonalize (5.53) and (5.54). On the other hand, if there are losses present, it is not possible, in general, to find suitable matrixes to simultaneously diagonalize $\mathbf{R}$, $\mathbf{L}$, and $\mathbf{C}$ ($\mathbf{G}$ is ignored) in the time domain. The situation is more manageable in the frequency domain. It can be shown that one can find transformation matrixes that decouple MTL equations diagonalizing two complex-valued impedance matrixes, $\mathbf{R} + j\omega \mathbf{L}$ and $\mathbf{G} + j\omega \mathbf{C}$, in most situations [14]. In the next sections, we will study two important cases, namely lossless coupled lines in homogenous and inhomogeneous media and describe how the transformation matrixes can be computed.

5.2.1 Lossless Lines in Homogenous Medium

In homogenous medium, the inductance and capacitance matrixes of the system satisfy the following important relationship:

$$\mathbf{LC} = \mathbf{CL} = \mu \varepsilon \, \mathbf{U}_{n \times n} \tag{5.69}$$

where $\mathbf{U}$ is the identity matrix. Then, one can find the required similarity transformation by following the steps given next [14].

1. Find a real orthogonal transformation such that

$$\mathbf{T}^t \mathbf{LT} = \mathbf{L}_m \tag{5.70}$$

where

$$\mathbf{L}_m = \begin{bmatrix} L_{11}^{(m)} & 0 & \cdots & 0 \\ 0 & L_{22}^{(m)} & \ddots & 0 \\ \vdots & \ddots & \ddots & \vdots \\ 0 & 0 & \cdots & L_{nn}^{(m)} \end{bmatrix} \qquad (5.71)$$

Note that the following relationship holds because $\mathbf{T}$ is an orthogonal matrix:

$$\mathbf{T}^{-1} = \mathbf{T}' \qquad (5.72)$$

2. Then, determine the current and voltage transformation matrixes as follows:

$$\mathbf{T}_I = \mathbf{T} \qquad (5.73)$$

$$\mathbf{T}_V = \mathbf{T} \qquad (5.74)$$

3. Finally, the mode characteristic impedances and propagation constants can be determined from

$$Z_{Ci}^{(m)} = \sqrt{\frac{1}{\mu\varepsilon}} L_{ii}^{(m)} \qquad (5.75)$$

$$\beta_{Ci}^{(m)} = \omega\sqrt{\mu\varepsilon} \qquad (5.76)$$

Since the medium is homogenous, only one phase velocity exists for the TEM modes.

As an example, the mode transformation matrix can be written as follows for a simple three-conductor system (i.e., a pair of coupled lines) [14]:

$$\mathbf{T} = \begin{bmatrix} \cos\theta & -\sin\theta \\ \sin\theta & \cos\theta \end{bmatrix} \qquad (5.77)$$

where

$$\theta = \frac{1}{2}\tan^{-1}\left(\frac{2L_{12}}{L_{11} - L_{22}}\right) \qquad (5.78)$$

If the lines are symmetrical (i.e., $L_{11} = L_{22}$), then

$$\mathbf{T} = \begin{bmatrix} 1/\sqrt{2} & -1/\sqrt{2} \\ 1/\sqrt{2} & 1/\sqrt{2} \end{bmatrix} \qquad (5.79)$$

5.2.2 Lossless Lines in Inhomogeneous Medium

If the medium is inhomogeneous, the relationship (5.69) does not hold, and the calculation of transformations becomes more involved. In this case, one can find the required similarity transformation by following the steps given next [14].

1. Find an orthogonal transformation $\mathbf{U}$ that diagonalizes $\mathbf{C}$ as

$$\mathbf{U'CU} = \mathbf{\theta}^2 \tag{5.80}$$

where

$$\mathbf{\theta}^2 = \begin{bmatrix} \theta_1^2 & 0 & \cdots & 0 \\ 0 & \theta_2^2 & \ddots & 0 \\ \vdots & \ddots & \ddots & \vdots \\ 0 & 0 & \cdots & \theta_n^2 \end{bmatrix} \tag{5.81}$$

and

$$\mathbf{U}^{-1} = \mathbf{U'} \tag{5.82}$$

2. Form the product $\mathbf{\theta U'LU\theta}$ and diagonalize it with another orthogonal transformation:

$$\mathbf{S}(\mathbf{\theta U'LU\theta})\mathbf{S} = \mathbf{\Lambda}^2 \tag{5.83}$$

where

$$\mathbf{\Lambda}^2 = \begin{bmatrix} \Lambda_1^2 & 0 & \cdots & 0 \\ 0 & \Lambda_2^2 & \ddots & 0 \\ \vdots & \ddots & \ddots & \vdots \\ 0 & 0 & \cdots & \Lambda_n^2 \end{bmatrix} \tag{5.84}$$

and

$$\mathbf{S}^{-1} = \mathbf{S'} \tag{5.85}$$

3. Define the matrix $\mathbf{T}$ as

$$\mathbf{T} = \mathbf{U\theta S} \tag{5.86}$$

4. Normalize the columns of $\mathbf{T}$ to a Euclidian length of unity as follows:

$$\mathbf{T}_{norm} = \mathbf{T\alpha} \tag{5.87}$$

where α is a diagonal matrix with diagonal terms given as

$$\alpha_{ii} = \frac{1}{\sqrt{\sum_{k=1}^{n} T_{ki}^2}} \tag{5.88}$$

5. Then, determine the current and voltage transformation matrixes as follows:

$$\mathbf{T}_I = \mathbf{T}_{norm} \tag{5.89}$$

$$\mathbf{T}_V = \left(\mathbf{T}_{norm}'\right)^{-1} \tag{5.90}$$

It can be shown that the modal inductance and capacitance matrixes are given as

$$\mathbf{L}_m = \mathbf{a}\Lambda^2\mathbf{a} \tag{5.91}$$

$$\mathbf{C}_m = \mathbf{a}^{-2} \tag{5.92}$$

6. Finally, the mode characteristic impedances and propagation constants can be determined from

$$Z_{Ci}^{(m)} = \sqrt{\frac{L_{ii}^{(m)}}{C_{ii}^{(m)}}} = \alpha_{ii}^2\Lambda_i \tag{5.93}$$

$$\beta_{Ci}^{(m)} = \omega\Lambda_i \tag{5.94}$$

5.3 Z-PARAMETERS OF COUPLED-LINE SECTIONS

In microwave engineering, four-port coupled-line sections (i.e., a pair of coupled lines) are extensively used in filter and matching networks [3, 38–44]. Therefore, circuit parameters and equivalent networks of such structures are important, and thus, they will be introduced in this section.

In most cases, the four-terminal coupled-line circuit shown in Figure 5.1 is utilized as a two-port circuit by terminating the other two terminals appropriately. Depending on the type of termination (short or open circuit) and the terminated ports, the resulting two-port circuit has different characteristics. Figure 5.8(a, b) shows the two most commonly used coupled-line sections and their transmission line equivalents [3]. The first type is a bandpass structure, whereas the second type is a lowpass structure. Note that the short circuit in the second type can be replaced by a quarter-wavelength open transmission line to eliminate via-holes. For narrowband applications, this will be accurate enough.

Transmission line equivalents given in Figure 5.8 are very convenient in transferring filter networks from lumped prototypes to distributed equivalents. The fundamental approach is to convert the lumped filter network into the transmission line equivalent containing line sections similar to the ones shown in the right-hand side of the figure. Then, the resulting transmission line network is converted to the coupled-line equivalents shown in the left-hand side. In this way, the lumped prototype filter is transformed into a distributed one that can easily be implemented using printed circuits. This method will be elaborated further when we explain microstrip filters in Chapter 6.

Before giving Z-parameters for each case in Figure 5.8, the Z-parameters of the general four-port coupled lines, whose port assignments are given in Figure 5.9, will be summarized first [44]:

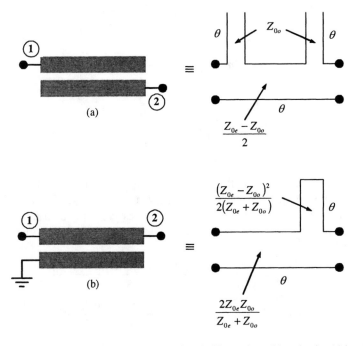

Figure 5.8 Two commonly used coupled-line sections in filter and matching circuits: (a) bandpass, and (b) lowpass structure. The short circuit in (b) is usually replaced by a quarter-wavelength open-circuited transmission line.

$$Z_{11} = Z_{22} = \frac{Z_{c1}\coth(\gamma_c l)}{(1 - R_c/R_\pi)} + \frac{Z_{\pi1}\coth(\gamma_\pi l)}{(1 - R_\pi/R_c)} \tag{5.95}$$

$$Z_{33} = Z_{44} = \frac{R_c^2 Z_{c1}\coth(\gamma_c l)}{(1 - R_c/R_\pi)} + \frac{R_\pi^2 Z_{\pi1}\coth(\gamma_\pi l)}{(1 - R_\pi/R_c)} \tag{5.96}$$

$$Z_{13} = Z_{31} = Z_{24} = Z_{42} = -\frac{Z_{c2}\coth(\gamma_c l)}{R_\pi(1 - R_c/R_\pi)} - \frac{Z_{\pi2}\coth(\gamma_\pi l)}{R_c(1 - R_\pi/R_c)} \tag{5.97}$$

$$Z_{14} = Z_{41} = Z_{23} = Z_{32} = \frac{R_c Z_{c1}}{(1 - R_c/R_\pi)\sinh(\gamma_c l)} + \frac{R_\pi Z_{\pi1}}{(1 - R_\pi/R_c)\sinh(\gamma_\pi l)} \tag{5.98}$$

$$Z_{12} = Z_{21} = \frac{Z_{c1}}{(1 - R_c/R_\pi)\sinh(\gamma_c l)} + \frac{Z_{\pi1}}{(1 - R_\pi/R_c)\sinh(\gamma_\pi l)} \tag{5.99}$$

$$Z_{34} = Z_{43} = \frac{R_c^2 Z_{c1}}{(1 - R_c/R_\pi)\sinh(\gamma_c l)} + \frac{R_\pi^2 Z_{\pi1}}{(1 - R_\pi/R_c)\sinh(\gamma_\pi l)} \tag{5.100}$$

where

$$R_c = \frac{1}{2b_1}\left[(a_2 - a_1) + \sqrt{(a_2 - a_1)^2 + 4b_1 b_2}\right]$$

$$R_\pi = \frac{1}{2b_1}\left[(a_2 - a_1) - \sqrt{(a_2 - a_1)^2 + 4b_1 b_2}\right]$$

$$a_1 = L_{11}C_{11} + L_m C_m$$
$$a_2 = L_{22}C_{22} + L_m C_m$$
$$b_1 = L_{11}C_m + L_m C_{22}$$
$$b_2 = L_{22}C_m + L_m C_{11}$$

In the above equations, γ_c and γ_e are the mode propagation constants. The mode characteristic impedances on each line are given by Z_{c1}, $Z_{\pi1}$, Z_{c2}, and $Z_{\pi2}$. For symmetrical lines, $R_c = -R_\pi = 1$, and the two modes correspond to the even and odd modes, which were discussed previously. In that case, the above equations can be simplified to

$$Z_{11} = Z_{22} = Z_{33} = Z_{44} = \frac{Z_{0e}\coth(\gamma_e l)}{2} + \frac{Z_{0o}\coth(\gamma_o l)}{2} \qquad (5.101)$$

$$Z_{13} = Z_{31} = Z_{42} = Z_{24} = \frac{Z_{0e}\coth(\gamma_e l)}{2} - \frac{Z_{0o}\coth(\gamma_o l)}{2} \qquad (5.102)$$

$$Z_{23} = Z_{32} = Z_{14} = Z_{41} = \frac{Z_{0e}}{2\sinh(\gamma_e l)} - \frac{Z_{0o}}{2\sinh(\gamma_o l)} \qquad (5.103)$$

$$Z_{12} = Z_{21} = Z_{43} = Z_{34} = \frac{Z_{0e}}{2\sinh(\gamma_e l)} + \frac{Z_{0o}}{2\sinh(\gamma_o l)} \qquad (5.104)$$

where Z_{0e} is the even-mode characteristic impedance, Z_{0o} is the odd-mode characteristic impedance, γ_e is the even-mode propagation constant, γ_o is the odd-mode propagation constant, and l is the length of the coupled section.

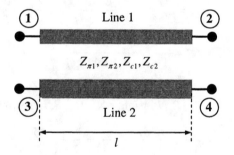

Figure 5.9 A pair of coupled lines and port assignments used in the circuit parameter definitions.

Then, Z-parameters of the two coupled-line sections shown in Figure 5.8(a, b) can be calculated from the above Z-parameters as follows (TEM propagation is assumed):

Type A:

$$Z_{11} = Z_{22} = \frac{Z_{0e} + Z_{0o}}{2} \coth(\gamma l) \tag{5.105}$$

$$Z_{21} = Z_{12} = \frac{Z_{0e} - Z_{0o}}{2} \frac{1}{\sinh(\gamma l)} \tag{5.106}$$

Type B:

$$Z_{11} = \frac{Z_{0e} + Z_{0o}}{2} \coth(\gamma l) - \frac{1}{2} \frac{(Z_{0e} - Z_{0o})^2}{(Z_{0e} + Z_{0o})} \coth(\gamma l)$$
$$= \frac{2Z_{0e}Z_{0o}}{Z_{0e} + Z_{0o}} \coth(\gamma l) \tag{5.107}$$

$$Z_{22} = \frac{Z_{0e} + Z_{0o}}{2} \coth(\gamma l) - \frac{1}{2} \frac{(Z_{0e} - Z_{0o})^2}{(Z_{0e} + Z_{0o})} \frac{1}{\sinh(\gamma l)\cosh(\gamma l)}$$
$$= \frac{2Z_{0e}Z_{0o}}{Z_{0e} + Z_{0o}} \frac{1}{\tanh(\gamma l)} + \frac{(Z_{0e} - Z_{0o})^2}{2(Z_{0e} + Z_{0o})} \tanh(\gamma l) \tag{5.108}$$

$$Z_{21} = Z_{12} = \frac{Z_{0e} - Z_{0o}}{2} \frac{1}{\sinh(\gamma l)} - \frac{1}{2} \frac{(Z_{0e} - Z_{0o})^2}{(Z_{0e} + Z_{0o})^2} \frac{1}{\sinh(\gamma l)}$$
$$= \frac{2Z_{0e}Z_{0o}}{Z_{0e} + Z_{0o}} \frac{1}{\sinh(\gamma l)} \tag{5.109}$$

where $\gamma_e = \gamma_o = \gamma$ (i.e., even- and odd-phase velocities are equal).

5.4 COUPLED MICROSTRIP LINES

So far, we have been concentrating on the analysis of coupled lines in general. The methods and equations presented up to this point are applicable to TEM coupled lines with any cross-section. In the following sections, we will focus on planar coupled-line structures used in printed microwave circuits.

Figure 5.10 shows a general printed coupled-line structure in a multilayer medium. Although one is shown in the picture, a large ground plane may or may not be present; strictly speaking, the ground plane is a return path for the RF currents and any conductor can serve this purpose if properly designed. In general, N conductor TEM lines (including the ground return) support $N - 1$ linearly independent TEM modes. If the medium is not homogenous (as in microstrip), then the propagation cannot be strictly TEM, as discussed before. Assuming quasi-TEM propagation, analysis of such structures can be carried out by first determining PUL parameters and then applying the MTL formulation. The PUL parameters are typically found through a 2D electromagnetic solver applied to the cross-section of the structure [45–49].

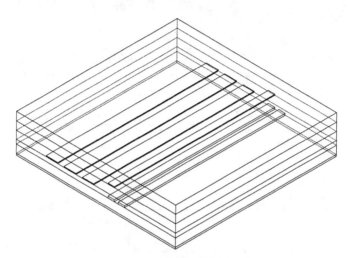

Figure 5.10 Coupled transmission lines in a multilayer medium. Study of coupling effects in such structures is especially important in fast digital and RF circuits.

For coupled microstrip lines with signal lines less than four, closed-form expressions for circuit parameters can be derived. Since they are used extensively in microwave engineering, we will now review two-conductor microstrip coupled lines, shown in Figure 5.11 in more depth. A cross-sectional view of the coupled microstrip lines is given in Figure 5.12. Note that the metallization is assumed to be infinitesimally thin. This is usually a valid assumption because, in most applications, all other physical dimensions are an order of magnitude greater than the metallization thickness. If this is not the case, the metallization thicknesses should be included in the analysis via numerical simulations. We will not address CPW coupled lines but interested readers can refer to literature [50].

Since the dominant mode of propagation for microstrip lines is not exactly TEM, the even- and odd-mode phase velocities in the coupled structure are not equal. This is especially important in microstrip coupled-line filters because it affects the filter transfer function. For directional couplers, the unequal phase velocities decrease the isolation relative to the TEM propagation. However, the TEM approximation is still useful if the requirements are not stringent. The coupled microstrip circuits designed using TEM analysis can always be optimized afterwards using computer simulations to compensate for the assumptions of the TEM analysis.

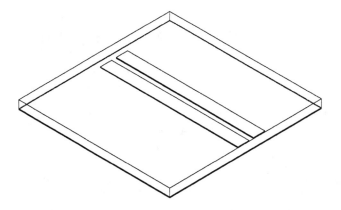

Figure 5.11 Coupled microstrip lines. See Figure 5.12 for the cross-sectional view.

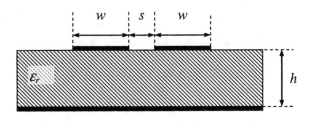

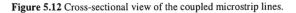

Figure 5.12 Cross-sectional view of the coupled microstrip lines.

Figure 5.13(a, b) shows the even and odd modes of a symmetrical microstrip coupled line. The even mode can be modeled by placing a magnetic wall vertically at the symmetry plane of the structure. Similarly, the odd mode can be modeled by placing an electric wall at the symmetry plane.

Note that the fields around the symmetrical coupled line can also be expressed as a linear combination of the even and odd modes. From another point of view, any linear combination of even and odd modes would, in fact, satisfy Maxwell's equations. For instance, summation and subtraction of the even and odd modes, as shown in the figure, would also be a valid solution. As an example, let's assume that only one of the lines is excited by a generator, and the terminals of the other line are shorted to ground. In this case, the field distribution would be closer to that shown in Figure 5.13(c, d) depending on the excited line. On the other hand, if it was assumed that two of the lines were simultaneously excited by two generators, then the fields would be like Figure 5.13(a, b) depending on phase of the generators. Typically, one of the lines is the driven line in most practical applications; resistive terminations are placed on the other terminals. In that case, the field distribution would be a superposition of Figure 5.13(a, b) as expected.

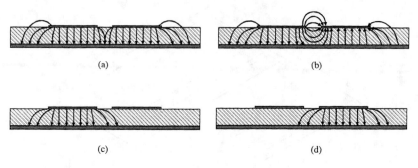

Figure 5.13 Electric field vectors for the even and odd modes of microstrip coupled lines: (a) even mode; (b) odd mode; (c) even mode plus odd mode; and (d) even mode minus odd mode.

This superposition property can also be seen by inspecting the eigenvector matrix given in (5.79) for symmetrical coupled TEM lines. The $1/\sqrt{2}$ factor appears because the columns of the current and voltage eigenvectors are orthonormal so that the total power is the sum of the powers of the individual modes [19].

We now provide the necessary design equations for microstrip coupled lines for the sake of completeness. The design equations that will be provided are based on TEM assumption and metallization thickness is ignored. Note that in MMIC design, the metal thickness can be comparable to the spacing and/or width of the lines. In that case, the designer should employ full-wave computer simulations to account for the metallization thickness properly. The even- and odd-mode characteristic impedance of microstrip coupled lines of zero-thickness metallization are given by the following expressions [51]:

$$Z_{0e} = Z_c \sqrt{\frac{\varepsilon_{eff}}{\varepsilon_{eff_e}}} \left[1 - \left(\frac{Z_0}{377}\right)\sqrt{\varepsilon_{eff}} Q_4 \right]^{-1} \tag{5.110}$$

$$Z_{0o} = Z_c \sqrt{\frac{\varepsilon_{eff}}{\varepsilon_{eff_o}}} \left[1 - \left(\frac{Z_0}{377}\right)\sqrt{\varepsilon_{eff}} Q_{10} \right]^{-1} \tag{5.111}$$

where

$$Q_1 = 0.8695 \cdot u^{0.194}$$

$$Q_2 = 1 + 0.7519 \cdot g + 0.189 \cdot g^{2.31}$$

$$Q_3 = 0.1975 + \left[16.6 + \left(8.4/g\right)^6\right]^{-0.387} + \ln\left[\frac{g^{10}}{1+\left(g/3.4\right)^{10}}\right]\frac{1}{241}$$

$$Q_4 = 2\frac{Q_1}{Q_2}\left[e^{-g} \cdot u^{Q_3} + \left(2 - e^{-g}\right)\cdot u^{-Q_3}\right]^{-1}$$

$$Q_5 = 1.794 + 1.14 \ln\left(1 + \frac{0.638}{g + 0.517 g^{2.43}}\right)$$

$$Q_6 = 0.2305 + \ln\left(\frac{g^{10}}{1 + (g/5.8)^{10}}\right)\frac{1}{281.3} + \ln\left(1 + 0.598 \cdot g^{1.154}\right)\frac{1}{5.1}$$

$$Q_7 = \left(10 + 190 \cdot g^2\right)/\left(1 + 82.3 \cdot g^3\right)$$

$$Q_8 = e^{-6.5 - 0.95\ln(g) - (g/0.15)^5}$$

$$Q_9 = \ln(Q_7) \cdot (Q_8 + 1/16.5)$$

$$Q_{10} = \frac{Q_2 Q_4 - Q_5 e^{\ln(u)Q_6 \cdot u^{-Q_9}}}{Q_2}$$

where $u = w/h$, $g = s/h$, and ε_{eff} and Z_c are the effective dielectric constant and characteristic impedance of a single microstrip line, respectively. The even- and odd-mode effective dielectric constants in the above expressions are given by [51]

$$\varepsilon_{eff_e} = 0.5(\varepsilon_r + 1) + 0.5(\varepsilon_r - 1) \cdot (1 + 10/v)^{-a_e b_e} \qquad (5.112)$$

where

$$v = u\left(20 + g^2\right)/\left(10 + g^2\right) + g \cdot e^{-g}$$

$$a_e = 1 + \ln\left[\frac{v^4 + (v/52)^2}{v^4 + 0.432}\right]\frac{1}{49} + \ln\left[1 + (v/18.1)^3\right]\frac{1}{18.7}$$

$$b_e = 0.564\left[(\varepsilon_r - 0.9)/(\varepsilon_r + 3)\right]^{0.053}$$

with

$$\varepsilon_{eff_o} = \left[0.5(\varepsilon_r + 1) + a_o - \varepsilon_{eff}\right] \cdot e^{-c_o \cdot g^{d_o}} + \varepsilon_{eff} \qquad (5.113)$$

where

$$a_o = 0.7287\left[\varepsilon_{eff} - 0.5(\varepsilon_r + 1)\right] \cdot \left(1 - e^{-0.179u}\right)$$

$$b_o = 0.747 \varepsilon_r / (0.15 + \varepsilon_r)$$

$$c_o = b_o - (b_o - 0.207)e^{-0.414 \cdot u}$$

$$d_o = 0.593 + 0.694 e^{-0.562u}$$

The range of validity of the above expressions can be given as follows:

$$0.1 \le u \le 10 \quad 0.1 \le g \le 10 \quad 1 \le \varepsilon_r \le 18$$

The above equations can easily be programmed into a digital computer to enable quick calculation.

An important parameter of the coupled-line circuits is the differential impedance when the lines are driven as a balanced transmission line. This differential impedance is related to the odd-mode impedance given in (5.111) by the following relationship:

$$Z_{diff} = 2Z_{0o} \qquad (5.114)$$

The two-factor appears because the odd-mode impedance is measured from strips to ground. Thus, the balanced excitation has twice of this impedance. Balanced configuration is commonly used in high-speed digital circuits and MMIC, where dielectric thickness is too large causing the single microstrip line to be prohibitively wide for 50-ohm characteristic impedance.

5.4.1 Microstrip Directional Couplers

Microstrip directional couplers are extensively used in microwave engineering to monitor forward- and backward-traveling waves on a microstrip line [3, 7, 52–56]. For instance, a directional coupler placed on the output of a power amplifier can detect if the load is disconnected (i.e., a large reflection), which can be further used to shut down the amplifier to prevent any damage. Figure 5.14 shows a typical microstrip directional coupler implemented using coupled microstrip lines. Note that the port transmission lines are bent 90° to provide coupling only through the length l, which is a quarter-wavelength at the operating frequency. The corners of the lines are also mitered to minimize reflections. This configuration of the directional couplers is contradirectional because the coupled port is the adjacent port to the input. The isolated port is the one next to the direct port. Ideally, very little power (theoretically zero for TEM couplers) should be directed to the isolated port when all the ports are terminated properly.

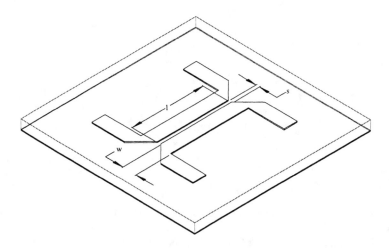

Figure 5.14 A microstrip directional coupler. Length l is typically quarter-wavelength long at the operating frequency.

Figure 5.15 shows port assignments for the symmetrical TEM coupler. The three important design parameters of a directional coupler can be defined as follows:

$$\text{Coupling} = 20\log\left|\frac{V_3}{V_1}\right| \tag{5.115}$$

$$\text{Directivity} = 20\log\left|\frac{V_4}{V_3}\right| \tag{5.116}$$

$$\text{Isolation} = 20\log\left|\frac{V_4}{V_1}\right| \tag{5.117}$$

The design of directional couplers is based on determination of the odd- and even-mode characteristic impedances according to the coupling coefficient, C, required. To express the characteristic impedance values in terms of the coupling coefficient and the normalization impedance, the following equations are used:

$$Z_{0e} = Z_0\sqrt{\frac{1+C}{1-C}} \tag{5.118}$$

$$Z_{0o} = Z_0\sqrt{\frac{1-C}{1+C}} \tag{5.119}$$

where Z_0 and C are the normalization impedance (usually 50 ohms) and coupling coefficient, respectively. Once the even- and odd-mode characteristic impedances are determined for a given coupling and normalization impedance, the width, w, and separation, s, of the lines can be found by means of (5.110) and (5.111).

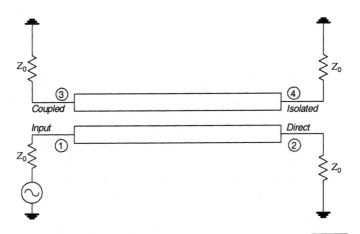

Figure 5.15 Port assignments of symmetrical TEM coupled lines. Note that $Z_0 = \sqrt{Z_{0o} \cdot Z_{0e}}$, where Z_{0o} and Z_{0e} are the odd- end even-mode impedances, respectively.

The scattering parameters of a symmetrical lossless TEM coupler can be given as follows:

$$S_{11} = S_{22} = S_{33} = S_{44} = 0 \qquad (5.120)$$

$$S_{14} = S_{41} = S_{23} = S_{32} = 0 \qquad (5.121)$$

$$S_{12} = S_{21} = S_{34} = S_{43} = \frac{\sqrt{1-C^2}}{\sqrt{1-C^2}\cos(\theta) + j\sin(\theta)} \qquad (5.122)$$

$$S_{13} = S_{31} = S_{24} = S_{42} = \frac{jC\sin(\theta)}{\sqrt{1-C^2}\cos(\theta) + j\sin(\theta)} \qquad (5.123)$$

where $\theta = \beta l$ denotes the electrical length of the coupler, and C is the coupling parameter as defined above. Note that there is a 90° phase shift between the direct and coupled ports. An interesting theory of lossless linear circuits states that any lossless four-port junction whose ports are all matched must be a directional coupler. The proof is based on the scattering matrix properties of lossless circuits and can be found in the literature [1]. By inspecting the above equations, one can see that the directivity of an ideal TEM coupler is infinity, as indicated before. The reason for this is that for a TEM coupler, the even- and odd-mode phase velocities are equal to each other. Therefore, perfect cancellation occurs at the isolated port. However, this is not the case for quasi-TEM couplers where microstrip-line couplers belong to that class. In that case, nonzero isolation occurs due to the difference between the phase velocities. In regular microstrip coupled lines, the phase velocity of the even mode is lower than the phase velocity of the odd mode because the electric field of the even mode is more concentrated in the dielectric (i.e., effective dielectric constant is higher), see Figure 5.16.

However, there are some means to compensate for the unequal phase velocities for microstrip line couplers as well, such as using lumped or interdigital distributed capacitors between the lines or using dielectric overlay [57, 58]. All of those methods are based on increasing the effective dielectric constant felt by the odd mode, so that its phase velocity can be reduced. Note that the mode characteristic impedances will also get affected during this process, but they will remain different so that the coupling (5.20) still becomes nonzero. In one of the most employed techniques for this purpose, lumped capacitors are placed on both ends of the coupled sections as demonstrated in Figure 5.17. Note that the compensating capacitors (C_p) are felt mainly by the odd mode increasing the effective dielectric constant for this mode.

Microstrip edge-coupled directional couplers described above are typically used for relatively small couplings (< -10 dB) because of the extremely narrow line spacing required. For higher couplings (e.g., -3 dB), different topologies other than simple edge-coupling must be employed. Hybrid couplers are the general name given for these kinds of structures that provide high couplings. In microstrip, hybrid couplers are usually implemented using interdigitated (Lange couplers) or branch-line configuration. In stripline, on the other hand, broadside coupling can be utilized by overlapping two striplines. Lange couplers are very popular in microwave engineering to design hybrids with good isolation. We will study the Lange couplers in more depth when we introduce the hybrids later.

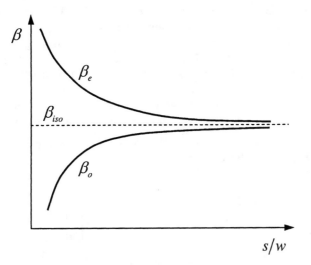

Figure 5.16 Qualitative plots of the even- and odd-mode propagation constants versus normalized strip spacing for microstrip coupled lines. The β_{iso} represents the propagation constant of an isolated strip with the same width.

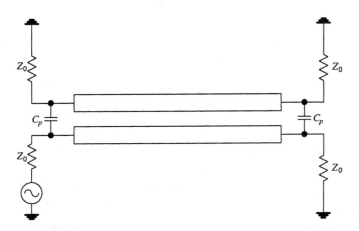

Figure 5.17 A method to improve the isolation in microstrip coupled lines. The value of capacitor C_p is determined such that the isolation is maximized.

Example 5.1

In this example, we will design a microstrip directional coupler using microstrip coupled lines. The required coupling amount is −20 dB at 15 GHz. The substrate parameters are $\varepsilon_r = 2.2$ and $h = 5$ mils.

We first determine the required even- and odd-mode characteristic impedances using (5.118) and (5.119) as $Z_{0e} = 55.3$ ohms and $Z_{0o} = 45.2$ ohms for the given coupling. This is the first step in any coupler design. Then, by using the formulae (5.110) through (5.113), we find that $s = 5$ mils and $w = 15$ mils, giving $Z_{0e} = 56.5$ ohms and $Z_{0o} = 44.8$ ohms. We decide to use these dimensions, although one could obtain more accurate dimensions using an optimization algorithm or synthesis equations. The even- and odd-mode effective dielectric constants are found as 1.96 and 1.76, respectively. Then, the physical length of the coupled section is determined as $\lambda_g/4 = 144$ mils. Note that the average of even- and odd-mode effective dielectric constants is used to determine the guided wavelength, λ_g. Finally, the coupler is realized based on these dimensions.

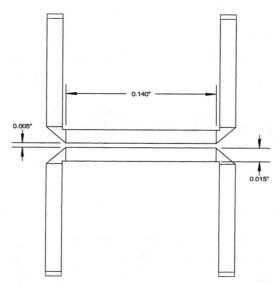

Figure 5.18 Full-wave electromagnetic simulation model of a microstrip directional coupler ($h = 5$ mils, $\varepsilon_r = 2.2$).

The full-wave model of the directional coupler implemented is shown in Figure 5.18, and simulation results are provided in Figures 5.19 and 5.20. Note that the length of the uniform coupled section is slightly less than the calculated length above. This modification was necessary to compensate for the effects of corners. The coupling is approximately −20 dB at 15 GHz, and 3-dB bandwidth is 10 GHz.

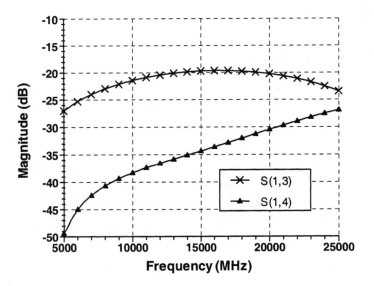

Figure 5.19 Magnitudes of S_{13} and S_{14} of the microstrip coupler shown in Figure 5.18. Note that S_{14} is nonzero and increases monotonically.

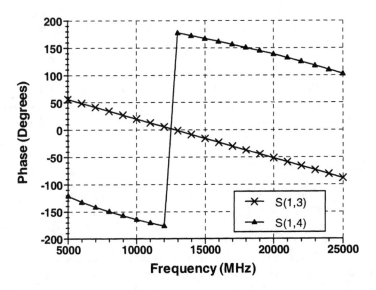

Figure 5.20 Phases of S_{13} and S_{14} of the microstrip coupler shown in Figure 5.18. Note that the phase difference between the isolated and coupled ports is 180°.

The important observation is that the isolation of the coupler is not very large, as expected. This nonideal behavior can be quite important, especially when the lines are electrically long, a situation can occur in long, close-proximity RF or high-speed digital signal lines. In that case, the resulting cross-talk distorts the pulse shape on the lines. To avoid this problem, physical separation between the lines can be increased. Using striplines instead of microstrip lines also improves the isolation significantly. Since the stripline geometry is homogeneous, the phase velocities are equal in that case.

Another observation is the quite small gap between the lines even for the modest coupling. One can easily predict that the gap between the lines becomes difficult to realize physically for couplings greater than −10 dB. For example, to design a −10-dB coupler using the same substrate, one would need 12.5-mils-wide lines with 0.5-mils separation, which is a small value for PCB implementation. This is the another limitation of microstrip directional couplers. It must be noted that using a thin-film process, on the other hand, one can obtain much finer dimensional resolution.

5.5 COUPLED STRIPLINES

Perhaps the chief advantage of coupled striplines is that even- and odd-mode phase velocities are equal in a stripline structure. This stems from the fact that because of its homogenous dielectric, the fundamental propagation mode is TEM in striplines (although higher-order modes can be excited if the frequency is high enough [59]). Practical implication of this is that it yields very good isolation compared to the microstrip case. Also, for small metallization thickness, one can obtain exact, closed-form expressions for even- and odd-mode characteristic impedance through conformal mapping technique.

Coupled striplines can be implemented as edge- or broadside-coupled configurations. When implemented in broadside configuration, they provide very high coupling, thus enabling the design of hybrids. These two configurations will be discussed in the next sections.

5.5.1 Edge-Coupled Striplines

Edge coupling in striplines provides moderate couplings as in the microstrip case. An edge-coupled stripline structure is shown in Figure 5.21. The cross-sectional view of the structure is given in Figure 5.22. Note that the metal strips are placed at the middle vertically.

The even- and odd-mode characteristic impedance of edge-coupled striplines for zero thickness metallization are given by the following expressions [60]:

$$Z_{0e} = \frac{1}{\sqrt{\varepsilon_r}} 30\pi \frac{K(k_e')}{K(k_e)} \tag{5.124}$$

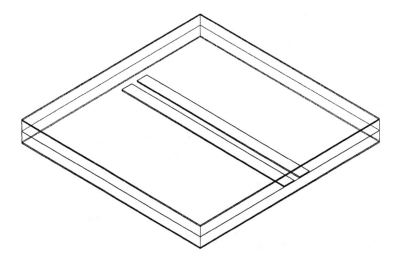

Figure 5.21 Edge-coupled striplines. See Figure 5.22 for the cross-sectional view.

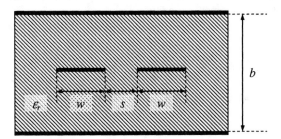

Figure 5.22 Cross-sectional view of edge-coupled striplines. Note that the metal strips are placed at the middle vertically.

$$Z_{0o} = \frac{1}{\sqrt{\varepsilon_r}} 30\pi \frac{K(k_o')}{K(k_o)} \qquad (5.125)$$

where

$$k_e = \tanh\left(\frac{\pi}{2}\frac{w}{b}\right)\tanh\left[\frac{\pi}{2}\frac{w+s}{b}\right]$$

$$k_o = \tanh\left(\frac{\pi}{2}\frac{w}{b}\right)\coth\left[\frac{\pi}{2}\frac{w+s}{b}\right]$$

and

$$k_{e,o}' = \sqrt{1 - k_{e,o}^2}$$

where w is the width of the lines, s is the separation of the lines, b is the total thickness of the dielectric slab, and $K(\ )$ is the complete elliptic integral of the first kind whose values can be found in mathematical handbooks. The above equations can easily be programmed into a digital computer to enable quick calculation.

5.5.2 Broadside-Coupled Striplines

An alternative way of achieving coupling in striplines is to use broadside coupling by placing the striplines in an overlapping fashion. This configuration considerably increases the coupling between the two transmission lines, thus enabling high-coupling directional couplers (i.e., hybrids). A broadside-coupled stripline structure is shown in Figure 5.23. The cross-sectional view of the structure is given in Figure 5.24. Usually, a broadside-coupled stripline is implemented using three dielectric layers as shown in the figure. Note that the metal strips are placed symmetrically in vertical direction and they are overlapping completely.

The even- and odd-mode impedances of the broadside-coupled striplines embedded in a homogenous medium for zero thickness metallization ($\varepsilon_1 = \varepsilon_2$) are given by the following expressions [61, 62]:

$$Z_{0e} = \frac{60\pi}{\sqrt{\varepsilon_r}} \frac{K(k')}{K(k)} \tag{5.126}$$

$$Z_{0o} = \frac{293.9}{\sqrt{\varepsilon_r}} \frac{s/b}{\tanh^{-1}(k)} \tag{5.127}$$

where

$$k' = \sqrt{1 - k^2}$$

and $K(\ )$ is the complete elliptic integral of the first kind. The factor k is the solution of the following transcendental equation:

$$\frac{W}{b} = \frac{1}{\pi}\left[\ln\left(\frac{1+R}{1-R}\right) - \frac{s}{b}\ln\left(\frac{1+R/k}{1-R/k}\right)\right]$$

$$R = \left[\left(k\frac{s}{b} - 1\right) \Big/ \left(\frac{1}{k}\frac{b}{s} - 1\right)\right]^{1/2}$$

where $2w$ is the width of the lines, $2s$ is the separation of the lines in the vertical direction, and $2b$ is the total thickness of the dielectric slab. The above equations can easily be programmed into a digital computer to enable quick calculation where the transcendental equation can be solved by means of numerical techniques.

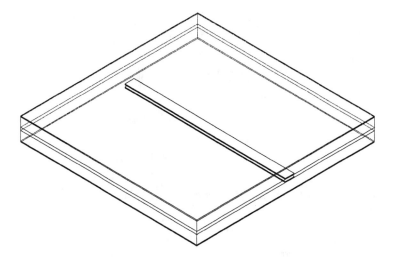

Figure 5.23 Broadside-coupled striplines. See Figure 5.24 for the cross-sectional view.

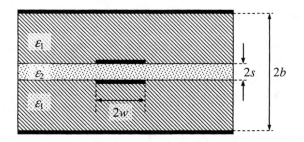

Figure 5.24 Cross-sectional view of broadside-coupled striplines. Note that it is also possible to offset the lines in the horizontal direction to control the amount of coupling.

An alternative explicit solution which doesn't require the solution of the transcendental equation can be given for the above formulation by first determining the odd-mode characteristic impedance as follows [62]:

$$Z_{0o} = \frac{Z_{0\infty}^a - \Delta Z_{0\infty}^a}{\sqrt{\varepsilon_r}} \qquad (5.128)$$

where

$$Z_{0\infty}^a = 60\ln\left[3\frac{s}{w} + \sqrt{\left(\frac{s}{w}\right)^2 + 1}\right]$$

$$\Delta Z_{0\infty}^a = \begin{cases} P & w/s \le 1/2 \\ P \cdot Q & w/s > 1/2 \end{cases}$$

$$P = 270 \left[1 - \tanh\left(0.28 + 1.2\sqrt{\frac{b-s}{s}} \right) \right]$$

$$Q = 1 - \tanh^{-1} \left[\frac{0.48\sqrt{2\dfrac{w}{s} - 1}}{\left(1 + \dfrac{b-s}{s}\right)^2} \right]$$

Then, the k-factor is found using the following formula:

$$k = \tanh\left(\frac{293.9\, s/b}{Z_{0o}\sqrt{\varepsilon_r}} \right)$$

After determining k, even-mode characteristic impedance is found using (5.126).

5.6 WIDE-BANDWIDTH DIRECTIONAL COUPLERS

As can be seen from the ongoing discussion, a single-section directional coupler is limited in bandwidth. However, by cascading many $\lambda/4$ sections as shown in Figure 5.25, one can increase the bandwidth and obtain more flat coupling characteristics. Due to more stringent phase-velocity-matching requirements, this kind of multisection directional couplers is usually constructed in stripline form.

The main assumption that we will make in the analysis of multisection couplers is that a small amount of total incident power is coupled at each coupler section. That is, most of the incident power is transferred to the direct arm. We will also assume there is an odd number of cascaded sections. Then, it can be shown that for a wideband coupler constructed using small-coupling directional couplers, the overall coupling can be expressed as follows [7]:

$$C_0 = \sin\theta \left[\sum_{i=1}^{M-1} 2C_i \cos(N - 2i + 1)\theta + C_M \right] \tag{5.129}$$

where $M = (N+1)/2$, N is the total number (odd) of sections in the coupler, θ is the electrical length, and C_i is the coupling contributed from each section ($i \ne 0$). The design process is based on the determination of couplings from each section to provide the overall desired coupling. Then, even- and odd-mode characteristic impedances of each section are calculated, and cascaded sections are physically realized. During the realization, transitions between the sections should be made as smooth as possible.

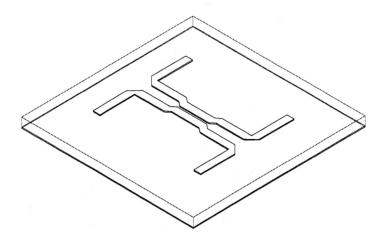

Figure 5.25 A multisection directional coupler. Note that an odd number of sections is usually employed. In the figure, a three-section coupler is shown.

Example 5.2

In this example, we will demonstrate the design of a multisection stripline directional coupler. The center frequency of the coupler will be $f = 15$ GHz, and the desired number of sections is three ($N = 3$). Substrate parameters are $\varepsilon_r = 2.2$ and $h = 20$ mils, where h is the total thickness of the stripline dielectric. We would like to have a maximally flat response and $C_0 = 20$-dB coupling at the center of the band.

The design process commences by determining the required couplings for each section. For a maximally flat response, the nth derivative of the total coupling, where n is the number of the section, must be equal to zero:

$$\frac{d^n}{d\theta^n} C(\theta) \bigg|_{\theta = \pi/2} = 0$$

where $n = 1, 2, \ldots, N - 1$. Using (5.129) and the above equation, we obtain the following set of equations:

$$-2C_1 + C_2 = 0.1$$
$$-10C_1 + C_2 = 0$$

Note that since the last section is the same as the first section, we don't need to include a separate equation for that. Solving the above equations for C_1 and C_2 gives $C_1 = C_3 = 0.0125$ and $C_2 = 0.125$. These are the couplings required for each section. Then, using (5.118) and (5.119), the even- and odd-mode characteristic impedances of each section are found as follows:

$$Z_{0e}^{(1)} = 50\sqrt{\frac{1+0.0125}{1-0.0125}} \cong 50.6 \quad \text{ohm}$$

$$Z_{0o}^{(1)} = 50\sqrt{\frac{1-0.0125}{1+0.0125}} \cong 49.4 \quad \text{ohm}$$

$$Z_{0e}^{(2)} = 50\sqrt{\frac{1+0.125}{1-0.125}} \cong 56.7 \quad \text{ohm}$$

$$Z_{0o}^{(2)} = 50\sqrt{\frac{1-0.125}{1+0.125}} \cong 44.1 \quad \text{ohm}$$

Finally, using the formulae (5.110) through (5.113), we find the spacing and line widths for each coupled section as $w_1 = 16.6$ mils, $s_1 = 19.1$ mils, $w_2 = 16.0$ mils, $s_2 = 4.9$ mils, $w_3 = 16.6$ mils, and $s_3 = 19.1$ mils. To test the design, the coupler shown in Figure 5.26 is analyzed using full-wave electromagnetic simulation. Note that some simplifications in the dimensions have been made to simplify the geometry so that the simulation time can be reduced. This is usually necessary if the simulation tool employed requires that the polygons representing the metallization must snap on a finite grid size. Simulation results are shown in Figures 5.27 and 5.28. From the figures, two important observations can be made: first, the coupling is more flat across the bandwidth compared to the single-section microstrip coupler analyzed previously. Note that the 3-dB bandwidth is nearly two octaves. Second, the isolation is better compared to the single-section microstrip coupler. This is expected because a stripline configuration is employed.

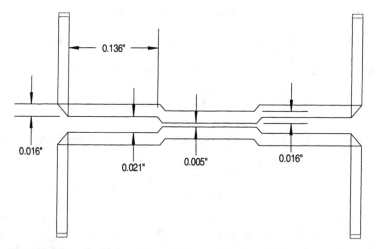

Figure 5.26 Full-wave electromagnetic simulation model of a stripline, three-section, directional coupler ($h = 20$ mils, $\varepsilon_r = 2.2$).

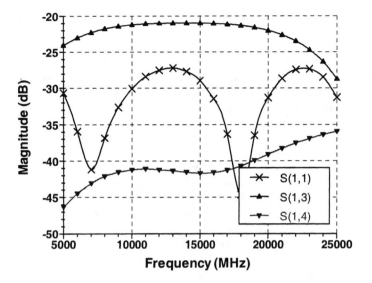

Figure 5.27 Magnitudes of S_{11}, S_{13}, and S_{14} of the stripline coupler shown in Figure 5.26.

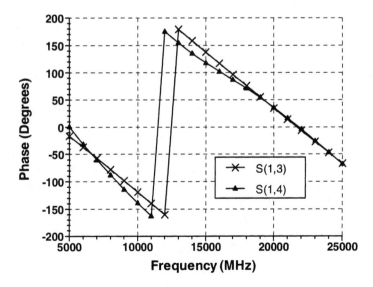

Figure 5.28 Phases of S_{13} and S_{14} of the stripline coupler shown in Figure 5.26.

5.7 HYBRID COUPLERS

Hybrid couplers are widely used in microwave circuits, such as balanced amplifiers, balanced mixers, phase shifters, circularly polarized patch antennas, and so forth, because of their unique features [63]. Hybrids can be designed to provide 180° or 90° coupling. To explain the use of the hybrids better, the design of balanced amplifiers will be discussed in more detail now.

In a balanced amplifier, the output power of many individual amplifiers is combined using hybrid couplers. A typical balanced amplifier configuration is shown in Figure 5.29. Note that the hybrids of a balanced amplifier are almost exclusively implemented using Lange couplers in MMIC design. Since the output power of each amplifier is added in phase, high output power levels can be achieved, which would otherwise be impossible to get from single amplifiers. The input and output reflection coefficients of the overall amplifier are also improved because reflected signals from the amplifiers are added out of phase and in phase at input and termination ports, respectively.

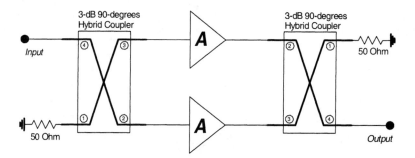

Figure 5.29 Structure of a balanced amplifier constructed with 3-dB hybrids. Balanced amplifiers are commonly used in microwave engineering.

Therefore, if the two amplifiers used in a hybrid block are identical, then, theoretically, unity input and output VSWR are obtained from the whole amplifier block. Note that the reflected powers from individual amplifiers are absorbed in the terminating resistor; therefore, this resistor must be selected with the proper power rating. It is important to design the individual amplifier modules as similarly as possible for optimum performance. It can be shown that idealized S-parameters of a balanced amplifier constructed with two hybrid couplers are given by

$$S_{11} = \frac{e^{-j\pi}}{2}\left(S_{11a} - S_{11b}\right) \tag{5.130}$$

$$S_{21} = \frac{e^{-j\pi/2}}{2}\left(S_{21a} + S_{21b}\right) \tag{5.131}$$

$$S_{12} = \frac{e^{-j\pi/2}}{2}\left(S_{12a} + S_{12b}\right) \quad (5.132)$$

$$S_{22} = \frac{e^{-j\pi}}{2}\left(S_{22a} - S_{22b}\right) \quad (5.133)$$

where the S-parameters in the parentheses represent the S-parameters of each individual amplifier. The cancellation of the reflection coefficients of each amplifier can be seen in the above equations.

For the sake of completeness, we will provide the idealized S-parameters of a 3-dB hybrid coupler as follows (port numbers are shown in Figure 5.29):

$$S_{11} = S_{22} = S_{33} = S_{44} = 0 \quad (5.134)$$

$$S_{14} = S_{41} = S_{23} = S_{32} = 0 \quad (5.135)$$

$$S_{12} = S_{21} = S_{34} = S_{43} = \frac{e^{-j\pi/2}}{\sqrt{2}} \quad (5.136)$$

$$S_{13} = S_{31} = S_{24} = S_{42} = \frac{e^{-j\pi}}{\sqrt{2}} \quad (5.137)$$

Note that there is a 90° phase difference between the direct and coupled arms. For this reason, this kind of coupler is also called a 90° hybrid.

There are many different ways of constructing printed hybrid couplers. They can be constructed in microstrip or stripline configuration. Figure 5.30 shows a branch-line microstrip hybrid coupler [64]. It is called branch-line because of the additional line segments connecting the two arms. This configuration can provide the necessary 3-dB coupling for a hybrid using microstrip lines. The electrical length of each arm is selected as $\lambda/4$. The design equations for a branch-line coupler are given as follows:

$$C = 10\log\frac{1}{1 - \left(Z_{01}/Z_0\right)^2} \quad (5.138)$$

$$Z_{02} = \frac{Z_{01}}{\sqrt{1 - \left(Z_{01}/Z_0\right)^2}} \quad (5.139)$$

where C is the coupling coefficient in decibels, and Z_{01} and Z_{02} are the characteristic impedance of the branch lines. For the 3-dB branch-line coupler and $Z_0 = 50$ ohms, Z_{01} and Z_{02} are found as 50 ohms and 35 ohms, respectively.

An interesting feature of branch-line couplers is that the coupling of a matched branch-line coupler at any frequency depends only on the impedance ratio of the shunt and series arms and is independent of frequency. Therefore, if one matches the coupler ports at some discrete frequencies using a proper matching network, then the coupling must be the same at those frequencies [65].

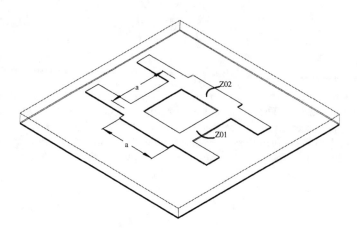

Figure 5.30 A branch-line 3-dB microstrip hybrid coupler. Electrical length of each arm is $\lambda/4$.

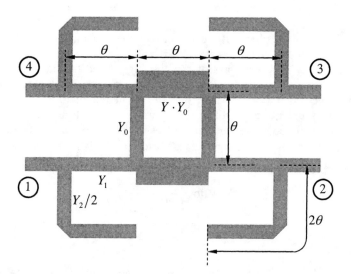

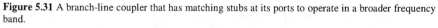

Figure 5.31 A branch-line coupler that has matching stubs at its ports to operate in a broader frequency band.

This fact can be used to design broadband hybrid couplers as demonstrated in Figure 5.31. The characteristic admittance of the arms and electrical length θ can be found from the solution of following expressions [65]:

$$\frac{\sin^2 \theta}{Y_1^2} + \left(1 + Y_2/Y_1\right)^2 \cos^2 \theta = \frac{\sin \theta}{Y_0 \sqrt{Y^2 - 1}} \qquad (5.140)$$

$$\frac{\sin\theta}{Y_1} - \left(1 + Y_2/Y_1\right)\left(Y_1\sin\theta - Y_2\frac{\cos^2\theta}{\sin\theta}\right) = -\left(\frac{Y+1}{Y-1}\right)^{1/2} \tag{5.141}$$

$$\theta = \cos^{-1}\left\{\sqrt{\frac{1}{2}}\cos\left[\frac{\pi}{2}\left(1 + \frac{\Delta f}{f}\right)\right]\right\} \tag{5.142}$$

where

$$Y = \sqrt{\left|\frac{S_{12}}{S_{13}}\right|^2 + 1} \tag{5.143}$$

In the above equations, Δf and f are the desired frequency bandwidth and center frequency, respectively. The factor Y is determined according to the required amount of coupling. For 3-dB coupling, Y becomes $\sqrt{2}$.

Broadside-coupled striplines can also be used to design hybrid couplers. Figure 5.32 shows a stripline hybrid coupler design using broadside-coupled striplines. Note that at least three dielectric layers (two metallization layers) are required to design a stripline hybrid coupler. In the structure shown in Figure 5.32, the top and bottom layers are covered with ground plane.

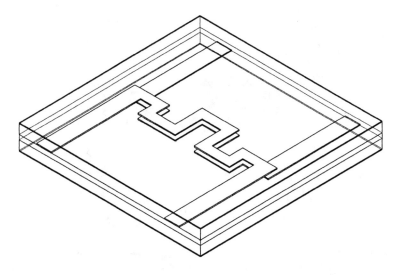

Figure 5.32 A stripline 3-dB hybrid coupler. Note that three dielectric layers are required to construct the stripline hybrid coupler. There are ground planes both on the top and bottom.

Another type of hybrid coupler which is frequently employed in microwave engineering (especially in MMIC design) is the Lange coupler [66–69]. A typical Lange coupler is shown in Figure 5.33. The coupling is enhanced by connecting the alternating fingers as depicted in the figure. The length of the coupled region is a quarter wavelength at the center frequency. Note that this concept can be generalized to interdigital couplers with an arbitrary number of fingers [67]. The design of the Lange coupler, like in other couplers, requires determination of even- and odd-mode characteristic impedance first. These can be found using the following expressions [69]:

$$Z_{0o} = Z_0 \left(\frac{1-C}{1+C}\right)^{1/2} \frac{(N-1)(q+1)}{(C+q)+(N-1)(1-C)} \tag{5.144}$$

$$Z_{0e} = Z_{0o} \frac{(C+q)}{(N-1)(1-C)} \tag{5.145}$$

where

$$q = \left[C^2 + (1-C^2)(N-1)^2\right]^{1/2}$$

In the above equations, C is the coupling coefficient, Z_0 is the terminating impedance, and N is the number of fingers (even). Once the Z_{0o} and Z_{0e} are determined, the width and separation of the fingers can be found using the standard coupled microstrip formulae.

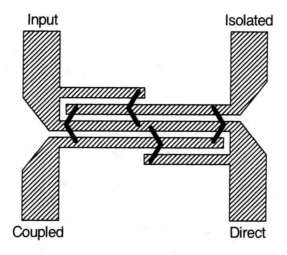

Figure 5.33 The Lange coupler. Note that how the alternating fingers are connected with wire bonds. The coupled section is quarter-wavelength long at the center frequency.

The main drawback of Lange couplers is that they need crossover connections to short the alternating fingers of the interdigitated structure, which requires wire bonds if the couplers are designed using PCB technology. In MMIC, these crossovers can be implemented conveniently using air bridges providing great flexibility.

Example 5.3

In this example, we will design a branch-line 3-dB microstrip hybrid coupler. The center frequency of the coupler is $f = 22$ GHz. Substrate parameters are $\varepsilon_r = 2.2$ and $h = 5$ mils.

We first determine characteristic impedances of the branch sections using (5.138) and (5.139) as $Z_{01} = 50$ ohms and $Z_{02} = 35$ ohms. Then, corresponding line widths of the branch sections are found as 15 mils and 25 mils, respectively. The effective dielectric constant is $\varepsilon_{eff} = 1.9$. The last step is to calculate the physical length of the branch sections. Using the effective dielectric constant, the length is found as $\lambda_g/4 = 100$ mils. The implemented branch-line coupler based on these numbers is shown in Figure 5.34, and simulation results are given in Figures 5.35 and 5.36. Note that the phase difference and power coupling between the coupled ports are nearly constant and equal to 90° and 3 dB, respectively. The maximum isolation obtained is −30 dB at $f = 23.25$ GHz. However, at the frequency point where the maximum isolation occurs, the power division between the coupled arms is not equal. Equal power division occurs around $f = 22$ GHz for this particular coupler. The change in the center frequency can easily be compensated for by utilizing an optimization algorithm.

One of the disadvantages of a branch-line hybrid coupler is that the size of the circuit can get quite large at lower frequencies due to quarter-wavelength transmission lines. To address this, lumped hybrid couplers can be used if the dimensions of the distributed network become unacceptable. The lumped model is extracted by replacing the transmission lines with their lumped Pi-network equivalent circuits over a narrow band of frequency.

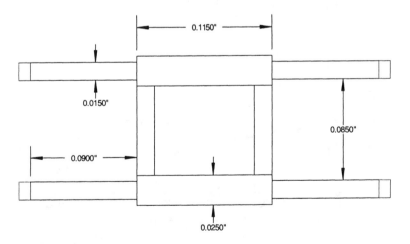

Figure 5.34 Full-wave electromagnetic simulation model of a branch-line hybrid line coupler ($h = 5$ mils, $\varepsilon_r = 2.2$). Electrical length of each arm is $\lambda/4$.

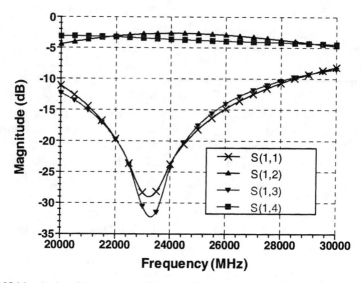

Figure 5.35 Magnitudes of *S*-parameters of the hybrid coupler shown in Figure 5.34. Note that the coupling to each coupled port is approximately −3 dB.

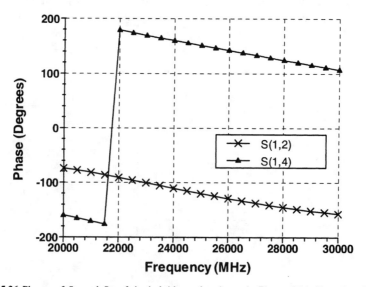

Figure 5.36 Phases of S_{12} and S_{14} of the hybrid coupler shown in Figure 5.34. Note that the phase difference between the coupled ports is 90° (or −270°).

It is important to note that the circuit dimensions should be judged based on the technology employed. For instance, a circuit that is acceptable for a PCB might be prohibitively large for an MMIC.

To demonstrate the approach, we will now design a lumped-circuit version of the branch-line coupler described above. The lumped Pi-network equivalents of the transmission lines of each branch can be found by comparing the *ABCD*-matrixes of the lines with the *ABCD*-matrix of a lumped *LC* Pi-network by assuming $\beta l \cong \pi/2$ at $f = 22$ GHz. Then, capacitance and inductance values of the equivalent lumped circuits are determined to be $C_h = 0.21$ pF, $L_h = 0.36$ nH, $C_l = 0.14$ pF, and $L_l = 0.25$ nH, where h and l subscripts refer to high- and low-impedance lines, respectively (see Figure 5.30). The circuit schematic of the resulting lumped branch-line coupler is shown in Figure 5.37, and simulation results are given in Figures 5.38 and 5.39. The first observation is that maximum isolation occurs at 22.25 GHz, approximately a 5% shift with respect to the microstrip counterpart. On the other hand, phase difference between coupled arms deteriorates relatively faster compared to the microstrip circuit, which is expected because lumped circuits can accurately approximate transmission lines only at a single frequency. However, the overall response of the lumped hybrid is quite acceptable. Between 20 GHz and 24 GHz, the power division is 3.5 ± 0.5 dB and return loss and isolation are better than −12 dB.

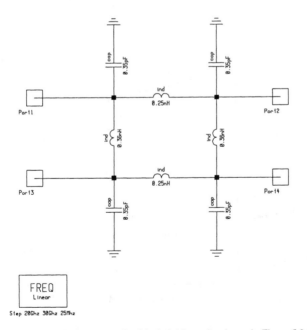

Figure 5.37 The lumped equivalent network of the hybrid coupler shown in Figure 5.34.

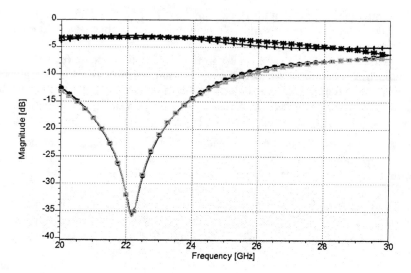

Figure 5.38 Magnitudes of S-parameters of the lumped 3-dB hybrid coupler (square: S_{11}; circle: S_{13}; plus: S_{12}; star: S_{14}).

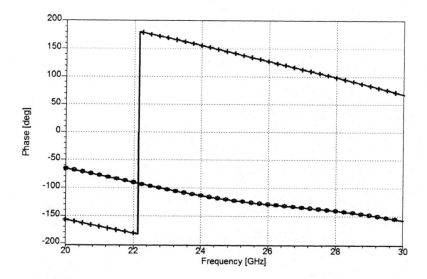

Figure 5.39 Phases of S_{12} and S_{14} of the lumped 3-dB hybrid coupler (circle: S_{12}; plus: S_{12}).

Example 5.4

In this example, we will design a 3-dB hybrid coupler using striplines. The center frequency and substrate dielectric constant are $f = 550$ MHz and $\varepsilon_r = 2.5$, respectively.

We first determine the required even- and odd-mode characteristic impedances using (5.118) and (5.119) as $Z_{0e} = 21$ ohms and $Z_{0o} = 121$ ohms for a 3-dB coupling. Then, by using (5.126) and (5.127), the required strip width, separation, and total dielectric thickness are found to be $2w = 1.1$ mm, $2s = 0.256$ mm, and $2h = 2.856$, respectively. From this, it is decided to use two different substrate materials with thickness $h_1 = 1.3$ mm and $h_2 = 0.256$ mm to form the stripline structure so that $2h_1 + h_2 = 2.856$.

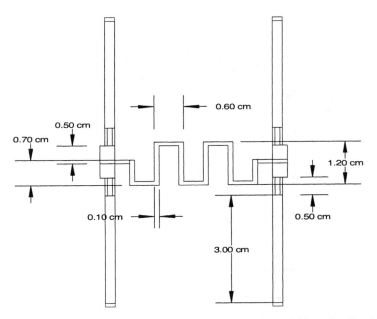

Figure 5.40 Full-wave electromagnetic simulation model of a stripline hybrid coupler ($h_1 = 1.3$ mm, $\varepsilon_{r1} = 2.5$, $h_2 = 0.256$ mm, $\varepsilon_{r2} = 2.48$, $h_3 = 1.3$ mm, $\varepsilon_{r3} = 2.5$).

The last step is to determine the physical length of the coupled section. Using the relative dielectric constant, the length of the coupled section is found to be $\lambda_g/4 = 8.6$ cm. The implemented stripline coupler is shown in Figure 5.40, and simulation results are given in Figures 5.41 and 5.42. Note that the coupled section is realized by using meander lines to reduce the size of the coupler. From the figures, it can be observed that power coupling and phase difference between the coupled ports are nearly constant and equal 3 dB and 90°, respectively, through the operation band.

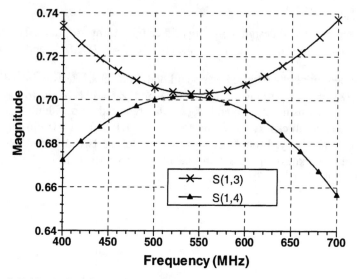

Figure 5.41 Magnitudes of S_{13} and S_{14} of the hybrid coupler shown in Figure 5.40. Note that the coupling to each coupled port is approximately −3 dB.

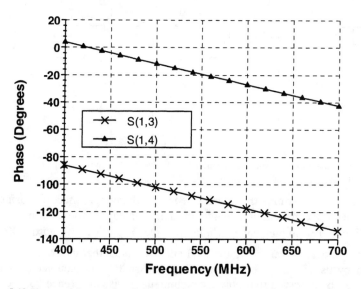

Figure 5.42 Phases of S_{13} and S_{14} of the hybrid coupler shown in Figure 5.40. Note that the phase difference between the coupled ports is 90°.

References

[1] C. G. Montgomery (Ed.), *Technique of Microwave Measurements*, Radiation Laboratory Series, New York: McGraw-Hill, 1947.

[2] C. G. Montgomery, R. H. Dicke, and E. M. Purcell (Eds.), *Principles of Microwave Circuits*, IEE Electromagnetic Waves Series, London, England: Peter Peregrinus Ltd., 1987.

[3] G. L. Matthaei, L. Young, and E. M. T. Jones, *Microwave Filters, Impedance Matching Networks and Coupling Structures*, Dedham, MA: Artech House, 1980.

[4] Rajesh Mongia, Inder Bahl, and Prakash Bhartia, *RF and Microwave Coupled-Line Circuits*, Norwood, MA: Artech House, 1999.

[5] T. C. Edwards and M. B. Steer, *Foundations of Interconnect and Microstrip Design*, New York: John Wiley and Sons, 2000.

[6] H. Howe, *Stripline Circuit Design*, Dedham, MA: Artech House, 1974.

[7] David M. Pozar, *Microwave Engineering*, New York: John Wiley and Sons, 1998.

[8] David E. Bockelman and William R. Eisenstadt, "Combined Differential and Common-Mode Scattering Parameters: Theory and Simulation," *IEEE Trans. Microwave Theory and Tech.*, vol. 43, pp. 1530–1539, July 1995.

[9] Dylan F. Williams, "Calibration in Multiconductor Transmission Lines," *NIST Publication*, 1996.

[10] Dylan F. Williams, "Multiconductor Transmission Line Characterization," *NIST Publication*, 1997.

[11] J. R. Carson and R. S. Hoyt, "Propagation of Periodic Waves over a System of Parallel Wires," *Bell System Tech. Journal*, vol. 4, pp. 495–545, July 1927.

[12] David A. Johns and Daniel Essig, "Integrated Circuits for Data Transmission over Twisted-Pair Channels," *IEEE Journal of Solid-State Circuits*, vol. 32, pp. 398–406, Mar. 1997.

[13] L. A. Pipes, "Matrix Theory of Multiconductor Transmission Lines," *Philosophical Magazine*, vol. 24, pp. 97–113, 1937.

[14] Clayton R. Paul, *Analysis of Multiconductor Transmission Lines*, New York: John Wiley and Sons, 1994.

[15] J. A. Brandao Faria, *Multiconductor Transmission-Line Structures*, New York: John Wiley and Sons, 1993.

[16] Fung-Yuel Chang, "Transient Analysis of Lossless Coupled Transmission Lines in a Nonhomogeneous Dielectric Medium," *IEEE Trans. Microwave Theory and Tech.*, vol. 18, pp. 616–626, Sept. 1970.

[17] K. D. Marx, "Propagation Modes, Equivalent Circuits, and Characteristic Terminations for Multiconductor Transmission Lines with Inhomogeneous Dielectrics," *IEEE Trans. Microwave Theory and Tech.*, vol. 21, pp. 450–457, July 1973.

[18] Clayton R. Paul, "On Uniform Multimode Transmission Lines," *IEEE Trans. Microwave Theory and Tech.*, pp. 556–558, Aug. 1973.

[19] Clayton R. Paul, "Decoupling the Multiconductor Transmission Line Equations," *IEEE Trans. Microwave Theory and Tech.*, vol. 44, pp. 1429–1440, Aug. 1996.

[20]   Mark K. Krage and George I. Haddad, "Characteristics of Coupled Microwave Transmission Lines-I: Coupled-Mode Formulation of Inhomogeneous Lines," *IEEE Trans. Microwave Theory and Tech.*, pp. 217–222, Apr. 1970.

[21]   Mark K. Krage and George I. Haddad, "Characteristics of Coupled Microwave Transmission Lines-II: Evaluation of Coupled-Line Parameters," *IEEE Trans. Microwave Theory and Tech.*, pp. 222–228, Apr. 1970.

[22]   Vijai K. Tripathi and Achim Hill, "Equivalent Circuit Modeling of Losses and Dispersion in Single and Coupled Lines for Microwave and Millimeter-Wave Integrated Circuits," *IEEE Trans. Microwave Theory and Tech.*, vol. 36, pp. 256–262, Feb. 1988.

[23]   Chih-Ming Tsai and Kuldip C. Gupta, "A Generalized Model for Coupled Lines and Its Applications to Two-Layer Planar Circuits," *IEEE Trans. Microwave Theory and Tech.*, vol. 40, pp. 2190–2199, Dec. 1992.

[24]   Krzystof Sachse, "The Scattering Parameters and Directional Coupled Analysis of Characteristically Terminated Asymmetric Coupled Transmission Lines in an Inhomogeneous Medium," *IEEE Trans. Microwave Theory and Tech.*, vol. 38, pp. 417–425, Apr. 1990.

[25]   Sophocles J. Orfanidis, *Electromagnetic Waves and Antennas*, ECE Department, Rutgers University, Piscataway, New Jersey, 2004.

[26]   Bernard M. Oliver, "Directional Electromagnetic Couplers," *Proceedings of the IRE*, pp. 1686–1692, Nov. 1954.

[27]   Lawrence Carin and Kevin J. Webb, "An Equivalent Circuit Model for Terminated Hybrid-Mode Multiconductor Transmission Lines," *IEEE Trans. Microwave Theory and Tech.*, vol. 37, pp. 1784–1793, Nov. 1989.

[28]   Vijai K. Tripathi and John B. Rettig, "A SPICE Model for Multiple Coupled Microstrip and Other Transmission Lines," *IEEE Trans. Microwave Theory and Tech.*, vol. 33, pp. 1513–1518, Dec. 1985.

[29]   G. G. Gentili and M. Salazar-Palma, "The Definition and Computation of Modal Characteristic Impedance in Quasi-TEM Coupled Transmission Lines," *IEEE Trans. Microwave Theory and Tech.*, vol. 42, pp. 338–343, Feb. 1995.

[30]   Jen-Tsai Kuo and Ching-Kuang C. Tzuang, "A Termination Scheme for High-Speed Pulse Propagation on System of Tightly Coupled Coplanar Strips," *IEEE Trans. Microwave Theory and Tech.*, vol. 42, pp. 1008–1015, June 1994.

[31]   Y. Y. Sun, "Comments on Propagation Modes, Equivalent Circuits, and Characteristic Terminations for Multiconductor Transmission Lines with Inhomogeneous Dielectrics," *IEEE Trans. Microwave Theory and Tech.*, vol. 26, pp. 915–918, Feb. 1978.

[32]   K. D. Marx, "Reply to Comments on Propagation Modes, Equivalent Circuits, and Characteristic Terminations for Multiconductor Transmission Lines with Inhomogeneous Dielectrics," *IEEE Trans. Microwave Theory and Tech.*, vol. 26, pp. 915–918, Feb. 1978.

[33]   Guang-Tsai Lei, Guang-Wen Pan, and Barry K. Gilbert, "Examination, Clarification, and Simplification of Modal Decoupling Method for Multiconductor Transmission Lines," *IEEE Trans. Microwave Theory and Tech.*, vol. 43, pp. 2090–2100, Sept. 1995.

[34]   Smain Amari, "Comments on Spectral-Domain Computation of Characteristic Impedances and Multiport Parameters of Multiple Coupled Microstrip Lines," *IEEE Trans. Microwave Theory and Tech.*, vol. 40, pp. 1733–1736, Aug. 1992.

[35] V. K. Tripathi and H. Lee, "Reply to Comments on Spectral-Domain Computation of Characteristic Impedances and Multiport Parameters of Multiple Coupled Microstrip Lines," *IEEE Trans. Microwave Theory and Tech.*, vol. 40, pp. 1733–1736, Aug. 1992.

[36] Josh G. Nickel et al., "Narrow-Band Matching Networks for Quasi-TEM Coupled Microstrip Lines," *IEEE Trans. Microwave Theory and Tech.*, vol. 50, pp. 1392–1399, May 2002.

[37] Smain Amari and Jens Bornemann, "Optimum Termination Networks for Tightly Coupled Microstrip Lines under Random and Deterministic Termination," *IEEE Trans. Microwave Theory and Tech.*, vol. 45, pp. 1785–1789, Oct. 1997.

[38] Adel A. M. Saleh, "Transmission-Line Identities for a Class of Interconnected Coupled-Line Sections with Application to Adjustable Microstrip and Stripline Tuners," *IEEE Trans. Microwave Theory and Tech.*, vol. 28, pp. 725–732, July 1980.

[39] Ersch Rotholz, "Transmission-Line Transformers," *IEEE Trans. Microwave Theory and Tech.*, vol. 29, pp. 327–331, Apr. 1981.

[40] Ralph Levy, "New Equivalent Circuits for Inhomogeneous Coupled Lines with Synthesis Applications," *IEEE Trans. Microwave Theory and Tech.*, vol. 36, pp. 1087–1094, June 1988.

[41] Ingo E. Losch and Johannes A. G. Malherbe, "Design Procedure for Inhomogeneous Coupled Line Sections," *IEEE Trans. Microwave Theory and Tech.*, vol. 36, pp. 1186–1190, July 1988.

[42] Vijai K. Tripathi, "Equivalent Circuits and Characteristics of Inhomogeneous Nonsymmetrical Coupled-Line Two-Port Circuits," *IEEE Trans. Microwave Theory and Tech.*, pp. 140–142, Feb. 1977.

[43] Vijai K. Tripathi, "On the Analysis of Symmetrical Three-Line Microstrip Circuits," *IEEE Trans. Microwave Theory and Tech.*, vol. 25, pp. 726–729, Sept. 1977.

[44] Vijai K. Tripathi, "Asymmetric Coupled Transmission Lines in an Inhomogeneous Medium," *IEEE Trans. Microwave Theory and Tech.*, vol. 23, pp. 734–739, Sept. 1975.

[45] Thomas G. Bryant and Jerald A. Weiss, "Parameters of Microstrip Transmission Lines and of Coupled Pairs of Microstrip Lines," *IEEE Trans. Microwave Theory and Tech.*, vol. 16, pp. 1021–1027, Dec. 1968.

[46] Cao Wei et al., "Multiconductor Transmission Lines in Multilayered Dielectric Media," *IEEE Trans. Microwave Theory and Tech.*, vol. 32, pp. 439–450, Apr. 1984.

[47] Kyung Suk Oh, Dmitri Kuznetsov, and Jose E. Schutt-Aine, "Capacitance Computations in a Multilayered Dielectric Medium Using Closed-Form Spatial Green's Functions," *IEEE Trans. Microwave Theory and Tech.*, vol. 42, pp. 1443–1453, Aug. 1994.

[48] S. Kai, D. Bhattacharya, and N. B. Chakraborti, "Empirical Relations for Capacitive and Inductive Coupling Coefficients of Coupled Microstrip Lines," *IEEE Trans. Microwave Theory and Tech.*, vol. 29, pp. 386–388, Apr. 1981.

[49] Vijai K. Tripathi and Hyuckjae Lee, "Spectral-Domain Computation of Characteristic Impedances and Multiport Parameters of Multiple Coupled Microstrip Lines," *IEEE Trans. Microwave Theory and Tech.*, vol. 37, pp. 215–221, Jan. 1989.

[50] Kwok-Keung M. Cheng, "Analysis and Synthesis of Coplanar Coupled Lines on Substrates of Finite Thickness," *IEEE Trans. Microwave Theory and Tech.*, vol. 44, pp. 636–643, July 1988.

[51] Manfred Kirschning and Rolf H. Jansen, "Accurate Wide-Range Design Equations for the Frequency-Dependent Characteristic of Parallel Coupled Microstrip Lines," *IEEE Trans. Microwave Theory and Tech.*, vol. 32, pp. 83–90, Jan. 1984 (Corrections: Mar. 1985).

[52] Edward G. Cristal, "Coupled Transmission-Line Directional Couplers with Coupled Lines of Unequal Characteristic Impedances," *IEEE Trans. Microwave Theory and Tech.*, vol. 14, pp. 337–346, July 1966.

[53] Risaburo Sato and Edward G. Cristal, "Simplified Analysis of Coupled Transmission-Line Networks," *IEEE Trans. Microwave Theory and Tech.*, vol. 18, pp. 122–131, Mar. 1970.

[54] James L. Allen and Marvin F. Estes, "Broadside-Coupled Strips in a Layered Dielectric Medium," *IEEE Trans. Microwave Theory and Tech.*, pp. 662–669, Oct. 1972.

[55] Sina Akhtarzad, Thomas R. Rowbotham, and Peter B. Johns, "The Design of Coupled Microstrip Lines," *IEEE Trans. Microwave Theory and Tech.*, pp. 486–492, June 1975.

[56] Ross A. Speciale, "Even- and Odd-Mode Waves for Nonsymmetrical Coupled Lines in Nonhomogeneous Media," *IEEE Trans. Microwave Theory and Tech.*, pp. 897–908, Nov. 1975.

[57] Steven L. March, "Phase Velocity Compensation in Parallel-Coupled Microstrip," *IEEE MTT Symposium Digest*, pp. 410–412, 1982.

[58] Michael Dydyk, "Accurate Design of Microstrip Directional Couplers with Capacitive Compensation," *IEEE MTT Symposium Digest*, pp. 581–584, 1990.

[59] Stelios Tsitsos et al., "Higher Order Modes in Coupled Striplines: Prediction and Measurement," *IEEE Trans. Microwave Theory and Tech.*, vol. 42, pp. 2071–2077, Nov. 1994.

[60] Seymour B. Cohn, "Shielded Coupled-Strip Transmission Line," *IEEE Trans. Microwave Theory and Tech.*, vol. 3, pp. 29–38, Oct. 1955.

[61] Seymour B. Cohn, "Characteristic Impedances of Broadside-Coupled Strip Transmission Lines," *IRE Trans. Microwave Theory and Tech.*, pp. 633–637, Nov. 1960.

[62] Prakash Bhartia and Protab Pramanick, "Computer-Aided Design Models for Broadside-Coupled Stripline and Millimeter-Wave Suspended Substrate Microstrip Lines," *IEEE Trans. Microwave Theory and Tech.*, vol. 36, pp. 1476–1481, Nov. 1988 (Corrections: Oct. 1989).

[63] M. D. Abouzahra and K. C. Gupta, "Hybrids and Power Dividers/Combiners," *Analysis and Design of Planar Microwave Components*, New York: IEEE Press, 1994.

[64] Vijai K. Tripathi, Hans B. Lunden, and J. Piotr Starski, "Analysis and Design of Branch-Line Hybrids with Coupled Lines," *IEEE Trans. Microwave Theory and Tech.*, vol. 32, pp. 425–432, Apr. 1984.

[65] Gordon P. Riblet, "A Directional Coupler with Very Flat Coupling," *IEEE Trans. Microwave Theory and Tech.*, vol. 26, pp. 70–74, Feb. 1978 (Corrections: Sept. 1978).

[66] Julius Lange, "Interdigitated Stipline Quadrature Hybrid," *IEEE Trans. Microwave Theory and Tech.*, pp. 1150–1151, Dec. 1969.

[67] Yusuke Tajima and Susumu Kamihashi, "Multiconductor Couplers," *IEEE Trans. Microwave Theory and Tech.*, pp. 795–801, Oct. 1978.

[68] Darko Kajfez, Zoja Paunovic, and Stane Pavlin, "Simplified Design of Lange Coupler," *IEEE Trans. Microwave Theory and Tech.*, pp. 806–808, Oct. 1978.

[69] R. M. Osmani, "Synthesis of Lange Couplers," *IEEE Trans. Microwave Theory and Tech.*, pp. 168–170, Feb. 1981.

Chapter 6

Microstrip Filters

The fundamental use of filters in electrical engineering is to shape signal spectrum, which is especially crucial in reducing input signal noise in receivers and spurious emissions in transmitters. It should be noted that electrical filters are not limited to shaping the magnitude response of the signal spectrum; they can be used to change the phase response as well. The fundamentals of filter theory are based on the works of the great French mathematician Jean Baptiste Joseph Fourier who showed that arbitrary functions can be represented by trigonometric series, called the spectrum of the signal. Broadly speaking, a filter is a continuous, translation-invariant, linear system. In electrical engineering, filtering can be intentional, as in the input stage of a receiver, or unintentional, as in the transmission path of a microwave signal. It is important to realize that almost every physical system has some sort of filtering action built in whenever a signal, an input, an output, and a transmission path can be defined in that object. The signals can be either electrical or mechanical vibrations, although many analogies exist between the two.

Microwave filters have traditionally been built using waveguide and coaxial lines. Following the enormous expansion in printed circuit technology and modeling techniques, many of them are now built using printed circuits, with few exceptions, which we will introduce in a moment. Printed circuit filters have advantages over rectangular or coaxial waveguide filters in terms of low cost, repeatability, high accuracy, and compact size. Repeatability and high accuracy are achieved through the use of photolithographic techniques. By selecting high-dielectric constant substrates, the size of the printed filters can be reduced significantly, which provides compact size. In addition, with the advances in full-wave electromagnetic simulation techniques, printed filters can be characterized accurately and rapidly. Another advantage of printed filters is their easy integration with active circuits; filters can be fabricated on the same substrate with transistor amplifiers, oscillators, and other circuits. Perhaps the main disadvantage of regular printed filters (i.e., nonsuperconducting) is the high insertion loss associated due to metallization and dielectric losses in some situations. For this reason, printed filters may not be suitable for applications where high power and very low loss are required. Output filter stages of base-station amplifiers are examples of such applications. In those cases, microwave cavities are still the primary choice for implementing the filters. Printed filters can be designed using microstrip or striplines. Stepped-impedance, interdigital, and coupled-line filters are the most commonly used forms of printed filters. In the first method, the widths of the transmission lines are changed in a periodic manner to replicate series inductance and shunt capacitance to implement lowpass filters. In the second method, quarter-wavelength-long transmission lines are arranged in an interdigital fashion where on one end the lines are shorted to ground in an alternating manner. In the last method, cascaded sections of quarter-wavelength-long coupled transmission lines are used to implement bandpass filters.

Like many other subjects in microwave engineering, major and rapid advances in microwave filters were made from 1940 to 1945 at U.S. and U.K. research laboratories. Some of the important laboratories in the United States working on microwave filters were M.I.T.'s Radiation Laboratory (specializing in waveguide-cavity filters), Harvard's Radio Research Laboratory (specializing in coaxial filters), and Bell Laboratories [1–7]. Among the scientists who worked at the Radiation Laboratory during that period, H. A. Bethe, N. Marcuvitz, E. M. Purcell, and J. Schwinger are the best known for their work in the theory of microwave filters, waveguides, and cavities. Just after 1945, a paper published by Paul I. Richards on commensurate circuits [8] showed how to apply the well-developed lumped-circuit filter theory to distributed circuits by using a simple transformation which would soon after be known as Richard's transformation. Then, in 1958, another important paper [9] was published describing the synthesis of stripline filters using a family of transformations discovered earlier by K. Kuroda in Japan. These transformations, known as Kuroda identities, enabled the easy design of coupled stripline filters starting from lumped prototype circuits.

Since lumped-circuit filter theory is an indispensable part of distributed filter synthesis, we will review the basics of this theory first. This will furnish us with the basic skills in the synthesis and analysis of microstrip filters. Then, we will gradually introduce distributed filters with an emphasis on microstrip filters.

6.1 BASIC FILTER THEORY

Filter networks can be designed using either the image parameter or insertion-loss method [10, 11]. In the image parameter method, *ABCD*-parameters of two port networks are selected appropriately so that when they are cascade-connected, the overall response will satisfy predetermined filter characteristics. The image parameter method is closely related to the analysis of infinite periodic structures. The main disadvantage of this approach is that one cannot realize an arbitrary transfer function. The insertion-loss method, on the other hand, is more flexible and allows the synthesis of arbitrary transfer functions provided that the transfer function is physically realizable. The insertion-loss method is more general and commonly used than the image parameter method. In this chapter, we will concentrate on only filter design by the insertion-loss method.

It is important to note that the most important physical parameter in any filter design is the Q-factor of resonators, which essentially determines how much energy is dissipated in the system. In filter design, one always would like to have resonators with Q-factors as high as possible. It can be shown that for any given regular resonator (i.e., nonsuperconducting), the Q of that resonator is proportional to $K \cdot b \cdot \sqrt{f}$, where K is a constant, b is a linear dimension of the cavity, and f is the frequency [2]. Thus, provided that all other parameters are kept constant, the larger the resonator dimensions, the higher the Q-factor. The crucial point here is that, for distributed resonators, this dependency is not only the result of lowering metallic losses by making the structure bigger; it is also a consequence of the field solutions. In most of the cases, it will be assumed that the resonators are lossless to simplify the design procedures. Once the initial design is obtained, one can always employ computer simulations to quantify and correct for (if possible) the losses.

Electric filters have a very well-established and extensive literature. One can find very good books and chapters on this subject, and a nonexhaustive list is provided in the references [10–24]. Matthaei et al. [10] have written the classic reference on microwave filter design. It provides tables, nomograms, and design techniques for distributed and lumped filters. Zverev [13] is another good reference on filters providing network transformations, extensive tables of component values, and the pole locations of various filters including elliptic ones. For network theory, Valkenburg [15] is a very good reference. It explains the basics of network synthesis, including Darlington's method, which is the anchor of filter design by insertion loss. Chen [16, 17] also explains network synthesis and impedance matching networks. Chen has a good discussion on elliptic functions and elliptic filter design, which is usually hard to find. Rhodes [18] is another fine reference on general filter theory. It separately addresses the amplitude and phase approximations of lumped and distributed networks. For linear algebra and handbook of various mathematical functions, Meyer [25] and Abramowitz [26] are very good references. It is strongly recommended that the reader review these references.

Filter design by the insertion-loss method starts with the selection of the transfer function that needs to be synthesized. Figure 6.1 shows the magnitude and phase characteristics of an ideal lowpass filter. Note that the insertion phase of an ideal filter must be a linear function to allow all components of the signal to travel with the same velocity. Obviously, such an idealized response cannot be achieved in practice. Therefore, it is the aim of filter synthesis to design a filter whose transfer function resembles ideal performance as close as possible (with the exception of equalizers, where filters are used to correct imperfections of the stages before them). Note that the magnitude and phase responses are not equally important in all of the applications. Hence, trade-offs are usually made to approximate one of them better than the other.

We will start our discussion by defining the transfer function of a minimum-phase system. The transfer function of a minimum-phase system is defined in the following form [18]:

$$H(s) = \frac{N(s)}{M(s)} \qquad (6.1)$$

where $N(s)$ and $M(s)$ are Hurwitz polynomials:

$$N(s) \neq 0 \quad M(s) \neq 0 \quad \text{Re}(s) > 0 \qquad (6.2)$$

From the above definition, it is seen that a Hurwitz polynomial has no zeros in the open right-half s plane. A strictly Hurwitz polynomial does not have zeros either in the open right-half s plane or on the imaginary axis. Examples of minimum-phase systems are ladder networks and waveguides, where energy can only travel from input to output along a single path.

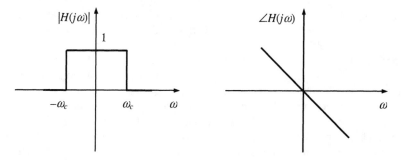

Figure 6.1 Magnitude and phase characteristics of an ideal lowpass filter.

Some lumped minimum and nonminimum phase networks are depicted in Figure 6.2(a–d). Minimum-phase systems have an important place in general filter theory because there exists a unique relationship between the magnitude and phase responses of a minimum-phase filter. It can be shown that the magnitude and phase of a stable minimum-phase system are related by a pair of Hilbert transforms as follows [14, 18]:

$$-\psi(\omega_0) = \frac{\omega_0}{\pi} \int_{-\infty}^{\infty} \frac{\alpha(\omega)}{\omega^2 - \omega_0^2} d\omega \qquad (6.3)$$

$$\alpha(\omega_0) = \alpha(0) + \frac{\omega_0^2}{\pi} \int_{-\infty}^{\infty} \frac{\psi(\omega)}{\omega(\omega^2 - \omega_0^2)} d\omega \qquad (6.4)$$

where

$$H(s) = e^{-\alpha(\omega) + j\psi(\omega)} \qquad (6.5)$$

This result tells us that for minimum-phase systems, the magnitude and phase of the transfer function are related and cannot be determined independently. In other words, for minimum-phase networks, once the magnitude response is determined, the phase response is automatically set, and vice versa. This is an important result because it puts boundaries on the magnitude and phase approximations that one can get from a minimum phase network, such as a simple ladder network. In most of the filter synthesis, only the magnitude of the transfer function is approximated; the resulting nonlinear phase response is tolerated. For instance, it is known that the human ear is relatively insensitive to phase error in the received signal. Thus, for analog voice transmission, magnitude response is the major consideration. However, this is not always the case. In some situations, very high-speed communication network filters for instance, phase linearity is also critical. In such cases, employing a nonminimum phase network is necessary to control the phase shift (alternatively, one could also use an equalizer, but we will not review those techniques here).

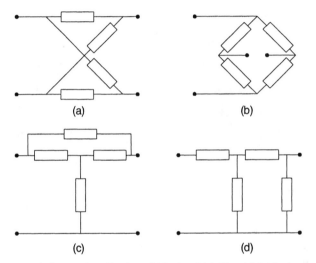

Figure 6.2 Four canonical network realizations: (a) lattice, (b) bridge, (c) bridged-tee, and (d) ladder. The networks shown in (a), (b), and (c) are nonminimum phase networks. The ladder network in (d) is a minimum phase network. (*After:* [14].)

Now, let's assume that the filter network that we would like to realize has the following magnitude response function:

$$|H(s)|^2 = \frac{1}{1+[F(s)]^2} \qquad (6.6)$$

By selecting the function $F(s)$ appropriately, one can realize different filter types. The most commonly used filter types are Butterworth, Chebyshev, Bessel, and elliptic filters. Butterworth and Chebyshev filters approximate the magnitude response of a transfer function, whereas Bessel filters provide maximally flat time delay. To distinguish them further, Chebyshev filters provide steeper stopband characteristic than Butterworth filters for the same filter order at the cost of ripple in the passband, thus it is more practically encountered. Butterworth and Chebyshev filters are also called all-pole filters since all zeros of the transfer function are at infinity or zero frequency (i.e., attenuation does not drop to zero at finite frequencies). Elliptic filters differ from those in the sense that they also place zeros at finite frequencies for maximum selectivity. From the ongoing discussion, it is not difficult to conclude that the pole and zero locations of the transfer function given in (6.6) are very important. For example, zeros on the imaginary axis would improve the filter selectivity and worsen the group delay. Zeros on the real axis, on the other hand, would improve the group delay but worsen selectivity [27, 28]. Essentially, filter synthesis can be thought of as pole-zero craftsmanship of the transfer function. Obviously, the physical feasibility of the resulting network is also a consideration. We will return to this point later in discussing cross-coupled filters.

After selecting the transfer function, a lowpass prototype ladder network as shown in Figure 6.3 can be realized if the transfer function is a minimum-phase one. In the figure, n is the order of the filter. Synthesizing the lowpass prototype network for a given transfer function is an important step because other response types (i.e., bandpass, highpass, and bandstop) can be derived directly from the lowpass prototype by using frequency mapping. Then, frequency transformation and impedance scaling are performed to obtain the actual filter network. Many design procedures for distributed filters start with the determination of lowpass prototype filter values.

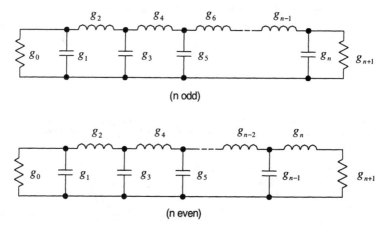

Figure 6.3 Lowpass filter prototype networks. Other filter types are generated from the lowpass prototype network.

At this point, it should again be stressed that one cannot find a ladder network equivalent to that shown in the Figure 6.3 for every $F(s)$. Sometimes usage of ideal transformers or cross-couplings between the circuit elements may be necessary. In that case, more advanced synthesis techniques must be employed [15–17]. In the next sections, we will introduce the basic filter responses and equations for calculating prototype component values.

Butterworth Response

For the Butterworth response, the nth-order filter-transfer function is expressed as follows [16]:

$$|H(j\omega)|^2 = \frac{H_0}{1+(\omega/\omega_c)^{2n}} \tag{6.7}$$

where ω_c is the 3-dB corner frequency. The constant H_0 is the dc attenuation, which may be equal to or less than unity for a passive network. The parameters of the Butterworth filter are given in Figure 6.4. This kind of filter is also known as maximally flat because of its flat passband characteristics. Figure 6.5 shows the insertion loss of the Butterworth filter for different filter orders.

The poles of the Butterworth response are given as [16]

$$s_k = \omega_c e^{j(2k+n-1)\pi/2n} \quad k = 1,2,\ldots,2n \tag{6.8}$$

Note that the poles of the Butterworth filter lie on a circle.

The component values of the lowpass Butterworth prototype network for equal resistive terminations (i.e., $H_0 = 1$) are determined using the following equations (see Figure 6.3):

$$g_0 = 1$$

$$g_k = 2\sin\left[\frac{(2k-1)\pi}{2n}\right], \quad k = 1,2,\ldots,n \tag{6.9}$$

$$g_{n+1} = 1$$

In spite of its simplicity, the Butterworth type of response is not used too often in practice due to insufficient selectivity.

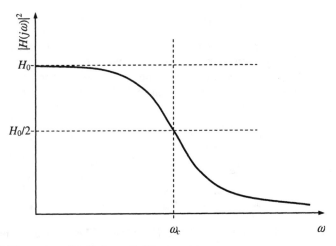

Figure 6.4 Parameters of the Butterworth filter response.

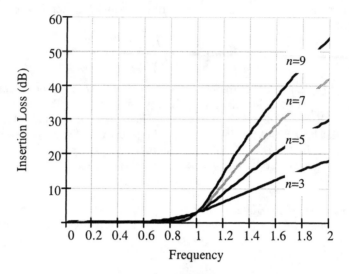

Figure 6.5 Insertion loss of the Butterworth prototype filter for n = 3, 5, 7, and 9.

Chebyshev Response

For the Chebyshev response, the nth-order filter-transfer function is expressed as follows [16]:

$$|H(j\omega)|^2 = \frac{H_0}{1+\varepsilon^2 C_n^2(\omega/\omega_c)} \quad (6.10)$$

$$C_n(\omega) = \cos\left[n\cos^{-1}(\omega)\right] \quad (6.11)$$

where ε and ω_c are ripple magnitude and 3-dB corner frequency, respectively. The constant H_0 is the dc attenuation, which may be equal to or less than unity for a passive network. The parameters of the Chebyshev filter are given in Figure 6.6. It can be shown that the transcendental equation in (6.11) is actually a polynomial whose first few orders are given as

$$\begin{aligned}
C_0(\omega) &= 1 \\
C_1(\omega) &= \omega \\
C_2(\omega) &= 2\omega^2 - 1 \\
&\vdots
\end{aligned} \quad (6.12)$$

$$C_{n+1}(\omega) = 2\omega C_n(\omega) - C_{n-1}(\omega)$$

Figure 6.7 shows the insertion loss of the Chebyshev filter for different filter orders.

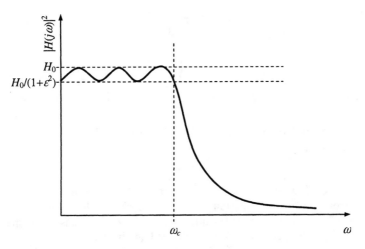

Figure 6.6 Parameters of the Chebyshev filter response.

The poles of the Chebyshev function are given as [16]

$$s_k = \omega_c(\sigma_k + j\omega_k) \quad k = 1,2,\ldots,2n \qquad (6.13)$$

where

$$\sigma_k = -\sinh(a)\sin\left[\frac{(2k-1)\pi}{2n}\right] \qquad (6.14)$$

$$\omega_k = \cosh(a)\cos\left[\frac{(2k-1)\pi}{2n}\right] \qquad (6.15)$$

$$a = \frac{1}{n}\sinh^{-1}\left(\frac{1}{\varepsilon}\right) \qquad (6.16)$$

Note that the poles of the Chebyshev filter lie on an ellipse.

The component values of the Chebyshev lowpass prototype network are determined using the following equations (see Figure 6.3):

$$g_0 = 1$$

$$g_1 = \frac{2a_1}{\gamma}$$

$$g_k = \frac{4a_{k-1}a_k}{b_{k-1}g_{k-1}}, \quad k = 2,3,\ldots,n \qquad (6.17)$$

$$g_{n+1} = 1, \quad \text{for } n \text{ odd}$$

$$g_{n+1} = \coth^2(\beta/4), \quad \text{for } n \text{ even}$$

where

$$a_k = \sin\left[\frac{(2k-1)\pi}{2n}\right], \quad k = 1,2,\ldots,n$$

$$b_k = \gamma^2 + \sin^2\left(\frac{k\pi}{n}\right), \quad k = 1,2,\ldots,n \tag{6.18}$$

$$\beta = \ln\left[\coth\left(\frac{A}{2 \cdot 8.686}\right)\right] \quad \gamma = \sinh\left(\frac{\beta}{2n}\right)$$

where A is the passband ripple in decibels. The relation between A and ε is given by the following expression:

$$A = 10\log\left(1 + \varepsilon^2\right) \tag{6.19}$$

Note that depending on the order of the filter, the termination resistance might be different from the source resistance. The Chebyshev filter is more commonly employed in practice than the Butterworth filter. The reason is the higher selectivity (i.e., sharper filter skirts) provided in the Chebyshev filter. Another observation is that the poles of the Chebyshev filter can be obtained from the poles of the Butterworth filter by multiplying the real and imaginary parts of the Butterworth poles with $\sinh(a)$ and $\cosh(a)$, respectively. There is also inverse Chebyshev response which has maximally flat passband and equal-ripple stopband characteristics.

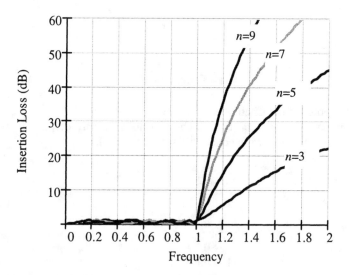

Figure 6.7 Insertion loss of the Chebyshev prototype filter for $n = 3, 5, 7$, and 9 ($\varepsilon = 0.5$).

Elliptic Response

Before explaining the elliptic filter response, it would be instructive to give definitions of elliptic integrals and functions that will be used in defining the elliptic response. The following integral is called the Legendre standard form of the elliptic integral of the first kind of modulus k [17]:

$$u \equiv F(k,\phi) = \int_0^x \frac{dx}{\left(1-x^2\right)^{1/2}\left(1-k^2x^2\right)^{1/2}} = \int_0^\phi \frac{d\phi}{\left(1-k^2\sin^2\phi\right)^{1/2}} \qquad (6.20)$$

Then, the Jacobian elliptic sine function of modulus k (i.e., the elliptic function) is defined as follows:

$$x \equiv sn(u,k) \qquad (6.21)$$

Thus, the result of an elliptic function is the x limit of the integral defined in (6.20), which produces the u for a given k. The values of the elliptic functions can be found in mathematics handbooks. Abramowitz and Stegun provide a good reference for mathematical functions and tables [26]. Alternatively, elliptic functions can easily be calculated using numerical algorithms. Since the application of Jacobian elliptic functions to filter design is first suggested by Cauer, elliptic filters are also called Cauer-type filters. Another function that will be used in the elliptic filter response is the complete elliptic integral of the first kind of modulus k, which is defined as follows:

$$K \equiv K(k) = F(k,\pi/2) \qquad (6.22)$$

For an elliptic response filter, the parameter k is a measure of the steepness of the attenuation in the transitional band and it is called the selectivity factor:

$$k = \frac{\omega_c}{\omega_s} \qquad (6.23)$$

where ω_c and ω_s are the cutoff frequency and edge of the stopband, respectively. The edge of the stopband is defined as the frequency where attenuation is dropped to the value of the maxima in the stopband. The parameters of the elliptic function are shown in Figure 6.8.

At this point, we are ready to define the transfer functions of an elliptic filter. The nth-order elliptic-filter transfer function is expressed as follows [16]:

$$|H(j\omega)|^2 = \frac{H_n}{1+\varepsilon^2 F_n^2(\omega/\omega_c)} \qquad (6.24)$$

where n is the order of the filter. The constants H_n and ε have the same interpretation as they do in the Chebyshev response. In addition to these parameters, one should also specify steepness of the elliptic filter, k, using (6.23). The function $F_n^2(\omega/\omega_c)$ can be expressed by the appropriate rational function whose zeros lie within the passband and whose poles lie in the stopband for odd and even filter orders as follows [17]:

n odd:

$$F_n(\omega) = H_0 \frac{\omega(\omega_1^2 - \omega^2) \cdot (\omega_2^2 - \omega^2) \cdot \ldots \cdot (\omega_q^2 - \omega^2)}{(1 - k^2 \omega_1^2 \omega^2) \cdot (1 - k^2 \omega_2^2 \omega^2) \cdot \ldots \cdot (1 - k^2 \omega_q^2 \omega^2)} \tag{6.25}$$

$$\omega_m = \mathrm{sn}(2mK/n, k), \quad m = 1, 2, \ldots, q, \quad q = \frac{1}{2}(n-1) \tag{6.26}$$

n even:

$$F_n(\omega) = H_0 \frac{(\omega_1^2 - \omega^2) \cdot (\omega_2^2 - \omega^2) \cdot \ldots \cdot (\omega_q^2 - \omega^2)}{(1 - k^2 \omega_1^2 \omega^2) \cdot (1 - k^2 \omega_2^2 \omega^2) \cdot \ldots \cdot (1 - k^2 \omega_q^2 \omega^2)} \tag{6.27}$$

$$\omega_m = \mathrm{sn}[(2m-1)K/n, k], \quad m = 1, 2, \ldots, q, \quad q = \frac{1}{2}n \tag{6.28}$$

where

$$H_0 = \left(\frac{k^n}{k_1} \right)^{1/2} \tag{6.29}$$

$$k_1 = k^n \left[\frac{(1 - \omega_1^2) \cdot (1 - \omega_2^2) \cdot \ldots \cdot (1 - \omega_q^2)}{(1 - k^2 \omega_1^2) \cdot (1 - k^2 \omega_2^2) \cdot \ldots \cdot (1 - k^2 \omega_q^2)} \right] \tag{6.30}$$

Figures 6.9 to 6.12 show the insertion loss of the elliptic filter for different filter orders.

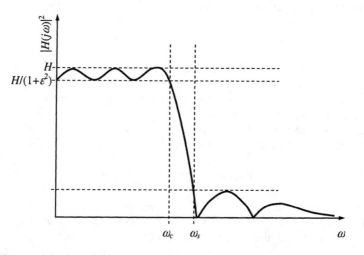

Figure 6.8 Parameters of the elliptic filter response.

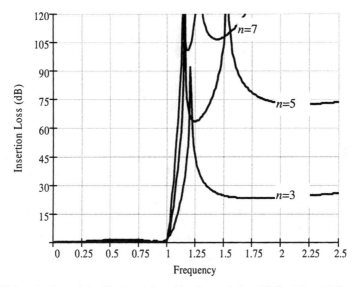

Figure 6.9 Insertion loss of the elliptic prototype filter for $n = 3, 5,$ and 7 ($k = 0.9$, $\varepsilon = 0.5$).

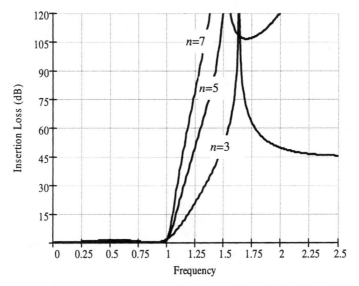

Figure 6.10 Insertion loss of the elliptic prototype filter for $n = 3, 5,$ and 7 ($k = 0.7$, $\varepsilon = 0.5$).

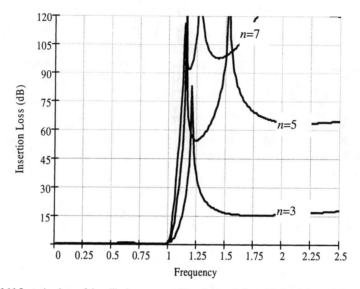

Figure 6.11 Insertion loss of the elliptic prototype filter for $n = 3$, 5, and 7 ($k = 0.9$, $\varepsilon = 0.3$).

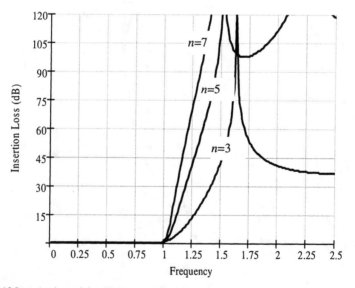

Figure 6.12 Insertion loss of the elliptic prototype filter for $n = 3$, 5, and 7 ($k = 0.7$, $\varepsilon = 0.3$).

6.1.1 Filter Transformations

In the previous sections, the design of the lowpass prototype networks for different transfer function responses was introduced. Those lowpass networks are normalized with respect to a 1-ohm load and source resistances and have a 1 rad/s cutoff frequency. Therefore, to obtain the desired frequency and impedance level, one needs to employ the proper transformations to scale the component values of the prototype networks. It will also be necessary to convert the lowpass networks into other types of filters (highpass, bandpass, and so forth) through frequency mapping. To accomplish the frequency mapping, a suitable function, $f(s)$, is found first, which maps the passband and stopband of the lowpass filter into the passband and stopband of the new filter. Then, s in the original filter is replaced by $f(s)$. Finally, components of the new filter are determined by inspecting the resulting impedance functions. Note that this frequency transformation is mathematically exact. However, the final component values may not always be realizable practically, especially for very narrow bandpass filters. In that case, additional network transformations might be necessary to convert extremely large or extremely small component values. Note that since the ideal resistors are frequency invariant, they are not affected by this frequency transformation.

Figure 6.13(a–d) shows four different fundamental ladder-type filter structures, namely, lowpass, highpass, bandpass, and bandstop. Simple ladder-type networks are very convenient in practice because they do not have bulky transformers. But, they are minimum-phase networks and have limitations in terms of realizable transfer function, as indicated before. The expressions for frequency mapping, impedance scaling, and frequency scaling to find the component values of each of these four fundamental filter structures are given in (6.31) through (6.34). In the expressions, the first equation in each set defines the required frequency mapping to obtain the desired response. Z_0 is the impedance level, g_k are the component values of the prototype lowpass filter, ω_c is the cutoff frequency of the prototype lowpass filter (which is 1 rad/s), and ω_0 is the frequency parameter of the new filter. If the new filter is lowpass or highpass, then ω_0 is the corner frequency; if the new filter is bandpass or bandstop, then ω_0 is the center frequency defined by two corner frequencies ω_1 and ω_2. A bandwidth parameter BW is also defined for the last two types of filters.

At this point, it would be beneficial to summarize what has been introduced so far. The first step in lumped-filter design is to select the desired filter function. Then, a proper filter topology is selected to implement the function. For the Chebyshev and Butterworth responses, one can utilize ladder networks where closed-form expressions are available for the component values. The next step is to determine the component values of the lowpass prototype network. Finally, impedance and frequency transformations are applied to the prototype network to obtain the desired filter. We will demonstrate the details of the design procedures in the next few examples.

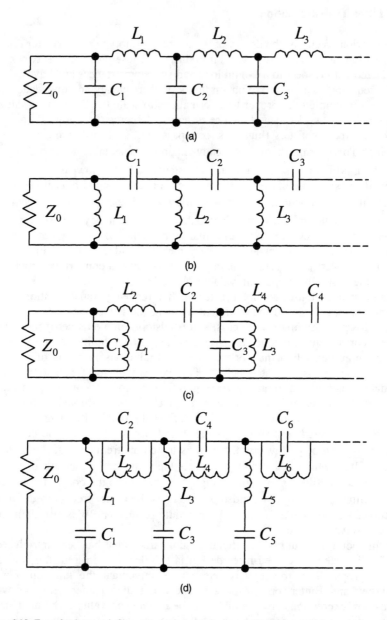

Figure 6.13 Four fundamental filter topologies: (a) lowpass, (b) highpass, (c) bandpass, and (d) bandstop. See (6.31) through (6.34) for obtaining component values of the lowpass prototype filter networks.

Lowpass filter:

$$\frac{\omega}{\omega_c} \leftarrow \frac{\omega}{\omega_0}$$

$$L_{k/2} = g_k \frac{Z_0}{\omega_0} \quad k = 2,4,6,\ldots \tag{6.31}$$

$$C_{(k+1)/2} = g_k \frac{1}{Z_0 \omega_0} \quad k = 1,3,5,\ldots$$

Highpass filter:

$$\frac{\omega}{\omega_c} \leftarrow -\frac{\omega_0}{\omega}$$

$$L_{k/2} = \frac{Z_0}{g_k \omega_0} \quad k = 2,4,6,\ldots \tag{6.32}$$

$$C_{(k+1)/2} = \frac{1}{g_k Z_0 \omega_0} \quad k = 1,3,5,\ldots$$

Bandpass filter:

$$\frac{\omega}{\omega_c} \leftarrow \frac{\omega_0}{BW}\left(\frac{\omega}{\omega_0} - \frac{\omega_0}{\omega}\right)$$

$$\omega_0 = \sqrt{\omega_1 \omega_2} \quad BW = \omega_2 - \omega_1 \tag{6.33}$$

$$L_k = g_k \frac{Z_0}{BW} \quad C_k = \frac{BW}{g_k Z_0 \omega_0^2} \quad k = 2,4,6,\ldots$$

$$L_k = \frac{BW Z_0}{g_k \omega_0^2} \quad C_k = g_k \frac{1}{BW Z_0} \quad k = 1,3,5,\ldots$$

Bandstop filter:

$$\frac{\omega_c}{\omega} \leftarrow -\frac{\omega_0}{BW}\left(\frac{\omega}{\omega_0} - \frac{\omega_0}{\omega}\right)$$

$$\omega_0 = \sqrt{\omega_1 \omega_2} \quad BW = \omega_2 - \omega_1 \tag{6.34}$$

$$L_k = g_k \frac{Z_0 BW}{\omega_0^2} \quad C_k = \frac{1}{g_k Z_0 BW} \quad k = 2,4,6,\ldots$$

$$L_k = \frac{Z_0}{g_k BW} \quad C_k = g_k \frac{BW}{Z_0 \omega_0^2} \quad k = 1,3,5,\ldots$$

Example 6.1

In this example, we will design a fifth-order Chebyshev lowpass filter with a corner frequency of 1.0 GHz and a passband ripple of 0.1 dB ($\varepsilon = 0.1526$). The input and output impedances are selected as 50 ohms.

The first step is to determine the prototype circuit values. Using expressions given in (6.17), the following prototype values are found:

$$g_1 = 1.147 \quad g_2 = 1.371$$
$$g_3 = 1.975 \quad g_4 = 1.371$$
$$g_5 = 1.147 \quad g_6 = 1.0$$

Then, using (6.31), we convert the prototype values to actual circuit component values:

$$C_1 = 3.65\,\text{pF} \quad L_1 = 10.91\,\text{nH}$$
$$C_2 = 6.29\,\text{pF} \quad L_2 = 10.91\,\text{nH}$$
$$C_3 = 3.65\,\text{pF}$$

To test the design, the filter circuit is entered into a circuit simulator as shown in Figure 6.14. The simulation results are shown in Figure 6.15 and 6.16. It can be seen that the return loss is better than 15 dB in the passband, and the insertion loss is 20 dB at 1.5 GHz in the stopband. One of the observations is that the group delay is not constant as expected. The consequence of this is that some frequency components travel at different speeds than others through the filter. This might cause problems for high-speed communication networks because it will affect transmitted pulse shapes.

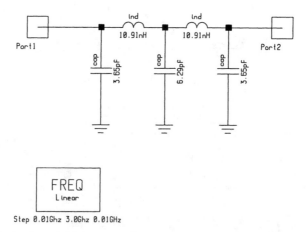

Figure 6.14 Circuit schematic of the lowpass filter given in Example 6.1.

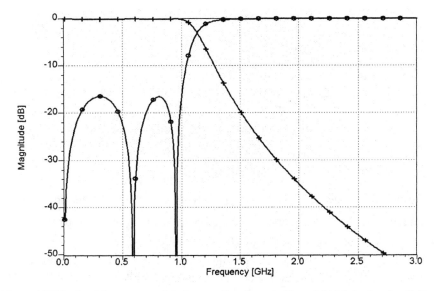

Figure 6.15 Simulated S-parameters of the lowpass filter shown in Figure 6.14 (circle: S_{11}; plus: S_{21}).

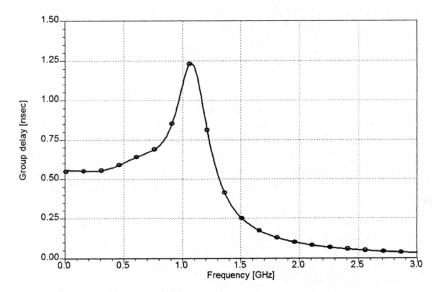

Figure 6.16 Simulated group delay of the lowpass filter shown in Figure 6.14.

Nevertheless, the plot given in Figure 6.16 is very typical for many practical filter circuits. Note that there are other filter polynomials that provide flat group delay response, which we will demonstrate.

It would also be informative to study the pole locations of the designed filter. Separation of the poles of the transfer function in the simulation results or measurements is one of the indications of a successful filter design. The poles of a fifth-order 0.1-dB ripple Chebyshev filter are found using (6.13):

$$p_{1,2} = -0.16653368 \pm j1.08037201$$

$$p_{3,4} = -0.43599085 \pm j0.66770662$$

$$p_5 = -0.53891143$$

Note that (6.13) gives the poles of $H(s) \cdot H(-s)$, or $|H(s)|^2$. But since we are interested only in stable systems, we assign poles with negative real parts to $H(s)$ and those with positive real parts to $H(-s)$. Thus, we only need to compute poles on the left-half of the s-plane. To determine the frequency points corresponding to the above poles, we first normalize the poles by $\cosh[1/n \cdot \sinh^{-1}(1/\varepsilon)] = 1.136$. Then, frequency transformation must be performed as follows:

$$f' = \frac{f}{f_0} \quad \Rightarrow \quad f = f_0 f'$$

By inserting imaginary parts of the poles into the above equation, the following frequency points are determined:

$$f_1 = 0$$
$$f_2 = 0.59 \text{ GHz}$$
$$f_3 = 0.95 \text{ GHz}$$

By examining Figure 6.15, one can verify that the return loss goes indeed to zero (i.e., $-\infty$ dB) at the above frequencies, indicating that the poles of the filter are correctly captured. The reader might also wonder why we have selected only the imaginary parts of the poles to determine the zeros of the reflection coefficient. The reason is as follows: since the order of the filter is odd, we know that the termination resistance equals the source resistance. This makes the dc attenuation of the filter unity. Then, we calculate magnitude square of the reflection coefficient from:

$$|\rho(j\omega)|^2 = 1 - |H(j\omega)|^2 = 1 - \frac{1}{1 + \varepsilon^2 C_n^2(\omega/\omega_c)} = \frac{\varepsilon^2 C_n^2(\omega/\omega_c)}{1 + \varepsilon^2 C_n^2(\omega/\omega_c)}$$

Now, we are interested in the roots of the polynomial $\varepsilon^2 C_n^2(\omega/\omega_c)$. It can be shown that the roots of $\varepsilon^2 C_n^2(\omega/\omega_c)$ are the imaginary parts of the roots of $1 + \varepsilon^2 C_n^2(\omega/\omega_c)$ normalized by $\cosh[1/n \cdot \sinh^{-1}(1/\varepsilon)]$. Since the roots of the latter term are the poles of the filter, the zeros of the reflection coefficient can readily be determined from the imaginary parts of the poles. Note that this is only applicable in the case of unity dc attenuation, that is, when the source and load terminations are equal. We will return to the evaluation of the reflection coefficient from the transfer function later when we introduce Darlington's method.

Example 6.2

In this example, we will design a fifth-order Chebyshev lowpass filter with a cutoff frequency of 1.0 GHz and passband ripple of 0.5 dB ($\varepsilon = 0.3493$). The input and output impedances are selected as 50 ohms.

The first step is to determine the prototype circuit values. Using expressions given in (6.17), the following prototype values are found:

$$g_1 = 1.706 \quad g_2 = 1.23$$
$$g_3 = 2.541 \quad g_4 = 1.23$$
$$g_5 = 1.706 \quad g_6 = 1.0$$

Then, using (6.31), we convert the prototype values into actual circuit component values:

$$C_1 = 5.43\,\text{pF} \quad L_1 = 9.79\,\text{nH}$$
$$C_2 = 8.1\,\text{pF} \quad L_2 = 9.79\,\text{nH}$$
$$C_3 = 5.43\,\text{pF}$$

To test the design, the filter circuit is entered to a circuit simulator as shown in Figure 6.17. The simulation results are shown in Figures 6.18 and 6.19. It can be seen that the return loss is better than 10 dB in the passband, and the insertion loss is 28 dB at 1.5 GHz in the stopband.

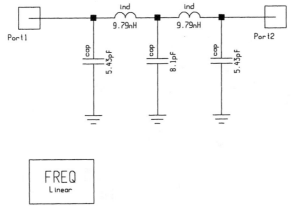

Figure 6.17 Circuit schematic of the lowpass filter given in Example 6.2.

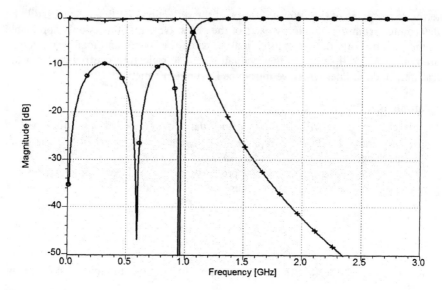

Figure 6.18 Simulated S-parameters of the lowpass filter shown in Figure 6.17 (circle: S_{11}; plus: S_{21}).

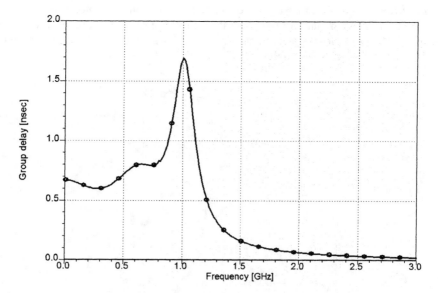

Figure 6.19 Simulated group delay of the lowpass filter shown in Figure 6.17.

The most important conclusion that can be drawn from this example is that this filter has worse group delay than that given in the previous example at the edge of the passband. This is because the filter here has sharper cutoff with respect to the previous filter. For minimum-phase networks, the group delay deteriorates more as the magnitude response of the filter approaches the ideal brick response.

Note that for a lossless two-port network, the amount of bandpass ripple and return loss is linked through the following equation:

$$|S_{11}|^2 + |S_{21}|^2 = 1$$

which directly follows from the properties of S-matrixes for lossless networks. In practice, the filters are typically designed for 20-dB return loss or better and the designer usually makes the trade-off between the return loss versus rejection rate.

Example 6.3

In this example, a seventh-order Chebyshev bandpass filter centered at 1.0 GHz with a 400-MHz bandwidth and 0.1-dB ($\varepsilon = 0.1526$) bandpass ripple is going to be realized. The input and output impedances are selected as 50 ohms.

The first step is to determine the prototype circuit values. Using expressions given in (6.17), the following prototype values are found:

$$g_1 = 1.181 \quad g_2 = 1.423$$
$$g_3 = 2.097 \quad g_4 = 1.573$$
$$g_5 = 2.097 \quad g_6 = 1.423$$
$$g_7 = 1.181 \quad g_8 = 1.0$$

Then, using (6.33), we convert the prototype values into actual circuit component values:

$$C_1 = 9.40 \text{ pF} \quad L_1 = 2.81 \text{ nH}$$
$$C_2 = 0.93 \text{ pF} \quad L_2 = 28.31 \text{ nH}$$
$$C_3 = 16.7 \text{ pF} \quad L_3 = 1.58 \text{ nH}$$
$$C_4 = 0.84 \text{ pF} \quad L_4 = 31.3 \text{ nH}$$
$$C_5 = 16.7 \text{ pF} \quad L_5 = 1.58 \text{ nH}$$
$$C_6 = 0.93 \text{ pF} \quad L_6 = 28.31 \text{ nH}$$
$$C_7 = 9.40 \text{ pF} \quad L_7 = 2.81 \text{ nH}$$

To test the design, the filter circuit is entered into a circuit simulator as shown in Figure 6.20. The simulation results are shown in Figures 6.21 and 6.22. It can be seen that the return is better than 15 dB in the passband, and insertion loss in the stopband is 40 dB at 0.72 GHz and 1.35 GHz. One might wonder why the group delay is worse at the lower edge of the passband than the upper edge. The answer is that because of the geometric symmetry (i.e., $\omega_0 = \sqrt{\omega_1 \omega_2}$) employed in the bandpass transformation, the filter has unequal attenuation slopes at the lower and upper edges of the response (the attenuation increases faster at the lower edge). This makes the group delay worse at the lower edge of the response [14]. This is one of the disadvantages of the frequency mapping employed. We will return to this point later when we introduce filter design by coupling.

We will study the pole locations of this filter as well. The poles of the seventh-order 0.1-dB ripple Chebyshev filter are found using (6.13):

$$p_{1,2} = -0.08384097 \pm j1.04183333$$

$$p_{3,4} = -0.23491716 \pm j0.83548546$$

$$p_{5,6} = -0.33946514 \pm j0.46365945$$

$$p_7 = -0.37677788$$

To determine the frequency points corresponding to the above poles, we first normalize the poles by $\cosh\left[1/n \cdot \sinh^{-1}(1/\varepsilon)\right] = 1.069$. Then, frequency transformation must be performed as follows:

$$f' = \frac{f_0}{B}\left(\frac{f}{f_0} - \frac{f_0}{f}\right) \quad \Rightarrow \quad f = \frac{f'B \pm \sqrt{(f'B)^2 + 4f_0^2}}{2}$$

By inserting the imaginary parts of the poles into the above equation, the following frequency points are determined:

$$f_1 = 1.194\,\text{GHz} \quad f_2 = 1.149\,\text{GHz}$$

$$f_3 = 1.071\,\text{GHz} \quad f_4 = 0.98\,\text{GHz}$$

$$f_5 = 0.897\,\text{GHz} \quad f_6 = 0.836\,\text{GHz}$$

$$f_7 = 0.804\,\text{GHz}$$

By examining Figure 6.21, one can verify that the return loss indeed goes to zero (i.e., $-\infty$ dB) at the above frequencies, indicating that the poles of the filter are correctly captured.

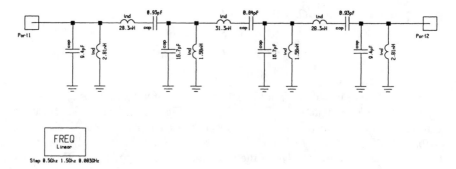

Figure 6.20 Circuit schematic of the bandpass filter given in Example 6.3.

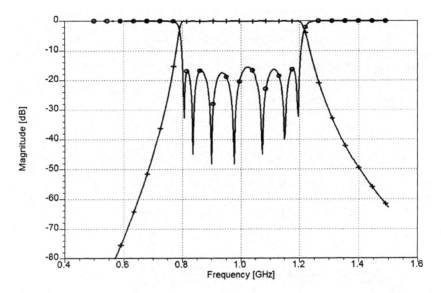

Figure 6.21 Simulated S-parameters of the bandpass filter shown in Figure 6.20 (circle: S_{11}; plus: S_{21}).

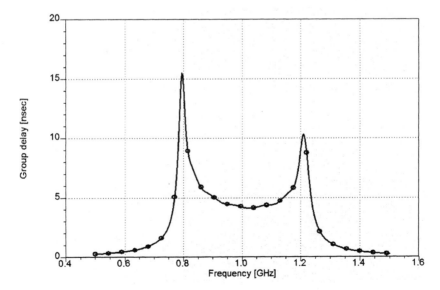

Figure 6.22 Simulated group delay of the bandpass filter shown in Figure 6.20.

Modern Microwave Circuits

Example 6.4

In this example, we will design a fifth-order maximally flat delay response lowpass filter with a cutoff frequency of 1.0 GHz. The input and output impedances are selected as 50 ohms.

The first step is to determine the prototype circuit values. Using tabulated values, the following prototype values are found [13]:

$$g_1 = 0.1743 \quad g_2 = 0.5072$$
$$g_3 = 0.8040 \quad g_4 = 1.1110$$
$$g_5 = 2.2582 \quad g_6 = 1.0$$

Then, frequency and impedance transformations are done to convert the prototype filter to the actual filter. To test the design, the filter circuit is entered into a circuit simulator as shown in Figure 6.23. Note that in this particular example, the frequency and impedance scaling are handled in the schematic by using simulator variables. This technique is very handy if it is required to have different designs with different cutoff frequencies based on the same topology.

Simulation results are shown in Figures 6.24 and 6.25. The first observation is the maximally flat group delay response, which is almost constant across the passband. If we compare the group delay of this filter with the group delay of the fifth-order Chebyshev filter, one can see that the first one is superior to the latter. However, the selectivity of this filter is very poor. In fact, in Figure 6.24, it is visually very difficult to distinguish the edge of the passband by just looking to the transmission characteristic (i.e., S_{21}).

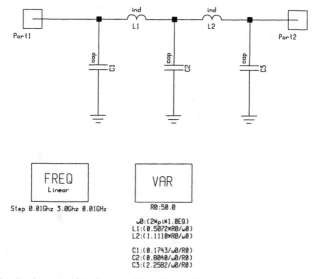

Figure 6.23 Circuit schematic of the lowpass filter given in Example 6.4.

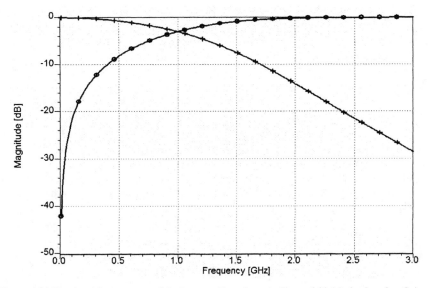

Figure 6.24 Simulated S-parameters of the lowpass filter shown in Figure 6.23 (circle: S_{11}; plus: S_{21}).

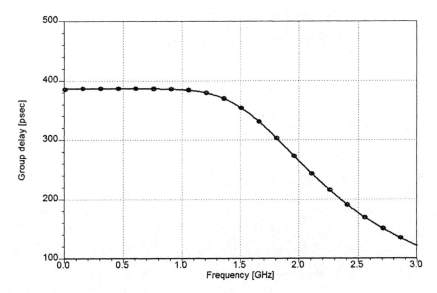

Figure 6.25 Simulated group delay of the lowpass filter shown in Figure 6.23.

This example clearly demonstrates the relationship between magnitude and phase responses for minimum-phase filters. As indicated before, it is not possible to control both responses independently for minimum-phase filters because of the Hilbert transform's relationship between magnitude and phase responses.

6.1.2 Norton's Transformations

Norton's transformations present two important network conversions that are extremely useful in altering the filter elements when necessary [13, 15]. These transformations, shown in Figure 6.26, are especially handy in reducing values of the filter elements to practically manageable levels at the expense of an increased number of components when the difference between the biggest and smallest elements in a filter network is relatively large. Note that depending on the value of K, some of the element values can be negative. However, it is usually possible to absorb these negative elements in the other filter elements.

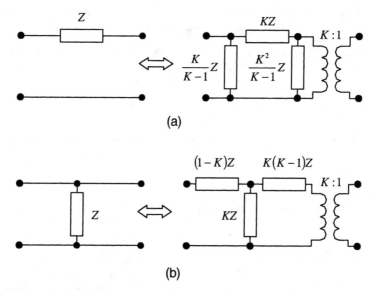

Figure 6.26 Norton's (a) first and (b) second transformations.

For the sake of demonstration, a typical application of Norton's transformation to a section of a bandpass filter is given in Figure 6.27. The transformation shown in the figure does take into account the absorption of negative elements in other components. Therefore, the constant K must be selected in a such a way that all component values are positive in the network. As can be seen from the figure, by selecting the K constant as less than unity, the inductor L_2 is scaled down, while the capacitor C_2 is scaled up. This helps in reducing the wide range of component values in narrow bandpass filters, which are obtained directly from lowpass prototype filters by frequency mapping. We will demonstrate this in the next example. Note that by proper selection of the K factor, the capacitor C_1 in the top circuit can also be annihilated completely.

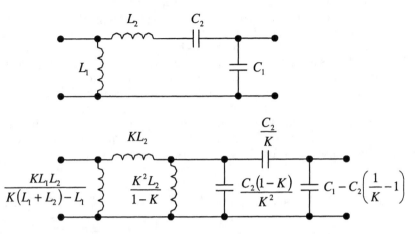

Figure 6.27 Application of Norton's transformations to a bandpass filter section. Top and bottom figures show the networks before and after transformation, respectively.

Example 6.5

In this example, we will design a narrowband bandpass filter and apply the Norton's transformation as depicted in Figure 6.27 to alter some of the component values to make the filter more practical. The filter that is going to be realized is a fifth-order Chebyshev bandpass filter centered at 1 GHz with a 100-MHz bandwidth and 0.1-dB ($\varepsilon = 0.1526$) bandpass ripple. The input and output impedances are selected as 50 ohms.

As usual, the first step is to determine the prototype circuit values. Using expressions given in (6.17), the following prototype values are found:

$$g_1 = 1.147 \quad g_2 = 1.371$$
$$g_3 = 1.975 \quad g_4 = 1.371$$
$$g_5 = 1.147 \quad g_6 = 1.0$$

Then, using (6.33), we convert the prototype values into actual circuit component values:

$$C_1 = 18.25 \text{ pF} \quad L_1 = 1.402 \text{ nH}$$
$$C_2 = 0.469 \text{ pF} \quad L_2 = 54.56 \text{ nH}$$
$$C_3 = 32.43 \text{ pF} \quad L_3 = 0.814 \text{ nH}$$
$$C_4 = 0.469 \text{ pF} \quad L_4 = 54.56 \text{ nH}$$
$$C_5 = 18.25 \text{ pF} \quad L_5 = 1.402 \text{ nH}$$

To test the design, the filter circuit is entered into a circuit simulator as shown in Figure 6.28. The simulation results of the filter is shown in Figure 6.30. From the figure, one can observe that return is better than 15 dB in the passband, and insertion loss in the stopband is 40 dB at 0.8 GHz and 1.25 GHz.

Now, as can be seen from Figure 6.28, the ratio of the biggest inductor value to the smallest one is more than 60, which might be quite impractical. Here, Norton's transformation can be applied to reduce this high ratio. The circuit given in Figure 6.29 shows the resulting filter after applying Norton's transformation. Simulation results of the new filter are given in Figure 6.31. In the transformation process, K is selected as 0.1. As can be seen from the circuit in Figure 6.29, the number of filter elements is increased. However, the biggest value of inductance is reduced by approximately tenfold. Although the biggest capacitor value is also increased, the increase is approximately 50%, which is not too important because implementing capacitors is typically easier than implementing inductors.

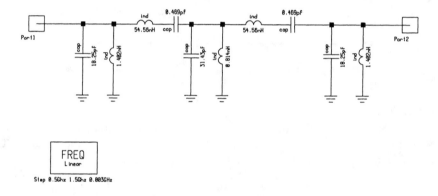

Figure 6.28 Circuit schematic of the bandpass filter given in Example 6.5. Note that inductor values in the series resonators are relatively large, a consequence of narrowband design.

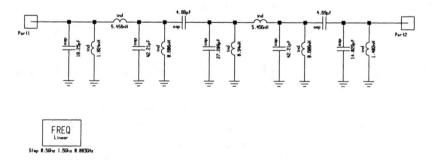

Figure 6.29 Circuit schematic of the bandpass filter given in Example 6.5 after applying Norton's transformation. Note the reduction in the maximum inductor values.

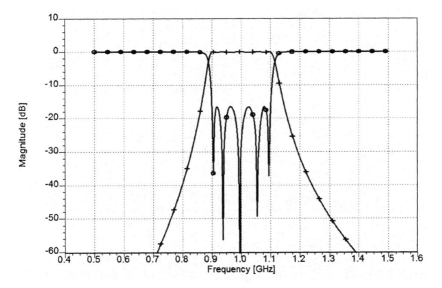

Figure 6.30 Simulated S-parameters of the bandpass filter shown in Figure 6.28 (circle: S_{11}; plus: S_{21}).

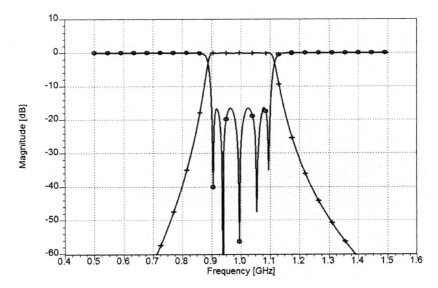

Figure 6.31 Simulated S-parameters of the bandpass filter shown in Figure 6.29 (circle: S_{11}; plus: S_{21}).

6.1.3 Darlington's Method

Darlington's method is a powerful technique for synthesizing arbitrary networks satisfying a given transfer function [15, 29]. Synthesis of matching networks using this technique was already demonstrated in Chapter 1. In this section, we will demonstrate the application of Darlington's method to filter synthesis.

Although the method is general, the resulting network may contain ideal transformers as shown in Chapter 1, which is not always practical. However, for all-pole transfer functions, which includes the Chebyshev and Butterworth responses, the method yields ladder networks after a continuous fraction expansion. Since closed-form expressions are available for these type of responses, the method's real power lies in the synthesis of arbitrary transfer functions, which we will leverage in explaining the design of cross-coupled filters. Nevertheless, application of the method to these filter classes, as will be demonstrated in the next example, will give us a good understanding of Darlington's synthesis method.

Example 6.6

In this example, we design a fifth-order Chebyshev lowpass filter with a 0.1-dB ripple using Darlington's synthesis method. Input and output impedances are selected as 1 ohm, and filter cutoff frequency is selected as 1 rad/s.

We will start by giving the fifth-order Chebyshev transfer function as follows:

$$|H(j\omega)|^2 = \frac{1}{1 + \varepsilon^2 \left[16\omega^5 - 20\omega^3 + 5\omega\right]^2} \qquad (6.35)$$

where $\varepsilon = 0.15262042$ for a 0.1-dB passband ripple. Then, $|H(s)|^2$ is found by replacing $j\omega$ with s:

$$|H(s)|^2 = \frac{1}{1 - \varepsilon^2 \left[-16s^5 - 20s^3 - 5s\right]^2} \qquad (6.36)$$

The next step is to determine the magnitude square of the reflection coefficient. This is achieved using the above transfer function specification as follows:

$$
\begin{aligned}
|\rho(s)|^2 &= \rho(s) \cdot \rho(-s) = 1 - |H(s)|^2 \\
&= \frac{\varepsilon^2 \left[-16s^5 - 20s^3 - 5s\right]^2}{1 - \varepsilon^2 \left[-16s^5 - 20s^3 - 5s\right]^2} \\
&= \frac{\left[-16s^5 - 20s^3 - 5s\right]^2}{\frac{1}{\varepsilon^2} - \left[-16s^5 - 20s^3 - 5s\right]^2}
\end{aligned}
\qquad (6.37)
$$

Now, to determine $\rho(s)$, we first find the zeros of the numerator and denominator polynomials. Then, by selecting the zeros on the left-half s-plane, an expression for $\rho(s)$ is found as follows:

$$\rho(s) = \pm \frac{16s(s+z_1)\cdot(s+z_2)\cdot(s+z_3)\cdot(s+z_4)}{16(s+y_1)\cdot(s+y_2)\cdot(s+y_3)\cdot(s+y_4)\cdot(s+y_5)} \qquad (6.38)$$

where

$$z_1 = 0.5878j \quad z_2 = -0.5878j$$
$$z_3 = 0.9511j \quad z_4 = -0.9511j$$
$$y_1 = 0.5389 \quad y_2 = 0.1665+1.0804j \quad y_3 = 0.1665-1.0804j$$
$$y_4 = 0.4360+0.6677j \quad y_5 = 0.4360-0.6677j$$

Then, input impedance $Z(s)$ can be found as follows:

$$Z(s) = \frac{1-\rho(s)}{1+\rho(s)} \qquad (6.39)$$

$$= \frac{27.9024s^4 + 24.3289s^3 + 38.3512s^2 + 17.9686s + 6.5523}{32s^5 + 27.9024s^4 + 64.3321s^3 + 38.3512s^2 + 27.9700s + 6.5523}$$

We know that it is possible to realize $Z(s)$ as the input impedance of a ladder network terminated with a resistive load because the transfer function given in (6.35) is an all-pole function.

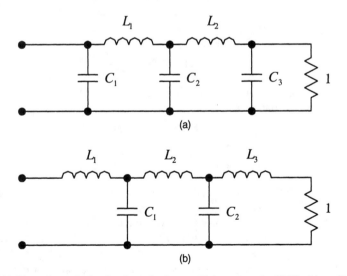

(a)

(b)

Figure 6.32 Network topologies for the reflection coefficient given in (6.38): (a) positive and (b) negative sign selected.

To find the component values of the ladder network, continuous fraction expansion of $Z(s)$ is carried out as follows:

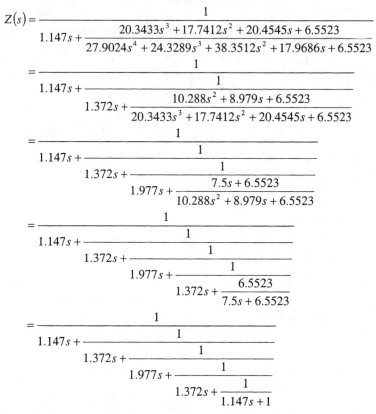

$$Z(s) = \cfrac{1}{1.147s + \cfrac{20.3433s^3 + 17.7412s^2 + 20.4545s + 6.5523}{27.9024s^4 + 24.3289s^3 + 38.3512s^2 + 17.9686s + 6.5523}}$$

$$= \cfrac{1}{1.147s + \cfrac{1}{1.372s + \cfrac{10.288s^2 + 8.979s + 6.5523}{20.3433s^3 + 17.7412s^2 + 20.4545s + 6.5523}}}$$

$$= \cfrac{1}{1.147s + \cfrac{1}{1.372s + \cfrac{1}{1.977s + \cfrac{7.5s + 6.5523}{10.288s^2 + 8.979s + 6.5523}}}}$$

$$= \cfrac{1}{1.147s + \cfrac{1}{1.372s + \cfrac{1}{1.977s + \cfrac{1}{1.372s + \cfrac{6.5523}{7.5s + 6.5523}}}}}$$

$$= \cfrac{1}{1.147s + \cfrac{1}{1.372s + \cfrac{1}{1.977s + \cfrac{1}{1.372s + \cfrac{1}{1.147s + 1}}}}}$$

Finally, required component values are determined from the last term of the above continuous fraction expansion as

$$C_1 = 1.147\,\text{F} \quad L_1 = 1.372\,\text{H}$$
$$C_2 = 1.977\,\text{F} \quad L_2 = 1.372\,\text{H}$$
$$C_3 = 1.147\,\text{F}$$

Note that one could also find the same component values by using the closed-form expression given in (6.17).

The applicable network topologies to the filter are given in Figure 6.32. The actual topology depends on the sign selected for the reflection coefficient in (6.38). For the plus sign selected here, the network shown in Figure 6.32(a) is applicable because (6.39) goes to zero as $\omega \to \infty$. If minus signs were selected, order of numerator of (6.39) would be greater than the order of numerator requiring a series ideal inductance as the first component, as in Figure 6.32(b).

6.1.4 Impedance Inverters

The frequency transformation introduced in Section 6.1.1 to obtain bandpass filters from lowpass prototypes results in a filter network containing both series and parallel resonators (see Figure 6.13). However, it is usually advantageous to use either series or parallel tuned circuits in designing bandpass filters to simplify the filter circuit. Besides, this would allow the realization of microstrip coupled-line filters, as we will demonstrate later. Therefore, a technique is required to convert networks containing both types of resonators obtained through frequency mapping to networks containing only one type of resonator.

Transformation of one resonator type to another is achieved by using circuit components called impedance or admittance inverters [13, 24]. Impedance inverters can be used to design bandpass filters using only parallel resonators. Similarly, admittance inverters are used to design bandpass filters using only series resonators. This is a very useful and powerful technique, which results in practical microstrip filter realizations starting from lumped-circuit filter prototypes. An ideal impedance inverter is a quarter-wave transformer whose input impedance is given as follows:

$$Z_{in} = \frac{K^2}{Z_L} \tag{6.40}$$

where Z_L and K are the load impedance and characteristic impedance, respectively. Figure 6.33 demonstrates the use of impedance and admittance inverters to convert series resonators into parallel resonators and vice versa. To explain how to select the characteristic impedance and admittance values of the transmission lines, let's consider the input impedance of a series LC network terminated with a load impedance Z_L:

$$Z_{in} = Z_L + j\omega L_1 + \frac{1}{j\omega C_1} \tag{6.41}$$

To convert this series resonator into a parallel resonator, we need to use two impedance inverters as depicted in Figure 6.33. The following is the input impedance of the parallel resonator with two impedance inverters on each side connected to the same load:

$$Z_{in} = \frac{K^2}{\dfrac{Z_L}{K^2} + j\omega C_2 + \dfrac{1}{j\omega L_2}} = Z_L + K^2 j\omega C_2 + K^2 \frac{1}{j\omega L_2} \tag{6.42}$$

where C_2 and L_2 are the component values of the parallel resonator. Our aim is to obtain the same input impedance from both equations so that the load transformation becomes the same. To accomplish this, one can have $K = \sqrt{L_1/C_2}$.

Note that for all of the resonators we have $LC = 1/\omega_0^2$, where ω_0 is the center frequency. A similar derivation can be carried out for the admittance inverter, yielding $J = \sqrt{C_2/L_1}$. It should be indicated that impedance and admittance inverters are, in fact, the same devices (i.e., quarter-wavelength-long transmission lines in the ideal case). However, depending on the requirements, we either specify the characteristic impedance or admittance of the line. Thus, the usage of K or J should not confuse the reader; both of them refer to the same device.

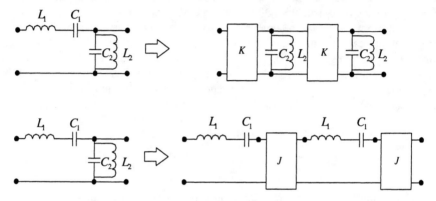

Figure 6.33 Use of inverters to convert a section of a bandpass filter into a network using only series or parallel resonators. Top figure demonstrates the impedance inverter ($K = \sqrt{L_1/C_2}$) and bottom figure demonstrates the admittance inverter ($J = \sqrt{C_2/L_1}$).

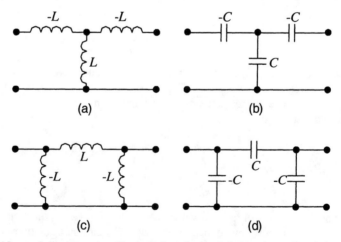

Figure 6.34 Lumped circuit inverters. The equivalent characteristic impedances (or admittances) of the inverter sections are (a) $K = \omega L$, (b) $K = 1/\omega C$, (c) $J = 1/\omega L$, and (d) $J = \omega C$.

It is also possible to have lumped equivalents of impedance inverters as shown in Figure 6.34(a–d). In this case, some of the component values are negative, and the circuit is frequency dependent. However, it is possible to absorb the negative components of a lumped inverter into other components of the filter. Also note that frequency dependent characteristics of the lumped inverters are less important for narrowband applications. It will later be demonstrated that lumped inverters are especially important in filter design using the coupling technique.

Example 6.7

In this example, a third-order Chebyshev bandpass filter centered at 2 GHz with a 400-MHz 3-dB bandwidth and 0.1-dB ($\varepsilon = 0.1526$) bandpass ripple is going to be realized using only series resonators. The input and output impedances are selected as 50 ohms.

As usual, the first step is to determine the prototype circuit value. Using expressions given in (6.17), the following prototype values are found:

$$g_1 = 1.032 \quad g_2 = 1.147$$
$$g_3 = 1.032 \quad g_4 = 1.0$$

Then, using (6.33), we convert the prototype values into actual circuit component values:

$$C_1 = 8.2 \text{ pF} \quad L_1 = 0.78 \text{ nH}$$
$$C_2 = 0.28 \text{ pF} \quad L_2 = 22.8 \text{ nH}$$
$$C_3 = 8.2 \text{ pF} \quad L_3 = 0.78 \text{ nH}$$

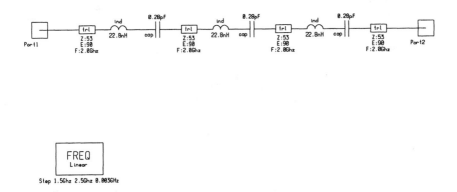

Figure 6.35 Circuit schematic of the bandpass filter given in Example 6.7. Notice that how the admittance inverters are being used.

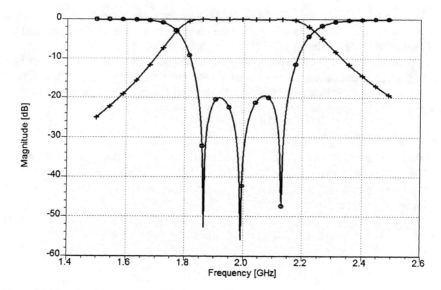

Figure 6.36 Simulated S-parameters of the bandpass filter shown in Figure 6.35 (circle: S_{11}; plus: S_{21}).

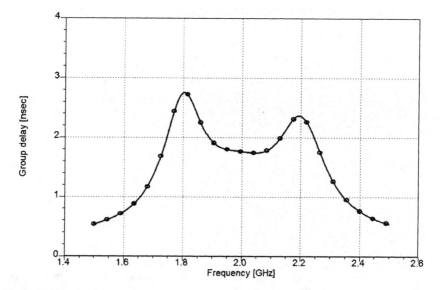

Figure 6.37 Simulated group delay of the bandpass filter shown in Figure 6.35.

Now, the first and last parallel resonators of the bandpass filter are converted into series resonators using quarter-wavelength-long transmission line admittance inverters whose characteristic admittance is selected as:

$$J = \sqrt{C_2/L_1} = \sqrt{0.28\times10^{-12}/0.78\times10^{-9}} = 0.0189$$

$$\Rightarrow Z_c = \frac{1}{Y_c} = \frac{1}{0.0189} \cong 53 \quad \text{ohm}$$

To test the design, the filter circuit is entered into a circuit simulator as shown in Figure 6.35. The simulation results of the filter are given in Figures 6.36 and 6.37. The circuit topology given in Figure 6.35 containing admittance inverters and series resonators is advantageous in designing distributed microstrip coupled-line filters because the series resonators can conveniently be converted into open-circuited coupled-line filter sections. We will demonstrate this point in the following sections.

6.2 RICHARD'S TRANSFORMATION

So far, we have been concentrating on the fundamentals of lumped filters and the main transformations used in implementing lumped filters, since those concepts are an indispensable part of general filter theory. However, little has been said about distributed filters. Starting from this section, we will shift our focus from lumped filters to distributed filters, which will eventually lead us to the microstrip filters.

Perhaps the most fundamental transformation used in designing distributed filters is Richard's transformation, which deals with the determination of the realizable impedance functions by resistor and transmission-line networks [8]. Richard's transformation is based on the following mapping:

$$\Omega = \tan(\beta l) \tag{6.43}$$

which maps the β plane to an Ω plane. Obviously, this mapping is not one-to-one because of the tangent function. The two main requirements of Richard's theory is that the lines must be lossless and commensurate (i.e., line lengths are divisible by a common integer). Both of these requirements can be approximated very well in practice. This theory enables the vast majority of lumped-circuit theories to be applicable to distributed networks by a simple change of variable. Thus, Richard's transformation holds a very important place in filter theory.

Synthesis of distributed filters from the lumped filters through Richard's transformation can be explained as follows: from (6.43) one can see that the reactance of an inductor and the susceptance of a capacitor can be written as:

$$jX_L = j\omega L = jL\tan(\beta l)$$
$$jB_C = j\omega C = jC\tan(\beta l) \tag{6.44}$$

This immediately suggests that an inductance can be replaced by a short-circuited transmission line with a characteristic impedance L, while a capacitor can be replaced by an open-circuited transmission line with a characteristic impedance of $1/C$. Pictorial representation of this transformation is shown in Figure 6.38. Note that the cutoff frequency of the transformed network occurs at $\Omega = 1$, which gives a stub length of $l = \lambda/8$. A disadvantage of Richard's transformation is that the frequency response of the resulting distributed network is periodic with $4\omega_c$ where ω_c is the cutoff frequency.

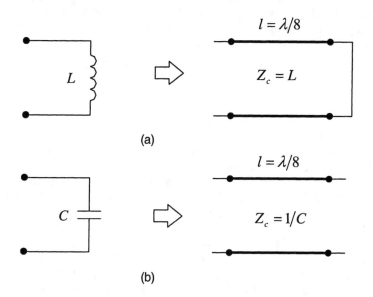

Figure 6.38 Richard's transformation: (a) for an inductor, and (b) for a capacitor.

Although Richard's transformation provides a simple way of converting lumped filters into distributed ones, there is difficulty in implementing distributed filters using microstrip circuits through this method: it is not easy to implement the series short-circuited stubs occurring due to series lumped inductors using printed transmission lines. Microstrip stubs are always easy to implement in shunt form. Thus, it would be very useful if all the microstrip stubs were implemented as shunt stubs. To circumvent this difficulty, Kuroda's identities shown in Figure 6.39 are used [9, 11]. As shown in the figure, the short-circuited stub is conveniently converted to a shunt open-circuited stub by the help of additional transmission line segment. The lengths of all of the lines are $\lambda/8$ at the cutoff frequency, ω_c. Note that application of Kuroda's identities will also physically separate the stubs by inserting additional transmission line segments between them, which is extremely helpful in practical realization of microstrip filters. A typical microstrip lowpass filter employing open-circuited stubs based on Richard's principle and Kuroda's identities is shown in Figure 6.40.

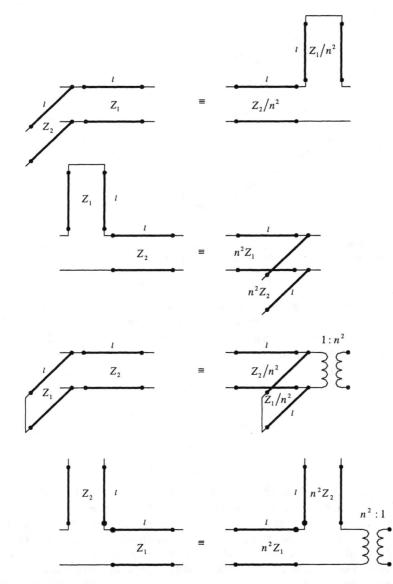

Figure 6.39 Kuroda's identities to convert short-circuited stubs into open-circuited stubs and vice versa $(n^2 = 1 + Z_2/Z_1)$.

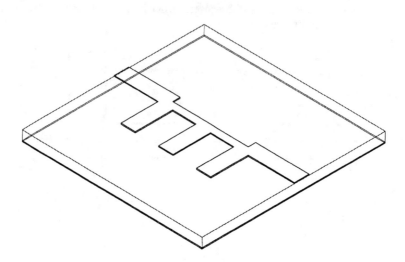

Figure 6.40 Microstrip lowpass filter using open-circuited stubs. Richard's transformation can be used to design microstrip lowpass filters like that shown in the figure.

Example 6.8

In this example, we will convert the lowpass filter design given in Example 5.2 into the microstrip equivalent with a cutoff frequency of 10 GHz using Richard's transformation. Substrate parameters are selected as $h = 5$ mils and $\varepsilon_r = 2.2$.

We first determine the characteristic impedances of the microstrip stubs using (6.44) and prototype values as follows:

$$Z_{c1} = 1/1.706 = 0.586 \quad \text{ohm}$$
$$Z_{c2} = 1.23 \quad \text{ohm}$$
$$Z_{c3} = 1/2.541 = 0.394 \quad \text{ohm}$$
$$Z_{c4} = 1.23 \quad \text{ohm}$$
$$Z_{c5} = 1/1.706 = 0.586 \quad \text{ohm}$$

Note that the above impedances are normalized with respect to 1 ohm. We will multiply these characteristic impedances by the normalization impedance (i.e., 50 ohms) during realization of the transmission lines. As discussed above, the application of Richard's transformation to microstrip filters requires the use of Kuroda's identities because series stubs are difficult to implement in microstrip. By applying a series of Kuroda's identities starting from the outside elements as shown in Figure 6.41(a–d), all short-circuited stubs will be converted into open-circuited stubs. The procedure is detailed in the following paragraph.

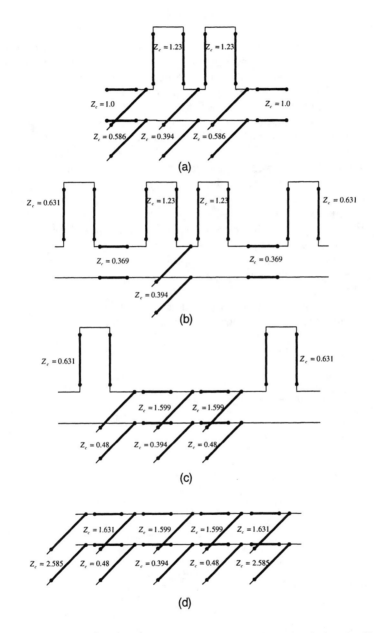

Figure 6.41 Application of Richard's transformation and Kuroda's identity to the lowpass filter given in Example 6.8: (a) the original filter after Richard's transformation, (b) and (c) successive applications of Kuroda's identities, and (d) the final network.

To start applying Kuroda's identities, we first add two transmission lines with a characteristic impedance of 1 ohm on both sides of the filter network obtained after Richard's transformation as shown in Figure 6.41(a). Note that these transmission lines will not affect magnitude response of the filter because they are matched with source and load impedances (they will only add a phase shift). Then, we apply the first Kuroda's identity in Figure 6.39 to the outside two elements on both sides to move the transmission line segments to the interior, as shown in Figure 6.41(b). Note that this creates additional series stubs but we will handle them at the end. Moving the transmission lines enables us to convert the stubs inside the network using the second identity of Figure 6.39, Figure 6.41(c). After converting the inside stubs, we again add unity characteristic impedance lines on both ends of the filter to convert the remaining two outside series stubs arriving the final network shown in Figure 6.41(d). Note that length of all the transmission lines is approximately 110 mils for $\omega_c = 10$ GHz.

To test the design, the filter circuit is entered into a circuit simulator as shown in Figure 6.42. The simulation results of the filter are given in Figures 6.43 and 6.44. One of the observations that can be made from the simulation results is that an attenuation pole occurs at $2\omega_c$. At this frequency, all the open-circuited stubs become short circuit and short-circuited stubs become open circuit at the plane of reference, attenuating transmission considerably. Thus, this is a normal result of Richard's transformation. Another observation is that the frequency response of the distributed circuit is periodic with $4\omega_c$. Again, this is an expected result due to periodicity of tangent function.

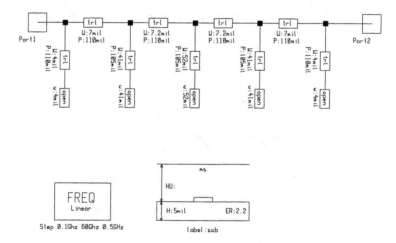

Figure 6.42 Circuit schematic of the lowpass filter given in Example 6.8.

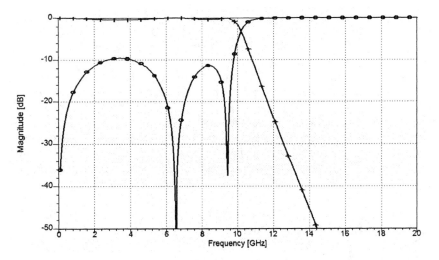

Figure 6.43 Simulated S-parameters of the lowpass filter shown in Figure 6.42 (circle: S_{11}; plus: S_{21}).

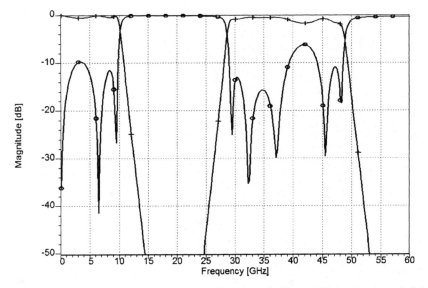

Figure 6.44 Simulated S-parameters of the lowpass filter shown in Figure 6.42 showing the periodicity in Richard's transformation (circle: S_{11}; plus: S_{21}).

It must be noted that the circuit shown in Figure 6.42 does not include discontinuity effects due to T-junctions because our aim was to demonstrate the application of the methodology. The effects of discontinuities, like T-junctions, impedance steps, and bends, are extremely pronounced in microwave frequencies. For accurate characterization, these must be included in the model through full-wave simulations. The most common approach to achieve this is to simulate the discontinuities separately using a full-wave simulation tool, then combine the resultant S-parameters with the transmission lines in a circuit simulator. This hybrid approach preserves the accuracy without reducing the simulation speed provided that the stubs are separated well so that the interactions between them are insignificant. Otherwise, full-wave simulation of the overall circuitry is required to account for all of the interactions between circuit elements properly.

6.3 STEPPED-IMPEDANCE MICROSTRIP FILTERS

Microstrip stub filters, implemented through Richard's transformation as introduced in Section 6.2, are not the only way of designing microstrip lowpass filters. Stepped-impedance microstrip filters are another way of achieving lowpass filter response [11, 21]. They are constructed by cascade connection of low- and high-impedance of electrically short transmission lines, as shown in Figure 6.45.

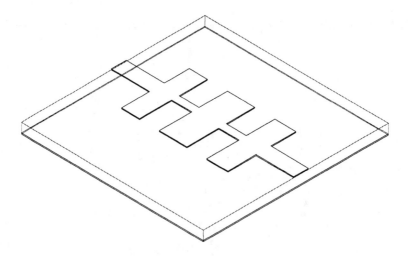

Figure 6.45 Stepped-impedance microstrip lowpass filter. Narrow and wide sections represent series inductors and shunt capacitors, respectively.

The design philosophy of stepped-impedance filters is based on lumped approximation of short transmission lines; a short low-impedance transmission line is approximated by a shunt capacitor to ground, and a short high-impedance transmission line is approximated by a series inductor. As one can notice, there is no mapping involved as in Richard's transformation. Thus, the frequency response is not periodic. In addition, the transmission lines in stepped-impedance filters do not need to be commensurate. Since there is no a rigorous mapping technique, this approach is limited to all-pole lowpass filters only. Nevertheless, due to their simple design and implementation, stepped-impedance microstrip filters are commonly used in practice.

In the design of stepped-impedance filters, one should ensure that the transmission lines are electrically short ($l < \lambda/8$) so that the lumped approximation will be valid. In addition, the low-impedance lines should not be too wide. Otherwise, they will act like open-circuited stubs.

The design of stepped-impedance filters starts with the determination of lowpass prototype circuit values for a given specification. Then, electrical lengths of low- and high-impedance line sections are determined according to the following equations:

For low-impedance lines:

$$\beta l = g_k \frac{Z_{low}}{Z_0} \quad k = 1,3,5,\ldots \tag{6.45}$$

For high-impedance lines:

$$\beta l = g_k \frac{Z_0}{Z_{high}} \quad k = 2,4,6,\ldots \tag{6.46}$$

where Z_{low} and Z_{high} are the characteristic impedances of low- and high-impedance lines, respectively. Z_0 is the normalization impedance, and g_k is the value of lowpass prototype components. In the design process, Z_{low} and Z_{high} are determined first by the designer. Then, required electrical lengths are found from the prototype values giving the required filter response. Note that Z_{low} and Z_{high} are usually set to the lowest and highest values that can be achieved with the used photolithographic technology, which limits realizable line widths. Alternatively, one could also fix the line lengths in advance, and vary the line widths (i.e., line characteristic impedances) to approximate prototype network values.

Example 6.9

In this example, we will design a fifth-order, stepped-impedance, microstrip line Chebyshev lowpass filter with cutoff frequency of 10 GHz and passband ripple of 0.1 dB ($\varepsilon = 0.1526$). The input and output impedances are selected as 50 ohm. The substrate parameters are $h = 5$ mils and $\varepsilon_r = 2.2$. The minimum and maximum line widths that will be used in the filter design should be 4 mils and 60 mils, respectively.

First, we find by using the design equations given in Chapter 2 that characteristic impedances corresponding to the minimum and maximum line widths are $Z_{low} = 17.5$ ($\varepsilon_{eff} = 1.75$, $\beta = 7.037 \times 10^{-3}$ rad/mils) and $Z_{high} = 105$ ($\varepsilon_{eff} = 2.02$, $\beta = 7.561 \times 10^{-3}$ rad/mils), respectively. The next step is to determine the prototype circuit values. Using expressions given in (6.17), the following prototype values are found:

$$g_1 = 1.147 \quad g_2 = 1.371$$
$$g_3 = 1.975 \quad g_4 = 1.371$$
$$g_5 = 1.147 \quad g_6 = 1.0$$

Then, the lengths of each transmission line segment are calculated using (6.45) and (6.46) as follows:

$$l_1 = \frac{1.147}{7.037 \times 10^{-3}} \frac{17.5}{50} = 57 \text{ mils} \quad l_2 = \frac{1.371}{7.561 \times 10^{-3}} \frac{50}{105} = 86 \text{ mils}$$

$$l_3 = \frac{1.975}{7.037 \times 10^{-3}} \frac{17.5}{50} = 98 \text{ mils} \quad l_4 = \frac{1.371}{7.561 \times 10^{-3}} \frac{50}{105} = 86 \text{ mils}$$

$$l_5 = \frac{1.147}{7.037 \times 10^{-3}} \frac{17.5}{50} = 57 \text{ mils}$$

Note that the guided wavelength at 10 GHz is approximately 850 mils in the substrate. So, all line lengths are shorter than $\lambda/8$, satisfying the short-line requirement. To test the design, the filter circuit is entered into a circuit simulator as shown Figure 6.46. The simulation results are shown in Figures 6.47 and 6.48. As one can see from the figures, the return loss is better than 15 dB in the passband, and the insertion loss in the stopband is 30 dB at 20 GHz. It must be noted that the circuit shown in Figure 6.46 does not include discontinuity effects due to changes in line widths. For accurate characterization, these must be included through full-wave simulations as well.

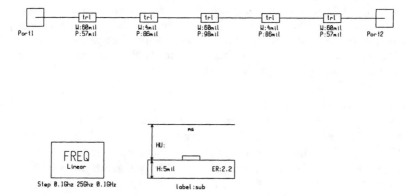

Figure 6.46 Circuit schematic of the lowpass filter given in Example 6.9.

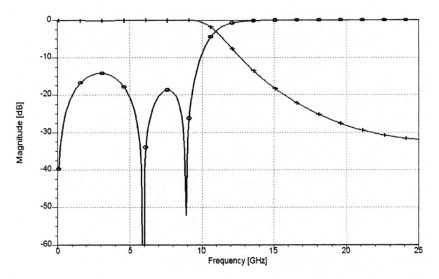

Figure 6.47 Simulated S-parameters of the lowpass filter shown in Figure 6.46 (circle: S_{11}; plus: S_{21}).

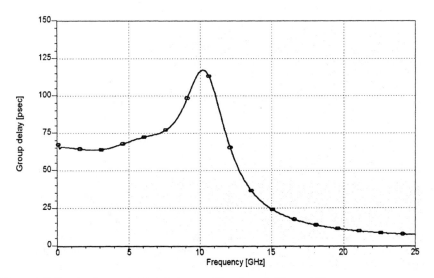

Figure 6.48 Simulated group delay of the lowpass filter shown in Figure 6.46.

6.4 COUPLED-LINE MICROSTRIP FILTERS

In this section, we introduce the design principles of coupled-line microstrip filters. Coupled-line microstrip filters are constructed using cascaded sections of coupled resonators as shown in Figure 6.49 [10, 21, 24, 30–33]. The filter response is synthesized by modifying the even- and odd-mode characteristic impedances of each resonator. Ideally, each resonator section is a quarter-wavelength long at the center frequency. However, it is usually necessary to adjust the lengths to compensate for fringing fields due to the open ends and quasi-TEM propagation.

Due to strict requirements on the resonator line widths and gaps, thin-film technology is best suited for this kind of filter. In designing edge-coupled microstrip coupled-line filters, it might sometimes be necessary to have small gaps (i.e., large couplings) between the conductors of some resonant elements, which would make the production of such filters difficult using PCB technology. This difficulty can be circumvented by implementing such high-coupling resonators in a stripline configuration with broadside coupling.

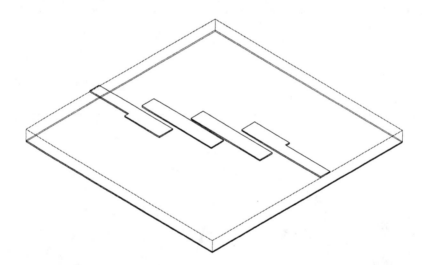

Figure 6.49 A typical bandpass filter constructed using coupled lines.

The design of coupled-line filters hinges on the equivalent models of open- and short-circuited coupled-line resonators. Figure 6.50 shows open- and short-circuited coupled-line resonators and their equivalent distributed models. Note that when the length of the resonator is selected to be a quarter-wavelength long at the center frequency, we have two stubs separated by an inverter as the equivalent model. The type of stubs used (i.e., series-open or shunt-short) depends on how the unused ports of the coupled lines are terminated.

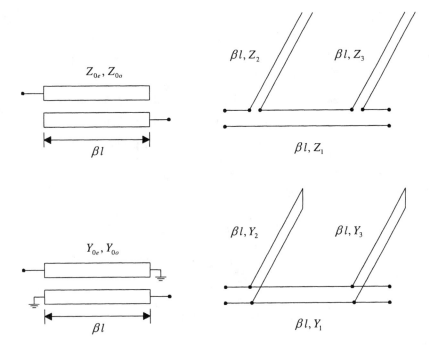

Figure 6.50 Open- and short-circuited microstrip coupled-line networks and their equivalent transmission-line circuits ($l = \lambda_g/4$).

Open-circuited coupled-line resonators are more commonly used in practice because they don't need ground via-holes. Then, the bandpass filter is designed using the equivalent distributed models according to the following principle: the open-circuited series stub is the approximate equivalent of a series-tuned *LC* resonator and the short-circuited shunt stub is the approximate equivalent of a parallel-tuned *LC* resonator. Since the design methods for bandpass filters using lumped-components are readily available, all one needs to do is convert the lumped bandpass filter that provides the desired response to a configuration containing only series or shunt resonators with the help of inverters as demonstrated in Section 6.1.4. The rest of the process is to utilize equivalent models to those given in Figure 6.50 to obtain coupled-line equivalents. Thus, the coupled-line microstrip filter can be designed methodically, starting from the lumped filter.

For the benefit of the reader, we will now show how the equivalent network of the open-circuited coupled-line section is obtained since this is the most used form in coupled-line bandpass filters. A similar derivation can be done for the short-circuited coupled-line section. The two-port *Z*-parameters of the open-circuited coupled-line sections shown in Figure 6.50 are given as follows [24]:

$$\begin{bmatrix} Z_{11} & Z_{21} \\ Z_{21} & Z_{22} \end{bmatrix} = -\frac{j}{2}\begin{bmatrix} Z_{0e}\cot(\theta_e) + Z_{0o}\cot(\theta_o) & Z_{0e}\csc(\theta_e) - Z_{0o}\csc(\theta_o) \\ Z_{0e}\csc(\theta_e) - Z_{0o}\csc(\theta_o) & Z_{0e}\cot(\theta_e) + Z_{0o}\cot(\theta_o) \end{bmatrix} \quad (6.47)$$

where the subscripts e and o designate even and odd modes, respectively. For $\theta_e \approx \theta_o \approx \beta l$, the above expression reduces to

$$\begin{bmatrix} Z_{11} & Z_{21} \\ Z_{21} & Z_{22} \end{bmatrix} = -\frac{j}{2}\begin{bmatrix} (Z_{0e} + Z_{0o})\cot(\beta l) & (Z_{0e} - Z_{0o})\csc(\beta l) \\ (Z_{0e} - Z_{0o})\csc(\beta l) & (Z_{0e} + Z_{0o})\cot(\beta l) \end{bmatrix} \quad (6.48)$$

The next step is to show that the two open stubs separated by an impedance inverter shown to the right in the figure are equivalent to the microstrip coupled line whose Z-parameters are given above. For this purpose, the Z-parameters of the stub-network are written as follows:

$$Z'_{11} = \frac{Z_1}{j\tan(\beta l)} + \frac{Z_2}{j\tan(\beta l)} \quad (6.49)$$

$$Z'_{22} = \frac{Z_1}{j\tan(\beta l)} + \frac{Z_3}{j\tan(\beta l)} \quad (6.50)$$

$$Z'_{12} = Z'_{21} = \sqrt{\left(Z'_{22} - \frac{1}{Y'_{22}}\right)Z'_{11}} \quad (6.51)$$

where

$$\frac{1}{Y'_{22}} = Z_1 \frac{\dfrac{Z_2}{j\tan(\beta l)} + jZ_1\tan(\beta l)}{Z_1 + j\dfrac{Z_2}{j\tan(\beta l)}\tan(\beta l)} + \frac{Z_3}{j\tan(\beta l)} \quad (6.52)$$

which leads to

$$Z'_{12} = \sqrt{-Z_1^2\left[\tan^2(\beta l) + 1\right]\frac{Z_1 + Z_2}{\tan^2(\beta l)(Z_1 + Z_2)}}$$

$$= -j\frac{Z_1}{\sin(\beta l)} \quad (6.53)$$

In these equations, Z_1 is the characteristic impedance of the inverter section, whereas Z_2 and Z_3 are the characteristic impedances of the stubs (see Figure 6.50). Note that the proper branch is selected for the square root. By comparing (6.49), (6.50), and (6.51) with (6.48), the characteristic impedances of the inverter and stubs are found in terms of the even- and odd-mode impedances of the coupled microstrip line as follows for open-circuited coupled lines:

$$Z_1 = \frac{Z_{0e} - Z_{0o}}{2}$$

(6.54)

$$Z_2 = Z_3 = Z_{0o}$$

One can also show that the following equations are obtained for short-circuited coupled lines:

$$Y_1 = \frac{Y_{0e} - Y_{0o}}{2}$$

(6.55)

$$Y_2 = Y_3 = Y_{0o}$$

The above equations are for symmetrical coupled lines. For asymmetrical coupled lines, the above expressions take the following forms [24]:

$$Z_1 = \frac{1}{2}\sqrt{\left(Z_{0e}^a - Z_{0o}^a\right)\left(Z_{0e}^b - Z_{0o}^b\right)}$$

$$Z_2 = \frac{1}{2}\left(Z_{0e}^a + Z_{0o}^a\right) - Z_1$$

(6.56)

$$Z_3 = \frac{1}{2}\left(Z_{0e}^b + Z_{0o}^b\right) - Z_1$$

for open-circuited coupled lines, and

$$Y_1 = \frac{1}{2}\sqrt{\left(Y_{0e}^a - Y_{0o}^a\right)\left(Y_{0e}^b - Y_{0o}^b\right)}$$

$$Y_2 = \frac{1}{2}\left(Y_{0e}^a + Y_{0o}^a\right) - Y_1$$

(6.57)

$$Y_3 = \frac{1}{2}\left(Y_{0e}^b + Y_{0o}^b\right) - Y_1$$

for short-circuited coupled lines, where the superscripts *a* and *b* denote the first and second lines (different widths).

In the above paragraphs, it was demonstrated that coupled-line microstrip resonators could be used in designing bandpass filters. The theory allows the transformation of a lumped bandpass filter with prescribed attenuation characteristics into a transmission line network containing stubs and inverters. Then, the coupled-line equivalent is derived from the stub network. Although the method is simple, it can be tedious, especially for a high number of sections. Fortunately, closed-form expressions are available for the design of coupled-line microstrip filters for most commonly used filter responses. Below, we will provide the closed-form expressions to determine even- and odd-mode characteristic impedances of the resonators in the case of Chebyshev-type attenuation characteristics [24, 31]:

$$Z_{0e}^1 = Z_{0e}^{N+1} = Z_0 \left[1 + \frac{K_{10}}{Z_0 \sqrt{S}} \sin\left(\frac{\pi \omega_1}{2\omega_0} \right) \right]$$

$$Z_{0o}^1 = Z_{0o}^{N+1} = Z_0 \left[1 - \frac{K_{10}}{Z_0 \sqrt{S}} \sin\left(\frac{\pi \omega_1}{2\omega_0} \right) \right] \tag{6.58}$$

$$Z_{0e}^{k+1} = Z_{0e}^{N-k+1} = \frac{Z_0}{S} \left(M_{k+1,k} + Y_0 K_{k+1,k} \right)$$

$$Z_{0o}^{k+1} = Z_{0o}^{N-k+1} = \frac{Z_0}{S} \left(M_{k+1,k} - Y_0 K_{k+1,k} \right) \tag{6.59}$$

where

$$K_{10} = K_{N+1,N} = \frac{Z_0}{\sqrt{g_0 g_1}} = \frac{Z_0}{\sqrt{g_{N+1} g_N}}$$

$$K_{k+1,k} = \frac{Z_0}{\sqrt{g_k g_{k+1}}}$$

$$M_{k+1,k} = \sqrt{\left(\frac{K_{k+1,k}}{Z_0} \right)^2 + \frac{1}{4} \tan^2\left(\frac{\pi \omega_1}{2\omega_0} \right)}$$

$$S = \frac{1}{2} \tan\left(\frac{\pi \omega_1}{2\omega_0} \right) + \left(\frac{K_{10}}{Z_0} \right)^2$$

$$Y_0 = \frac{1}{Z_0}$$

In these equations, Z_0 is the reference impedance (usually 50 ohms), Z_{0e} is the even-mode characteristic impedance, Z_{0o} is the odd-mode characteristic impedance, ω_0 is the center frequency, ω_1 is the frequency at the lower edge of the passband, N is the order of the filter, and g_k is the value of the components of lowpass prototype filter calculated using (6.17). All coupled-line sections are a quarter-wavelength long at the center frequency. Note that the number of coupled-line sections is $N + 1$. If the bandwidth of the filter is specified, then ω_1 should be calculated using the bandwidth and center frequency. These equations can easily be programmed into a computer for quick calculation.

It should be stated that the above equations have been derived by assuming that the even- and odd-mode phase velocities are equal. Therefore, they are strictly valid for TEM structures like stripline. For quasi-TEM structures, like microstrip lines for instance, the resultant filter response will deviate somewhat from the ideal response due to different phase velocities. In that case, the resonator lengths should be optimized to get the desired response. The optimization can easily be carried out using a circuit or full-wave electromagnetic simulator.

Example 6.10

In this example, we will design a third-order coupled-line stripline Chebyshev bandpass filter with edge frequencies of $f_1 = 9.8$ GHz and $f_2 = 10.2$ GHz, and a passband ripple of 0.5 dB ($\varepsilon = 0.349$). The substrate parameters are $h = 20$ mils and $\varepsilon_r = 2.2$.

The first step is to obtain prototype component values. Using the expressions given in (6.17), the following prototype values are found:

$$g_1 = 1.032 \quad g_2 = 1.147$$
$$g_3 = 1.032 \quad g_4 = 1.0$$

Then, even- and odd-mode impedance of each coupled section are determined using (6.58) and (6.59) as follows:

$$Z_{0e}^1 = 59.7 \quad Z_{0o}^1 = 40.3$$
$$Z_{0e}^2 = 50.5 \quad Z_{0o}^2 = 45.9$$
$$Z_{0e}^3 = 50.5 \quad Z_{0o}^3 = 45.9$$
$$Z_{0e}^4 = 59.7 \quad Z_{0o}^4 = 40.3$$

By using the above impedance values, the width and separation of each coupled stripline section are determined using the formulas for coupled striplines as

$$w_1 = 15.7 \text{ mil} \quad s_1 = 2.4 \text{ mil}$$
$$w_2 = 17.6 \text{ mil} \quad s_2 = 11.2 \text{ mil}$$
$$w_3 = 17.6 \text{ mil} \quad s_3 = 11.2 \text{ mil}$$
$$w_4 = 15.7 \text{ mil} \quad s_4 = 2.4 \text{ mil}$$

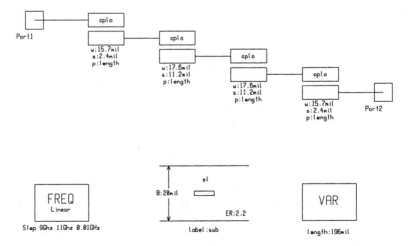

Figure 6.51 Circuit schematic of the bandpass filter given in Example 6.10.

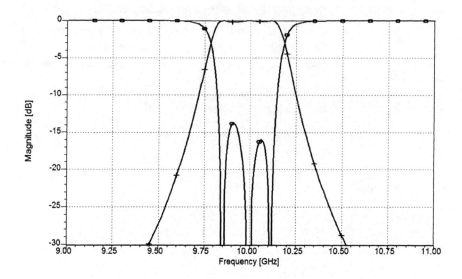

Figure 6.52 Simulated S-parameters of the bandpass filter shown in Figure 6.51 (circle: S_{11}; plus: S_{21}).

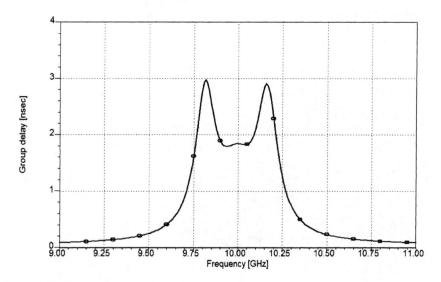

Figure 6.53 Simulated group delay of the bandpass filter shown in Figure 6.51.

To test the design, the filter circuit is entered into a circuit simulator as shown in Figure 6.51. Note that the number of coupled-line sections is one more than the order of the filter. The simulation results are shown in Figures 6.52 and 6.53. The return loss is better than 14 dB in the passband, and the insertion loss in the stopband is 28 dB at 9.5 GHz and 10.5 GHz.

It is important to indicate that the solution is not unique for edge-coupled, combline, and interdigital filters. In other words, different combinations of line widths and gaps will give the same filter response. In practice, the line widths are usually forced to be the same and required resonator gaps are determined. This can be done by optimization after the initial design is found.

Example 6.11

In this example, we will implement the microstrip version of the stripline bandpass filter designed in the previous example. By using the same even- and odd-mode impedance values, the width and separation of each coupled microstrip section are determined using the formulas for coupled microstrip lines as follows:

$$w_1 = 14.7 \text{ mil} \quad s_1 = 1.9 \text{ mil}$$
$$w_2 = 16.2 \text{ mil} \quad s_2 = 12.4 \text{ mil}$$
$$w_3 = 16.2 \text{ mil} \quad s_3 = 12.4 \text{ mil}$$
$$w_4 = 14.7 \text{ mil} \quad s_4 = 1.9 \text{ mil}$$

To test the design, the filter circuit is entered into a circuit simulator as shown in Figure 6.54. Note that in this case, it was necessary to optimize the length of the resonators due to unequal even- and odd-mode phase velocities.

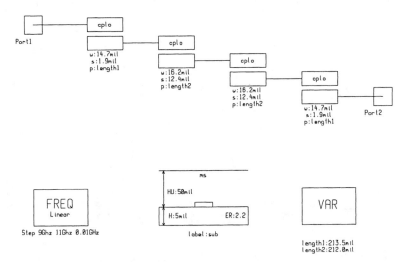

Figure 6.54 Circuit schematic of the bandpass filter given in Example 6.11.

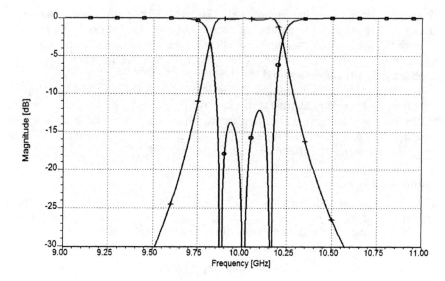

Figure 6.55 Simulated S-parameters of the bandpass filter shown in Figure 6.54 (circle: S_{11}, plus: S_{12}).

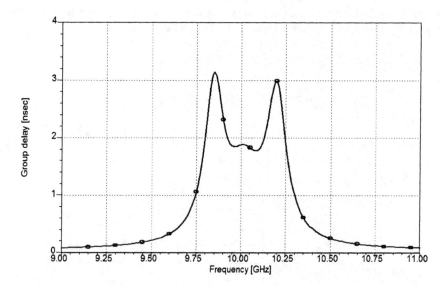

Figure 6.56 Simulated group delay of the bandpass filter shown in Figure 6.54.

The simulation results are shown in Figures 6.55 and 6.56. The response is similar to the one obtained for the stripline case in the previous example as expected. The return loss is better than 12 dB in the passband, and the insertion loss in the stopband is 25 dB at 9.5 GHz and 10.5 GHz.

Example 6.12

In this example, a multilayer three-resonator bandpass filter with broadside coupling is designed using the procedures described in previous sections. The full-wave electromagnetic simulation model and cross-section of the filter are shown in Figures 6.57 and 6.58, respectively. Note that the coupling of the center two coupled sections is implemented as edge-coupled whereas coupling in the first and last coupled sections is implemented as broadside coupled. The simulation results of this filter are given in Figures 6.59 and 6.60.

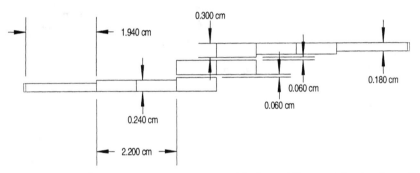

Figure 6.57 Full-wave electromagnetic simulation model of a multilayer stripline bandpass filter ($h_1 = 2.06$ mm, $\varepsilon_{r1} = 2.62$, $h_2 = 0.97$ mm, $\varepsilon_{r2} = 2.62$, $h_3 = 2.06$ mm, $\varepsilon_{r3} = 2.62$).

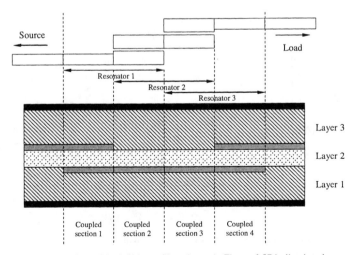

Figure 6.58 Cross-section view of the bandpass filter shown in Figure 6.57 indicating the resonators.

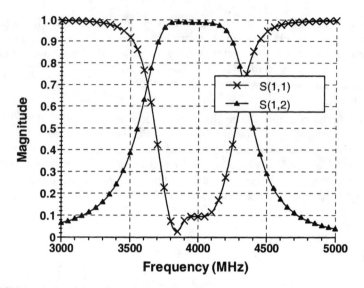

Figure 6.59 Magnitudes of S_{11} and S_{12} of the bandpass filter shown in Figure 6.57.

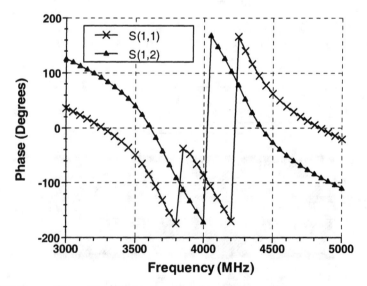

Figure 6.60 Phases of S_{11} and S_{12} of the bandpass filter shown in Figure 6.57.

6.5 BANDSTOP FILTERS

The coupled-line sections introduced in Section 6.4 to implement bandpass filters can be used to design bandstop filters as well [10, 34]. A typical bandstop filter constructed using microstrip coupled-lines is shown in Figure 6.61. As in the bandpass filters, bandstop filters are implemented starting from either all-series or all-shunt lumped resonators separated by quarter-wavelength-long transmission lines (i.e., inverters) as depicted in Figure 6.62. Then, the filter is converted to microstrip using the equivalent circuits in Figure 6.63. In practice, the topology with series resonators can be implemented conveniently with coupled-line sections resulting in the circuit shown in Figure 6.61.

The design of bandstop filters using coupled-line sections first requires calculation of slope parameters for each distributed resonator section that will be realized [10]. The slope parameters can be obtained directly from the lowpass prototype filter values, which are determined for a desired filter response. Then, the lumped circuit is converted to a transmission line equivalent using short- or open-circuited stubs and slope parameters. Finally, the transmission line circuit with stubs is converted into the coupled-line microstrip equivalent through a similar approach described in the previous section. Note that the designer also has the option of implementing stub resonators directly in microstrip to implement the filter (instead of converting them into coupled lines). The disadvantage of this approach is that the resulting stub characteristic impedances might be difficult to physically realize.

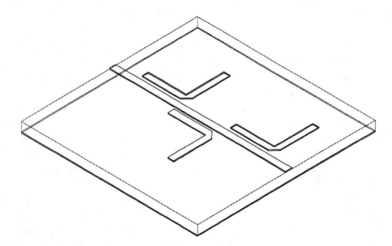

Figure 6.61 A microstrip bandstop filter implemented with coupled lines.

At this point, we will elaborate the design procedure in detail. As indicated above, the first step in the design procedure is to determine slope parameters based on the given filter bandwidth and center frequency. The slope parameters establish a relationship between the lumped-circuit resonators and distributed resonators around the resonance frequency, and they are introduced in Chapter 1. The slope parameter of each resonator section for the filters shown in Figure 6.62 are given by the following equations [10]:

Filter with shunt branches:

$$x_i\Big|_{i=even} = \omega_0 L_i = \frac{1}{\omega_0 C_i} = Z_0 \frac{g_0 f_0}{g_i BW}$$

$$x_i\Big|_{i=odd} = \omega_0 L_i = \frac{1}{\omega_0 C_i} = \frac{Z_0 f_0}{g_0 g_i BW} \qquad (6.60)$$

$$Z_i = Z_0 \quad i = 1,2,\ldots,n-1$$

Filter with series branches:

$$b_i\Big|_{i=even} = \omega_0 C_i = \frac{1}{\omega_0 L_i} = Y_0 \frac{g_0 f_0}{g_i BW}$$

$$b_i\Big|_{i=odd} = \omega_0 C_i = \frac{1}{\omega_0 L_i} = \frac{Y_0 f_0}{g_0 g_i BW} \qquad (6.61)$$

$$Y_i = Y_0 \quad i = 1,2,\ldots,n-1$$

where g_i are the prototype lowpass filter values, C_i are the resonator capacitors, L_i are the resonator inductors, BW is the filter bandwidth, f_0 is the center frequency, Z_i is the characteristic impedance of the inverters, n is the order of the filter, and Z_0 is the reference impedance. In the above equations, it is assumed that the filter has an odd number of sections. This causes the inverter impedances to be equal to the reference impedance. Note that in general, it is not necessary to have the inverter impedances equal the reference impedance, although having an odd number of filter sections simplifies the design equations.

After obtaining the slope parameters of each resonator section, the lumped circuit is converted into a transmission line equivalent by replacing the lumped resonators with appropriate stub resonators, using the slope parameters. The characteristic impedances of the stubs required in this process are determined using the equations given in Chapter 1; thus, the reader is encouraged to review those sections if necessary.

The final step is to replace the resulting distributed circuit with the equivalent coupled-line circuit. As indicated before, although the designer has the option of implementing stubs directly using microstrip lines, coupled-line equivalents are usually preferred because they are easy to implement in terms of line characteristic impedance. This point will be clearer in the next example. For microstrip coupled-line bandstop filters, the equivalent circuit shown in Figure 6.63 can be used to transform simple transmission line resonators into coupled lines.

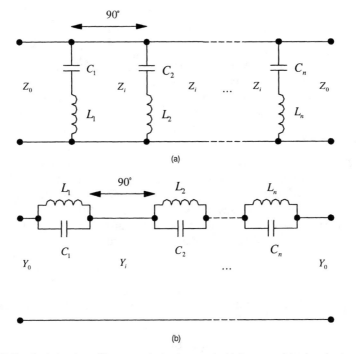

(a)

(b)

Figure 6.62 Two basic bandstop filter networks implemented with inverters: (a) using shunt resonators, and (b) using series resonators. (*After:* [10].)

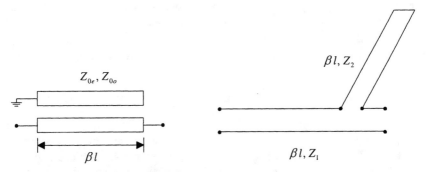

Figure 6.63 Short-circuited bandstop coupled-line resonator and its distributed equivalent circuit.

After some simple calculations, it can be shown that relations between the stub circuit and coupled-line circuit impedances for bandstop filters are given as follows:

$$Z_{0o} = Z_1 + Z_2 \pm \sqrt{Z_2^2 + Z_1 \cdot Z_2} \tag{6.62}$$

$$Z_{0e} = Z_1 \frac{Z_1 + Z_2 \pm \sqrt{Z_2^2 + Z_1 \cdot Z_2}}{Z_1 + 2Z_2 \pm 2\sqrt{Z_2^2 + Z_1 \cdot Z_2}} \qquad (6.63)$$

where Z_{0e} is the even-mode characteristic impedance of the coupled-lines, Z_{0o} is the odd-mode characteristic impedance of the coupled lines, Z_1 is the characteristic impedance of the inverter transmission line, and Z_2 is the characteristic impedance of the short-circuited stub.

Example 6.13

In this example, a third-order Chebyshev bandstop filter centered at 10 GHz with a 400-MHz bandwidth and 0.1-dB ($\varepsilon = 0.1526$) ripple is going to be realized using microstrip coupled lines. The input and output impedances are selected as 50 ohms. The substrate parameters are $h = 10$ mils and $\varepsilon_r = 6.0$.

The first step is to determine the prototype circuit values. Using the expressions given in (6.17), the following prototype values are found:

$$g_1 = 1.032 \quad g_2 = 1.147$$
$$g_3 = 1.032 \quad g_4 = 1.0$$

Then, using (6.61), we determine the slope parameters of the series resonators as

$$b_1 = 0.4845$$
$$b_2 = 0.4359$$
$$b_3 = 0.4845$$

The series resonators described by the above slope parameters will be separated by admittance inverters with characteristic admittance of $Y_0 = 0.02$ (i.e., $1/50$). The next step is to determine the characteristic admittances of stub transmission lines that will serve as series resonators, using the equations given in Chapter 1:

$$Y_1 = 0.6169$$
$$Y_2 = 0.5555$$
$$Y_3 = 0.6169$$

Note that although the characteristic impedances of the stub resonators are very low, we are not planning to implement the stub resonators in transmission line form. We will be transferring each inverter and resonator section into a coupled-line section using the approach described previously. To accomplish this, even- and odd-mode impedances of the required coupled-line sections are found using (6.62) and (6.63) as follows:

$$Z_{0e1} = 60.7 \quad Z_{0o1} = 42.5$$
$$Z_{0e2} = 61.5 \quad Z_{0o2} = 42.1$$
$$Z_{0e3} = 60.7 \quad Z_{0o3} = 42.5$$

As one can see, the stubs with very low characteristic impedances are converted to a circuit that is easily realizable by virtue of coupled lines. The rest is to physically realize the coupled-line sections, which have the above given even- and odd-mode characteristic impedances.

To test the design, the filter circuit is entered into a circuit simulator as shown in Figure 6.64. Note that one terminal of the coupled-line resonators is short circuited using an ideal ground. In practice, an RF short circuit can be accomplished by using an additional quarter-wavelength-long open transmission line connected to this point. It must be noted that this approach will narrow the stopband of the filter because the open-circuited stub is exactly a short-circuit only at a given frequency. However, since this approach eliminates the usage of grounding via-holes, it is commonly used in practice. The simulation results of the filter are given in Figures 6.65 and 6.66. The first figure shows the response around the center frequency, whereas the second one presents a wideband view.

For the sake of completeness, we will now study transmission zeros of the bandstop filter. To determine frequency points corresponding to the zeros, we first normalize poles of the third-order Chebyshev response as usual. Then, frequency transformation must be performed as follows:

$$\frac{1}{f'} = -\frac{f_0}{B}\left(\frac{f}{f_0} - \frac{f_0}{f}\right) \quad \Rightarrow \quad f = \frac{-B/f' \pm \sqrt{(B/f')^2 + 4f_0^2}}{2}$$

By inserting imaginary parts of the poles into the above equation, the following frequency points are determined:

$$f_1 = 9.77 \text{ GHz} \quad f_2 = 10.23 \text{ GHz}$$

which can be seen captured in Figure 6.65. Note that there is also an additional zero at dc. The other zeros are due to the fact that the coupled-line sections are acting as a half-wave transformer at those frequencies.

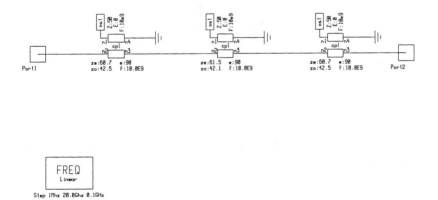

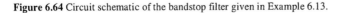

Figure 6.64 Circuit schematic of the bandstop filter given in Example 6.13.

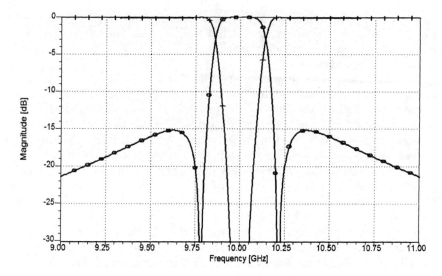

Figure 6.65 Simulated S-parameters of the bandstop filter shown in Figure 6.64 (circle: S_{11}; plus: S_{21}).

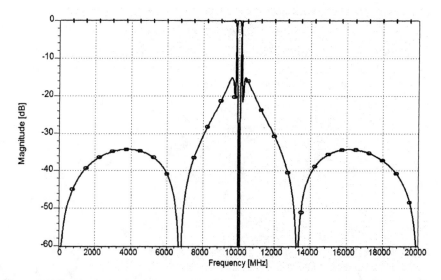

Figure 6.66 Simulated wide-band S-parameters of the bandstop filter shown in Figure 6.64 (circle: S_{11}; plus: S_{21}).

6.6 FILTER DESIGN BY COUPLING

As demonstrated in Section 6.1, the first step in designing bandpass filters is to determine the lowpass prototype network. Then, frequency mapping and impedance scaling are used to obtain the desired bandpass filter response. The filters obtained this way can be designed both for narrow and wide band. The frequency mapping effectively replaces every series inductor by a series inductor and capacitor, and every shunt capacitor by a parallel inductor and capacitor, to transform the lowpass network into a bandpass network. Although the frequency mapping is mathematically exact, the resultant component values may be too small or large for narrowband designs, which was demonstrated in Example 5.5. Besides, series arm resonators must have high Q in order to have low insertion loss at high frequencies. These problems make the usage of the frequency-mapping technique practically difficult in high-frequency narrowband bandpass filters [13]. The remedy to this problem, at the cost of an increased number of components, is to use resonators of one type coupled to each other through either inductances or capacitances, or both. Removing the selectivity function from series arms eliminates the aforementioned difficulties significantly. Essentially, this is what has been done using inverters introduced in Section 6.1.4. Now, if one replaces the inverters with their equivalent lumped circuits as shown in Figure 6.34, then we end up with a filter circuit containing only one type of resonator coupled to each other by lumped components, which is the main topology that we will study in this section. Thus, the inverter theory is closely linked to coupled filters. This technique is called filter design by coupling, and it was first introduced in the literature by Milton Dishal [35, 36]. Note that the concepts that we will introduce here are not just limited to lumped filters. By proper definition, the method becomes quite general and can be applied to coupled resonator filters, greatly expanding the applicability of the method in modern microwave filter design [13]. A good review on this subject can be found in Puglia [37, 38].

Figure 6.67(a–c) shows the narrowband approximation of a typical bandpass filter using the coupling method. The new design scheme requires the determination of the Q-factors of each resonator in general (in case of lossless resonators, only determination of Q-factors of the first and last resonators is sufficient) and the coupling amount between the resonators based on the specifications. The selection of the coupling method (i.e., capacitive or inductive) between the resonators may be important in some applications. For instance, the attenuation rates can be made more equal if inductive coupling is used between the resonators rather than capacitive coupling [13]. It should be noted that a more linear phase response can be obtained for minimum-phase filters if the amplitude response of the filter has more arithmetical symmetry then the geometrical symmetry in the linear frequency scale. In other words, the attenuation rates on the high and low sides of the filters should be the same to have better phase linearity. This means that the filter amplitude response should be geometrically asymmetrical in the linear frequency scale (see Figure 6.68; also Example 6.3).

Note that the geometric symmetry is the result of center frequency definition $\omega_0 = \sqrt{\omega_1 \omega_2}$ done in the frequency mapping. For sufficiently narrowband filters, this can be approximated by $\omega_0 = (\omega_1 + \omega_2)/2$.

There are two main methods in designing the coupled filters: the synchronously tuned method and the symmetrically detuned method. In the first method, all of the resonators are tuned to the desired center frequency, whereas in the second method, the resonators are symmetrically detuned about the center frequency. Although both methods may arrive at the same amplitude response, the passband insertions loss of the synchronous filter design is lower; for this reason, that method is more commonly used in practice [13]. It should be noted that there are two important assumptions in designing the filter with the coupling method. First, the coupling reactance is independent of frequency. This assumption can easily be justified for narrowband designs. Second, each resonator is coupled only with the resonator adjacent to it. A more general theory governing filters that have cross-couplings will be reviewed in Section 6.10.

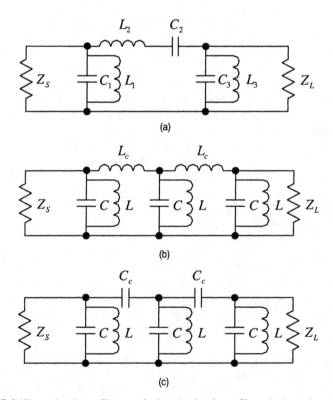

Figure 6.67 Different bandpass filter topologies: (a) bandpass filter obtained through frequency mapping technique, (b) bandpass filter with inductive coupling, and (c) bandpass filter with capacitive coupling. Note that mixed coupling between the resonators is also possible.

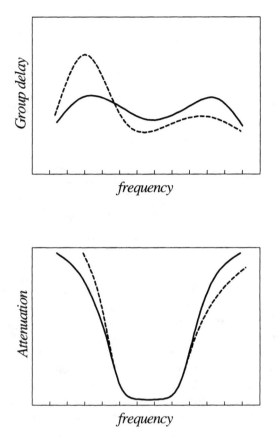

Figure 6.68 Effects of arithmetical (solid) and geometrical (dashed) symmetries on group delay and attenuation in bandpass filter transformation. Note that group delay is the negative of the derivative of the insertion phase with respect to frequency.

We start our explanation of the design technique by introducing two types of coupled-resonator ladder networks as depicted in Figure 6.69(a, b): a nodal circuit filter and a mesh circuit filter. In the figure, nodes or mesh loops are indicated by underlined numbers. In the nodal circuit, there are parallel resonators coupled with inductors or capacitors. In the mesh circuit, on the other hand, there are series resonators coupled with inductors or capacitors. Mutual coupled inductors between inductors are also possible in either kind of topology. Depending on the circuit topology, different expressions must be used for the evaluation of coupling elements, although the topologies are dual of each other. From practical point of view, the mesh network inductances are an order of magnitude greater than the optimum value of nodal network inductances for the same filter response. Therefore, the mesh network is more practical when the source and load resistances are relatively low [14].

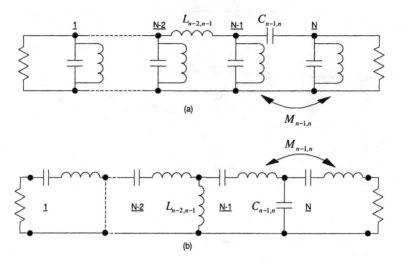

Figure 6.69 Two main circuit topologies used in filter design with coupling: (a) basic nodal circuit with inductive and capacitive coupling, and (b) basic mesh circuit with inductive and capacitive coupling. (*After:* [13].)

The design process of bandpass filters using the coupling technique first requires the determination of the loaded Q of each resonator (assuming each resonator has finite Q) and of the coupling between the resonators. We will be assuming that all the resonators have the same resonance frequency (i.e., are synchronously tuned). When the bandwidth, Δf, the center frequency, f_m, and the resonator Q-factor due to parasitic losses, Q_0, are given, then the normalized resonator Q-factor can be determined from the following formula:

$$q_0 = \frac{\Delta f}{f_m} Q_0 \tag{6.64}$$

Then, by using the resonator Q-factor (i.e., q_0), degree of the filter (i.e., n), and type of polynomial approximation as design parameters, the normalized loaded quality factors of each resonator, q_i, and coupling coefficients between the resonators, $k_{i,k}$, can be determined from the tables that are available in the literature [13]. Once the normalized quantities are obtained, they are unnormalized using the following expressions:

$$K_{i,k} = k_{i,k} \frac{\Delta f}{f_m} \tag{6.65}$$

$$Q_i = q_i \frac{f_m}{\Delta f} \tag{6.66}$$

The final step is to determine the coupling elements of the bandpass filter. Depending on the topology of the ladder network (i.e., nodal or mesh), the following expressions are used [13]:

Nodal circuit design:

1. For capacitive coupling:

$$C_{i,k} = K_{i,k} \sqrt{C_i C_k} \qquad (6.67)$$

where $C_{i,k}$ is the coupling capacitance between the ith and kth node, $K_{i,k}$ is the unnormalized coupling coefficient, and C_i is the total shunt capacitance of the ith node when all other nodes are shorted to ground.

2. For inductive coupling:

$$L_{i,k} = \frac{\sqrt{L_i L_k}}{K_{i,k}} \qquad (6.68)$$

where $L_{i,k}$ is the coupling inductance between the ith and kth node, $K_{i,k}$ is the unnormalized coupling coefficient, and L_i is the total shunt inductance of the ith node when all other nodes are shorted to ground.

Mesh circuit design:

1. For capacitive coupling:

$$C_{i,k} = \frac{\sqrt{C_i C_k}}{K_{i,k}} \qquad (6.69)$$

where $C_{i,k}$ is the coupling capacitance between the ith and kth mesh, $K_{i,k}$ is the unnormalized coupling coefficient, and C_i is the total series capacitance of the ith mesh when all other meshes are open circuited.

2. For inductive coupling:

$$L_{i,k} = K_{i,k} \sqrt{L_i L_k} \qquad (6.70)$$

where $L_{i,k}$ is the coupling inductance between the ith and kth mesh, $K_{i,k}$ is the unnormalized coupling coefficient, and L_i is the total series inductance of the ith mesh when all other meshes are open circuited.

Note that the resonance frequency of each node or mesh is given by the following equation:

$$f_m = \frac{1}{2\pi \sqrt{L_i C_i}} \qquad (6.71)$$

To determine C_i and L_i, all nodes in the nodal circuit, except the ith node, must be short-circuited to ground, and all meshes in the mesh circuit, except the ith mesh, must be open-circuited. This shows that the immediate coupling elements of the ith resonator are, in fact, contributing to the resonance frequency. However, if the coupling is relatively weak, this contribution might be negligible.

If the filter circuit contains transformers, then they can be converted into a suitable equivalent circuit with the help of the transformations given in Figure 6.70.

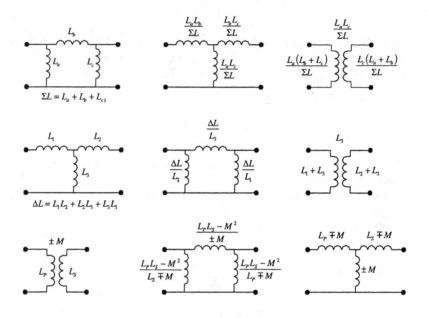

Figure 6.70 Equivalent circuits for the coupled inductors. (*After:* [15].)

Although this procedure is quite general for narrowband filters, it requires the use of tables to determine coupling coefficients, Q-factors, and component values for each resonator. In some applications, it can be assumed that resonators are lossless for the sake of simplification. In that case, the coupling coefficients can be conveniently derived directly from the lowpass prototype values, which will be demonstrated in the next sections.

Example 6.14

In this example, we will demonstrate filter design using the coupling technique by implementing a fifth-order Chebyshev bandpass filter. The center frequency and bandwidth are 5 GHz and 200 MHz, respectively. The ripple in the bandpass should be 0.1 dB. It is decided to employ capacitive coupling between the resonators. The input and output impedances are selected as 50 ohms, and the unloaded Q of the internal resonators will be assumed to be very large (i.e., ideal resonators).

The design process commences by determining loaded Q-factors and coupling coefficients for ideal resonators from the tables available in the literature [13]:

$$q_1 = 1.3013 \quad q_5 = 1.3013$$

$$k_{12} = 0.7028 \quad k_{23} = 0.5355$$

$$k_{34} = 0.5355 \quad k_{45} = 0.7028$$

If the resonators are not ideal, then there will be some insertion loss due to the loss associated with the resonators. Therefore, if the filter insertion loss is specified, then the tables will also provide the required unloaded Q-factor of each resonator. Alternatively, one can resort to computer optimization to take losses into account.

The next step of the process is to decide on the resonator capacitance and inductance values. These values are selected in such a way that they are easily realizable at the given frequency with low loss. Besides, the component values of the resonators are important to achieve the required loaded Q-factor of the input and output resonators. Although it may be possible to select component values such that they produce the required Q when they are terminated with the source and load impedances, this is not usually done in practice. The common approach is to use transformer stages at the input and output. Thus, it is also important to select the resonator component values in such a way that the transformer ratios required to produce the necessary Q are not peculiar. By taking these issues into consideration, the capacitance and inductance values of the resonators are selected as 5.0 pF and 0.203 nH, respectively. These values will provide a tank circuit, which resonates at 5 GHz (note that all resonators will use the same tank circuit). This selection will result in the following input/output transformer ratio:

$$n = \left[\frac{Z_s}{q_{1,5} \frac{f_m}{\Delta f} \frac{1}{\omega C}} \right]^{1/2} = \left[\frac{50}{1.3013 \frac{5 \times 10^9}{200 \times 10^6} \frac{1}{2\pi \cdot 5 \times 10^9 \cdot 5 \times 10^{-12}}} \right]^{1/2} \cong 0.49$$

Then, coupling capacitances between the resonators are obtained by using the above coupling coefficients as follows:

$$C_{12} = k_{12} \frac{\Delta f}{f_m} \sqrt{C_{\mathrm{I}} C_{\mathrm{II}}} = 0.7028 \frac{5.0 \times 10^{-12}}{25} = 0.1406 \quad \mathrm{pF}$$

$$C_{23} = k_{23} \frac{\Delta f}{f_m} \sqrt{C_{\mathrm{II}} C_{\mathrm{III}}} = 0.7028 \frac{5.0 \times 10^{-12}}{25} = 0.1071 \quad \mathrm{pF}$$

$$C_{34} = k_{34} \frac{\Delta f}{f_m} \sqrt{C_{\mathrm{III}} C_{\mathrm{IV}}} = 0.7028 \frac{5.0 \times 10^{-12}}{25} = 0.1071 \quad \mathrm{pF}$$

$$C_{45} = k_{45} \frac{\Delta f}{f_m} \sqrt{C_{\mathrm{IV}} C_{\mathrm{V}}} = 0.7028 \frac{5.0 \times 10^{-12}}{25} = 0.1406 \quad \mathrm{pF}$$

where the numbering convention of Zverev is adopted. In this convention, capacitances with Roman numerals refer to total nodal capacitance at each node when all of the other nodes are shorted to ground. Note that the total nodal capacitance of each node should be equal to 5.0 pF to provide the same tank resonance frequency. This requires modification of the capacitance of the original tank circuit for each resonator because the coupling capacitors also contribute to the nodal capacitance. The modified resonator capacitances are found in the following way by taking the coupling capacitors into consideration:

$$C_a = C_I - C_{12} = 4.859 \quad \text{pF}$$
$$C_b = C_{II} - C_{12} - C_{23} = 4.752 \quad \text{pF}$$
$$C_c = C_{III} - C_{23} - C_{34} = 4.786 \quad \text{pF}$$
$$C_d = C_{IV} - C_{34} - C_{45} = 4.752 \quad \text{pF}$$
$$C_e = C_V - C_{45} = 4.859 \quad \text{pF}$$

where the capacitances with letters identify the value of the capacitors in each tank circuit. Note that inductance values are not affected. This completes the design of the bandpass filter.

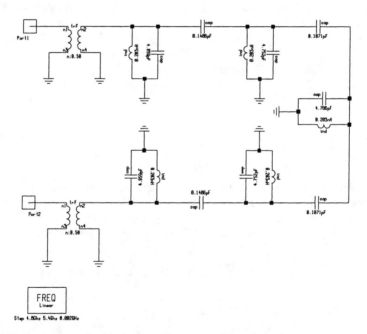

Figure 6.71 Circuit schematic of the bandpass filter given in Example 6.14.

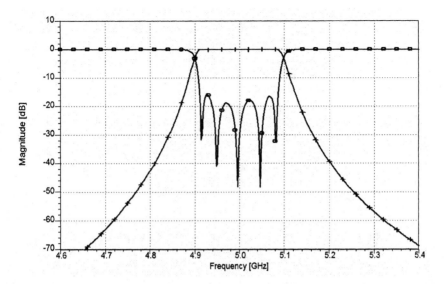

Figure 6.72 Simulated S-parameters of the bandpass filter shown in Figure 6.71 (circle: S_{11}; plus: S_{21}).

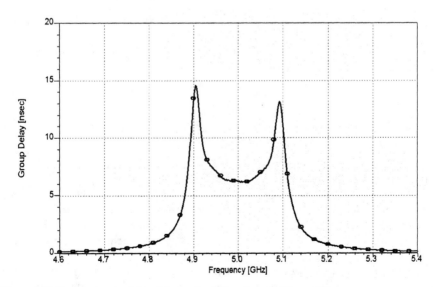

Figure 6.73 Simulated group delay of the bandpass filter shown in Figure 6.71.

To test the design, the filter circuit is entered into a circuit simulator as shown in Figure 6.71. Note that input and output transformers are implemented using ideal transformers for the sake of demonstration. In practice, one may need to employ a lumped approximation to the ideal transformer. Also note that there are no large variations between the component values, which would happen in a narrowband filter design based on the frequency mapping, as demonstrated before. Again, this is one of the biggest advantages of filter design using the coupling technique. The cost is the increased number of components. The simulation results are shown in Figures 6.72 and 6.73. The center frequency and bandwidth of the filter are 5 GHz and 200 MHz, respectively, as prescribed. The five poles of the transfer function are also clearly separable in the response. The return loss is better than 15 dB in the passband, and the insertion loss in the stopband is 30 dB at 4.85 GHz and 5.15 GHz.

So far, we have demonstrated how to design bandpass filters using the coupling method. The technique is general and can be applied to lossy resonators as well. However, it would be practically advantageous if one could determine the coupling coefficients without resorting to tabulated values. Fortunately, one can calculate the coupling coefficient directly from the lowpass prototype values in the case of narrowband filters with lossless (or practically low-loss) resonators. It can be shown that the coupling coefficients of narrowband lossless bandpass filters can be found using the following formula:

$$K_{n,n+1} = \frac{BW}{f_0\sqrt{g_n g_{n+1}}} \quad n = 1,2,\ldots,N \tag{6.72}$$

where BW is the 3-dB bandwidth, f_0 is the center frequency, N is the filter order, and g_n's are the lowpass prototype filter values. $K_{n,n+1}$ is the coupling between the nth and $(n + 1)$st resonators. The loaded Q of the first and last resonators are found using the following expressions:

$$Q_1 = \frac{f_0}{BW} g_0 g_1$$

$$Q_N = \frac{f_0}{BW} g_N g_{N+1} \tag{6.73}$$

Once the coupling coefficients and loaded Q-factors are determined, the remaining task is to physically implement the filter using coupled resonators.

The design procedure of coupled-resonator bandpass filter has been introduced using lumped components where coupling between the resonators is facilitated by lumped capacitors, inductors, transformers, or any combination of them. However, in designing microstrip distributed filters, the coupling is achieved by placing the transmission line resonators in close proximity; no lumped components are used. Thus, some other means must be employed to characterize coupling between two resonators in microstrip distributed filters. Fortunately, full-wave electromagnetic simulations are very handy to address this problem.

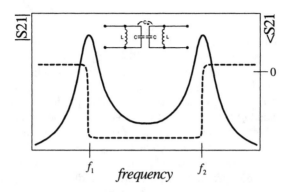

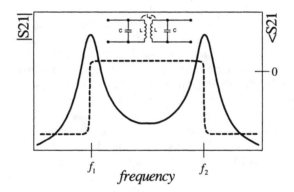

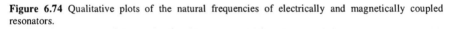

Figure 6.74 Qualitative plots of the natural frequencies of electrically and magnetically coupled resonators.

The actual coupling coefficient (i.e., unnormalized) for a pair of resonators separated by a distance is determined from the natural frequencies as follows:

$$K = \pm \frac{f_2^2 - f_1^2}{f_2^2 + f_1^2} \cong \pm \frac{f_2 - f_1}{f_0} \qquad (6.74)$$

where f_2 and f_1 are the two natural frequencies obtained from full-wave simulations [see Figure 6.74(a, b)]. The sign of (6.74) is selected depending on the type of the coupling (i.e., magnetic or electric). It is important to note that the absolute sign of a particular coupling is not important in the filter realization; it is important, however, to use different signs for couplings with different transmission phases. Equation (6.74) readily suggests that one can determine the coupling coefficient between two microstrip resonators by simulating them with different separations and inspecting the locations of the natural frequencies. Note that some means of input/output coupling (i.e., tapped or capacitive) to the resonators must be employed to facilitate the signal transfer.

This characterization is carried out in all the desired resonator separations. After performing the experiment, the designer then adjusts the separations to get the coupling coefficients found from (6.72). Sometimes it might be necessary to revisit the specifications or change the resonator topology because it turns out that it wouldn't be practically possible to obtain some of the coupling coefficients. Similarly, the loaded Q-factors given by (6.73) for the input and output resonators can be determined efficiently and accurately using full-wave simulations.

At this point, it would be instructive to explain how (6.72) and (6.73) are derived. For this purpose, let's consider the filter circuits shown in Figure 6.75. The first circuit in the figure shows a third-order bandpass filter, which is obtained directly from a lowpass prototype using the frequency-mapping technique, which was visited extensively before. Our aim is to replace the series arm resonator with a shunt resonator and then determine the required coupling coefficients between each resonator. To achieve this, we will use inverters. Converting the series arm resonator to a shunt resonator with the help of inverters is depicted in the second circuit. However, as one can notice, not all of the shunt resonators in the second circuit are the same; the last resonator is different from the first two. To convert the last resonator, we utilize a series of transformations using inverters as depicted in the third and fourth circuits. After converting all of the resonators to the same type, we eliminate the last inverter by transforming the load resistor as shown in the fifth circuit. The remaining task is to obtain coupling coefficients from the characteristic impedances of the inverters. For this purpose, the distributed inverters are replaced by an appropriate lumped equivalent (see Figure 6.34). Then, the coupling coefficients are obtained as follows:

$$K_{12} = \frac{C_m^{(12)}}{C_1} \quad K_{23} = \frac{C_m^{(23)}}{C_1}$$

where $C_m^{(ij)}$ is the coupling capacitance between the ith and jth resonators which can be obtained from (assuming $C_m \ll C_i$)

$$C_m^{(12)} = \frac{1}{\omega}\sqrt{\frac{C_1}{L_2}} \quad C_m^{(23)} = \frac{1}{\omega}\sqrt{\frac{C_1^2}{L_2 C_3}}$$

Then, by combining the above two equations, one can obtain

$$K_{12} = \frac{1}{\omega}\sqrt{\frac{C_1}{L_2}}\frac{1}{C_1} = \frac{B}{\omega}\sqrt{\frac{1}{g_1 g_2}}$$

$$K_{23} = \frac{1}{\omega}\sqrt{\frac{C_1^2}{L_2 C_3}}\frac{1}{C_1} = \frac{B}{\omega}\sqrt{\frac{1}{g_2 g_3}}$$

which are the same as (6.72).

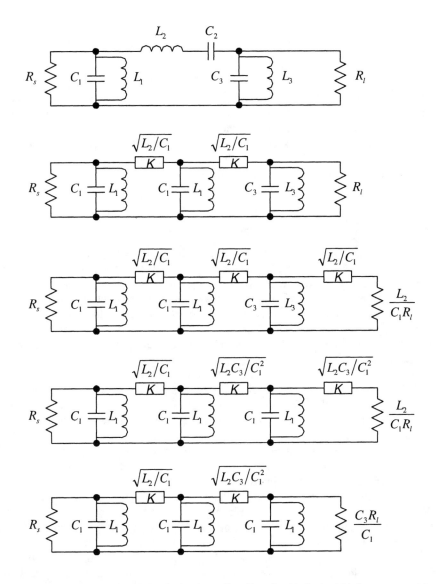

Figure 6.75 Conversion of a third-order bandpass filter (first circuit) obtained by frequency mapping into a coupled resonator filter (last circuit). The circuit blocks represented by K are impedance inverters. Coupling coefficients can be determined from the characteristic impedances of the inverters.

The final step is to verify the Q-factors of the input and output resonators. By inspecting Figure 6.75, one can see that

$$Q_1 = \omega C_1 R_s = \frac{\omega}{B} g_0 g_1$$

$$Q_N = \omega C_1 \frac{R_l C_3}{C_1} = \frac{\omega}{B} g_N g_{N+1}$$

which is the same as (6.73). This procedure can be generalized for coupled filters with higher order. Thus, it was demonstrated that filter design by coupling can also be formulated using the inverters and their lumped equivalences.

In the following sections, the two most commonly used microstrip bandpass filters are introduced. The design of these filters can be successfully carried out using the coupling technique explained above.

6.7 INTERDIGITAL FILTERS

Interdigital filters are very popular in microwave engineering since they use the available circuit area efficiently and can be designed both for narrow and wider bandwidths. Another advantage of interdigital filters is that they have perfectly arithmetical symmetry providing better phase and delay characteristics. However, this latter statement is more easily achievable in the coaxial rod type of structures; microstrip interdigital filers will have relatively more distortion. Perhaps the main disadvantage of interdigital filters is the spurious bandpass response appearing around the third harmonic of center frequency [2].

Figure 6.76(a, b) shows the two common forms of interdigital filters. The tapped-line configuration [Figure 6.76(b)] has the advantage of reducing the total number of lines by two. It also has the advantage in situations where couplings between the first and last pair of lines of the original direct-fed configuration [Figure 6.76(a)] become very tight (hence, there is very small line separation). The disadvantage of the tapped version is that no exact relationship exists to the original configuration; only approximate models are available. However, for narrowband filters, the approximation works well. Besides, by the addition of two lumped capacitors at the open ends of the input and output lines, as shown in Figure 6.76(b), the response can be compensated for further. The approximate design formulas for the original interdigital configuration were given by Matthaei [39]. An exact synthesis approach was presented by Wenzel [40], who showed that the interdigital structures can be modeled by using the capacitance matrix of the structure. The design approach of tapped-line interdigital filters was demonstrated by Cristal and others [41–46].

There are different techniques to design interdigital filters. The method given by Matthaei is perhaps the most commonly employed one for the direct-fed configuration [39]. One can also use the approach presented in Section 6.6 to design interdigital filters, provided that the coupling between two lines is characterized accurately and properly. Once this is done, the required couplings (thus, the line separations) can be determined using the prototype filter values as usual [43]. Another important point is the determination of input and output loaded Q-factors. For a given Q, the required tapping position of quarter-wavelength-long resonators is given as follows:

$$Q_L = \frac{Z_0}{R} \frac{\pi}{4\sin^2\left(\dfrac{\pi l}{2L}\right)} \qquad (6.75)$$

where Z_0 and R are the source and filter internal impedances, respectively. The tapping position is given by l. Derivation of this expression can be found in Chapter 1.

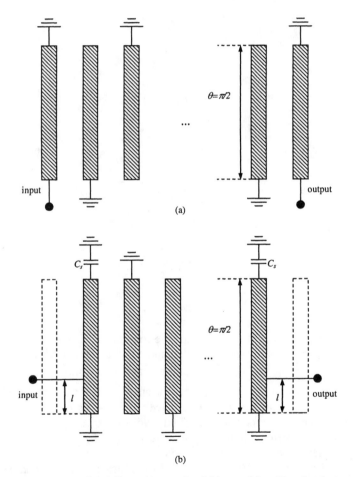

Figure 6.76 Microstrip interdigital filters: (a) normal and (b) tapped line. Note that the tapped version reduces the total number of lines by two. Tapping position determines the Q-factor of input and output resonators.

Closed-form design equations for the tapped-line interdigital filters are also available in the literature [45]. For Figure 6.76(b), the self and mutual capacitances of the lines can be found from the following expressions:

$$C_1 = \frac{376.7\varepsilon}{\sqrt{\varepsilon_r}}(Y_1 - Y_{1,2}) \tag{6.76}$$

$$C_N = \frac{376.7\varepsilon}{\sqrt{\varepsilon_r}}(Y_1 - Y_{N-1,N}) \tag{6.77}$$

$$C_{i,i+1} = \frac{376.7\varepsilon}{\sqrt{\varepsilon_r}}Y_{i,i+1} \quad i = 1,2,...,N-1 \tag{6.78}$$

$$C_i = \frac{376.7\varepsilon}{\sqrt{\varepsilon_r}}(Y_1 - Y_{i-1,i} - Y_{i,i+1}) \quad i = 1,2,...,N-1 \tag{6.79}$$

where

$$Y_{i,i+1} = J_{i,i+1}\sin(\theta_1) \quad i = 1,2,...,N-1 \tag{6.80}$$

$$J_{i,i+1} = \frac{Y}{\sqrt{g_i g_{i+1}}} \quad i = 1,2,...,N-1 \tag{6.81}$$

$$Y = \frac{Y_1}{\tan(\theta_1)} \tag{6.82}$$

$$\theta_1 = \frac{\pi}{2}\left(1 - \frac{FBW}{2}\right) \tag{6.83}$$

$$\theta_2 = \frac{\sin^{-1}\left\{\sqrt{\frac{Y\sin^2(\theta_1)}{g_0 g_1 Y_0}}\right\}}{1 - \frac{FBW}{2}} \tag{6.84}$$

In the above equations, N is the order of the lowpass prototype filter, g_i are the values of the lowpass prototype filter element, FBW is the fractional bandwidth, which is given by $(\omega_2 - \omega_1)/\omega_0$, Y_0 is the source and load admittance, and Y_1 is the resonator line admittance. The tapping distance, l, is determined using the electrical length, θ_2. Note that the capacitance matrix is a block-diagonal matrix because it has been assumed that a given line couples only to its immediate neighbors. After determining the capacitance matrix, the remaining task is to physically implement the interdigital structure.

The lumped capacitance, C_s [Figure 6.76(b)], which is used for compensating the resonance frequency shift, is given as follows:

$$C_s = \frac{Y_0^2}{Y_t \omega_0} \frac{\cos(\theta_2)\sin^3(\theta_2)}{1 + Y_0^2/Y_t^2 \cdot \sin^2(\theta_2)\cos^2(\theta_2)} \tag{6.85}$$

where

$$Y_t = Y_1 - Y_{1,2}^2/Y_1 \tag{6.86}$$

In the above equations, ε and ε_r are the permittivity of free space and relative permittivity, respectively. Alternatively, one can also make the input and output lines little longer to achieve the same effect.

6.8 HAIRPIN FILTERS

Hairpin filters belong to the class of microstrip filters that use quarter-wavelength-long coupled resonators [43, 46]. Instead of forming a straight cascade connection as demonstrated in Section 6.4, however, the lines are folded in a U-shape and arranged in a fashion so that the overall filter can be compacted. Alternatively, one can also view the structure as half-wavelength-long resonators coupled to each other. Note that it is assumed that a given resonator couples only to its immediate neighbors.

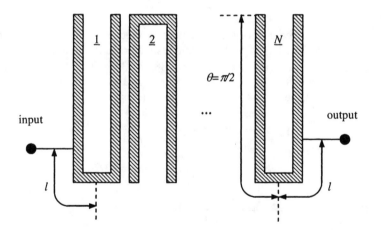

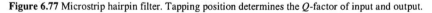

Figure 6.77 Microstrip hairpin filter. Tapping position determines the Q-factor of input and output.

Figure 6.77 shows a tapped-line, hairpin, microstrip line filter. Note that it could also be possible to use line coupling, instead of tapping, employing additional coupled lines at the input and output as in the interdigital case. The design of hairpin filters can be carried out successfully using the filter design by coupling technique introduced in Section 6.6. For a given Q, the required tapping position of half-wavelength-long resonators is given by

$$Q_L = \frac{Z_0}{R} \frac{\pi}{2\sin^2\left(\dfrac{\pi l}{2L}\right)} \tag{6.87}$$

where Z_0 and R are the source and filter internal impedances, respectively. The tapping position is given by l. Derivation of this expression can be found in Chapter 1. Another interesting point is that the Q-factor of the half-wavelength resonator is twice the Q of a quarter-wavelength resonator employed in interdigital filters. The design process of hairpin filters will now be demonstrated by an example.

Example 6.15

In this example, we will design a hairpin bandpass filter using the coupling technique. The selected filter response is a fifth-order Chebyshev response with a 0.1-dB ($\varepsilon = 0.1526$) ripple. The center frequency and filter bandwidth are 5.0 GHz and 200 MHz, respectively. The substrate parameters are $h = 10$ mils and $\varepsilon_r = 2.2$.

Design of the filter starts with determination of the lowpass prototype values, using (6.17) for the selected response as follows:

$$g_1 = 1.147 \quad g_2 = 1.371$$
$$g_3 = 1.975 \quad g_4 = 1.371$$
$$g_5 = 1.147 \quad g_6 = 1.0$$

The characteristic impedance of the resonators is selected as 50 ohms resulting in 30-mils wide lines. The length of the hairpin resonators is found to be 860 mils by using the effective dielectric constant, $\varepsilon_{eff} = 1.88$, for 30-mils wide lines. The next step is to determine the Q-factors of the input and output resonators and the coupling coefficients between the resonators as follows:

$$k_{12} = k_{45} = \frac{200 \times 10^6}{5 \times 10^9 \sqrt{1.147 \cdot 1.371}} = 0.032$$

$$k_{23} = k_{34} = \frac{200 \times 10^6}{5 \times 10^9 \sqrt{1.975 \cdot 1.371}} = 0.024$$

$$Q_1 = Q_2 = \frac{5 \times 10^9}{200 \times 10^6} 1.147 = 28.7$$

After determining the main design parameters of the filter, one continues the design procedure by simulating a pair of hairpin resonators to find coupling coefficient versus resonator separation, obtaining the plot shown in Figure 6.78. By using this coupling plot, the following required resonator separations can be found to realize k_{12}, k_{23}, k_{34}, and k_{45}:

$$k_{12} = k_{45} = 0.032 \rightarrow s_{12} = s_{45} \cong 19 \quad \text{mils}$$
$$k_{23} = k_{34} = 0.024 \rightarrow s_{23} = s_{34} \cong 24 \quad \text{mils}$$

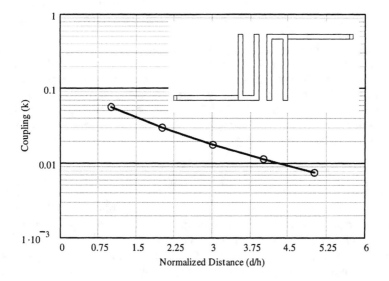

Figure 6.78 Coupling coefficient versus normalized resonator spacing for a pair of hairpin resonators. Note that the distance is normalized to substrate thickness. The inset figure shows the simulated geometry.

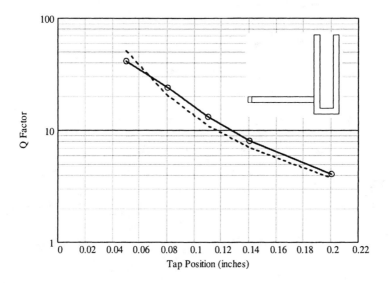

Figure 6.79 Q-factor of the hairpin resonator versus tapping position, l, obtained through full-wave electromagnetic simulations (solid) and equation (6.87) (dashed). The inset figure shows the simulated geometry.

The final step is to calculate the tapping positions of the input and output resonators to obtain the input and output Q-factors. For this purpose, another set of full-wave simulations of a single resonator are performed for different tapping positions, resulting in the plot shown in Figure 6.79.

Note that determination of the Q-factor from simulation results needs particular attention. The usual 3-dB-points method may not be accurate for resonators with relatively low Q. Thus, an alternative method based on the derivative of susceptance is employed here [47]. In this method, the resonator is modeled as a shunt conductance-inductance-capacitor (GLC) in the vicinity of resonance point. Then, Q is calculated from

$$Q = \frac{\omega}{2(G+Y_0)} \frac{\partial B}{\partial \omega} \bigg|_{\omega=\omega_0}$$

The full-wave electromagnetic simulation model of the filter is shown in Figure 6.80 and simulation results are given in Figures 6.81 and 6.82. Through the passband, the insertion loss is around 4 dB and the return loss is better than 15 dB. The poles of the filter are also resolved, which can be seen from the return loss. The reason for not having very distinct pole resolution is two-fold: One is the accuracy of the circuit dimensions. Due to the fact that all shapes must snap on a finite grid size in the computer simulation, the dimensions are usually quantized while the circuit is prepared for simulation, resulting in slight variations from the actual dimensions. The second is the loss of each resonator. The coupling theory outlined above assumes ideal lossless resonators.

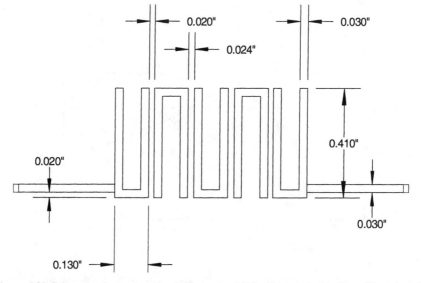

Figure 6.80 Full-wave electromagnetic simulation model for the hairpin bandpass filter given in Example 6.15.

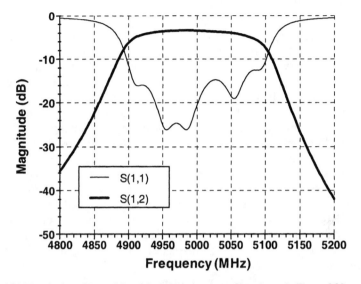

Figure 6.81 Magnitudes of S_{11} and S_{12} of the hairpin bandpass filter shown in Figure 6.80.

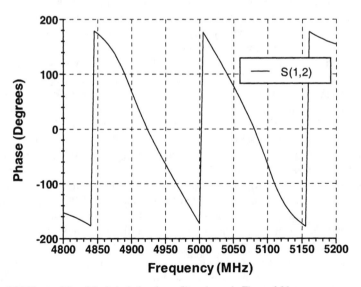

Figure 6.82 Phase of S_{12} of the hairpin bandpass filter shown in Figure 6.80.

6.9 CROSS-COUPLED FILTERS

For some applications, the elliptic filter response is the optimum choice in the sense that it provides a much steeper transition band for a given order and relatively small insertion losses as compared to Butterworth and Chebyshev filters. This is because the latter have attenuation poles at infinity and/or zero frequency; which is not practically required. Elliptic filters alleviate this by placing the attenuation poles at finite frequencies thus providing steeper transitions (or higher selectivity). Elliptic filters can be constructed by employing cross-coupling between the filter resonators. It has been shown that the most general form of bandpass transfer functions of symmetrical networks can be realized by using either single- or dual-mode coupled resonators. Optimum performance is obtained when the cavities are synchronously tuned [13]. Then, the desired transfer function is obtained by adjusting the amount and sign of the coupling between each resonators. Note that depending on the transfer function, it may not be necessary to couple a given resonator to every other resonator in the filter. In fact, it is the key point in cross-coupled filter design to reduce the coupling matrix so that the minimum possible number of resonators are coupled. This is quite important in terms of the practical implementation of the resultant filter.

Traditionally, synchronously tuned rectangular or circular waveguide cavities have been used to design low-loss cross-coupled filters. However, microstrip resonators have increasingly been used to design cross-coupled filters where very high Q is not the main design parameter. Figure 6.83 shows a typical cross-coupled resonator network. Although only some of the couplings are shown, there is coupling between each resonator in the most general form. The design of cross-coupled filters also has a very wide literature. Usage of dual-mode resonators is demonstrated in the classic paper by Lin [48]. Rhodes showed the design of linear phase filters using cross-coupled waveguide cavities [49]. Then, Atia et al. published a series of papers on the synthesis of dual-mode bandpass filters based on Darlington's theory [50–53]. Zaki et al. demonstrated usage of different coupling mechanisms other than the simple iris between the resonators [54–56]. Recent works also showed that microstrip networks can be used to design cross-coupled filters [57–65]. One of the intriguing points of the general theory is that it is possible to design filters by facilitating the coupling between two linearly independent modes of a given cavity and then coupling such cavities to other ones. This way, the number of resonators can be reduced, thus providing significant advantages in applications where weight is critical, such as satellites. The original work of Atia and others was concentrated heavily on such filters. Of course, the tricky part here is to adjust mode coupling in each metal cavity, which is done by inserting metal screws at suitable locations. One could imagine that this approach can be extended to overmoded cavities supporting more than two modes. However, the adjustment of coupling between each mode becomes extremely difficult in the case of more than two modes. In this section, we will concentrate on single-mode cavities only.

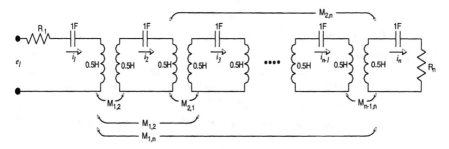

Figure 6.83 Typical cross-coupled resonator network. Note that only some of the couplings are shown. In the most general form, there is coupling between each resonator.

In cross-coupled filters, the design of the bandpass filter requires determination of the coupling coefficients, as well as the termination resistors, according to a given voltage transfer ratio through an appropriate synthesis procedure. Here, we will introduce the synthesis procedure based on Darlington's theory. The synthesis procedure starts by writing the loop equations for the given resonator circuit topology as follows [51]:

$$
\begin{bmatrix} e_1 \\ 0 \\ 0 \\ 0 \\ \vdots \\ 0 \end{bmatrix} =
\begin{bmatrix}
\lambda + R_1 & jM_{12} & jM_{13} & jM_{14} & \cdots & & jM_{1n} \\
jM_{12} & \lambda & jM_{23} & & & & \\
jM_{13} & jM_{23} & \lambda & & & & \vdots \\
jM_{14} & & & \lambda & & & \\
\vdots & & & & \lambda & & jM_{n-1,n} \\
jM_{1n} & & \cdots & & & jM_{n-1,n} & \lambda + R_n
\end{bmatrix}
\begin{bmatrix} i_1 \\ i_2 \\ i_3 \\ i_4 \\ \vdots \\ i_n \end{bmatrix}
\tag{6.88}
$$

where

$$
\lambda = j\left(\omega - \frac{1}{\omega}\right) \tag{6.89}
$$

$$
jM_{ij} \approx j\omega M_{ij} \approx j\omega_0 M_{ij}
$$

In the above equations, ω_0 is the center frequency of the filter. The coupling between the ith and jth resonators is indicated by M_{ij}. Note that it is assumed that the coupling coefficients are frequency independent. For narrowband applications, this assumption holds up well. The loop equations can be written in a more compact form as

$$
\mathbf{E} = (\lambda\,\mathbf{U} + j\mathbf{M} + \mathbf{R})\times\mathbf{I} \tag{6.90}
$$

The $\mathbf{U}$ matrix is the identity matrix, and the $\mathbf{R}$ matrix accounts for the source and load terminations. $\mathbf{M}$ is called the coupling matrix, and it is assumed that its entries are frequency independent. Then, the voltage transfer ratio of the filter network is defined as follows:

$$\frac{e_n}{e_1} = K \frac{P(\lambda)}{Q(\lambda)} \tag{6.91}$$

where K is a constant, $Q(\lambda)$ is a Hurwitz polynomial of degree n, and $P(\lambda)$ is an even polynomial whose degree is $m \le n-2$. A Hurwitz polynomial has no zeroes in the open right half of the s-plane. A polynomial that does not have zeros either in the open right half of the s-plane or on the $j\omega$-axis is called strictly Hurwitz. The coupling matrix $\mathbf{M}$ has general entries of M_{ij} for $i \ne j$ and 0 for $i = j$. For even $P(\lambda)$ polynomial, the coupling matrix can be reduced to the following form [51]:

$$\mathbf{M} = \begin{bmatrix} 0 & M_{1,2} & 0 & \cdots & 0 & M_{1,n} \\ M_{1,2} & 0 & M_{2,3} & \cdots & M_{2,n-1} & 0 \\ 0 & M_{2,3} & 0 & \cdots & & M_{3,n} \\ \vdots & & & & & \vdots \\ 0 & M_{2,n-1} & 0 & \cdots & 0 & M_{n-1,n} \\ M_{1,n} & 0 & M_{3,n} & \cdots & M_{n,n-1} & 0 \end{bmatrix} \tag{6.92}$$

Now, the remaining task is to determine the coupling matrix entries and the values of the termination resistors. For this purpose, we first note that the denominator of the transfer functions is the determinant of the current coefficient matrix [51]:

$$Q(\lambda) = \sum_{k=0}^{n} q_k \lambda^{n-k} = \det(\lambda \mathbf{U} + j\mathbf{M} + \mathbf{R}) \tag{6.93}$$

Then, using the Leverrier-Souriau-Frame algorithm [25], it can be shown that

$$q_0 = 1$$

$$q_k = -\frac{\text{trace}\{(j\mathbf{M} + \mathbf{R}) \times \mathbf{B}_{k-1}\}}{k} \quad k = 1, 2, \ldots, n$$

where

$$\mathbf{B}_0 = \mathbf{U}$$

$$\mathbf{B}_k = -\frac{\text{trace}\{(j\mathbf{M} + \mathbf{R}) \times \mathbf{B}_{k-1}\}}{k} \mathbf{U} + (j\mathbf{M} + \mathbf{R}) \times \mathbf{B}_{k-1} \quad k = 1, 2, \ldots, n$$

which immediately results in

$$q_1 = R_1 + R_n \tag{6.94}$$

The above equation determines the termination resistors from the denominator of the transfer function when the ratio of R_1/R_n is given. The next step is to determine the coupling matrix entries. To accomplish this, consider the cross-coupled resonator network as a two-port structure as shown in Figure 6.84.

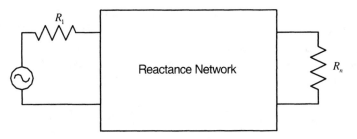

Figure 6.84 Two-port equivalent of the cross-coupled resonator network.

The voltage-current relationship of the two-port network shown in the figure can be written using Y-parameters as follows:

$$\begin{bmatrix} I_1 \\ I_2 \end{bmatrix} = \begin{bmatrix} y_{11} & y_{12} \\ y_{21} & y_{22} \end{bmatrix} \begin{bmatrix} V_1 \\ V_2 \end{bmatrix} \tag{6.95}$$

The short-circuit admittances can be obtained from the short-circuit currents as

$$y_{11} = \left. (\lambda \mathbf{U} + j\mathbf{M})^{-1} \right|_{1,1}$$
$$y_{21} = \left. (\lambda \mathbf{U} + j\mathbf{M})^{-1} \right|_{n,1} \tag{6.96}$$

Since $\mathbf{M}$ is a real symmetric matrix, its eigenvalues are real, and there exists an orthogonal matrix $\mathbf{T}$ such that

$$\mathbf{M} = \mathbf{T}\mathbf{\Lambda}\mathbf{T}^T \tag{6.97}$$

$$\mathbf{\Lambda} = \mathrm{diag}(\lambda_1, \lambda_2, ..., \lambda_n) \tag{6.98}$$

where λ_k are the eigenvalues of $\mathbf{M}$. Superscript T denotes the transpose operation.

Then, the following relationships are obtained between the Y-parameters and the entries of the $\mathbf{T}$ matrix [51]:

$$y_{11} = \sum_{k=1}^{n} \frac{T_{1,k}^2}{\lambda - \lambda_k}$$
$$y_{21} = \sum_{k=1}^{n} \frac{T_{1,k} \cdot T_{n,k}}{\lambda - \lambda_k} \tag{6.99}$$

Since the admittance functions y_{11} and y_{12} can be obtained from a given network transfer function through Darlington's method (see Chapter 1 for details), it is possible to identify the residues and form the first and last rows of the orthogonal matrix $\mathbf{T}$. The remaining rows of matrix $\mathbf{T}$ can be constructed using the Gram-Schmidt orthogonalization procedure. After forming the orthogonal matrix, the coupling matrix can be constructed using (6.97), and the synthesis procedure is completed.

For symmetrical networks, the coupling matrix can be simplified further. The assumption of a symmetrical network ($R_1 = R_n$, and $\mathbf{M}$ is symmetrical about the diagonal) provides convenience in designing the bandpass filters. In that case, it is computationally advantageous to use the even-mode coupling matrix of the system, which is obtained from the original coupling matrix by folding it along the centerline of rows and columns [51]:

$$
\mathbf{M}_e = \begin{bmatrix} M_{1,n} & M_{1,2} & M_{1,n-2} & M_{1,4} & \cdots & M_{1,n/2} \\ M_{1,2} & M_{2,n-1} & M_{2,3} & \cdots & \\ M_{1,n-2} & M_{2,3} & M_{3,n-2} & \cdots & & \vdots \\ \vdots & & & & & \\ & & & \cdots & & \\ M_{1,n/2} & & & \cdots & & M_{n/2,n/2} \end{bmatrix}
\tag{6.100}
$$

The eigenvalues of the even-mode matrix $\mathbf{M}_e$ are identical in magnitude to the $n/2$ distinct eigenvalues of the original matrix $\mathbf{M}$. The even-mode matrix corresponds to the excitation of the unterminated network by two identical voltage sources at both ends. Note that the new even-mode orthogonal matrix $\mathbf{T}_e$ can be constructed only by knowing the first row elements obtained through the even-mode driving point admittance.

The coupling matrix obtained using the method outlined above is not unique since an infinite number of $\mathbf{T}$ matrixes can be constructed using the first and last rows. However, all of those coupling matrixes would produce the same transfer function. On the other hand, not all of them can be easily realized. Thus, it is important to obtain a coupling matrix that is easy to realize. This is achieved by reducing the desired matrix entries (hence, the couplings between the particular cavities) to zero through similarity transformations. To perform the similarity transformation, one must first construct an orthogonal matrix, which is identical to the unity matrix except for the (i, i), (i, j), (j, i), and (j, j) entries:

$$
\mathbf{O} = \begin{bmatrix} 1 & 0 & & \cdots & & 0 & 0 \\ 0 & \ddots & & & & & 0 \\ & & \cos\theta & 0 & \sin\theta & & \\ \vdots & & 0 & \ddots & 0 & & \vdots \\ & & -\sin\theta & 0 & \cos\theta & & \\ 0 & & & & & \ddots & 0 \\ 0 & 0 & & \cdots & & 0 & 1 \end{bmatrix}
\tag{6.101}
$$

where θ is the rotation angle. Then, the new coupling matrix is given by

$$
\mathbf{M}' = \mathbf{O}^T \times \mathbf{M} \times \mathbf{O}
\tag{6.102}
$$

Now, the rotation angle must be chosen in a such a way that the required matrix entries are minimized in absolute value after the similarity transformation. This procedure should be continued until the particular entries are sufficiently minimized. If the aimed for topology is physically realizable, the iteration should converge. It is important to note that during the similarity transformation, the first and the last rows of the coupling matrix, $\mathbf{M}$, should not be used (i.e., $i \neq 1,n$ and $j \neq 1,n$) [66]. The same rule applies to the first row of the even-mode coupling matrix, $\mathbf{M}_e$.

Example 6.16

In this example, we will demonstrate the cross-coupled filter synthesis procedure, which is outlined above. This example is the detailed version of one of the designs that can be found in Atia et al. [50].

Let's consider the following six-order elliptic transfer function:

$$|t(s)|^2 = \frac{1}{1 + \varepsilon^2 \dfrac{s^4 \left(s^2 + z_1^2\right)^2 \left(s^2 + z_1^2\right)^2}{\left(s^2 + p_1^2\right)^2 \left(s^2 + p_1^2\right)^2}} \tag{6.103}$$

where $s = j\lambda$. The following poles and zeros are chosen to produce an elliptic function response having a selectivity of 0.63:

$$z_1 = 0.725591 \quad z_2 = 0.971439$$
$$p_1 = 1.592692 \quad p_2 = 2.132335 \tag{6.104}$$

For a ripple of 0.05 dB, ε is equal to 21.997276. Plots of this transfer function are given in Figures 6.85 and 6.86. To commence the synthesis process, we first place the transfer function in the form of $t(s) \cdot t(-s)$, which results in the following:

$$t(s) = \frac{1}{\varepsilon} \frac{P(s)}{Q(s)}$$

where

$$P(s) = s^4 + 7.083520s^2 + 11.533853$$
$$Q(s) = s^6 + 2.175887s^5 + 3.837417s^4 + 4.217645s^3$$
$$+ 3.392813s^2 + 1.794499s + 0.524331$$

From this, the resistive terminations can be determined as $R_1 + R_6 = 2.175887$ (i.e., $R = 1.0879435$). The next step is to find the Y-parameters of the two-port network through Darlington's synthesis method so that we will be able to determine the coupling matrix of the filter.

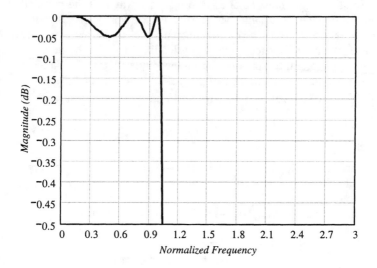

Figure 6.85 Plot of the transfer function given in (6.103) showing the passband ripple. The ripple is 0.05 dB as selected.

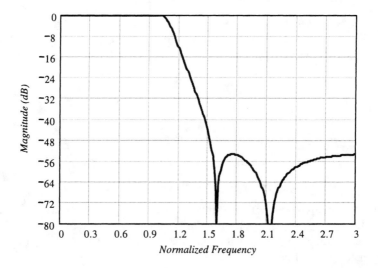

Figure 6.86 Plot of the transfer function given in (6.103) showing overall response. Note the two zeros in the stopband.

To find the *Y*-parameters, we first determine the input impedance of the two-port network which gives the prescribed insertion loss as follows:

$$\left|\rho_1\left(s\right)\right|^2 = 1 - \left|t\left(s\right)\right|^2 = \frac{\left|R\left(s\right)\right|^2}{\left|N\left(s\right)\right|^2}$$

where numerator and denominator are given as

$$\left|R\left(s\right)\right|^2 = 483.88\,s^{12} + 1422.78\,s^{10} + 1526.69\,s^8 + 706.89\,s^6 + 119.44\,s^4$$

$$\left|N\left(s\right)\right|^2 = 483.88\,s^{12} + 1422.78\,s^{10} + 1526.69\,s^8 + 721.06\,s^6 + 192.69\,s^4$$
$$+ 163.04\,s^2 + 133.03$$

To find the reflection coefficient $\rho_1(s)$, we should find the zeros of the numerator and the denominator:

$$R(s)\cdot R(-s) = s^4\left(s - j0.7256\right)^2\left(s + j0.7256\right)^2\left(s - j0.9714\right)^2\left(s + j0.9714\right)^2$$

$$N(s)\cdot N(-s) = \left(s + 0.1003 - j0.7256\right)^2\left(s + 0.1003 + j0.7256\right)^2$$
$$\cdot\left(s + 0.3507 - j0.8617\right)^2\left(s + 0.3507 + j0.8617\right)^2$$
$$\cdot\left(s + 0.6369 - j0.3414\right)^2\left(s + 0.6369 + j0.3414\right)^2$$

Then, the polynomials $R(s)$ and $N(s)$ are constructed as follows:

$$R(s) = s^6 + 1.4702\,s^4 + 0.4968\,s^2$$
$$N(s) = s^6 + 2.1759\,s^5 + 3.8374\,s^4 + 4.2176\,s^3 + 3.3928\,s^2 + 1.7945\,s + 0.5243$$

Note that complex zeros are used in conjugate pairs in forming the reflection coefficient. After this step, input impedance of the two-port network is found from the reflection coefficient:

$$Z(s) = \frac{1 + \dfrac{R(s)}{N(s)}}{1 - \dfrac{R(s)}{N(s)}} = \frac{N(s) + R(s)}{N(s) - R(s)}$$

$$Z(s) = \frac{2s^6 + 2.1759\,s^5 + 5.3076\,s^4 + 4.2176\,s^3 + 3.8897\,s^2 + 1.7945\,s + 0.5243}{2.1759\,s^5 + 2.3672\,s^4 + 4.2176\,s^3 + 2.8960\,s^2 + 1.7945\,s + 0.5243}$$

This completes the extraction of the input impedance. Now, we need to find the *Y*-parameters of the network from the input impedance. For this purpose, the numerator and denominator of the input impedance are separated into even and odd parts first:

$$Z(s) = \frac{m_1 + n_1}{m_2 + n_2}$$

where m and n refer to even and odd parts of the polynomials, respectively:

$$m_1 = 2s^6 + 5.3076s^4 + 3.8897s^2 + 0.5243$$

$$n_1 = 2.1759s^5 + 4.2176s^3 + 1.7945s$$

$$m_2 = 2.3672s^4 + 2.8960s^2 + 0.5243$$

$$n_2 = 2.1759s^5 + 4.2176s^3 + 1.7945s$$

Then, the Y-parameters can be calculated from

$$y_{11} = \frac{n_2}{m_1}$$

$$y_{12} = \frac{\sqrt{n_1 n_2 - m_1 m_2}}{m_1}$$

which results in

$$y_{11} = \frac{2.175886s^5 + 4.217644s^3 + 1.794499s}{2s^6 + 5.307592s^4 + 3.889651s^2 + 0.524331}$$

$$y_{12} = \frac{j \cdot 0.04546\left(s^4 + 7.083520s^2 + 11.533855\right)}{2s^6 + 5.307592s^4 + 3.889651s^2 + 0.524331}$$

Then, we expand y_{11} and y_{12} into partial fractions:

$$y_{11} = \frac{C_1}{s+p_1} + \frac{C_2}{s+p_2} + \frac{C_3}{s+p_3} + \frac{C_4}{s+p_4} + \frac{C_5}{s+p_5} + \frac{C_6}{s+p_6}$$

$$y_{12} = \frac{C_1}{s+p_1} - \frac{C_2}{s+p_2} - \frac{C_3}{s+p_3} + \frac{C_4}{s+p_4} + \frac{C_5}{s+p_5} - \frac{C_6}{s+p_6}$$

where the poles and residues are given as

$$p_1 = -j \cdot 1.179722 \quad C_1 = 0.094344$$

$$p_2 = j \cdot 1.179722 \quad C_2 = 0.094344$$

$$p_3 = -j \cdot 1.043595 \quad C_3 = 0.196596$$

$$p_4 = j \cdot 1.043595 \quad C_4 = 0.196596$$

$$p_5 = -j \cdot 0.415888 \quad C_5 = 0.253032$$

$$p_6 = j \cdot 0.415888 \quad C_6 = 0.253032$$

With above Y-parameters, the even-mode driving point admittance becomes

$$Y_e(s) = \frac{1}{2}[y_{11}(s) + y_{12}(s)] = \frac{C_{e1}}{s+p_1} + \frac{C_{e2}}{s+p_2} + \frac{C_{e3}}{s+p_3}$$

where the poles and residues are given as

$$p_1 = -j \cdot 1.179722 \quad C_{e1} = 0.094344$$
$$p_2 = j \cdot 1.043595 \quad C_{e2} = 0.196596$$
$$p_3 = -j \cdot 0.415888 \quad C_{e3} = 0.253032$$

The square root of the residues of the even-mode driving point impedance are the first row of the orthogonal matrix, $\mathbf{T}_e$. The rest of the orthogonal matrix is constructed using the Gram-Schmidt orthonormalization process using the basis vectors $(0,1,0)$ and $(0,0,1)$:

$$\mathbf{T}_e = \begin{bmatrix} 0.682023 & 0.601174 & 0.416455 \\ -0.513083 & 0.799119 & -0.313297 \\ -0.521143 & 0 & 0.85347 \end{bmatrix}$$

Then, the even-mode coupling matrix is found as

$$-\mathbf{M}_e = \begin{bmatrix} -0.020892 & 0.800809 & -0.271491 \\ 0.800809 & 0.44115 & 0.204241 \\ -0.521143 & 0.204241 & -0.972273 \end{bmatrix}$$

Now, it is difficult to realize the above coupling matrix practically because it includes all the possible couplings for a even-order network. However, using the reduction technique that was explained before, $\mathbf{M}_{13}$ and $\mathbf{M}_{31}$ can be reduced to zero, yielding the following coupling matrix:

$$-\mathbf{M}_e = \begin{bmatrix} -0.020893 & 0.845578 & 0.0 \\ 0.845578 & 0.171237 & 0.591914 \\ 0.0 & 0.591914 & -0.702359 \end{bmatrix}$$

The final step in the design is the realization of resonators and couplings between the resonators according to the found coupling matrix. Input and output transformers are also necessary to realize the loaded Q_e $(=1/R)$ of the first and last resonators.

For the sake of demonstration, we will now design a cross-coupled bandpass filter using the above transfer function with a center frequency and bandwidth of 5.0 GHz and 30 MHz, respectively. As a first step, the coupling coefficients $(\mathbf{M}_e)$ and the input and output Q-factors should be scaled according to the selected center frequency and bandwidth as follows:

$$K = k \frac{BW}{f_0} \tag{6.105}$$

$$Q = q \frac{f_0}{BW} \tag{6.106}$$

where k is the normalized coupling coefficient (entries of $\mathbf{M}_e$), q is the normalized Q-factor, K is the actual coupling coefficient, Q is the actual Q-factor $(1/R)$, f_0 is the center frequency, and *BW* is the 3-dB bandwidth. After determining the actual coupling coefficients and *Q*-factors, the filter circuit is implemented using lumped capacitors and ideal coupled inductors, resulting in the circuit shown in Figure 6.87. Note that for the sake of demonstration, the inductance of each resonator is divided into three equal parts. Although this is not necessary, it makes the schematic easy to follow. One could also use a single six-coupled inductor model to draw the overall circuit. However, that would slightly complicate the circuit schematic. Nevertheless, both approaches would yield the same electrical response. One important point when using divided inductors as depicted in the schematic is that the coupling coefficients between the resonators must also be modified accordingly to keep the voltage excitation due to mutual coupling the same.

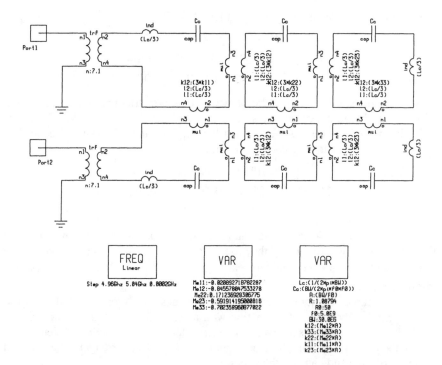

Figure 6.87 Circuit schematic of the bandpass filter given in Example 6.16.

The simulation results are shown in Figures 6.88 and 6.89. As can be seen from the *S*-parameter plots, the transfer function is correctly synthesized, which is evident from having six separate poles in S_{11} and four zeros in S_{21}. One might have noticed that the input and output transformer ratios are slightly higher than those dictated by theoretical calculations. The reason for this is that the transformer ratios had to be optimized to obtain the proper pole separation.

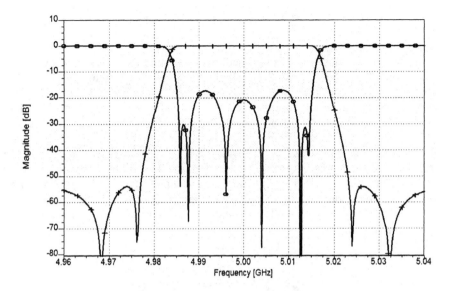

Figure 6.88 Simulated S-parameters of the bandpass filter shown in Figure 6.87 (circle: S_{11}; plus: S_{21}).

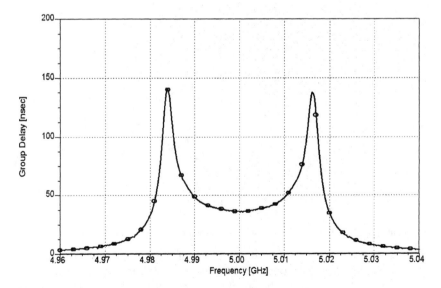

Figure 6.89 Simulated group delay of the bandpass filter shown in Figure 6.87.

6.10 CASCADE QUADRUPLETS AND TRIPLETS

The cross-coupled filters in canonical form are optimum in the sense that they can synthesize any realizable transfer function by adjusting the couplings between the resonators. However, the resulting filters might be difficult to align and tune because all of the couplings are responsible in a collective way to produce the required poles and zeros. Cascaded sections of quadruplets (filters containing four resonators) and triplets (filters containing three resonators) can address this problem by specifically pairing each transmission zero to a section of the filter, thus making the filters easy to tune. Although the resulting filters constructed in this manner may not be the optimum, they are extensively used in practice because of this advantage [67–77]. Employing cross-couplings between three and four resonators to produce finite transmission zeros appeared in the literature by Kurzrok [67–69]. Figure 6.90(a, b) shows the quadruplet and triplet sections involving only one cross-coupling.

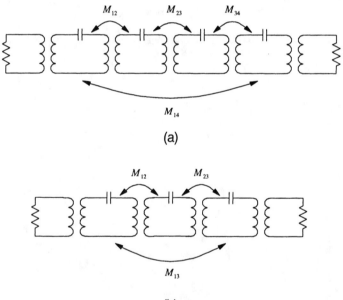

Figure 6.90 (a) Quadruplet and (b) triplet filter sections with only one cross-coupling.

It would be instructive to further elaborate on the zero-producing mechanisms of quadruplet and triplet sections shown in Figure 6.90. In a quadruplet with one cross-coupling, the zeros must be on either the real or imaginary axis; no complex transmission zeros are allowed. The placement of the zeros is determined by the polarity of the cross-coupling. When the cross-coupling has the same sign as the direct couplings, then the finite transmission zeroes are produced on the real axis, providing improved passband delay characteristics. On the contrary, when the cross-coupling has the opposite sign with respect to the main couplings, then imaginary-axis zeros (i.e., real frequency) are produced, resulting in sharper filter skirt. The sign of the cross-coupling plays a different role in the triplets. A finite transmission zero is produced on the upper and lower stopband when the couplings have the same and opposite signs with respect to main couplings, respectively. Therefore, the main advantage of a triplet is the capability of producing asymmetric frequency response, which is desirable in some applications. The zero-creating mechanisms explained above are summarized in Figure 6.91(a–d). To iterate once more, the real advantage of the quadruplet with same-sign couplings is to improve group-delay characteristics.

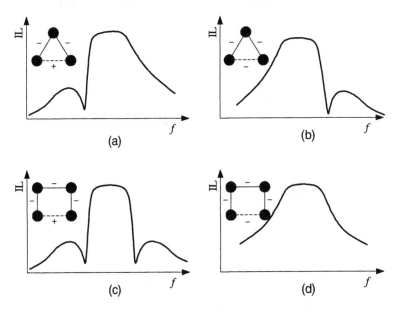

Figure 6.91 Qualitative frequency responses (insertion loss) for the triplet and quadruplet filters. Inset figures show the filter configuration. Each resonator is depicted by a filled circle. Direct and indirect coupling paths are depicted by solid and dashed lines, respectively. Note that the location of the finite transmission zeros depends on the cross-coupling sign.

A cascaded quadruplet (CQ) filter consists of cascaded sections of four cavities each with only one cross-coupling. Similarly, a cascaded triplet (CT) filter consists of cascaded sections of three cavities. Figure 6.92(a–c) shows the basic construction of cascaded sections with a comparison to the canonical cross-coupled filter. The canonical topology corresponds to the filters obtained through Darlington's method as described in the previous sections. In the figure, all cross-couplings are shown by dashed lines. Note that in the CT configuration, the overall filter can be optimized both for magnitude and phase characteristics by selecting different signs for cross-couplings in each resonator. Another advantage of cascaded CQ and CT filters over the canonical structure is that the input and output of the filter are on opposite sides, which may be helpful to assembly in some applications.

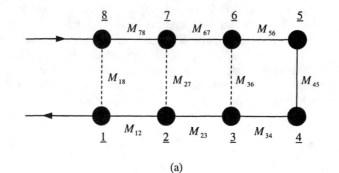

(a)

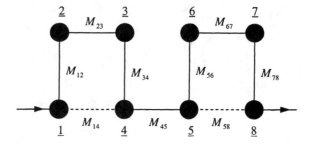

(b)

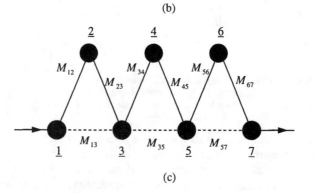

(c)

Figure 6.92 Three different cross-coupled filters: (a) canonical, (b) cascaded quadruplets (CQ), and (c) cascaded triplets (CT).

One could also employ microstrip resonators to construct CQ or CT filters. The two fundamental resonator-coupling techniques used in these kinds of filters are depicted in Figure 6.93(a, b). Note that the nature of coupling between the two resonators depends on how the resonators are placed with respect to each other. If the open ends of the resonators are on the same side, as in Figure 6.93(b), the coupling is mainly electric. On the contrary, if the open ends are on opposite sides, as in Figure 6.93(a), the coupling is mainly magnetic. These two types of placements provide the sign reversal of coupling that is necessary in CT and CQ filters. Triplet and quadruplet filters constructed using these resonators are shown in Figure 6.94(a–c). Note that both tapped and resonator-coupled input/output can be utilized. Hong and Lanchester reported extensive work on microstrip CQ and CT filters [57–60].

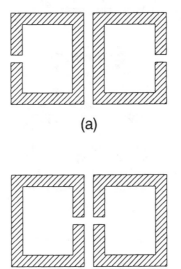

(a)

(b)

Figure 6.93 Two fundamental resonator couplings used in microstrip CT and CQ filters: (a) mainly magnetic, and (b) mainly electric. Note that signs of coupling between the two cases are opposite, thus enabling construction of CT and CQ filters.

Full-wave simulations can be used effectively in the design of microstrip CQ and CT filters by following the technique described for hairpin filters in Section 6.8. Essentially, there are two fundamental steps in designing such filters in which the full-wave simulators become quite useful: The first one is to determine the coupling coefficient between two resonators with different coupling polarities (see Figure 6.93). As described, a function for the coupling versus resonator separation can easily be found using computer simulations by observing the two natural frequencies. The second step is the analysis of a single resonator to determine the necessary tapping position for input/output coupling.

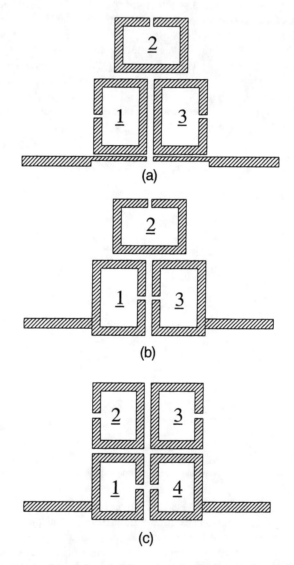

Figure 6.94 Microstrip implementation of triplet and quadruplet filter sections: (a) triplet where the cross-coupling has the same sign as the main coupling, (b) triplet where the cross-coupling has the opposite sign with respect to the main coupling, and (c) quadruplet where the cross-coupling has the opposite sign with respect to the main couplings. Note the different input/output coupling mechanisms for the triplets. (*After:* [19].)

The important point in this step is to use the correct method to calculate loaded Q of the resonators [47]. Since the loaded Q-factors of regular microstrip resonators at microwave frequencies tend to be relatively low (~ 100 or less), the standard 3-dB method would yield inaccurate results. Thus, the method based on the derivative of reactance or admittance should be utilized [47].

Example 6.17

In this example, a four-pole bandpass filter realized using a microstrip quadruplet section (see Figure 6.95) is analyzed, and the effect of the resonator spacing, s, on the filter response is investigated. This structure was originally reported by Hong and Lancester [57]. Note that aligning the resonator gaps next to each other and on opposite sides, as shown in the figure, provides coupling coefficients with different signs, a requirement in the coupled filter design.

The simulation results of the filter shown in Figure 6.95, which is designed based on this approach, are provided in Figures 6.96 and 6.97. Note that the optimum separation of the resonators is close to 0.175 cm. The characteristic finite-transmission zeros of a quadruplet section on both sides of the passband are clearly visible in Figure 6.97. One can also distinguish the four poles of the filter in Figure 6.96. For the sake of demonstration, the resonator separations are changed by 20% in each direction (equally on every side) and the effect of this is investigated. As one can expect, bandwidth of the filter is decreased as the couplings between the resonators decrease and vice versa. However, this deteriorates the return loss because the resonators become over- or undercoupled.

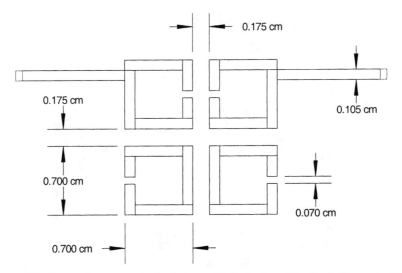

Figure 6.95 Full-wave electromagnetic simulation model of a bandpass filter ($h = 1.27$ mm, $\varepsilon_r = 10.8$).

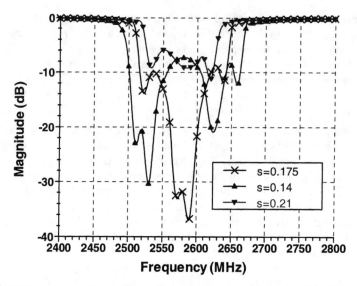

Figure 6.96 Magnitude of S_{11} of the bandpass filter shown in Figure 6.95 for different resonator spacings, s (mm).

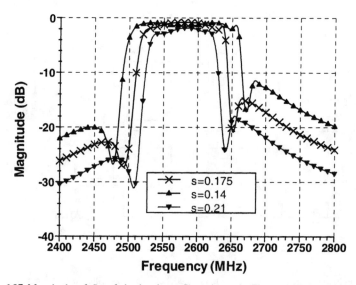

Figure 6.97 Magnitude of S_{21} of the bandpass filter shown in Figure 6.95 for different resonator spacings, s (mm).

6.11 FILTER SYNTHESIS BY OPTIMIZATION

In Section 6.9, the synthesis of cross-coupled filters based on Darlington's method was introduced. As demonstrated, extraction of the coupling matrix from the transfer-function specification is the key point in such filters. Another approach that could be used to extract the coupling matrix to synthesize a given transfer function is to utilize computer optimization [23, 78–85]. However, computer optimization of filter networks can be notoriously difficult if one blindly employs the optimization algorithms. The first step in a successful filter optimization is to select the right circuit topology so that the desired zeros and poles can be realized. This requires a basic understanding of the underlying physics. Second, a proper error function should be employed so that fast and optimal convergence will be achieved. Essentially, formulation of the error function itself is quite as important as the optimization method employed.

The standard approach in the optimization of microwave networks is to define an error function based on the difference of simulated scattering parameters and a target response in a given frequency band. Then, using an appropriate optimization method, one can try to extract the variables (coupling matrix entries in this case) by minimizing the error function. The difficulty of this approach is that the error function is linked to the filter transfer function by the whole circuit response. That is, the circuit response is evaluated at many frequency points in the band of interest at every iteration. Therefore, convergence to the desired transfer function can be quite slow. Even when convergence is achieved, the resulting filter topology may not be optimum. An alternative approach is to formulate the error function so that the scattering parameters are calculated only at the zeros and poles of the transfer function [27, 28, 78, 81]. This approach significantly improves the convergence because the poles and zeros of the transfer function are directly used in the optimization. Besides, the number of function evaluations is significantly lower than in the regular approach, providing an additional advantage.

To demonstrate the technique, let's assume a given insertion loss ratio defined as follows [27, 78]:

$$|S_{21}|^2 = \frac{1}{1 + \varepsilon^2 \Phi^2(\lambda)} \tag{6.107}$$

where

$$\Phi(\lambda) = \frac{\displaystyle\prod_{i=1}^{N} (\lambda - A_i)}{\displaystyle\prod_{j=1}^{M} (\lambda - B_j)} \tag{6.108}$$

$$\lambda = \frac{f_0}{BW} \left(\frac{f}{f_0} - \frac{f_0}{f} \right) \tag{6.109}$$

Then,

$$|S_{11}|^2 = 1 - |S_{21}|^2 = \frac{\varepsilon^2 \Phi^2(\lambda)}{1 + \varepsilon^2 \Phi^2(\lambda)} \tag{6.110}$$

Note that A_i's and B_i's are the zeros of S_{11} and S_{21}, respectively. This immediately suggests the frequency points where S_{11} and S_{21} need to be evaluated. However, to be able to synthesize the coupling matrix, one also needs the expressions linking the scattering parameters to the coupling matrix. This can be done mainly in two ways, which will be demonstrated next.

The first way of achieving this is to calculate the scattering parameters using the following equations [81]:

$$S_{21} = -2j\sqrt{R_1 R_2}\, \mathbf{A}^{-1}\big|_{n1} \tag{6.111}$$

$$S_{11} = 1 + 2jR_1 \mathbf{A}^{-1}\big|_{11} \tag{6.112}$$

where R_1 and R_2 are the input and output impedances, respectively. The matrix $\mathbf{A}$ is evaluated from the coupling, termination, and frequency matrixes as follows:

$$\mathbf{A} = \lambda \mathbf{U} - j\mathbf{R} + \mathbf{M} \tag{6.113}$$

where $\mathbf{U}$ is the unitary matrix. Note that numerical matrix inversion is required to calculate the scattering parameters whenever the frequency or coupling matrix is changed.

The second way of linking the scattering parameters to the coupling matrix is to express the scattering parameters by the ratio of two polynomials in λ as follows [27, 78]:

$$S_{21} = \frac{-2j\sqrt{R_1 R_2}\, P_{12}}{D + P_0 R_1 R_2 - j(P_{11} R_1 + P_{22} R_2)} \tag{6.114}$$

$$S_{11} = \frac{D - P_0 R_1 R_2 + j(P_{11} R_1 + P_{22} R_2)}{D + P_0 R_1 R_2 - j(P_{11} R_1 + P_{22} R_2)} \tag{6.115}$$

where P_{11}, P_{12}, P_{22}, P_0, and D are polynomials in λ related to the coupling matrix $\mathbf{M}$ by the following:

$$[\lambda \mathbf{U} - \mathbf{M}]^{-1}\big|_{11} = \frac{P_{11}}{D} \tag{6.116}$$

$$[\lambda \mathbf{U} - \mathbf{M}]^{-1}\big|_{nn} = \frac{P_{22}}{D} \tag{6.117}$$

$$[\lambda \mathbf{U} - \mathbf{M}]^{-1}\big|_{1n} = \frac{P_{12}}{D} \tag{6.118}$$

$$P_0 D = P_{12}^2 - P_{11} P_{22} \tag{6.119}$$

To evaluate the above polynomials, we invoke the Leverrier-Souriau-Frame algorithm once more [25]:

$$[\lambda \mathbf{U} - \mathbf{M}]^{-1} = \frac{\text{adj}(\lambda \mathbf{U} - \mathbf{M})}{\det(\lambda \mathbf{U} - \mathbf{M})} = \frac{\text{adj}(\lambda \mathbf{U} - \mathbf{M})}{\lambda^n + c_1 \lambda^{n-1} + c_2 \lambda^{n-2} + \ldots + c_n}$$

where

$$c_k = -\frac{\text{trace}\{\mathbf{M} \times \mathbf{B}_{k-1}\}}{k} \quad k = 1, 2, \ldots, n$$

$$\mathbf{B}_0 = \mathbf{U}$$

$$\mathbf{B}_k = -\frac{\text{trace}\{\mathbf{M} \times \mathbf{B}_{k-1}\}}{k} \mathbf{U} + \mathbf{M} \times \mathbf{B}_{k-1} \quad k = 1, 2, \ldots, n$$

and the adj( ) operator stands for the matrix adjoint (i.e., complex-conjugate transpose). The advantage of the above algorithm, which is essentially based on a symbolic matrix inversion, is that the scattering parameters are obtained as functions of frequency (i.e., λ) for a given coupling matrix. Thus, at each iteration, the error function can be evaluated extremely quickly at different frequencies without resorting to the numerical matrix inversion once the polynomials defining the scattering matrixes are determined.

The remaining task is to define the error function that will be used in the optimization:

$$\text{Error} = \sum_{i=1}^{N} |S_{11}(A_i)|^2 + \sum_{i=1}^{M} |S_{21}(B_i)|^2 + |\varepsilon - \hat{\varepsilon}|^2 \tag{6.120}$$

where S_{11}, S_{21}, and $\hat{\varepsilon}$ are evaluated from the current trial matrix, $\mathbf{M}$, using either of the approaches above. To minimize the error function, a gradient-based algorithm can be used. Note that the error function becomes zero only if the coupling matrix, $\mathbf{M}$, reproduces the reflection zeros, transmission zeros, and scale factor. A salient feature of using zeros and poles of the transfer function, $\Phi(\lambda)$, is that the convergence is almost independent of the initial value of the matrix, $\mathbf{M}$. Thus, one can choose the topology matrix as the starting value. After determining the coupling matrix parameter, which satisfies the given transfer function, the actual filter circuit can be implemented using the coupled resonators.

The above approaches essentially describe how to find the coupling parameters using optimization instead of resorting to Darlington's method. Alternatively, one can easily design CT and CQ filters by direct optimization. This is doable because each coupling in this type of filter is responsible for a particular transmission zero (unlike the general cross-coupled filters where all couplings contribute to every pole/zero in a complicated manner). Therefore, the designer can physically place the zeros where they are desired by adjusting the individual couplings. More information on this can be found in Swanson [23].

6.12 DC-BLOCK CIRCUITS

The dc-block circuits are used in microwave circuits to isolate the dc bias levels of active circuits. They are most commonly employed at the input and output of MMIC amplifiers. Since they are on the RF signal path, accurate design and characterization of such components are important. Note that a simple lumped metal-insulator-metal (MIM) capacitor for dc-blocking might be prohibitive at high frequencies because of the associated parasitics. Besides, MIMs can be vulnerable to static discharge due to their thin layer of dielectrics. Thus, microstrip coupled lines are frequently used to implement dc-block circuits performing both dc blocking and filtering functions.

Figure 6.98 shows a typical coupled-line dc-block circuit [86, 87]. Although only a pair of coupled lines is depicted in the figure, they can also be designed with multiple fingers.

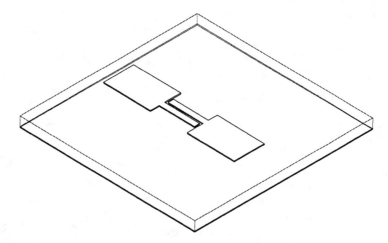

Figure 6.98 A typical dc-block implemented using coupled lines. Line lengths are a quarter-wavelength long at the center frequency.

For the two-finger version shown in the figure, the approximate design equations for a rippled response are given as follows [87]:

$$Z_{0e} = \sqrt{R_1 R_2} \left\{ \begin{array}{l} \sqrt{S} \left[1 + \left(1 + \dfrac{1 + \sqrt{1 + \Omega_c^2}}{\Omega_c^2} \left(1 - \dfrac{1}{S} \right) \right)^{1/2} \right] \\[4mm] \dfrac{1}{\sqrt{S}} \left[1 + \left(1 + \dfrac{-1 + \sqrt{1 + \Omega_c^2}}{\Omega_c^2} (S - 1) \right)^{1/2} \right] \end{array} \right. \qquad (6.121)$$

$$Z_{0o} = \sqrt{R_1 R_2} \left\{ \begin{array}{l} \sqrt{S} \left[-1 + \left(1 + \frac{1 + \sqrt{1 + \Omega_c^2}}{\Omega_c^2} \left(1 - \frac{1}{S} \right) \right)^{1/2} \right] \\ \frac{1}{\sqrt{S}} \left[-1 + \left(1 + \frac{-1 + \sqrt{1 + \Omega_c^2}}{\Omega_c^2} (S - 1) \right)^{1/2} \right] \end{array} \right. \qquad (6.122)$$

where

$$\Omega_c = \cot \left[\frac{\pi}{2} \left(1 - \frac{f_2 - f_1}{f_2 + f_1} \right) \right]$$

In the above equations, R_1 and R_2 are the input and output terminations, respectively (typically 50 ohms). The lower and upper edge frequencies are represented by f_1 and f_2, respectively. S is the standing wave ratio that needs to be realized. The length of the coupled section is a quarter-wavelength at the center frequency. Once the even- and odd-mode characteristic impedances are determined, the coupled-line section is implemented using the standard coupled-line synthesis equations. Note that there are two sets of solutions in the above equations for a given specification. The selection of the solution depends on the feasibility of each solution.

Note that due to process variations, it is not always possible to realize dimensions of the coupled-line dc block accurately; there will be variations caused by the over- or underetching, which can affect the circuit performance.

Example 6.18

In this example, an interdigital dc-block filter designed for the frequency bandwidth of 70 to 80 GHz is analyzed. Figure 6.99 shows the full-wave electromagnetic simulation model of the interdigital filter with equal finger widths and spacing.

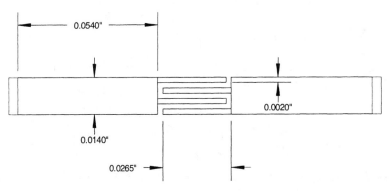

Figure 6.99 Full-wave electromagnetic simulation model of a dc-block filter ($h = 5$ mils, $\varepsilon_r = 2.2$).

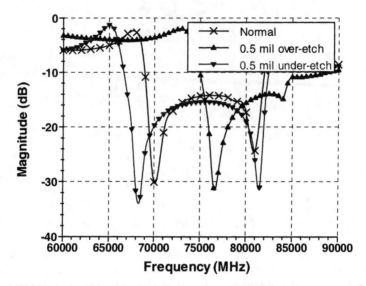

Figure 6.100 Magnitudes of S_{11} of the dc-block filter shown in Figure 6.99 for over- and underetching.

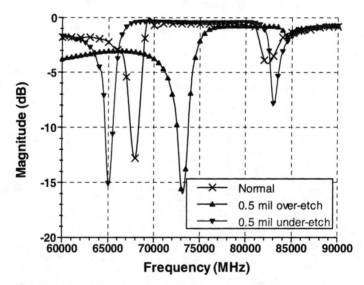

Figure 6.101 Magnitudes of S_{21} of the dc-block filter shown in Figure 6.99 for over- and underetching.

Figures 6.100 and 6.101 show the changes in the circuit response for a ± 0.5-mil variation in the filter dimensions, which is a tight specification for a high-volume PCB process with today's state-of-the-art technology. Note that the original response is nicely centered at 75 GHz. However, the altered responses due to over- and underetching place unwanted resonance points in the frequency band of interest. Therefore, one should take etching variations into consideration when using the multiple-finger dc-block circuits at millimeter-wave frequencies.

Note that the finger lengths do not change appreciably in over- or underetching (because approximately the same amounts of metal are removed/added from/to either end of the lines); only the finger widths and spacing change. However, this can be enough to deteriorate the response of the filter at high frequencies as demonstrated here.

References

[1]   R. Levy and S. B. Cohn, "A History of Microwave Filter Research, Design and Development," *IEEE Trans. on Microwave Theory Tech.*, vol. 32, pp. 1055–1067, Sept. 1984.

[2]   R. Levy, Richard V. Snyder, and George Matthaei, "Design of Microwave Filters," *IEEE Trans. on Microwave Theory Tech.*, vol. 50, pp. 783–793, Mar. 2002.

[3]   R. M. Fano and A. W. Lawson, "The Theory of Microwave Filters," *M.I.T. Rad. Lab. Series*, vol. 9, pp. 540–612, 1948.

[4]   A. W. Lawson and R. M. Fano, "The Design of Microwave Filters," *M.I.T. Rad. Lab. Series*, vol. 9, pp. 613–716, 1948.

[5]   A. A. Oliner, "Historical Perspectives on Microwave Field Theory," *IEEE Trans. on Microwave Theory Tech.*, vol. 32, pp. 1022–1045, Sept. 1984.

[6]   Vitold Belevitch, "Recent Developments in Filter Theory," *IRE Trans. on Circuit Theory.*, pp. 236–252, Dec. 1958.

[7]   Leo Young, "Microwave Filters—1965," *IEEE Trans. on Microwave Theory Tech.*, vol. MTT-13, pp. 489–508, Sept. 1965.

[8]   Paul I. Richards, "Resistor-Transmission-Line Circuits," *Proceedings of the I.R.E.*, pp. 217–220, Feb. 1948.

[9]   H. Ozaki and J. Ishii, "Synthesis of a Class of Stripline Filters," *I.R.E. Trans. on Circuit Theory*, pp. 104–109, June 1958.

[10] Matthaei, G. L., L. Young, and E. M. T. Jones, *Microwave Filters, Impedance Matching Networks and Coupling Structures*, Dedham, MA: Artech House, 1980.

[11] David M. Pozar, *Microwave Engineering*, New York: John Wiley and Sons, 1998.

[12] Ronald M. Foster, "Academic and Theoretical Aspects of Circuit Theory," *Proceedings of the IRE*, pp. 866–871, May 1961.

[13] Anatol I. Zverev, *Handbook of Filter Synthesis*, New York: John Wiley and Sons, 1967.

[14] Herman J. Blinchikoff and Anatol I. Zverev, *Filtering in the Time and Frequency Domains*, Noble Publishing, 2001.

[15]   Van Valkenburg, *Introduction to Modern Network Synthesis*, New York: John Wiley and Sons, 1960.

[16]   Wai-Kai Chen, *Passive and Active Filters: Theory and Implementations*, New York: John Wiley and Sons, 1986.

[17]   Wai-Kai Chen, *Theory and Design of Broadband Matching Networks*, Pergamon International Library, 1976.

[18]   J. D. Rhodes, *Theory of Electrical Filters*, New York: John Wiley and Sons, 1976.

[19]   Jia-Shen Hong and M. J. Lancaster, *Microstrip Filter for RF/Microwave Applications*, New York: John Wiley and Sons, 2001.

[20]   Leo Young (Ed.), *Microwave Filters Using Parallel Coupled Lines*, Dedham, MA: Artech House, 1972.

[21]   J. A. G. Malherbe, *Microwave Transmission Line Filters*, Dedham, MA: Artech House, 1979.

[22]   H. Howe, *Stripline Circuit Design*, Dedham, MA: Artech House, 1974.

[23]   Daniel G. Swanson, Jr., and Wolfgang J. R. Hoefer, *Microwave Circuit Modeling Using Electromagnetic Field Simulation*, Norwood, MA: Artech House, 2003.

[24]   Robert E. Collin, *Foundations for Microwave Engineering*, IEEE Series on Electromagnetic Wave Theory, New York: IEEE Press, 2001.

[25]   Carl D. Meyer, *Matrix Analysis and Applied Linear Algebra*, Philadelphia: SIAM Publications, 2000.

[26]   Milton Abramowitz and Irene A. Stegun (Eds.), *Handbook of Mathematical Functions*, New York: Dover, 1972.

[27]   A. E. Atia, "Multiple Coupled Resonator Filters Synthesis by Optimization," *IEEE-MTTS Filters Workshop*, 2000.

[28]   Kawthar Zaki, "Coupling Conversion to Physical Filter Dimensions by EM Simulations," *IEEE-MTTS Filters Workshop*, 2001.

[29]   Sydney Darlington, "Synthesis of Reactance 4-Poles Which Produce Prescribed Insertion Loss Characteristics," *Journal of Mathematics and Physics*, vol. 17, pp. 257–353, 1938.

[30]   Seymour B. Cohn, "Parallel-Coupled Transmission-Line-Resonator Filters," *IRE Trans. on Microwave Theory Tech.*, pp. 223–231, Apr. 1958.

[31]   George L. Matthaei, "Design of Wide-Band (and Narrow-Band) Band-Pass Microwave Filters on the Insertion Loss Basis," *IRE Trans. on Microwave Theory Tech.*, pp. 580–593, Nov. 1960.

[32]   Edward G. Cristal, "Design Equation for a Class of Wide-Band Bandpass Filters," *IEEE Trans. on Microwave Theory Tech.*, pp. 696–699, Oct. 1972 (Corrections: pp. 598, Sept. 1973).

[33]   Edward G. Cristal, "New Design Equations for a Class of Microwave Filters," *IEEE Trans. on Microwave Theory Tech.*, pp. 486–490, May 1971.

[34]   B. M. Schiffman and G. L. Matthaei, "Exact Design of Band-Stop Microwave Filters," *IEEE Trans. on Microwave Theory Tech.*, pp. 6–15, Jan. 1964.

[35]   Milton Dishal, "Design of Dissipative Band-Pass Filter Producing Desired Exact Amplitude-Frequency Characteristics," *Proceedings of the I.R.E.*, pp. 1050–1069, Sept. 1949.

[36]   Milton Dishal, "Alignment and Adjustment of Synchronously Tuned Multiple-Resonant-Circuit Filters," *Proceedings of the I.R.E.*, pp. 1448–1455, Nov. 1951.

[37] K. V. Puglia, "A General Design Procedure for Bandpass Filters Derived from Low Pass Prototype Elements: Part I," *Microwave Journal*, Dec. 2000.

[38] K. V. Puglia, "A General Design Procedure for Bandpass Filters Derived from Low Pass Prototype Elements: Part II," *Microwave Journal*, Jan. 2001.

[39] George L. Matthaei, "Interdigital Band-Pass Filters," *IRE Trans. on Microwave Theory Tech.*, pp. 479–491, 1962.

[40] R. J. Wenzel, "Exact Theory of Interdigital Band-Pass Filters and Related Coupled Structures," *IEEE Trans. on Microwave Theory Tech.*, pp. 559–575, Sept. 1965.

[41] R. J. Wenzel, "Synthesis of Combline and Capacitively Loaded Interdigital Bandpass Filters of Arbitrary Bandwidth," *IEEE Trans. on Microwave Theory Tech.*, vol. MTT-19, pp. 678–686, Aug. 1971.

[42] J. D. Rhodes, "The Stepped Digital Elliptic Filter," *IEEE Trans. on Microwave Theory Tech.*, vol. MTT-17, pp. 178–184, Apr. 1969.

[43] Joseph S. Wong, "Microstrip Tapped-Line Filter Design," *IEEE Trans. on Microwave Theory Tech.*, pp. 44–50, Jan. 1979.

[44] Edward G. Cristal, "Tapped-Line Coupled Transmission Lines with Applications to Interdigital and Combline Filters," *IEEE Trans. on Microwave Theory Tech.*, pp. 1007–1012, Dec. 1975.

[45] Shimon Caspi and J. Adelman, "Design of Combline and Interdigital Filters with Tapped-Line Input," *IEEE Trans. on Microwave Theory Tech.*, vol. 36, pp. 759–763, Apr. 1988.

[46] Edward G. Cristal and Sidney Frankel, "Hairpin-Line and Hybrid Hairpin-Like/Half-Wave Parallel-Coupled-Line Filters," *IEEE Trans. on Microwave Theory Tech.*, pp. 719–728, Nov. 1972.

[47] J. Michael Drozd and William T. Joines, "Determining Q Using S-Parameter Data," *IEEE Trans. on Microwave Theory Tech.*, vol. 44, pp. 2123–2127, Nov. 1996.

[48] Wei-Guan Lin, "Microwave Filters Employing a Single Cavity Excited in More Than One Mode," *Journal of Applied Physics*, vol. 22, pp. 989–1001, Aug. 1951.

[49] John D. Rhodes, "The Generalized Direct-Coupled Cavity Linear Phase Filter," *IEEE Trans. on Microwave Theory Tech.*, vol. 6, pp. 308–313, June 1970.

[50] A. E. Atia and A. E. Williams, "New Types of Waveguide Bandpass Filters for Satellite Transponders," *Comsat Technical Review*, vol. 1, pp. 21–43, 1971.

[51] A. E. Atia and A. E. Williams, "Narrow-Bandpass Waveguide Filters," *IEEE Trans. on Microwave Theory Tech.*, pp. 258–265, Apr. 1972.

[52] A. E. Atia and A. E. Williams, "Nonminimum-Phase Optimum-Amplitude Bandpass Waveguide Filters," *IEEE Trans. on Microwave Theory Tech.*, pp. 425–431, Apr. 1974.

[53] A. E. Atia, A. E. Williams, and R. W. Newcomb, "Narrow-Band Multiple-Coupled Cavity Synthesis," *IEEE Trans. on Circuits and Systems*, pp. 649–655, Sept. 1974.

[54] Kawthar A. Zaki, Chunming Chen, and Ali E. Atia, "Canonical and Longitudinal Dual-Mode Dielectric Resonator Filters Without Iris," *IEEE Trans. on Microwave Theory Tech.*, vol. 35, pp. 1130–1135, Dec. 1987.

[55] Xiao-Peng Liang, Kawtar A. Zaki, and Ali E. Atia, "Dual Mode Coupling by Square Corner Cut in Resonators and Filters," *IEEE Trans. on Microwave Theory Tech.*, vol. 40, pp. 2294–2302, Dec. 1992.

[56]   Hsin-Chin Chang and Kawthar A. Zaki, "Evanescent-Mode Coupling of Dual-Mode Rectangular Waveguide Filters," *IEEE Trans. on Microwave Theory Tech.*, vol. 39, pp. 1307–1312, August 1991.

[57]   J. Hong and M. J. Lancaster, "Couplings of Microstrip Square Open-Loop Resonators for Cross-Coupled Planar Microwave Filters," *IEEE Trans. on Microwave Theory Tech.*, vol. 44, pp. 2099–2109, Dec. 1996.

[58]   J. Hong and M. J. Lancaster, "Cross-Coupled Microstrip Hairpin-Resonator Filters," *IEEE Trans. on Microwave Theory Tech.*, vol. 46, pp. 118–122, Jan. 1998.

[59]   J. Hong and M. J. Lancaster, "Design of Highly Selective Microstrip Bandpass Filters with a Single Pair of Attenuation Poles at Finite Frequencies," *IEEE Trans. on Microwave Theory Tech.*, vol. 48, pp. 1098–1107, July 2000.

[60]   Kenneth S. K. Yeo, M. J. Lancaster, and Jia-Sheng Hong, "The Design of Microstrip Six-Pole Quasi-Elliptic Filter with Linear Phase Response Using Extracted Pole Technique," *IEEE Trans. on Microwave Theory Tech.*, vol. 49, pp. 321–327, Feb. 2001.

[61]   Shen-Yuan Lee and Chih-Ming Tsai, "New Cross-Coupled Filter Design Using Improved Hairpin Resonators," *IEEE Trans. on Microwave Theory Tech.*, vol. 48, pp. 2482–2490, Dec. 2000.

[62]   J. A. Curtis and S. J. Fiedziuszko, "Miniature Dual Mode Microstrip Filters," *IEEE MTT-S Digest.*, pp. 443–446, 1991.

[63]   J. A. Curtis and S. J. Fiedziuszko, "Multi-Layered Planar Filters Based on Aperture Coupled, Dual Mode Microstrip or Stripline Resonators," *IEEE MTT-S Digest.*, pp. 1203–1206, 1992.

[64]   Wolfgang Schwab and Wolfgang Menzel, "Compact Bandpass Filters with Improved Stop-band Characteristics Using Planar Multilayer Structures," *IEEE MTT-S Digest.*, pp. 1207–1209, 1992.

[65]   W. Schwab, F. Boegelsack, and W. Menzel, "Multilayer Suspended Stripline and Coplanar Line Filters," *IEEE Trans. on Microwave Theory Tech.*, vol. MTT-42, pp. 1403–1407, July 1994.

[66]   Richard J. Cameron and J. D. Rhodes, "Asymmetric Realization for Dual-Mode Bandpass Filters," *IEEE Trans. on Microwave Theory Tech.*, vol. MTT-29, pp. 51–58, Jan. 1981.

[67]   R. M. Kurzrok, "Couplings in Direct-Coupled Waveguide Band-Pass Filters," *I.R.E Trans. on Microwave Theory Tech.*, pp. 389–390, Sept. 1962.

[68]   R. M. Kurzrok, "General Four-Resonator Filters at Microwave Frequencies," *IEEE Trans. on Microwave Theory Tech.*, pp. 295–296, June 1966.

[69]   R. M. Kurzrok, "General Three-Resonator Filters in Waveguide," *IEEE Trans. on Microwave Theory Tech.*, pp. 46–47, Jan. 1966.

[70]   Ralph Levy, "Synthesis of General Asymmetric Singly and Doubly Terminated Cross-Coupled Filters," *IEEE Trans. on Microwave Theory Tech.*, vol. 42, pp. 2468–2471, Dec. 1994.

[71]   Ralph Levy, "Direct Synthesis of Cascaded Quadruplet (CQ) Filters," *IEEE Trans. on Microwave Theory Tech.*, vol. 43, pp. 2940–2945, Dec. 1995.

[72]   Ralph Levy and Peter Petre, "Design of CT and CQ Filters Using Approximation and Optimization," *IEEE Trans. on Microwave Theory Tech.*, vol. 49, pp. 2350–2356, Dec. 2001.

[73]   Ralph Levy, "Filters with Single Transmission Zeros at Real or Imaginary Frequencies," *IEEE Trans. on Microwave Theory Tech.*, vol. 24, pp. 172–180, Apr. 1976.

[74] Gerhard Pfitzenmaier, "Synthesis and Realization of Narrow-Band Canonical Microwave Bandpass Filters Exhibiting Linear Phase and Transmission Zeros," *IEEE Trans. on Microwave Theory Tech.*, vol. 30, pp. 1300–1310, Sept. 1982.

[75] John D. Rhodes, "The Design and Synthesis of a Class of Microwave Bandpass Linear Phase Filters," *IEEE Trans. on Microwave Theory Tech.*, vol. 17, pp. 189–204, Apr. 1969.

[76] J. Brian Thomas, "Cross-Coupling in Coaxial Cavity Filters — A Tutorial Overview," *IEEE Trans. on Microwave Theory Tech.*, vol. MTT-51, pp. 1368–1376, Apr. 2003.

[77] Nevzat Yildirim et al., "A Revision of Cascade Synthesis Theory Covering Cross-Coupled Filters," *IEEE Trans. on Microwave Theory Tech.*, vol. MTT-50, pp. 1536–1543, June 2002.

[78] Walid A. Atia, Kawtar A. Zaki, and Ali E. Atia, "Synthesis of General Topology Multiple Coupled Resonator Filters by Optimization," *IEEE MTT-S Digest.*, pp. 821–824, 1998.

[79] Ralph Levy, "Direct Synthesis of Cross-Coupled Filters Based on formation of Transfer Matrices," *IEEE-MTTS Filters Workshop*, 2000.

[80] Wai-Cheung Tang, "The Art and Science of Selecting Physical Filter Configuration for a Given Transfer Function," *IEEE-MTTS Filters Workshop*, 2000.

[81] Smain Amari, "Synthesis of Cross-Coupled Resonator Filters Using an Analytical Gradient-Based Optimization Technique," *IEEE Trans. on Microwave Theory Tech.*, vol. 48, pp. 1559–1564, Sept. 2000.

[82] Heng-Tung Hsu et al., "Parameter Extraction for Symmetric Coupled-Resonator Filters," *IEEE Trans. on Microwave Theory Tech.*, vol. MTT-50, pp. 2971–2978, Dec. 2002.

[83] Tao Shen et al., "Full-Wave Design of Canonical Waveguide Filters by Optimization," *IEEE Trans. on Microwave Theory Tech.*, vol. MTT-51, pp. 504–511, Feb. 2003.

[84] Alejandro Garcia-Lamperez et al., "Efficient Electromagnetic Optimization of Microwave filters and Multiplexers Using Rational Models," *IEEE Trans. on Microwave Theory Tech.*, vol. MTT-52, pp. 508–521, Feb. 2004.

[85] John W. Bandler, Radoslaw M. Biernacki, Shao Hua Chen, Daniel G. Swanson Jr., and Shen Ye, "Microstrip Filter Design Using Direct EM Field Simulation," *IEEE Trans. on Microwave Theory Tech.*, vol. MTT-42, pp. 1353–1359, July 1994.

[86] David Lacombe and Jerome Cohen, "Octave-Band Microstrip dc Blocks," *IEEE Trans. on Microwave Theory Tech.*, pp. 555–556, Aug. 1972.

[87] Darko Kajfez and B. Sarma Vidula, "Design Equations for Symmetric Microstrip dc Blocks," *IEEE Trans. on Microwave Theory Tech.*, vol. MTT-28, pp. 974–981, Sept. 1980.

Chapter 7

Microwave Lumped Elements

Microwave circuits had traditionally been built using the distributed approach in which impedance-matching and resonator networks were constructed by stub-loaded transmission lines or waveguide cavities. With the invention of integrated circuits (ICs), however, the distributed approach became a big obstacle in the path to the integration of passive microwave circuits with active components, directly affecting the cost, volume, and weight of the circuits. Lumped elements (i.e., capacitors, inductors, resistors, and so forth) would provide a unique solution to these problems because they could be manufactured directly on the same substrate with active circuits using photolithographic techniques. Prompted with these requirements, work on microwave lumped elements was reported as early as 1967, when the authors demonstrated spiral inductors and metal-insulator-metal (MIM) capacitors at frequencies up to 2 GHz with Q-factors around 50 [1]. In the following years, more applications of lumped elements for MMICs were reported [2–7].

Despite the clear advantages in terms of integration, perhaps the biggest two problems with using lumped elements in microwave frequencies have been the lack of accurate modeling tools and relatively low Q-factors compared to their distributed counterparts because of the associated metallization and dielectric losses. Unlike waveguide and coaxial circuits, where closed-form analytical solutions can be obtained in many cases, it is quite difficult to get accurate analytical models for lumped elements because of their intricate construction. The first models created for lumped elements were based on quasi-static approximations and semiempirical formulations [8–12]. This started to change in the mid-1980s with the advent of full-wave electromagnetic (EM) computer simulators, where the circuit parameters of microwave lumped elements can be extracted very accurately [13–16]. Sonnet was the first successful, commercially available EM simulation software for planar geometries that could be used for this purpose [17]. In terms of the Q-factor, however, lumped elements still had a disadvantage. Fortunately, in most practical applications, except for filters with very low insertion-loss requirements, this would have been tolerated. The fast development speed of radio-frequency integrated circuits (RFICs) in recent years for consumer products (mainly cellular phones) further increased the interest in microwave lumped elements.

There is an extensive literature on the subject, and a nonexhaustive list is provided in the references [18–24]. A recent book by Bahl [19] provides a detailed review of lumped elements that are used in microwave circuits including resistors, transformers, wire bonds, and via-holes. Bahl also provides examples of microwave circuits (e.g., switches, phase shifters) constructed using lumped elements. The books by Hoffmann [21] and Edwards [20] are general references on microwave integrated circuits with a broader view. Fundamental knowledge of numerical analysis [25–29] and network theory [30–34] is also useful if one would like to extract the equivalent parameters of lumped elements.

7.1 BASIC LUMPED ELEMENTS

In this section, we will review the three basic lumped elements that are frequently used in microwave circuits: MIM capacitors, interdigital capacitors, and spiral inductors. Approximate closed-form formulas and equivalent circuits for each case will be provided for the sake of completeness. Note that the equivalent circuits that will be given are accurate only up to the first self-resonance frequency or so, although in some cases they can be used at frequencies slightly higher than that. It must be stressed that the closed-form expressions provided here are used today only for preliminary calculations; full-wave EM simulators are almost always employed to analyze and design lumped elements. Once the circuit is analyzed, the results can be used directly in circuit simulations, or equivalent circuits can be extracted. We will return to equivalent-circuit extraction using full-wave simulators later in this chapter.

7.1.1 MIM Capacitors

MIM capacitors are formed by sandwiching a dielectric layer between two layers of metal, resulting in a simple parallel-plate capacitor (see Figure 7.1). Note that in the figure, the input and output lines are on the same layer. The top plate is constructed using a second layer of metal deposition as shown. The connection between the top plate and bottom metal layer is usually called an air-bridge connection. The thickness of the dielectric layer between the top and bottom plates is typically on the order of a few hundred angstroms. Due to this very thin dielectric layer, high capacitance values can be attained.

At frequencies lower than the self-resonance frequency, the simple model shown in Figure 7.2 containing loss and resonance components is sufficient to model MIM capacitors [5, 6].

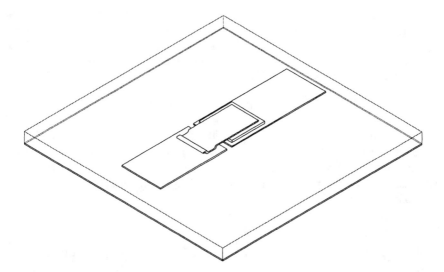

Figure 7.1 A typical MIM capacitor. Note that the dielectric thickness of the parallel-plate section is exaggerated for easy visualization.

In this model, the series capacitor accounts for the parallel plate and the fringe-field capacitances, whereas the series inductor is due to the capacitor plates and air-bridge connection. The parallel and series resistors represent dielectric and metallization losses, respectively, and they are frequency dependent. The capacitors connected to ground are the parasitic capacitances, which usually have very low values with respect to the series capacitance in a MIM capacitor.

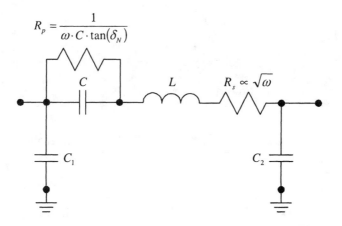

Figure 7.2 Simple lumped model for a MIM capacitor. See Figure 7.3 for the layout of a MIM capacitor.

The layout of a typical MIM capacitor is given in Figure 7.3 showing the important dimensions and reference planes for the equivalent circuit. Note that the effective area of the capacitor is determined by the dimensions l and w.

The capacitance of a MIM capacitor can be calculated using the following simple parallel-plate formula:

$$C = \varepsilon_r \varepsilon_0 \frac{(l + \Delta s)(w + \Delta s)}{h} \qquad (7.1)$$

where h is the thickness of the capacitor dielectric (usually nitride for MMIC), l is the length of the effective area, w is the width of the effective area, and ε_r is the relative dielectric constant of the dielectric layer (~ 7 for nitride). In the above expression, Δs is a factor to include the fringe-field effects, and it can be approximated by

$$\Delta s \cong \frac{4h \ln(2)}{\pi} \qquad (7.2)$$

The series resistance of a MIM capacitor can be calculated by using the following formula [5]:

$$R_s = \frac{2}{3} \frac{l}{w} R_F \qquad (7.3)$$

where R_F is the frequency-dependent surface resistivity of the conductors [21]. Note that lengths of the reference planes and air-bridge connection should be included in the above expression if they are comparable to the top-plate length, l.

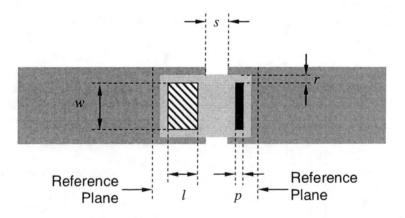

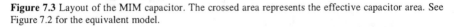

Figure 7.3 Layout of the MIM capacitor. The crossed area represents the effective capacitor area. See Figure 7.2 for the equivalent model.

Although this simple model is sufficient at low frequencies, it fails when the dimensions become comparable with the wavelength. In that case, one can use a distributed model where the MIM capacitor is modeled as two broadside-coupled transmission lines with necessary shunt capacitances to ground [11].

One of the most important points in determining the equivalent models of MIM capacitors, and other lumped elements in general, is the extraction of loss factors (i.e., dielectric and metallization losses). For accurate loss simulations, computer models must first be calibrated with the measurement results. The reason for this is that the loss models of commercially available full-wave simulators do not automatically take into account surface roughness and the thickness of metallizations that are made by depositing various metals in a layered fashion (which is typical for a MMIC process), although dielectric losses can be modeled relatively well once the loss tangent is known. MoM formulation, which is the most commonly used technique in the analysis of planar circuits, assumes that all currents are confined to the surface of metallizations (i.e., metal thickness is an order of magnitude greater than the skin depth). Fortunately, all of the metallization imperfections can be approximated quite well by defining an equivalent conductivity, which can be obtained from measurements. However, separation of conductor and dielectric losses from the measurement results can also be problematic because all loss effects will essentially be lumped together in the measured data. To facilitate separation of conductor and dielectric losses, different size capacitors should be measured in a wide band of frequency (at least a decade) so that the functional dependency of each loss mechanism can be identified. One of the ways of achieving this is to measure the self-resonance frequency of the capacitors with different sizes. Then, the loss factors can be identified from the overall Q-factor, which can be written as follows [5]:

$$\frac{1}{Q} = \frac{1}{Q_c} + \frac{1}{Q_d} \tag{7.4}$$

where

$$Q_c = \frac{3}{2\omega R_s (C/A) l^2} \tag{7.5}$$

$$Q_d = \frac{1}{\tan(\delta)} \tag{7.6}$$

where R_s and l are the surface resistivity and electrode length, respectively. A is the area of the parallel-plate section. Note that the loss-tangent can be determined from the Q-factor of the small capacitors ($l \ll 1$) at low frequencies where Q_c is very high. Once the Q_d is determined, Q_c can be extracted from the high-frequency resonators formed using capacitors. Another way of determining the loss factors is to utilize an optimization algorithm to optimize the parameters of the model given in Figure 7.2 until the simulated and measured scattering parameters agree. Note that usage of an optimization algorithm would allow the extraction of frequency-dependent resistive terms as well.

Although these approaches are simple, they may still lack the required accuracy because of the assumptions made (especially for the conductor losses). Perhaps the most accurate way of determining the loss factors is to utilize full-wave EM simulations. For this purpose, the loss-tangent of the capacitor dielectric and the conductivity of metallization are varied until the simulated Q-factors of various capacitors match the experimental data. Note that conductivity obtained this way would be an equivalent conductivity due to surface roughness and other imperfections, as described above. It should also be stated that the relative permittivity of the capacitor dielectric can be extracted by measuring the capacitance of the large capacitors where the fringe-field effects are insignificant compared to the parallel-plate capacitance.

7.1.2 Interdigital Capacitors

Interdigital capacitors are realized, as the name implies, by arranging coupled transmission line sections in an interlaced manner (see Figure 7.4). The resulting capacitance is mainly due to the fringe fields between the fingers. For this reason, the capacitance values obtained from interdigital capacitors are an order of magnitude lower than the MIM capacitors for the same MMIC area. Nevertheless, interdigital capacitors are very useful in MMIC design whenever low capacitance and high accuracy (no dielectric thickness variation) are required. They are also extensively used in integrated circuits where nitride MIM capacitors are not available.

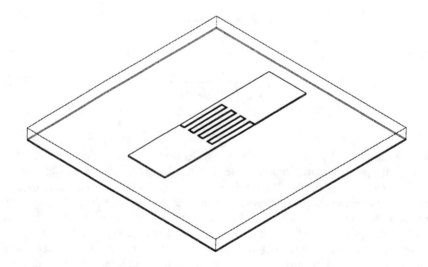

Figure 7.4 A typical interdigital capacitor.

At frequencies lower than the self-resonance frequency, a simple model shown in Figure 7.5 containing loss and resonance components is sufficient to model an interdigital capacitor [6–10]. In this model, the series capacitor accounts for the capacitance between the fingers, whereas the series inductor is due to the finger lengths and reference planes. The series resistor represents metallization losses, and it is frequency dependent in general. The capacitors connected to ground are the parasitic capacitances, which may have comparable values with respect to the series capacitance for an interdigital capacitor, depending on the substrate thickness.

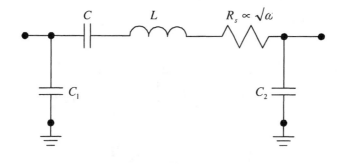

Figure 7.5 Simple lumped model for an interdigital capacitor. See Figure 7.6 for the layout of an interdigital capacitor.

The layout of a typical interdigital capacitor is given in Figure 7.6 showing the important dimensions and reference planes for the equivalent circuit.

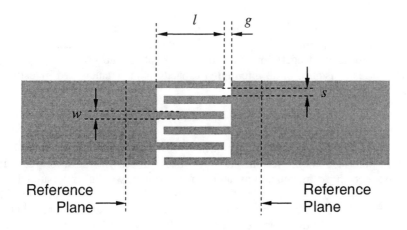

Figure 7.6 Layout of the interdigital capacitor. See Figure 7.5 for the equivalent model.

The capacitance of an interdigital capacitor can be calculated by using the following formula [9, 18]:

$$C = \varepsilon_{eff} \frac{1 \times 10^{-3}}{18\pi} \frac{K(k)}{K'(k)} (N-1) \cdot l \tag{7.7}$$

where

$$\frac{K(k)}{K'(k)} = \begin{cases} \left[\dfrac{1}{\pi} \ln\left(2\dfrac{1+\sqrt{k'}}{1-\sqrt{k'}} \right) \right]^{-1} & 0 \le k \le 0.7 \\[4mm] \dfrac{1}{\pi} \ln\left(2\dfrac{1+\sqrt{k}}{1-\sqrt{k}} \right) & 0.7 \le k \le 1 \end{cases} \tag{7.8}$$

$$k = \tan\left[\frac{w \cdot \pi}{4 \cdot (w+s)} \right]^2 \tag{7.9}$$

$$k' = \sqrt{1-k^2} \tag{7.10}$$

and

$$\varepsilon_{eff} = \frac{\varepsilon_r + 1}{2} + \frac{\varepsilon_r - 1}{2} F - \frac{\varepsilon_r - 1}{4.6} \frac{t/h}{\sqrt{w/h}} \tag{7.11}$$

where

$$F = \begin{cases} \left[\left(1 + 12\dfrac{h}{w}\right)^{-1/2} + 0.04\left(1 - \dfrac{w}{h}\right)^2 \right] & \dfrac{w}{h} \le 1 \\[4mm] \left(1 + 12\dfrac{h}{w}\right)^{-1/2} & \dfrac{w}{h} \ge 1 \end{cases} \tag{7.12}$$

In the above expressions, w is the width of the fingers, l is the length of the fingers, N is the number of fingers, t is the thickness of metallization, s is the spacing between the fingers, and h is the thickness of the substrate.

The series resistance of an interdigital capacitor can be calculated by using the following formula [7]:

$$R_s = \frac{4}{3} \frac{l}{wN} R_F \tag{7.13}$$

where R_F is the frequency-dependent surface resistivity of the conductors [21].

Note that (7.7) does not include the effect of the metallization thickness directly. For thick metallizations and narrow finger widths, the parallel-plate capacitance due to the side-walls of the fingers can be a substantial amount of the total series capacitance. This capacitance can be accounted for either by calculating the resulting parallel-plate capacitance due to side-walls and then adding it to (7.7), or by increasing the width of the fingers to an effective width to accommodate the metal thickness. The effective finger width due to metal thickness, t, is given by [10]:

$$w_{eff} = w + \left(\frac{t}{\pi}\right) \cdot \left[1 + \ln\left(\frac{8\pi \cdot w}{2 \cdot t}\right)\right]$$ (7.14)

where w and w_{eff} are the physical and effective finger widths, respectively.

7.1.3 Spiral Inductors

Spiral inductors are usually implemented in circular, square, or octagonal forms (see Figure 7.7). The square form has the advantage of being easily simulated and laid out although shapes close to circular have somewhat better Q-factor (~ 10% higher).

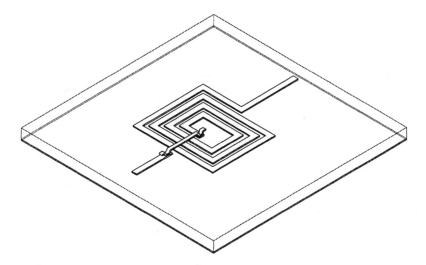

Figure 7.7 A typical spiral inductor. Note the air-bridge connection that connects the left terminal to the innermost turn of the spiral.

At frequencies lower than the self-resonance frequency, a simple model shown in Figure 7.8 containing loss and resonance components is sufficient to model spiral inductors [6, 35–37]. In this model, the series inductor represents the inductance of windings, and the parallel capacitor accounts for the capacitance due to interwinding gaps. The series resistor represents metallization losses, and it is frequency dependent. The capacitors connected to ground are the parasitic capacitances due to substrate. If the substrate is conductive, then appropriate resistive terms should also be added in parallel to these capacitors. It should be stressed that this approximate model will lose accuracy quite quickly after the first self-resonance frequency because of the fact that spiral inductors usually contain more closely spaced poles in the transfer function than interdigital or MIM capacitors due to complex interactions between the turns. The layout of a typical spiral inductor is given in Figure 7.9 showing the important dimensions and reference planes for the equivalent circuit.

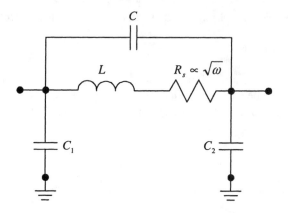

Figure 7.8 Simple lumped model for a spiral capacitor. See Figure 7.9 for the layout of a spiral inductor.

Multilayer inductors taking advantage of modern RFIC processes, which support three or more metallization layers, are also possible to increase the inductance per area of the circuit. In designing multilayer inductors, one must be careful not to align windings of different layers. Doing so will increase the interwinding capacitance considerably, causing the self-resonance frequency to drop. To avoid this, each layer must be offset such that the windings align with the gaps in the next layer of the spiral.

For spiral inductors, we also define the average diameter, D_{avg}, and fill-ratio, F, as follows (see Figure 7.9) [36]:

$$D_{avg} = \frac{D_{out} + D_{in}}{2} \tag{7.15}$$

$$F = \frac{D_{out} - D_{in}}{D_{out} + D_{in}} \tag{7.16}$$

Also worth mentioning are the parameters D_{in} and D_{out}, which represent the inner and outer diameters of the spiral inductor, respectively. Depending on the fill ratio given by (7.16), the spiral is called either hollow ($F \cong 0$) or full ($F \gg 1$). A hollow spiral will have a slightly higher inductance than a full one with the same average diameter because in the full spiral, the turns close to the center of the inductor contribute more negative mutual inductance. The fill ratio is also important in terms of maximizing the center area of the spiral to allow magnetic fluxes to concentrate. For full inductors, magnetic fluxes have to pass through the conductors, increasing losses due to eddy currents induced on the conductors (because of Faraday's law). Thus, providing relatively large, conductor-free space at the center of spirals increases the *Q*-factor by both reducing losses due to eddy currents and increasing the inductance. Because of these reasons, designing high-*Q* inductors can be quite challenging for voltage controlled oscillators (VCO) where low phase noise is a paramount consideration.

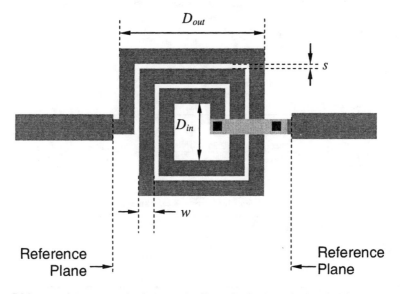

Figure 7.9 Layout of the square-spiral inductor. See Figure 7.8 for the equivalent model.

The inductance of a spiral inductor can be calculated by using the following formulas [36]:

Modified Wheeler Formula

$$L = \mu_0 K_1 \frac{n^2 D_{avg}}{1 + K_2 F} \tag{7.17}$$

where n is the number of turns. The coefficients K_1 and K_2 are layout-dependent parameters and are given in Table 7.1.

Current-Sheet Approximation

$$L = \mu_0 \frac{n^2 D_{avg}}{2} C_1 \left[\ln\left(\frac{C_2}{F}\right) + C_3 F + C_4 F^2 \right] \tag{7.18}$$

where n is the number of turns. The coefficients C_1, C_2, C_3, and C_4 are layout-dependent parameters and are given in Table 7.2.

The series resistance of a circular spiral inductor can be calculated by using the following formula [7]:

$$R_s = \pi \frac{K n D_{avg}}{2w} R_F \tag{7.19}$$

where R_F is the frequency-dependent surface resistivity [21]. In the above expression, K is a parameter to take into account the current-crowding effect. The typical value of K is around 1.5 [2].

Table 7.1

Coefficients for Modified Wheeler Expression

Layout	K_1	K_2
Square	2.34	2.75
Hexagon	2.33	3.82
Octagon	2.25	3.55

Source: [36].

Table 7.2

Coefficients for Current Sheet Expression

Layout	C_1	C_2	C_3	C_4
Square	1.27	2.07	0.18	0.13
Hexagon	1.09	2.23	0.00	0.17
Octagon	1.07	2.29	0.00	0.19
Circle	1.00	2.46	0.00	0.20

Source: [36].

Note that neither (7.17) nor (7.18) depends on the line width, w, of the spiral. This is because these formulas are based on static approximations and are valid when the guided wavelength is orders of magnitude larger than the maximum circuit dimensions (i.e., the lumped approximation). Under this assumption, the line width dominantly affects the series resistance of the spiral, hence the Q-factor. Line separation, s, affects the interwinding capacitance, hence the self-resonance frequency. Similarly, the inductance does not depend on the substrate dielectric constant either because the electric and magnetic fields are assumed to be decoupled. These assumptions hold well for RFIC applications where frequencies and maximum circuit dimensions are around 1 GHz and a couple of millimeters, respectively. However, for microwave frequencies (~ 10 GHz or greater), the distributed nature of the spirals starts to be pronounced; the inductance will be dependent on w. Although circuit dimensions can still be reduced to go into the lumped domain again, usually this cannot be done because reducing the dimensions after a limit will increase the conductor losses prohibitively. Thus, accurate modeling of spiral inductors at microwave frequencies without resorting to measurements can only be achieved using full-wave electromagnetic simulators, where all the distributed and coupling effects are taken into account rigorously.

7.2 MODEL EXTRACTION OF LUMPED ELEMENTS

It has been indicated that full-wave EM simulations should be used to analyze lumped elements whenever possible because of their accuracy instead of employing static and/or quasi-empirical equations. Although measurement is essentially the most direct way of characterization, time and cost issues do not always permit a measurement-based lumped-element characterization. Employing full-wave simulations, however, has a caveat: it may not be efficient to directly use computer generated *S*-parameters in the subsequent circuit simulations with the active circuits, especially in the time domain. The reason is that it is notoriously difficult to make time-domain circuit simulations using frequency-domain (i.e., *S*-parameters) models because the circuit simulator must first determine the impulse response of the circuit so that convolution can be employed. Determinations of both the impulse response and convolution are not the most efficient processes, and sometimes they can cause convergence problems. Note that one can still use the *S*-parameter data obtained through full-wave simulations directly in frequency-domain simulators (e.g., harmonic balance) without any problem (though it is important to obtain broadband data for dc convergence and proper harmonic response). Thus, although EM simulations can be used to analyze the lumped elements, a secondary effort, which we call model extraction, is sometimes necessary to determine the equivalent models of lumped elements so that they can be used efficiently in time-domain circuit simulators. This is the idea that we will explore in this section.

The extracted models of lumped elements can be as simple as the ones introduced in Section 7.1, or more elaborate ones can be constructed. In the case of simple models, the designer uses various curve-fitting and/or optimization techniques to determine the component values by matching the response of the equivalent model to the circuit parameters obtained from the full-wave EM simulations. Once the component values are determined, the equivalent model can then be used in time-domain circuit simulations. Another advantage of extracting equivalent models is that it is relatively safer to use equivalent models at frequencies where the circuit was not originally simulated, instead of extrapolating the simulated frequency response. Because the model is physical, it is guaranteed that nonphysical results will not occur (such as having a gain for a passive lumped element) due to possible erroneous extrapolation of simulated data. This point is especially important in harmonic-balance simulations where circuit response is required at the harmonics of the input tone. Finally, the dc response of the components is readily available, which is important in the biasing of active components. Without proper dc information, lumped-element models may create problems during dc analysis (to alleviate this problem, the designer sometimes enters the dc *S*-parameters manually if frequency-domain data is to be used).

Although the approach described in the above paragraph seems relatively straightforward, difficulties arise because the model topology has to be determined by intuition first. For a simple component (e.g., MIM capacitor) on an insulating by intuition first. For a simple component (e.g., MIM capacitor) on an insulating substrate, this may not be a problem. However, for components that have multiple resonance points (e.g., spiral inductors) and/or are on semi-insulating substrates, determination of the physical equivalent model may be tricky. This step is crucial because if the model is not suitable to approximate the component in the frequency range of interest, then subsequent component extraction will fail. Let's suppose for a moment that we have determined the circuit topology of the model correctly. The next question will be how we should extract the component values of the model. There are many issues associated with this problem, and one must be cautious. First, there is always some level of numerical noise in the simulations, forcing the designer to use more than the necessary number of frequency points, making the algebraic equations overdetermined. Second, because of the inadequacy of the equivalent model, it is possible to lump the effects of an unaccounted-for component in with those of the other components. Finally, extraction of loss figures is usually difficult, as was briefly mentioned in explaining MIM capacitors. The answer to the first question is relatively easy; one can use a least-squares or optimization algorithm to determine component values, therefore properly utilizing the frequency redundancy in full-wave simulations. The second issue, however, is not so easy to address and will require some trial and error. Note that conductive substrates further complicate the equivalent network of the components. The third issue can be resolved by extracting the resistive terms around the self-resonance frequencies and optimizing. Since near or at resonance the loss-parameters of components start to be pronounced, it will be numerically easier to extract them. In conclusion, although modeling based on equivalent physical networks has important advantages, the aforementioned challenges make the automatic generation of models difficult using this approach.

A completely different methodology for modeling lumped elements, which could provide ease of automation (at least in theory), would be to employ network synthesis based on the poles of the circuit parameters [42–45]. In that approach, the circuit parameters (usually Y-parameters) are expressed as a summation of the poles with associated residues through a process called parameter estimation. Then, a mathematically equivalent network, such as the Cauer type, is directly generated from these extracted poles. This resulting network might contain ideal transformers, and it might not be intuitive. However, it would be mathematically exact. Besides, model-reduction techniques could also be employed to simplify the models of electrically large circuits [46–48]. Yet another technique would be to devise a state-space representation once the poles and residues are determined [34]. It can be shown that one can generate equivalent SPICE models directly from the state-space realization using only controlled sources, capacitors, and resistors. This latter method is extensively used in signal integrity to model PCB traces, ground planes, and so forth [46, 48].

7.2.1 Modeling Based on Equivalent Physical Networks

In this section, we will provide details on the model extraction of lumped elements using equivalent physical networks. The technique is based on the extraction component values of an equivalent physical network, which is devised by intuition, for the lumped element that needs to be modeled. To devise the equivalent model, the designer inspects the circuit and tries to model direct current paths and discontinuities by inductors and capacitors, respectively. For simple circuits, this approach works well. Note that the success of the technique hinges on the selection of the equivalent model. The overall process can be summarized as follows [38, 41]:

1. Generate broadband S-parameter data of the lumped circuit using full-wave EM simulator. Most simulators support appropriate rational interpolation schemes to generate broadband data without simulating many frequency points. If the simulator does not have this feature, then a rational approximation for the circuit parameters can be used to generate broad-band data [49–52].

2. Assume a simple Pi- or T-equivalent circuit for the lumped element as depicted in Figure 7.10. Based on the equivalent model, convert the S-parameters to Y- or Z-parameters at each frequency point. Then, evaluate the admittance or impedance of each arm from the appropriate circuit parameters using Figure 7.10. Examples of the forms that each arm can have are shown in Figures 7.11 and 7.12.

3. Plot the imaginary versus real parts of the impedances for each arm. Then, identify the correct topology with the help of Figures 7.11 and 7.12. Note that the responses given in the figures are representative; the actual locus may vary from the ones depicted depending on the component values and/or frequency range.

4. Extract the component values of each arm of the equivalent circuit by fitting the impedance values across the frequency using either the linear least-squares or Gauss-Newton algorithm. The Gauss-Newton method is utilized if the impedance is a nonlinear function of the component values.

Note that the above procedure is most efficient in case of simple Pi- or T-networks, where admittance or impedance of each arm can easily be determined from the two-port circuit parameters, and the lumped circuit under investigation is not electrically large. If a more complicated or distributed equivalent model were to be devised, then a full-blown optimization would be required. For circuits of more than two ports, one can also generate equivalent models by assuming that each port is connected to any other one through a simple admittance that does not have any internal ground connection, as demonstrated in Figure 7.13 [41]. Then, the admittance can be computed directly from the N-port Y-parameter matrix as follows:

$$y_{ii} = \sum_{n=1}^{N} Y_{i,n} \tag{7.20}$$

$$y_{ik} = -Y_{i,k}$$

where N is the total number of ports. After calculating the admittances, one can fit a parallel *RLCG* network, as shown in Figure 7.14 to each admittance to complete the model [41]. Note that this approach also has the same disadvantage of being valid for electrically small circuits. To cope with the problem of modeling electrically large circuits, one can divide the frequency band of interest into many narrow bands and generate separate equivalent models for each of them. However, the resulting models obtained this way quickly become too cumbersome to handle. Nevertheless, this approach is still useful in many cases.

Since they are used in many modeling schemes, a good understanding of linear least-squares and Gauss-Newton algorithms would be helpful. For this purpose, they are summarized below for the benefit of the reader.

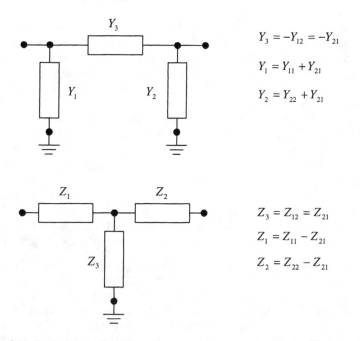

Figure 7.10 Simple Pi- and T-networks that are used in the modeling of lumped circuits and their component values in terms of Y- and Z-matrixes.

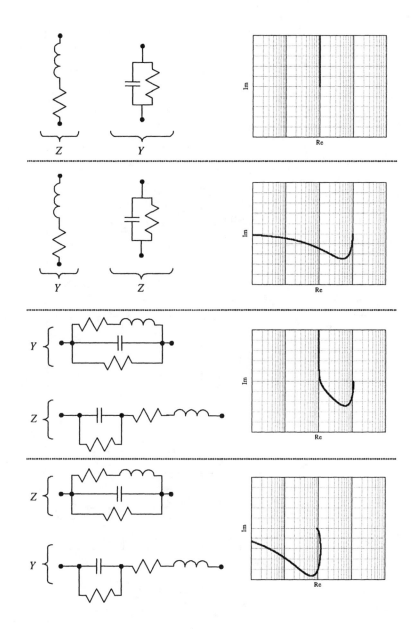

Figure 7.11 Qualitative impedance or admittance locus of various lumped circuits (frequency-independent loss terms: R and G are constant).

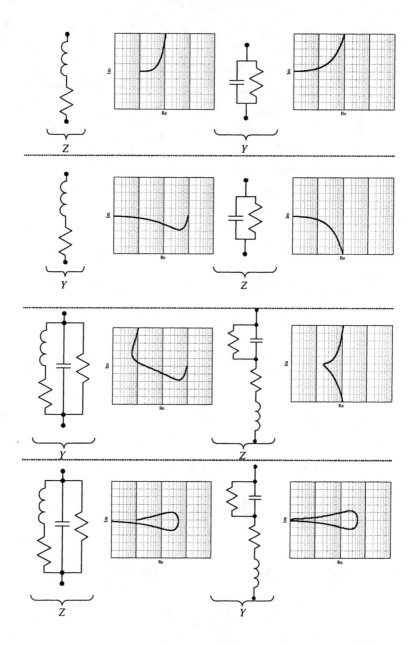

Figure 7.12 Qualitative impedance or admittance locus of various lumped circuits (frequency-dependent loss terms: $R = \alpha + \beta\sqrt{\omega}$, $G = \chi\omega$).

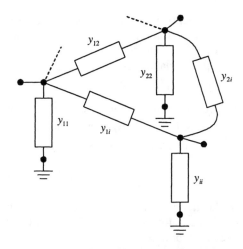

Figure 7.13 Generalization of the simple model extraction of multiple ports where it is assumed that a port is connected to any other port through a simple impedance with no internal ground.

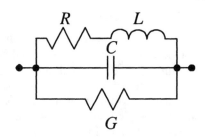

Figure 7.14 The simple parallel *RLCG* network that can be used for each admittance term in the topology given in Figure 7.13.

Linear Least-Squares Algorithm

The linear least-squares algorithm provides the optimum solution to a system of equations when there are more data points than the number of unknowns (i.e., an overdetermined system). Let's assume that we have following linear system of equations:

$$
\begin{bmatrix}
A_{1,1} & A_{1,2} & \cdots & A_{1,n} \\
A_{2,1} & \ddots & \ddots & A_{2,n} \\
A_{3,1} & \ddots & \ddots & \vdots \\
\vdots & \ddots & \ddots & A_{m-1,n} \\
A_{m,1} & \cdots & A_{m,n-1} & A_{m,n}
\end{bmatrix}
\times
\begin{bmatrix}
x_1 \\
x_2 \\
\vdots \\
x_n
\end{bmatrix}
=
\begin{bmatrix}
y_1 \\
y_2 \\
y_3 \\
\vdots \\
y_m
\end{bmatrix}
\tag{7.21}
$$

where A are the known coefficients, y are the data points, and x are the unknowns to be determined. For an overdetermined system, m is greater than n. Then, according to the least-squares theory, the set of solutions to the above system is the solution of the following matrix equation [26, 28]:

$$\mathbf{A}^T \mathbf{A} \mathbf{x} = \mathbf{A}^T \mathbf{y} \qquad (7.22)$$

where the superscript T denotes the complex-conjugate transpose. Now, if $\text{rank}(\mathbf{A}) = n$ (i.e., the matrix $\mathbf{A}$ has linearly independent columns, which is the case in the majority of applications), then there is a unique least-squares solution which is given by

$$\mathbf{x} = \left(\mathbf{A}^T \mathbf{A}\right)^{-1} \mathbf{A}^T \mathbf{y} \qquad (7.23)$$

Note that evaluating the matrix inverse in the above equation may not be numerically stable. Instead, the following equation based on QR factorization can be used to improve numerical stability:

$$\mathbf{R} \mathbf{x} = \mathbf{Q}^T \mathbf{y} \qquad (7.24)$$

where the $\mathbf{Q}$ and $\mathbf{R}$ matrixes are found from the QR factorization of the $\mathbf{A}$ matrix:

$$\mathbf{A} = \mathbf{Q} \mathbf{R} \qquad (7.25)$$

The QR factorization expresses any matrix $\mathbf{A}_{m \times n}$ with linearly independent columns as a product of a $\mathbf{Q}_{m \times n}$ matrix with orthonormal columns, and an upper triangular $\mathbf{R}_{n \times n}$ matrix with positive diagonal entries. Since $\mathbf{R}_{n \times n}$ is upper triangular, (7.24) can be solved efficiently by back-substitution. The QR factorization is readily available in most of the matrix-algebra computer software.

Alternatively, one can also employ singular value decomposition (SVD) to solve (7.21). It can be shown that the solution of (7.21) is given as [26, 28]

$$\mathbf{x} = \mathbf{A}^\dagger \mathbf{B} \qquad (7.26)$$

where $\dagger$ represents the generalized inverse (or Moore-Penrose pseudoinverse), which can be found using

$$\mathbf{A}^\dagger = \mathbf{V} \mathbf{\Sigma}^{-1} \mathbf{U}^T \qquad (7.27)$$

where the matrixes $\mathbf{U}$, $\mathbf{\Sigma}$, and $\mathbf{V}$ are obtained after the SVD:

$$\mathbf{A} = \mathbf{U} \mathbf{\Sigma} \mathbf{V}^T \qquad (7.28)$$

The real, diagonal matrix $\mathbf{\Sigma}$ obtained after the SVD is called singular-value matrix and plays an important role in the model synthesis. If the SVD is applied to an appropriate system matrix, then the number of nonzero (or above a numerical threshold) diagonal entries of $\mathbf{\Sigma}$ indicates the number of dominant modes in the system and can be used to inspect the system order. Similar to QR decomposition, the SVD is readily available in many matrix-algebra computer software packages.

The least-squares algorithm is believed to have been invented by Carl F. Gauss, who is considered to be the greatest mathematician of all time by many [28]. In 1801, astronomers of the time were puzzled by the new star observed in the Taurus constellation. The object was actually an asteroid and would soon be lost. All the efforts of famous astronomers to find the lost celestial object were in vain. Gauss decided to find the lost asteroid. In a very short time, Gauss shocked the scientific community by predicting both the location and future positions of the asteroid from apparently insufficient data. His calculations were based on his theory of least squares.

To demonstrate the application of the least-squares technique in the model extraction, consider the impedance of a series *RLC* circuit:

$$Z(\omega) = R + j\omega L + \frac{1}{j\omega C}$$

We want to determine *R*, *L*, and *C* using the least-squares algorithm. The above equation can be rewritten as follows:

$$Z(\omega) = A_1 x_1 + A_2 x_2 + A_3 x_3$$

where

$$A_1 = 1 \quad A_2 = j\omega \quad A_3 = \frac{1}{j\omega}$$

which are the terms of the **A** matrix given in (7.21). The unknowns represented by *x* are the *R*, *L*, and *C* to be determined. Then, the following matrix equation is obtained:

$$\begin{bmatrix} 1 & j\omega_1 & 1/j\omega_1 \\ 1 & j\omega_2 & 1/j\omega_2 \\ 1 & j\omega_3 & 1/j\omega_3 \\ \vdots & \vdots & \vdots \\ 1 & j\omega_m & 1/j\omega_m \end{bmatrix} \times \begin{bmatrix} x_1 \\ x_2 \\ x_3 \end{bmatrix} = \begin{bmatrix} Z'(\omega_1) \\ Z'(\omega_2) \\ Z'(\omega_3) \\ \vdots \\ Z'(\omega_m) \end{bmatrix}$$

where ω_1, ω_2, ..., ω_m are the frequency points. Primed quantities on the right-hand side indicate the simulated (or measured) impedance values. The solution to the above system can now be obtained through the least-squares technique described above. One may wonder why one should resort to the least-squares method in the above example when the unknowns can readily be obtained by just using two frequency samples (utilizing both real and imaginary parts). The answer is that if the model approximately represents the actual circuit and/or if there exists noise in the data (either experimental or numerical), one needs more samples to determine the unknowns in an optimal fashion.

Note that it is usually necessary to scale the component values when applying the least-squares method to modeling problems so that numerical accuracy is preserved. For instance, working with nH and pF, instead of Henries and Farads, would scale the ω and $1/\omega$ terms on the left-hand side such that they would have a close order of magnitude at microwave frequencies. Another important note is that the above system is usually solved by separating the real and imaginary parts. This would guarantee that no imaginary parts for the component values could result due to finite floating-point resolution.

Gauss-Newton Method

The Gauss-Newton method is an iterative technique used in solving nonlinear data-fitting problems. Let's assume that we have the following data and nonlinear function vectors:

$$\mathbf{y} = \begin{bmatrix} y_1 \\ y_2 \\ y_3 \\ \vdots \\ y_m \end{bmatrix}, \quad \mathbf{f}(x) = \begin{bmatrix} f_1(x_1, x_2, \ldots, x_n) \\ f_2(x_1, x_2, \ldots, x_n) \\ f_3(x_1, x_2, \ldots, x_n) \\ \vdots \\ f_m(x_1, x_2, \ldots, x_n) \end{bmatrix} \qquad (7.29)$$

where x are the unknowns to be determined. We also define a Jacobian matrix constructed using the partial derivatives of the function f with respect to each variable:

$$Df(\xi) = \begin{bmatrix} \dfrac{\partial f_1}{\partial x_1} & \dfrac{\partial f_1}{\partial x_2} & \cdots & \dfrac{\partial f_1}{\partial x_n} \\ \dfrac{\partial f_2}{\partial x_1} & \ddots & \ddots & \vdots \\ \vdots & \ddots & \ddots & \dfrac{\partial f_{m-1}}{\partial x_n} \\ \dfrac{\partial f_m}{\partial x_1} & \ddots & \dfrac{\partial f_m}{\partial x_{n-1}} & \dfrac{\partial f_m}{\partial x_n} \end{bmatrix}_{x=\xi} \qquad (7.30)$$

Then, one can construct the following linear least-squares problem using the Jacobian, $Df(\xi)$, and data vector, $\mathbf{y}$ [26]:

$$Df(\mathbf{x}^{(i)})\mathbf{s} = \mathbf{y} - f(\mathbf{x}^{(i)}) \qquad (7.31)$$

where $\mathbf{x}^{(i)}$ is the initial estimate of the unknowns, and $\mathbf{s}$ is an error vector that needs to be solved. Once $\mathbf{s}$ is solved, a new estimate of the unknowns is determined using

$$\mathbf{x}^{(i+1)} = \mathbf{x}^{(i)} + 2^{-k}\mathbf{s} \qquad (7.32)$$

where k is a constant that will be defined in a moment. After the new approximation, $\mathbf{x}^{(i+1)}$, is calculated, $\mathbf{x}^{(i)}$ is replaced with $\mathbf{x}^{(i+1)}$ in (7.31) and iteration continues until convergence is reached. The constant k affects the convergence of the algorithm. If the initial estimate is close enough to the solution, then k can be chosen as zero. Otherwise, the following method should be used to improve the convergence [26]. To find the constant k, we first define a function in the following form:

$$\varphi(\tau) := \left\| \mathbf{y} - f\left(\mathbf{x}^{(i)} + \tau \mathbf{s}\right) \right\|^2 \tag{7.33}$$

Then, the constant k can be selected as the minimum integer $k \geq 0$, which satisfies

$$\varphi\left(2^{-k}\right) < \varphi(0) \tag{7.34}$$

This would reduce the step size, s, if necessary, so that the new function value is always smaller than the previous one.

To demonstrate the application of the Gauss-Newton method in the model extraction, consider the impedance of a series *RLCG* circuit:

$$Z(\omega) = R + j\omega L + \frac{1}{j\omega C + G}$$

We want to determine R, L, C, and G using the Gauss-Newton method. To do this, we first write the Jacobian matrix for this function as follows:

$$DZ(\xi) = \begin{bmatrix} 1 & j\omega_1 & -j\omega_1 / \left(j\omega_1 C^{(i)} + G^{(i)}\right)^2 & -1 / \left(j\omega_1 C^{(i)} + G^{(i)}\right)^2 \\ 1 & j\omega_2 & -j\omega_2 / \left(j\omega_2 C^{(i)} + G^{(i)}\right)^2 & -1 / \left(j\omega_2 C^{(i)} + G^{(i)}\right)^2 \\ \vdots & \vdots & \vdots & \vdots \\ 1 & j\omega_m & -j\omega_m / \left(j\omega_m C^{(i)} + G^{(i)}\right)^2 & -1 / \left(j\omega_m C^{(i)} + G^{(i)}\right)^2 \end{bmatrix}_{x=\xi}$$

Note that the first column is the partial derivative with respect to R, the second column is the partial derivative with respect to L, and so on. Then, we construct the linearized system as

$$DZ\left(\mathbf{x}^{(i)}\right)\mathbf{s} = \mathbf{Z}' - \mathbf{Z}\left(\mathbf{x}^{(i)}\right)$$

where ω_1, ω_2, ..., ω_m are the frequency points. The superscript (i) denotes the estimates of R, L, C, and G at ith iteration. The primed vector on the right-hand side is the simulated (or measured) impedance data. The solution of R, L, C, and G can now be obtained through the iterative method described above.

If the analytical derivatives could not be obtained, then numerical derivatives can be used to determine the Jacobian matrix. An important advantage of the Gauss-Newton method is that it is quite general. For example, one can also extract the component values of more complicated equivalent networks other than the simple Pi- or T-networks using the Gauss-Newton method. It can be applied to nonlinear problems as well. Another advantage of the method is that it is quadratically convergent.

Example 7.1

In this example, we will model a MIM capacitor in the frequency range 0.08 to 80 GHz, using the equivalent network approach. The parallel-plate dimensions of the MIM capacitor are 16 μm and 24 μm. Thickness and relative permittivity of the capacitor dielectric are 0.066 μm and 7, respectively. It will be assumed that the loss-tangent of the dielectric is 0.01.

We have decided to use the Pi-model shown in Figure 7.10 to model the capacitor. Since the selected equivalent model is best represented in terms of Y-parameters, the first step is to convert the given S-parameters of the capacitor to two-port Y-parameters. Then, using the simple equations given in Figure 7.10, we determine the admittance of each arm from the two-port Y-parameters. The results are plotted in Figure 7.15. Note that magnitude of Y_3 is orders of magnitude greater than that of Y_2 and Y_1. Besides, Y_2 and Y_1 show both series and parallel resonances, which are relatively difficult to model. Thus, it is appropriate to apply the modeling scheme only to the series arm of the model. The parameters of the shunt arm will be approximated using S-parameters directly, as described later. Note that a very simple network containing a parallel capacitor and resistor to model the shunt arms is usually sufficient for MIM and interdigital capacitors, provided that the circuits are electrically small.

The next step is to plot the imaginary versus real parts of the admittance of the series arm (i.e., Y_3), as shown in Figure 7.16, so that the type of network can be identified. By inspecting Figures 7.11 and 7.12, it can be seen that Y_3 can be represented best using a series LC with loss components (preferably frequency dependent). It is now necessary to extract the component values of this $RLCG$ network using an appropriate algorithm. Note that since we knew the loss-tangent of the capacitor dielectric, the parallel resistor to the capacitor can already be determined using this value. This leaves us to determine the inductor, capacitor, and series resistor. It will also be assumed that the series resistance is proportional to the square root of the frequency to model the decreasing skin depth with frequency. It will further be assumed that the dc component of this resistor is zero, although this is not strictly true. However, this is usually a reasonable assumption to improve the accuracy if we are trying to extract loss parameters using only a single set of given data. Since we are assuming frequency-dependent loss terms, it is best to use the Gauss-Newton algorithm. For the benefit of the reader, the admittance function that will be optimized is given below:

$$Y(\omega) = \cfrac{1}{\underbrace{K \cdot 10^{-6}\sqrt{\omega}}_{R} + j\omega L \cdot 10^{-9} + \cfrac{1}{j\omega C \cdot 10^{-12} + \omega C \cdot 10^{-12}\tan(\delta)}}$$

Note the scaling factors used to improve the numerical convergence. After optimization, the following component values are found for the series arm:

$$L = 0.032 \quad \text{nH}$$

$$C = 0.379 \quad \text{pF}$$

$$R = 0.688 \cdot 10^{-6}\sqrt{2\pi f / \text{Hz}} \quad \text{ohm}$$

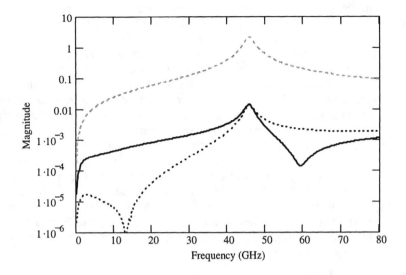

Figure 7.15 Admittance magnitudes of the MIM capacitor (Y_1: solid; Y_2: dotted; Y_3: dashed).

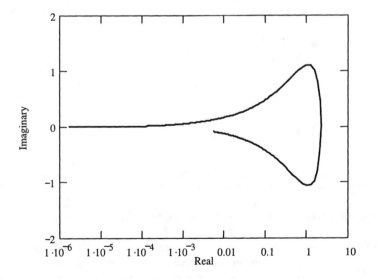

Figure 7.16 Imaginary versus real parts of Y_3 for the MIM capacitor. This plot tells us that Y_3 can be represented by a series LC resonator (see Figures 7.11 and 7.12).

As a sanity check, we can calculate the simple parallel-plate capacitance of the MIM, which is found to be 0.361 pF. The difference between this value and the value found using optimization is due to fringe electric fields.

The last step is to determine the admittances of the shunt arms. For this purpose, the model S-parameters are compared with the simulated results, and the shunt capacitance and conductance values are adjusted until a good match is obtained. A manual optimization is sufficient, although a second step of computer optimization could also be used. After this, it is found that a parallel combination of 4000 ohms and 0.003 pF is appropriate to approximate the shunt arms. The resistance is due to the finite conductivity of the substrate. Although the value of the shunt capacitors is quite small compared to the series arm, they are necessary to shift the frequency response so that simulated and modeled results will agree well. The final network is shown in Figure 7.17. Note that nonideal L and C have been used in the circuit schematic such that the loss is entered through the Q-factor definitions. Comparisons of the model and simulated data are given in Figures 7.18 and 7.19. As can be seen from the plots, the model and simulation results agree quite well; the curves are visually indistinguishable.

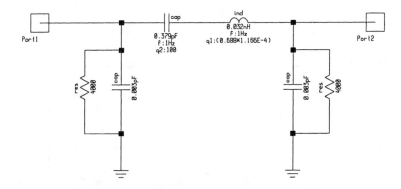

Figure 7.17 Equivalent model of the MIM capacitor. Note that nonideal L and C have been used in the model such that the loss is entered through the Q-factor definitions.

7.2.2 Modeling Based on Parameter Estimation

The modeling technique based on equivalent physical networks becomes difficult to apply when the circuit under investigation is electrically large and/or has multiple ports. This is because, under such circumstances, it can be quite difficult to determine the broadband equivalent physical model (either lumped or distributed) by intuition. In that case, the designer needs to reduce the bandwidth until the match between the equivalent model and simulations is acceptable.

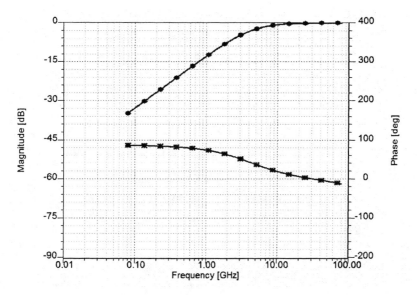

Figure 7.18 Comparison of S_{11} for the simulated and modeled S-parameters for the MIM capacitor (circle: model magnitude; square: model phase; plus: simulated magnitude; star: simulated phase).

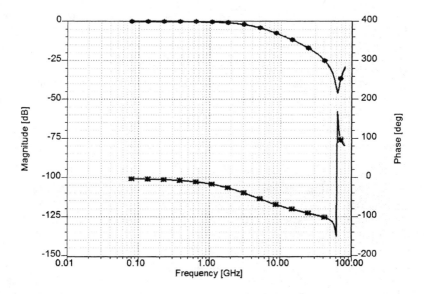

Figure 7.19 Comparison of S_{12} for the simulated and modeled S-parameters for the MIM capacitor (circle: model magnitude; square: model phase; plus: simulated magnitude; star: simulated phase).

Although the problem of modeling electrically large circuits can be addressed to some extent by reducing the modeling bandwidth, the situation becomes quickly unmanageable in the case of multiple-port circuits, since the equivalent physical model gets quite complicated to consider all electrically dominant interactions between the ports. This is especially important in modeling structures such as electronic circuit packages or multiple electrical interconnects.

To address these issues, an alternative technique based on mathematical modeling instead of physical modeling, can be employed. In this approach, an equivalent mathematical model is derived from the pole-residue information of the circuit parameters [34, 48, 53, 54]. Although the resulting model might not always be intuitive, it can correctly model the electrically large and multiport circuits. Another unique advantage of this approach is the possibility of model reduction for the electrically large circuits in which only the dominant eigenvalues are selected such that the complexity of the equivalent model is reduced.

We start our discussion by indicating that the admittance of any linear, passive network can be represented by

$$\mathbf{Y}(s) = \mathbf{A}^{(0)} + \sum_{n=1}^{N} \frac{A_0^{(n)}}{s - \alpha_n} \mathbf{A}^{(n)} + \mathbf{A}^{(\infty)} s \qquad (7.35)$$

where α_n and $\mathbf{A}_n$ are the poles and residue matrixes of the multiport circuit, respectively. Many features of this representation should be mentioned at this point. First of all, complex poles should appear as complex-conjugate pairs, and the real parts of the poles should be negative for a stable system. The residues must be real for real poles. However, they can be complex for complex poles. The first and last terms in (7.35) represent the zero frequency and asymptotic response as the frequency goes to infinity, respectively. Depending on the circuitry, either or both of them may be zero. Note that all terms of $\mathbf{Y}(s)$ (i.e., Y_{11}, Y_{12}, and so forth) are assumed to have the same poles. Once a representation in the form of (7.35) is obtained, an equivalent network can be deduced using either the Cauer topology [42–44] or state-space approach [34].

Another important property of (7.35) is the residue condition, which can be stated for lossless two-port networks as follows:

$$k_{11}k_{22} - k_{12}^2 \geq 0 \qquad (7.36)$$

where k_{11}, k_{22}, and k_{12} are the residues. This condition is important to satisfy because it determines whether or not the system is physically realizable. If the residue condition is satisfied with the equality sign, the corresponding pole is called compact. For more thorough information on the properties of pole-residue representation and residue condition, see Guillemin [33].

The critical step in the overall modeling, as one might expect, is the determination of the poles and residues of the circuit that needs to be modeled. Blind application of an optimization algorithm to determine the parameters in (7.35) may yield nonphysical representations. Thus, the poles and residues are usually determined in separate steps. There are many ways of obtaining the poles and residues of circuit parameters [55–66]. Two of the frequency-domain methods that are used frequently will be summarized next.

Frequency-Domain Prony Method

The frequency-domain Prony method is used to extract pole-residue information directly from the frequency-domain data [55–57]. Both time- and frequency-domain versions of the Prony technique have been extensively used in the literature for target identification using the late-time electromagnetic transient response of conducting bodies. Prony's technique can also be utilized in the characterization of passive microwave circuits [45]. Although it is relatively simple, the main drawback of the method is that it requires the determination of polynomial roots, which can be inaccurate if the system contains a large number of (> 20) and/or closely spaced poles. Besides, a priori information on the number of system poles is required. If the system needs to be modeled in the time domain, then a better alternative to the Prony method is the generalized pencil-of-function (GPOF) method, and interested readers can refer to the literature for more information on that technique [62–65]. GPOF has better noise immunity than Prony's technique. Another advantage of the GPOF is that the number of system poles is determined from the nonzero singular values of an information matrix, which eliminates the necessity of a priori information on the number of poles.

The frequency-domain Prony method starts by assuming that the transfer function of an electromagnetic system can be represented by

$$F(s) = \sum_{n=1}^{N} \frac{R_n}{s - s_n} \tag{7.37}$$

where s_n and R_n are the poles and residues of the system, respectively. N represents the number of poles, which we need to assign an initial value. We will visit this point later in more detail. Then, (7.37) is put into the following form:

$$F(s) = \sum_{n=1}^{N} R_n \frac{\displaystyle\prod_{\substack{i=1 \\ i \neq n}}^{N} (s - s_i)}{\displaystyle\prod_{k=1}^{N} (s - s_k)} \tag{7.38}$$

One can also expand the products in the numerator and denominator as follows:

$$\prod_{k=1}^{N} (s - s_k) = \sum_{k=0}^{N} \alpha_k s^k, \quad \alpha_N = 1 \tag{7.39}$$

$$\prod_{\substack{i=1 \\ i \neq n}}^{N} (s - s_i) = \sum_{i=0}^{N-1} \beta_i^{(n)} s^i, \quad \beta_{N-1}^{(n)} = 1 \tag{7.40}$$

Note that roots of the polynomial given in (7.39) are the poles of the system. Thus, we need to determine α_k first to find the roots. By inserting (7.39) and (7.40) into

(7.38), then changing the order of the summation in the numerator, one can obtain the following expression:

$$F(s) = \frac{\sum_{i=0}^{N-1} s^i \sum_{n=1}^{N} R_n \beta_i^{(n)}}{\sum_{k=0}^{N} \alpha_k s^k}$$

After some algebraic operations, the following equation is obtained:

$$\sum_{i=0}^{N-1} \left[\alpha_i F(s) - \tilde{\beta}_i \right] s^i = -F(s) s^N \qquad (7.41)$$

where

$$\tilde{\beta}_i = \sum_{n=1}^{N} R_n \beta_i^{(n)} \qquad (7.42)$$

Now, by sampling $F(s)$ at different frequency points, one can form a linear system of equations using (7.41) whose solution will provide the polynomial coefficients:

$$\begin{bmatrix} F(s_1) & \cdots & s_1^{N-1}F(s_1) & -1 & \cdots & -s_1^{N-1} \\ F(s_2) & \cdots & s_2^{N-1}F(s_2) & -1 & \cdots & -s_2^{N-1} \\ \vdots & \vdots & \vdots & \vdots & \vdots & \vdots \\ F(s_M) & \cdots & s_M^{N-1}F(s_M) & -1 & \cdots & -s_M^{N-1} \end{bmatrix} \begin{bmatrix} \alpha_0 \\ \vdots \\ \alpha_{N-1} \\ \tilde{\beta}_0 \\ \vdots \\ \tilde{\beta}_{N-1} \end{bmatrix} = \begin{bmatrix} -F(s_1)s_1^N \\ -F(s_2)s_2^N \\ \vdots \\ -F(s_M)s_M^N \end{bmatrix} \qquad (7.43)$$

where s_i equals $j\omega_i$, and M is the total number of samples. Note that the above equation is best solved by separating the real and imaginary parts. For this purpose, (7.43) can be written as

$$(\mathbf{A}_r + j\mathbf{A}_i) \times (\mathbf{x}_r + j\mathbf{x}_i) = (\mathbf{B}_r + j\mathbf{B}_i)$$

where subscripts r and i refer to real and imaginary parts, respectively. After some simple matrix algebra, it can be shown that

$$\begin{bmatrix} \mathbf{A}_r & -\mathbf{A}_i \\ \hline \mathbf{A}_i & \mathbf{A}_r \end{bmatrix} \begin{bmatrix} \mathbf{x}_r \\ \hline \mathbf{x}_i \end{bmatrix} = \begin{bmatrix} \mathbf{B}_r \\ \hline \mathbf{B}_i \end{bmatrix} \qquad (7.44)$$

Note that (7.44) contains two times more unknowns than the original equation (because we separated the real and imaginary parts). However, it can be solved by using only real arithmetic, which can be advantageous. Another important point is to enforce the conjugate property of the complex poles (i.e., $F^*(s) = F(s^*)$). This requires that all coefficients of the polynomial (7.39) be real, which can be simply achieved by picking only the first half of the solution vector in (7.44). However, there should also be a mechanism so that the remainder of the solution vector (i.e., imaginary parts) becomes zero, or very close to zero. This can be enforced by mirroring the data to negative frequencies around the zero frequency such that the real and imaginary parts of the data become even and odd functions of the frequency, respectively.

The system given in (7.44) can be solved using many different ways to obtain the polynomial coefficients. If the number of frequency samples is greater than the unknowns, then the system can be solved using the least-squares approach already described. After determining the polynomial coefficients, α, the poles can be found from the roots of (7.39). The roots of polynomials can be determined in a very robust manner using the companion matrix approach instead of employing the classical Newton-Raphson algorithm. It is known that the roots of the polynomial given in (7.39) are the eigenvalues of the following matrix:

$$\begin{bmatrix} -\alpha_0/\alpha_N & -\alpha_1/\alpha_N & -\alpha_2/\alpha_N & \cdots & -1 \\ 1 & 0 & 0 & \cdots & 0 \\ 0 & 1 & 0 & \ddots & 0 \\ \vdots & \ddots & \ddots & \ddots & \vdots \\ 0 & 0 & \cdots & 1 & 0 \end{bmatrix} \qquad (7.45)$$

where α_n are the polynomial coefficients. Although it would be more CPU intensive, this approach is more numerically robust in determining closely spaced poles. After the poles are found, they are inserted into (7.37) and residues can be determined directly from the solution of the resulting linear system. Note that there will be some restrictions on the residues, such as the residue condition of lossless networks [see (7.36)] so that the resulting network function can represent a physical network. If these restrictions are not satisfied, one can still get a relatively good match to the data but fail to generate an equivalent network. In that case, the model becomes merely a mathematical curve fitting. For this reason, an appropriate optimization algorithm with proper constraints can be employed so that the signs and magnitudes of the residues stay within the required limits. We will return to this point later in explaining how to interface with circuit simulators.

At this point, it would also be instructive to elaborate on how the number of poles, N, should be selected to begin with. To find the optimum value, one can increase the number of poles from a starting value and apply Prony's method iteratively until the error between the actual samples and model does not improve any further. Unfortunately, there is no easy way of determining the actual number of poles in Prony's method. Thus, some trial and error is required in this step. One should also utilize broadband frequency data so that all dominant poles are sampled. Prony's method works best if the frequency band of sampling covers all the dominant poles of the system.

Cauchy Method

The Cauchy method is another technique used to approximate a frequency-domain system function in terms of rational polynomials [60, 61]. If available, it utilizes derivatives of the function that needs to be approximated for higher accuracy. The method assumes a system function in the following form:

$$F(s) = \frac{A(s)}{B(s)} = \frac{\displaystyle\sum_{k=0}^{P} a_k s^k}{\displaystyle\sum_{k=0}^{Q} b_k s^k} \tag{7.46}$$

where P and Q are the orders of the numerator and denominator, respectively. The aim is to determine a_k, b_k, P, and Q from the samples of $F(s)$ and its N_j derivatives at some frequency points s_j, where $j = 1, 2, \ldots, J$; that is,

Given $\quad F^{(n)}(s_j)\quad$ for $\quad j = 1, 2, \ldots, J \quad n = 0, 1, \ldots, N_j$

Find $\quad P, Q, \{a_k, k = 0, 1, \ldots, P\} \quad$ and $\quad \{b_k, k = 0, 1, \ldots, Q\}$

where the superscript (n) denotes the nth derivative. The solutions for a_k and b_k are unique if the total number of samples, N, is greater than or equal to the total number of unknown coefficients, $P + Q + 2$ [60]:

$$N \equiv \sum_{j=1}^{J}(N_j + 1) \geq P + Q + 2 \tag{7.47}$$

Rearranging (7.46) and taking the derivative of both sides n times results in the following binomial expression:

$$A^{(n)}(s_j) = \sum_{i=0}^{n} C_{i,n} H^{(n-i)}(s_j) B^{(i)}(s_j) \tag{7.48}$$

where

$$C_{i,n} = \frac{n!}{(n-i)! \, i!} \tag{7.49}$$

Then, using the polynomial expressions for $A(s)$ and $B(s)$, the above expression can be written as [60]

$$\sum_{k=0}^{P} a_k A_{j,n,k} = \sum_{k=0}^{Q} b_k B_{j,n,k} \tag{7.50}$$

where

$$A_{j,n,k} = \frac{k!}{(k-n)!} s_j^{k-n} u(k-n)$$

$$B_{j,n,k} = \sum_{i=0}^{n} C_{i,n} H^{(n-i)}(s_j) \frac{k!}{(k-i)!} s_j^{k-i} u(k-i)$$

where $u(l)=0$ for $l<0$, and $u(l)=1$ otherwise. Note that the indexes j and n represent the sample points and order of the derivatives at the sample points, respectively. If only values of the function are used (i.e., no derivatives are available), then n becomes zero for all sample points. The above equations can be put into matrix form for easy visualization as follows:

$$\begin{bmatrix} A_{0,0,0} & A_{0,0,1} & \cdots & A_{0,0,P} \\ A_{0,1,0} & A_{0,1,1} & \cdots & A_{0,1,P} \\ \vdots & \vdots & \vdots & \vdots \\ A_{0,N_0,0} & A_{0,N_0,1} & \cdots & A_{0,N_0,P} \\ A_{1,0,0} & A_{1,0,1} & \cdots & A_{1,0,P} \\ A_{1,1,0} & A_{1,1,1} & \cdots & A_{1,1,P} \\ \vdots & \vdots & \vdots & \vdots \\ A_{1,N_1,0} & A_{1,N_1,1} & \cdots & A_{1,N_1,P} \\ \vdots & \vdots & \vdots & \vdots \\ A_{J,N_J,0} & A_{J,N_J,1} & \cdots & A_{J,N_J,P} \end{bmatrix} \begin{bmatrix} a_0 \\ a_1 \\ a_2 \\ \vdots \\ a_P \end{bmatrix} = \begin{bmatrix} B_{0,0,0} & B_{0,0,1} & \cdots & B_{0,0,Q} \\ B_{0,1,0} & B_{0,1,1} & \cdots & B_{0,1,Q} \\ \vdots & \vdots & \vdots & \vdots \\ B_{0,N_0,0} & B_{0,N_0,1} & \cdots & B_{0,N_0,Q} \\ B_{1,0,0} & B_{1,0,1} & \cdots & B_{1,0,Q} \\ B_{1,1,0} & B_{1,1,1} & \cdots & B_{1,1,Q} \\ \vdots & \vdots & \vdots & \vdots \\ B_{1,N_1,0} & B_{1,N_1,1} & \cdots & B_{1,N_1,Q} \\ \vdots & \vdots & \vdots & \vdots \\ B_{J,N_J,0} & B_{J,N_J,1} & \cdots & B_{J,N_J,Q} \end{bmatrix} \begin{bmatrix} b_0 \\ b_1 \\ b_2 \\ \vdots \\ b_Q \end{bmatrix} \quad (7.51)$$

By taking the right-hand side to the left, one obtains

$$[\mathbf{A} \mid -\mathbf{B}]\begin{bmatrix} \mathbf{a} \\ \overline{\mathbf{b}} \end{bmatrix} = \mathbf{0} \tag{7.52}$$

The order of matrix $\mathbf{A}$ is $N \times (P+1)$, and that of $\mathbf{B}$ is $N \times (Q+1)$. The solution of (7.52) gives the desired polynomial coefficients of (7.46).

As in Prony's method, one needs to have estimates of P and Q a priori. Initial selection of P and Q can be done according to (7.47). Then, one can employ SVD on the following matrix:

$$\mathbf{C} \equiv [\mathbf{A} \mid -\mathbf{B}] = \mathbf{U}\mathbf{\Sigma}\mathbf{V}^T$$

where $\mathbf{\Sigma}$ is a diagonal matrix whose entries are the singular values of matrix $\mathbf{C}$. The SVD operation can easily be performed using linear-algebra computer software libraries, and it is a built in function in many numerical software packages. The number of nonzero singular values provides us the information on how much redundancy there exists in the system. Specifically, we would like to have [60]

$$R+1 = P+Q+2 \tag{7.53}$$

where R is the number of nonzero singular values. Therefore, once the singular values are determined, the initial estimates of P and Q are updated according to (7.53), and a new system is formed using (7.52) based on updated P and Q.

One can also enforce some specific conditions on the rational function in (7.46). For instance, we know from Chapter 1 that the driving point function (impedance or admittance) of passive networks should be a positive real function. This puts some restrictions on the possible forms of rational functions that we can use to approximate such networks. The necessary (but not sufficient) conditions for a rational function to be positive real are repeated here again for convenience:

1. All polynomials' coefficients should be real and positive.
2. Degrees of numerator polynomial, P, and denominator polynomial, Q, differ at most by one.
3. Numerator and denominator terms of lowest degree differ at most by one.
4. Imaginary axis poles and zeros are simple.
5. There should be no missing terms in the numerator or denominator polynomials, unless all even or all odd terms are missing.

It should be stressed once more that the above conditions apply to driving point functions (i.e., Z_{11} or Y_{11}); transfer functions (i.e., Z_{12} or Y_{12}) do not need to be positive real. In conclusion, the Cauchy method equipped with SVD to determine the optimum values of P and Q, with the above restrictions, can effectively be used to characterize microwave lumped elements.

Example 7.2

In this example, we will demonstrate the application of Prony's technique. For this purpose, we are going to extract the poles and residues of the Y-parameters of a two-layer, 4.5 turn, MMIC square-spiral inductor with outer dimensions of 114 μm by 114 μm. Line width and separation of the inductor are both 6 μm. Note that two-layers of turns are utilized to increase the inductance for a given area. We will extract the poles and residues of this microwave lumped element in the frequency range of 0.08 to 80 GHz. Broad band modeling of spiral inductors is notoriously difficult with simple lumped circuits like the one shown in Figure 7.8. However, extraction poles of the Y-parameters and then realizing a Cauer type of network using these poles will address this difficulty.

To apply the technique, we first convert the given two-port S-parameters to Y-parameters. For the sake of information, the S-parameter matrix contains 300 frequency samples in the frequency range of interest obtained through an adaptive frequency-sampling technique. Then, we mirror the Y-parameters around the zero frequency and use the following odd and even properties of the admittance functions to create information for the negative frequencies as well:

$$\mathrm{Re}\{Y\} = \frac{1}{2}[Y(s) + Y(-s)]$$

$$\mathrm{Im}\{Y\} = \frac{1}{2}[Y(s) - Y(-s)]$$

(7.54)

One more final step before start applying Prony's technique is to utilize normalized frequencies to improve the numerical solution. For this purpose, we use frequencies in units of gigahertz (e.g., 1 instead of 1×10^9) so that the dynamic range of the numbers in (7.43) does not fall below the number of significant digits in the computer arithmetic.

The next step is forming the system given in (7.43) using Y_{12} and the frequency points. The number of poles, N, is initially selected as 5. Then, we separate the system into real and imaginary parts as shown in (7.44), and SVD is employed to solve the resulting real, linear matrix equation. Finally, the roots of the polynomial (7.39) are determined by employing the companion matrix approach, as described before, resulting in the following system poles:

$$\alpha = \begin{bmatrix} -2.627 - j76.797 \\ -2.627 + j76.797 \\ -1.353 + j30.602 \\ -1.353 - j30.602 \\ -0.279 \end{bmatrix} \tag{7.55}$$

A quick inspection of the poles confirms that all poles are on the left-hand side of the s-plane as expected. Besides, complex poles appear as conjugate pairs, and there is a pole at the zero imaginary frequency, which is also expected because an inductor passes dc (i.e., zero imaginary frequency) with much less attenuation provided that it has low dc resistivity. Then, we continue with our process and extract the residues corresponding to Y_{11}, Y_{22}, and Y_{12} through a linear least-squares algorithm, using the obtained poles. The result is as follows:

$$R_{11} = \begin{bmatrix} 0.241 - j0.024 \\ 0.241 + j0.024 \\ 0.107 + j6.014 \times 10^{-3} \\ 0.107 - j6.014 \times 10^{-3} \\ 0.044 \end{bmatrix} \quad R_{22} = \begin{bmatrix} 0.2 - j0.027 \\ 0.2 + j0.027 \\ 0.084 + 9.762 \times 10^{-3} \\ 0.084 - 9.762 \times 10^{-3} \\ 0.044 \end{bmatrix} \tag{7.56}$$

$$R_{12} = \begin{bmatrix} -0.204 + j9.778 \times 10^{-3} \\ -0.204 - j9.778 \times 10^{-3} \\ -0.097 - j5.36 \times 10^{-3} \\ -0.097 + j5.36 \times 10^{-3} \\ -0.044 \end{bmatrix}$$

By inspecting the extracted residues, one can also confirm that all of the real parts of R_{11} and R_{22} are greater than zero, an indication of positive real behavior. Note that the transfer admittances do not need to be positive real, so it is allowable to have negative real parts in R_{12}. Finally, all of the complex residues appear as conjugate pairs.

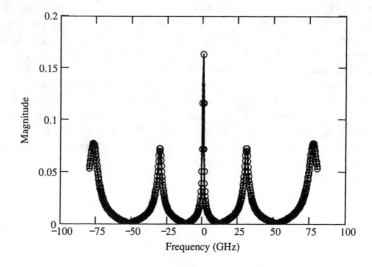

Figure 7.20 Magnitude of Y_{12} for the modeled and given Y-parameters of the square-spiral inductor (circle: data; line: model).

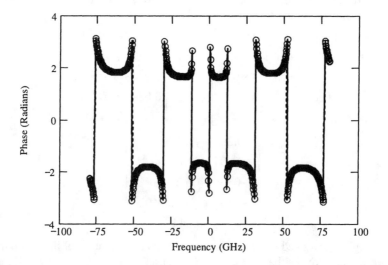

Figure 7.21 Phase of Y_{12} for the modeled and given Y-parameters of the square-spiral inductor (circle: data; line: model).

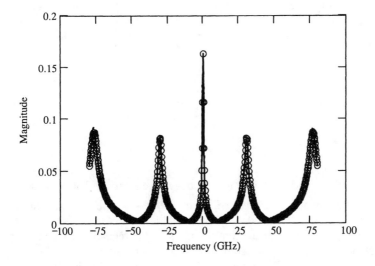

Figure 7.22 Magnitude of Y_{11} for the modeled and given Y-parameters of the square-spiral inductor (circle: data; line: model).

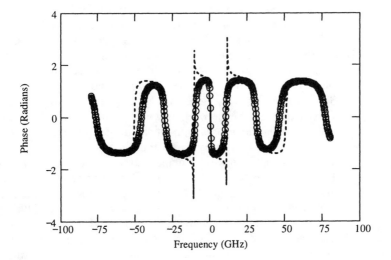

Figure 7.23 Phase of Y_{11} for the modeled and given Y-parameters of the square-spiral inductor (circle: data; line: model).

After obtaining the poles and residues and verifying their mathematical properties, it is necessary to compare the model with the input data. For this purpose, the mathematical model of the inductor is constructed as follows:

$$Y_{11}^{(m)} = \sum_{n=1}^{5} \frac{R_n^{(11)}}{s - \alpha_n}$$

$$Y_{12}^{(m)} = \sum_{n=1}^{5} \frac{R_n^{(12)}}{s - \alpha_n} \qquad (7.57)$$

$$Y_{22}^{(m)} = \sum_{n=1}^{5} \frac{R_n^{(22)}}{s - \alpha_n}$$

where $R^{(11)}$, $R^{(12)}$, and $R^{(22)}$ are the residues given above. The superscript (m) represents the model Y-parameters and the poles are designated by α_n. Comparisons of the above model with the data are given in Figures 7.20 to 7.23 for Y_{12} and Y_{11}. As can be seen from the plots, the agreement between the model and input data is very good. Similar behavior was also observed for Y_{22}. This immediately proves that five poles were sufficient to model this inductor in the given frequency range. Thus, no further iteration is required. In fact, it is always a good idea to plot Y_{12} versus frequency as a first step to get an idea about the required number of poles.

7.2.3 Interfacing with Circuit Simulators

In Section 7.2.2, we described how to represent microwave lumped elements using pole-residue information of the circuit parameters. That approach, however, might have somewhat limited value by itself in modeling the lumped elements because it is not in a universal form that can be used in every circuit simulator. To complete the modeling, therefore, these representations should be transferred into equivalent circuits so that they can be employed in time- and/or frequency-domain circuit simulators (e.g., SPICE) with active components. It should also be noted that some circuit simulators may allow direct entry of the s-domain information; in that case the pole-residue representation can be used without any further processing.

There are fundamentally two ways of transferring the pole-residue representation into an equivalent passive, lumped circuit. The first way is to employ a network synthesis approach to obtain a Cauer-type of network [42–45]. A Cauer network essentially contains parallel or series connection of simple pole-producing circuits, which are properly isolated using ideal transformers. Although Cauer-type networks are usually difficult to implement in practice due to ideal transformers, they are acceptable in our case because we are only interested in an equivalent network that could be used in a circuit simulator. The second way of converting the pole-residue representation into an equivalent network is through the state-space representation [34, 46, 48]. In that approach, coefficient matrixes of an equivalent state-space system is extracted from the poles and residues of the network. Once this is done, an equivalent network is generated for this state-space system using controlled sources. Both of these approaches will be elaborated in more detail in the following sections.

Cauer Network Representation

This method was originally developed by Wilhelm Cauer for the synthesis of two-terminal pair *LC*, *RC*, and *RL* networks using the partial-fraction expansion of the network parameters [30]. Generalization of this method for multiport *RLCG* networks can be found in the literature [42, 43]. Here, we will demonstrate the application of the method to two-port circuits. For multiport networks, the state-space approach that is described in the next section might be more suitable in terms of automation.

To demonstrate the technique, first consider a simple Pi-network with an ideal transformer as shown in Figure 7.24. It can easily be shown that the admittances can be obtained in terms of the two-port *Y*-parameters as follows:

$$Y_a = Y_{11} + \frac{1}{a}Y_{12}$$

$$Y_b = \frac{1}{a^2}Y_{22} + \frac{1}{a}Y_{12} \qquad (7.58)$$

$$Y_c = -\frac{1}{a}Y_{12}$$

where a is the transformer ratio. What is the purpose of the ideal transformer in the network? First, we already know from the network theory that in order to have a passive network, the admittances Y_a, Y_b, and Y_c must be positive-real functions. However, the transfer admittance Y_{12} is not necessarily positive real. Nor is the sum of a positive-real function and a transfer admittance positive real. To make the Y_a, Y_b, and Y_c positive real, we need to select the transformer ratio and sign properly in (7.58).

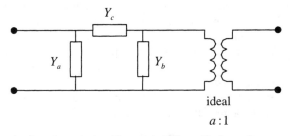

Figure 7.24 The simple pole-producing Pi-network with an ideal transformer used in the Cauer synthesis method.

Second, as it will be clear in a moment, the transformer will help to isolate each pole so that different circuits as shown in Figure 7.24 representing individual poles in the partial-fraction expansion can be connected in parallel to approximate the network parameters. Note that under certain circumstances, the transformer ratios become unity. In that case, it will be possible to eliminate the corresponding transformers [30].

Then, we assume that the partial-fraction expansions for the short-circuit impedance functions of the two-port network are given as follows:

$$Y_{11} = \frac{k_{11}^{(0)}}{s} + \frac{2\,\mathrm{Re}\{k_{11}^{(1)}\}s - 2\,\mathrm{Re}\{k_{11}^{(1)}\alpha_1^*\}}{s^2 - 2\,\mathrm{Re}\{\alpha_1\}s + |\alpha_1|^2} + \frac{2\,\mathrm{Re}\{k_{11}^{(2)}\}s - 2\,\mathrm{Re}\{k_{11}^{(2)}\alpha_2^*\}}{s^2 - 2\,\mathrm{Re}\{\alpha_2\}s + |\alpha_2|^2}$$
$$+ \ldots + \frac{2\,\mathrm{Re}\{k_{11}^{(n)}\}s - 2\,\mathrm{Re}\{k_{11}^{(n)}\alpha_n^*\}}{s^2 - 2\,\mathrm{Re}\{\alpha_n\}s + |\alpha_n|^2} + \ldots + k_{11}^{(\infty)}s \qquad (7.59)$$

$$Y_{22} = \frac{k_{22}^{(0)}}{s} + \frac{2\,\mathrm{Re}\{k_{22}^{(1)}\}s - 2\,\mathrm{Re}\{k_{22}^{(1)}\alpha_1^*\}}{s^2 - 2\,\mathrm{Re}\{\alpha_1\}s + |\alpha_1|^2} + \frac{2\,\mathrm{Re}\{k_{22}^{(2)}\}s - 2\,\mathrm{Re}\{k_{22}^{(2)}\alpha_2^*\}}{s^2 - 2\,\mathrm{Re}\{\alpha_2\}s + |\alpha_2|^2}$$
$$+ \ldots + \frac{2\,\mathrm{Re}\{k_{22}^{(n)}\}s - 2\,\mathrm{Re}\{k_{22}^{(n)}\alpha_n^*\}}{s^2 - 2\,\mathrm{Re}\{\alpha_n\}s + |\alpha_n|^2} + \ldots + k_{22}^{(\infty)}s \qquad (7.60)$$

$$Y_{12} = \frac{k_{12}^{(0)}}{s} + \frac{2\,\mathrm{Re}\{k_{12}^{(1)}\}s - 2\,\mathrm{Re}\{k_{12}^{(1)}\alpha_1^*\}}{s^2 - 2\,\mathrm{Re}\{\alpha_1\}s + |\alpha_1|^2} + \frac{2\,\mathrm{Re}\{k_{12}^{(2)}\}s - 2\,\mathrm{Re}\{k_{12}^{(2)}\alpha_2^*\}}{s^2 - 2\,\mathrm{Re}\{\alpha_2\}s + |\alpha_2|^2}$$
$$+ \ldots + \frac{2\,\mathrm{Re}\{k_{12}^{(n)}\}s - 2\,\mathrm{Re}\{k_{12}^{(n)}\alpha_n^*\}}{s^2 - 2\,\mathrm{Re}\{\alpha_n\}s + |\alpha_n|^2} + \ldots + k_{12}^{(\infty)}s \qquad (7.61)$$

where α_n and k are the complex-conjugate poles and residues, respectively, which can be found using Prony's method or another similar technique. If we assume that each term in the above expansion represents a simple Pi-circuit in the form of Figure 7.24, then the overall network can be represented by the parallel connection of these (hence, the summation of partial-fraction expansion terms) Pi-circuits as depicted in Figure 7.25. Each Pi-network shown in Figure 7.25 can be one of those networks indicated in Figure 7.26 depending on the admittance-function term. The remaining task is to determine the admittances of each arm of the Pi-network from the partial-fraction expansion.

For the sake of demonstration, we will now present how the components of the last network shown in Figure 7.26 are obtained. After grasping the methodology, it would be simple to do the same thing for the first two topologies. To accomplish this, we first insert (7.59), (7.60), and (7.61) into (7.58). Then, by comparing the terms of (7.58) with the admittances in Figure 7.26, one can find that

$$L = \frac{1}{A_1} \qquad R = \frac{1}{A_1}(A_3 - A_2/A_1) \qquad (7.62)$$

$$C = \frac{A_1^2}{A_1 A_4 - A_2(A_3 - A_2/A_1)} \qquad G = \frac{A_1 A_2}{A_1 A_4 - A_2(A_3 - A_2/A_1)}$$

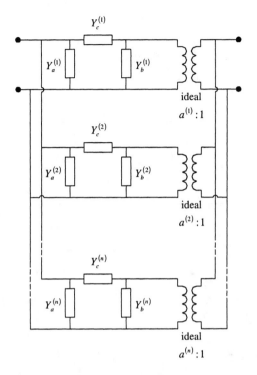

Figure 7.25 General form of the Cauer-type *RLCG* network obtained by parallel connection of different Pi-networks. Each Pi-network is responsible for producing a specific pole and can be one of the circuits represented in Figure 7.26. Note that under specific conditions, it is possible to eliminate the ideal transformers.

where the positive coefficients A_1, A_2, A_3, and A_4 are given for each arm as follows:

$$A_1 = \begin{cases} 2\,\mathrm{Re}\{k_{11}^{(n)}\} + \dfrac{2}{a}\,\mathrm{Re}\{k_{12}^{(n)}\} & \text{for arm } a \\[2mm] \dfrac{2}{a^2}\,\mathrm{Re}\{k_{22}^{(n)}\} + \dfrac{2}{a}\,\mathrm{Re}\{k_{12}^{(n)}\} & \text{for arm } b \\[2mm] -\dfrac{2}{a}\,\mathrm{Re}\{k_{12}^{(n)}\} & \text{for arm } c \end{cases} \tag{7.63}$$

$$A_2 = \begin{cases} -2\,\mathrm{Re}\{k_{11}^{(n)}\alpha_n^*\} - \dfrac{2}{a}\,\mathrm{Re}\{k_{12}^{(v)}\alpha_n^*\} & \text{for arm } a \\[2mm] -\dfrac{2}{a^2}\,\mathrm{Re}\{k_{22}^{(n)}\alpha_n^*\} - \dfrac{2}{a}\,\mathrm{Re}\{k_{12}^{(n)}\alpha_n^*\} & \text{for arm } b \\[2mm] \dfrac{2}{a}\,\mathrm{Re}\{k_{12}^{(n)}\alpha_n^*\} & \text{for arm } c \end{cases} \tag{7.64}$$

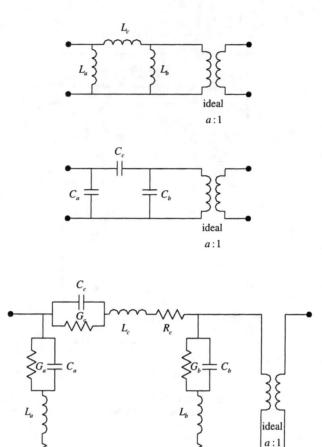

Figure 7.26 The three canonical networks that are used in the synthesis of Cauer-type network. Admittances of the arms from the top to bottom are as follows: $Y(s) = s$; $Y(s) = 1/s$; or $Y(s) = (A_1 s + A_2)/(s^2 + A_3 s + A_4)$. The turn ratios of the ideal transformers are determined during the synthesis.

$$A_3 = -2\,\mathrm{Re}\{\alpha_n^*\} \quad \text{for arms } a, b, \text{ and } c \tag{7.65}$$

$$A_4 = |\alpha_n|^2 \quad \text{for arms } a, b, \text{ and } c \tag{7.66}$$

Note that A_4 is greater than zero by definition. So is A_3 because we are only interested in the poles on the left half plane. The signs of A_1 and A_2 will depend on the transformer ratio and polarity, as well as the residues. But we already know that A_1 and A_2 need to be positive also (from the positive real requirement of any driving point admittance), because they are the coefficients of the rational admittance function. Then one needs to have the following additional conditions to ensure that the component values in (7.62) are all positive:

$$A_1 A_4 > A_2 \left(A_3 - \frac{A_2}{A_1} \right) \qquad A_3 > \frac{A_2}{A_1}$$

It can be shown that for lossless circuits (i.e., *LC* networks), the above conditions with $A_1 > 0$, $A_2 > 0$, $A_3 > 0$, and $A_4 > 0$, are reduced to the well-known residue condition. To ensure that these conditions are satisfied, one needs to select the sign and magnitude of the transformer ratio a properly. To satisfy the above requirement, an optimization algorithm with constraints (such as Levenberg-Marquardt, which is known as for its robust convergence [29]) is usually used to determine the transformer ratios and adjust the residues. Note that if an optimization algorithm is used, then it is a good idea to use the residues obtained from the linear least-squares solution as an initial guess so that the convergence is accelerated.

State-Space Representation

The equivalent circuit extracted using the Cauer synthesis method contains resistors, inductors, capacitors, and ideal transformers, which make it very convenient to implement in circuit simulators. However, for circuits that have a high number of ports (>4), the method becomes difficult to automate. Although microwave lumped elements can typically be represented by two ports in microstrip circuits, lumped elements that are used in differential circuits will require four-port representation for proper modeling. Note that differential circuits are extensively employed in high-speed communication circuits.

To address the model automation for circuits containing a high number of ports, one can generate a state-space representation for the admittance matrix, which will yield compact models for *N*-ports as an alternative to the Cauer synthesis method [34, 46, 48]. In this scheme, the resulting network will contain controlled sources. Thus, the circuit simulator must be able to handle controlled sources to use the equivalent model. To demonstrate the method, let's assume that we are given the *s*-domain relationship between the terminal voltages and currents of a linear, passive, *N*-port circuit such that

$$\mathbf{I} = \mathbf{Y}(s)\mathbf{V} \qquad (7.67)$$

where $\mathbf{I}$ and $\mathbf{V}$ are the current and voltage vectors, respectively, and $\mathbf{Y}(s)$ is an $N \times N$ admittance matrix whose entries are represented as follows:

$$Y_{ij}(s) = \sum_{l=1}^{Q} \frac{k_l^{(ij)}}{s - \alpha_l} \qquad (7.68)$$

where Q is the number of poles. It is assumed that all of the *Y*-parameters have the same poles but different residues. Let's also assume that all of the poles are real for the time being. Then, it can be shown that the following state-space representation can characterize the circuit described by (7.67):

$$\frac{dx}{dt} = \mathbf{A}\mathbf{x} + \mathbf{B}\mathbf{v}$$

$$\mathbf{i} = \mathbf{C}\mathbf{x}$$
(7.69)

In the above equation, the vector $\mathbf{x}$ represents the state of the circuit. The vectors $\mathbf{v}$ and $\mathbf{i}$ are the Laplace transforms of the s-domain voltage, $\mathbf{V}$, and current, $\mathbf{I}$, vectors, respectively. The state matrix $\mathbf{A}$ has the size of $NQ \times NQ$. It can be shown that the following relationship exists between the $\mathbf{A}$, $\mathbf{B}$, and $\mathbf{C}$ matrixes and the $\mathbf{Y}(s)$ matrix:

$$\mathbf{Y}(s) = \mathbf{C}(s\mathbf{U} - \mathbf{A})^{-1}\mathbf{B}$$
(7.70)

where $\mathbf{U}$ is the identity matrix. Then, by taking the Laplace transform of (7.69) and comparing the result with (7.67), one can readily obtain the $\mathbf{A}$, $\mathbf{B}$, and $\mathbf{C}$ matrixes as follows:

$$\mathbf{A} = \begin{bmatrix} \mathbf{P}_1 & 0 & \cdots & 0 \\ 0 & \mathbf{P}_2 & \ddots & \vdots \\ \vdots & \ddots & \ddots & 0 \\ 0 & \cdots & 0 & \mathbf{P}_Q \end{bmatrix}$$
(7.71)

$$\mathbf{C} = \begin{bmatrix} \mathbf{k}_{11} & \mathbf{k}_{12} & \cdots & \mathbf{k}_{1Q} \\ \mathbf{k}_{21} & \mathbf{k}_{22} & \ddots & \vdots \\ \vdots & \ddots & \ddots & \mathbf{k}_{(N-1)Q} \\ \mathbf{k}_{N1} & \cdots & \mathbf{k}_{N(Q-1)} & \mathbf{k}_{NQ} \end{bmatrix}$$
(7.72)

$$\mathbf{B} = \begin{bmatrix} \mathbf{U}_1 & \mathbf{U}_2 & \cdots & \mathbf{U}_Q \end{bmatrix}^T$$
(7.73)

where $\mathbf{U}_i$ is the $N \times N$ identity matrix. The $N \times N$ matrixes $\mathbf{P}_i$ and the $1 \times N$ vectors $\mathbf{k}_{ij}$ are given in terms of poles and residues as

$$\mathbf{P}_i = \begin{bmatrix} \alpha_i & 0 & \cdots & 0 \\ 0 & \alpha_i & \ddots & \vdots \\ \vdots & \ddots & \ddots & 0 \\ 0 & \cdots & 0 & \alpha_i \end{bmatrix}$$
(7.74)

$$\mathbf{k}_{ij} = \begin{bmatrix} k_j^{(i1)} & k_j^{(i2)} & \cdots & k_j^{(iN)} \end{bmatrix}$$
(7.75)

As an example, consider a two-port lumped circuit whose admittance matrix is approximated by three real poles and the corresponding residues:

$$\begin{bmatrix} I_1 \\ I_2 \end{bmatrix} = \begin{bmatrix} \sum_{l=1}^{3} \dfrac{k_l^{(11)}}{s-\alpha_l} & \sum_{l=1}^{3} \dfrac{k_l^{(12)}}{s-\alpha_l} \\ \sum_{l=1}^{3} \dfrac{k_l^{(21)}}{s-\alpha_l} & \sum_{l=1}^{3} \dfrac{k_l^{(22)}}{s-\alpha_l} \end{bmatrix} \begin{bmatrix} V_1 \\ V_2 \end{bmatrix}$$

Then, the resultant matrixes can be written as follows:

$$\mathbf{A} = \begin{bmatrix} \alpha_1 & 0 & 0 & 0 & 0 & 0 \\ 0 & \alpha_1 & 0 & 0 & 0 & 0 \\ 0 & 0 & \alpha_2 & 0 & 0 & 0 \\ 0 & 0 & 0 & \alpha_2 & 0 & 0 \\ 0 & 0 & 0 & 0 & \alpha_3 & 0 \\ 0 & 0 & 0 & 0 & 0 & \alpha_3 \end{bmatrix}$$

$$\mathbf{C} = \begin{bmatrix} k_1^{(11)} & k_1^{(12)} & k_2^{(11)} & k_2^{(12)} & k_3^{(11)} & k_3^{(12)} \\ k_1^{(21)} & k_1^{(22)} & k_2^{(21)} & k_2^{(22)} & k_3^{(21)} & k_3^{(22)} \end{bmatrix}$$

$$\mathbf{B} = \begin{bmatrix} 1 & 0 & 1 & 0 & 1 & 0 \\ 0 & 1 & 0 & 1 & 0 & 1 \end{bmatrix}^T$$

So far, we have assumed real poles so that all the matrixes in the state-space representation become real. In the case of complex poles, we need one more step to convert complex matrixes into real ones since complex matrixes in the time domain do not have any direct interpretation. For this purpose, we employ a nonsingular similarity transformation matrix, $\mathbf{J}$, such that the two realizations shown next are equivalent [34]:

$$\frac{d\mathbf{x}}{dt} = \mathbf{A}\mathbf{x} + \mathbf{B}\mathbf{v} \quad \Leftrightarrow \quad \frac{d\hat{\mathbf{x}}}{dt} = \mathbf{J}\mathbf{A}\mathbf{J}^{-1}\hat{\mathbf{x}} + \mathbf{J}\mathbf{B}\mathbf{v} \tag{7.76}$$
$$\mathbf{i} = \mathbf{C}\mathbf{x} \qquad\qquad \mathbf{i} = \mathbf{C}\mathbf{J}^{-1}\hat{\mathbf{x}}$$

where the transformation matrix $\mathbf{J}$ is selected such that all the matrixes become real after the transformation shown in (7.76). One can choose such a matrix as follows:

$$\mathbf{J} = \begin{bmatrix} \mathbf{U}_{N\cdot\frac{Q_c}{2}\times N\cdot\frac{Q_c}{2}} & \mathbf{U}_{N\cdot\frac{Q_c}{2}\times N\cdot\frac{Q_c}{2}} & \mathbf{0} \\ j\mathbf{U}_{N\cdot\frac{Q_c}{2}\times N\cdot\frac{Q_c}{2}} & -j\mathbf{U}_{N\cdot\frac{Q_c}{2}\times N\cdot\frac{Q_c}{2}} & \mathbf{0} \\ \mathbf{0} & \mathbf{0} & \mathbf{U}_{N\cdot Q_r \times N\cdot Q_r} \end{bmatrix} \tag{7.77}$$

where Q_c and Q_r are the number of complex (which are in conjugate pairs) and real poles, respectively. Note that the state matrix should be arranged so that the states corresponding to complex poles start from the upper-left section.

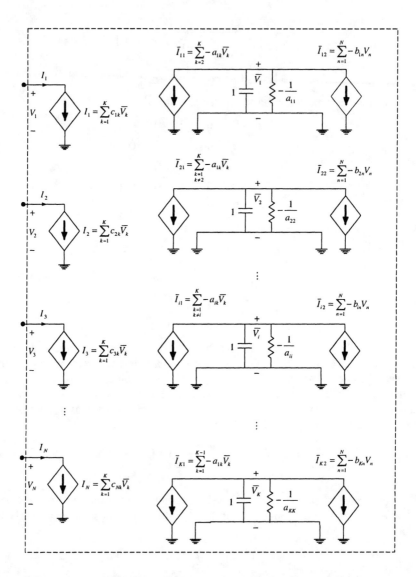

Figure 7.27 State-space representation of an N-port linear circuit using controlled current sources, resistors, and capacitors. Note that K equals $N \cdot Q$, where Q is the number of poles. Ports are shown on the left-hand side.

This completes the state-space representation of an N-port linear system described by the admittance matrix. The final step is to generate a SPICE model, which implements the equations given by (7.69). This can easily be done by assigning the state variables to capacitor voltages and setting capacitance values to

1 Farad. Then, the current through the capacitor becomes derivative of the state variable. The overall process can be generalized as depicted in Figure 7.27. Note that the matrixes **A**, **B**, and **C**, which are referenced in the figure, should be modified according to the similarity transformation given in (7.76) in case of complex poles.

Example 7.3

In this example, we will generate an equivalent model of the spiral inductor given in Example 6.2 using Cauer's equivalent network. Since the number of poles is five, we will need three Pi-sections as shown in Figure 7.28 (one for each complex pair and one for the pole at zero frequency) connected in parallel. Note that the model depicted in the figure is the most general representation. Depending on the residues, some of the shunt arms might be removed. When a particular pole pair is represented by only a series arm (i.e., the corresponding shunt arms are removed), then that pole is called compact.

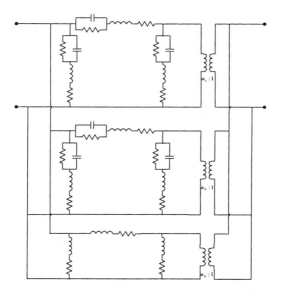

Figure 7.28 General equivalent model of an inductor, which can be represented by five poles (two complex pairs, one at the origin).

Once the topology is determined, the next step is to extract the component values of each arm using the approach described. Since the poles and residues are already calculated in Example 6.2, what is left is to optimize residues and transformer ratios so that all the coefficients in (7.62) are positive. For this purpose, we have utilized the Levenberg-Marquardt [29] method optimizing all three residues (Y_{11}, Y_{22}, Y_{12}) and transformer ratios simultaneously. The initial values of the residues are obtained from linear least-squares solutions, and the

initial values of the transformer ratios are set to unity. After the synthesis, it turns
out that the second pair and dc poles are compact. Thus, they are represented by
only a series arm (see Figure 7.29). The first pair is nearly compact and has only
the left shunt arm to correct phase at higher frequencies.

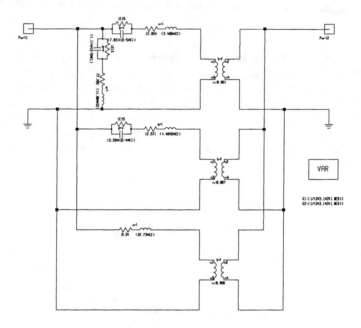

Figure 7.29 Equivalent model of the inductor based on pole-residue extraction and Cauer synthesis.

The resultant component values and transformer ratios are given below:

$$L_a^{(1)} = 14.80/2\pi \cdot 10^9 \quad \text{H} \qquad L_c^{(1)} = 2.41/2\pi \cdot 10^9 \quad \text{H}$$

$$C_a^{(1)} = 1.11 \cdot 10^{-5}/2\pi \cdot 10^9 \quad \text{F} \quad C_c^{(1)} = 7.03 \cdot 10^{-5}/2\pi \cdot 10^9 \quad \text{F}$$

$$R_a^{(1)} = 77.8 \quad \text{ohm} \qquad R_c^{(1)} = 12.7 \quad \text{ohm}$$

$$L_c^{(2)} = 4.47/2\pi \cdot 10^9 \quad \text{H} \qquad L_c^{(3)} = 22.73/2\pi \cdot 10^9 \quad \text{H}$$

$$C_c^{(2)} = 2.29 \cdot 10^{-4}/2\pi \cdot 10^9 \quad \text{F} \quad C_c^{(3)} \to \infty$$

$$R_c^{(2)} = 12.6 \quad \text{ohm} \qquad R_c^{(3)} = 6.34 \quad \text{ohm}$$

$$a_1 = 0.981 \quad a_2 = 0.887 \quad a_3 = 0.998$$

Note that we have to scale down component values because we have calculated the
poles and residues using scaled frequency units (gigahertz) to avoid loss of
numerical accuracy. Comparisons of the model with simulated results are
presented in Figures 7.30 and 7.31. As can be seen from the plots, the simulation
and model results agree quite well.

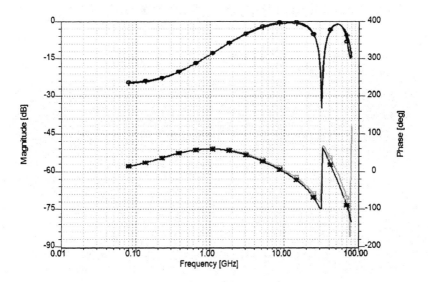

Figure 7.30 Comparison of S_{11} for the simulated and modeled S-parameters for the spiral inductor (circle: model magnitude; square: model phase; plus: simulated magnitude; star: simulated phase).

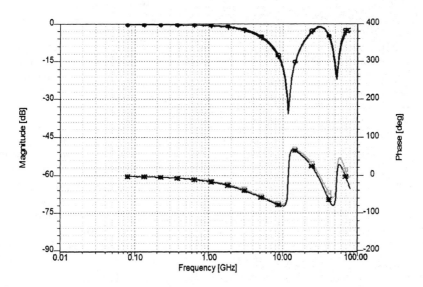

Figure 7.31 Comparison of S_{12} for the simulated and modeled S-parameters for the spiral inductor (circle: model magnitude; square: model phase; plus: simulated magnitude; star: simulated phase).

7.3 SCALABLE MODELS OF LUMPED ELEMENTS

So far, we have been concentrating on different techniques of model extraction for microwave lumped elements. Not much has been said on how to extract scalable (or parameterized) models of lumped elements. Even though the analysis speed of full-wave simulations of typical lumped elements, like MIM capacitors or spiral inductors, is on the order of minutes or shorter for today's engineering workstations, incorporating full-wave simulations directly into optimization and Monte Carlo analyses is still prohibitive because of the vast number of circuit simulations that are necessary. The purpose of parameterized models, then, is to provide interpolated response of passive components for any geometric and/or electrical variations, provided that the variations stay inside some predetermined limits. In this way, the speed of circuit simulations is significantly increased without resorting to approximate, quasi-static models, which reduces accuracy. Figure 7.32 summarizes the concept of generating and using scalable models based on computer simulations. Preferably, the model should generate the required output based on a relatively small number of initial full-wave simulations. Although the methodology is general, we will limit ourselves to parameterized model generation based on full-wave EM simulations only.

At this point, it would be worthwhile to explore the underlying philosophy of parameterized model generation. As stated above, the aim of parameterized models is to supply full-wave simulated circuit parameters for any geometrical parameter combination for a given passive circuit (e.g., MIM capacitor, spiral inductor). Clearly, it wouldn't be feasible to simulate the passive circuit in advance using very small step sizes and generating a database of circuit parameters for later use. Therefore, it is desirable to utilize some limited number of simulations (~ 5–10 steps for each parameter to be varied), then to use an interpolation scheme in such a way that when the user asks for a parameter combination that does not exist in the original set, then an approximation to actual circuit parameters can be supplied. Note that interpolation can be performed either directly on the circuit parameters or on the equivalent model parameters which are extracted from the circuit parameters at each step. The latter usually provides better accuracy. However, it includes an additional step of model extraction. On the other hand, the first method is very easy to apply, but it may fail to provide accurate interpolation near resonance points. It is also important to employ a suitable rational approximation for the circuit parameters for each parameter combination so that broadband response is constructed using a small number of EM simulations (~ 10 discrete frequency simulations for multiple octaves, typically). These are the basic concepts of parameterized modeling. The actual implementation of how the full-wave EM simulations are used to build parameterized models varies significantly from one method to another.

There are many ways of creating parameterized models, and they have been studied extensively in the literature [51–52]. Space Mapping and OSA90 were the first commercially successful tools that could be used for optimization and parameter sweep of passive microwave circuits using EM simulations [13, 14]. The Space Mapping technique forms a transformation between the so-called coarse and fine EM models so that overall response can be constructed using a relatively small number of finely meshed models. Since for large circuits the total solution time increases by approximately N^3 in MoM, where N is the number of unknowns, this method can drastically cut the CPU time required. Thus, one could, for example, optimize response of a microstrip bandpass filter using EM simulations in a relatively short time, compared to the case where every simulations is performed using the same fine meshing. Artificial neural networks have also been used to extract parameterized models [67–69].

Essentially, any of the techniques introduced so far in this chapter can also be used in parameterized models by first extracting models at predetermined parameter values covering the design space, then interpolating the resultant components or coefficients using a multidimensional interpolation algorithm.

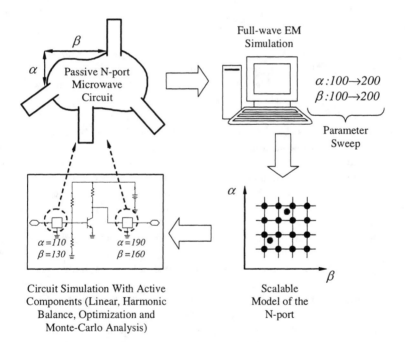

Figure 7.32 Concept of generating and using scalable models. Scalable models significantly decrease the MMIC design time by eliminating resimulation of passive components for different geometrical variations.

As an example, for bivariate interpolation with data points arranged on a Cartesian grid, it can be shown that a simple polynomial interpolation function can be found by using the following rule [15, 16]:

$$F(x, y) = \sum_{i=1}^{p} \sum_{j=1}^{q} f(x_i, y_j) \cdot v_j(y) \cdot u_i(x) \qquad (7.78)$$

where $F(x, y)$ is the polynomial interpolant, and $f(x_i, y_j)$ are the data points (e.g., component values of a Pi-network that represents a MIM capacitor, such as series and shunt capacitances). Then, the functions $u_i(x)$ and $v_j(y)$ are given by the following expressions:

$$u_i(x) = \prod_{\substack{k=1 \\ k \neq i}}^{p} \frac{x - x_k}{x_i - x_k} \quad (1 \leq i \leq p) \qquad (7.79)$$

$$v_j(y) = \prod_{\substack{k=1 \\ k \neq j}}^{q} \frac{y - y_k}{y_j - y_k} \quad (1 \leq j \leq q) \qquad (7.80)$$

In the above expressions, x_i and y_j are the predefined parameter values (e.g., length and width of the MIM capacitor) where full-wave simulation results exist. Note that most of the other parameters are fixed by the MMIC process employed (like dielectric thickness, dielectric permittivity, dielectric loss tangent). It is important to comment that the above interpolation scheme is usually used in the vicinity of a desired data point. Therefore, p and q are usually selected as some small number (e.g., $p = q = 4$ generates a cubic interpolant), and then interpolation is applied around the required data point; the rest of the data set is not used. This method is also called the look-up table approach. This eliminates the necessity of finding an interpolant for the whole parameter space. Although it is possible to find an interpolant function covering the whole parameter space, this would require very large p and q, reducing the speed and accuracy of the method.

The approach of interpolating circuit parameters will be reviewed now in more detail. This method does not have the advantages of equivalent models described at the beginning of this chapter but is extremely simple to implement. Besides, it can be automated with minimal effort. Since it directly uses the circuit parameters, loss factors are automatically included, yielding accurate determination of quality factors. Finally, interpolation accuracy can be made very good if enough parameter samples are taken during the full-wave simulations. There are, however, two important disadvantages of this method. First, one needs relatively fine parameter steps for the initial full-wave simulations to interpolate near resonance points accurately later. Second, if the circuit parameters are to be stored directly, then the resulting data files can be very large, causing slow initialization in the subsequent circuit simulations (because the circuit simulator must first load the stored data to interpolate them). The latter point can be avoided by storing the coefficients of rational polynomials that fit the frequency response at each step, provided that the full-wave simulator has such an option. Since the method of direct interpolation (i.e., storing the circuit parameters) can be employed very easily, it is beneficial to summarize the overall procedure as follows:

1. Prepare geometry of the lumped element that needs to be parameterized in a full-wave electromagnetic simulator.

2. Assign parameters to dimensions that you want to vary. Select the step size and number of steps for a particular dimension to be not more than 10% of the maximum value of that particular dimension and 10, respectively. Try to limit simultaneous parameter sweeps to two (e.g., length and width). If a third parameter is required (e.g., substrate thickness), then it is probably more efficient to generate another set of simulations for that parameter instead of sweeping three parameters at the same time. This makes sense because, for example, it is not required to have a continuous variation on the substrate thickness of lumped elements.

3. Simulate the geometry by sweeping the parameters with the steps that are selected above. It is very important to obtain broadband circuit parameters. Try to set the low-frequency limit to less than 100 MHz so that dc parameters can be extrapolated safely. For dc S-parameters, simulating the circuit at a very low frequency is usually sufficient in most cases to safely extrapolate the low-frequency point to dc. To use this approach, the full-wave simulator must be able to simulate at low frequencies (< 100 MHz) accurately. Note that this is not trivial because as the frequency gets smaller, the MoM matrix becomes more ill-conditioned. Alternatively, one can also extract the dc parameters manually and add in the simulation results afterwards.

4. Convert the output of the full-wave simulator to a suitable file format that supports parameterization (e.g., MDIF for ADS) so that it can be loaded into the particular circuit simulator for subsequent interpolation. In generalized MDIF file format, one can have multiple data blocks corresponding to each parameter combination. The data blocks themselves can be anything from S-parameters to component values of the equivalent model.

5. Prepare a circuit model in the circuit simulator to read the circuit parameters file and interpolate the parameters for any given parameter combination (e.g., using equation-based linear components and DAC in ADS). The DAC component will automatically perform multidimensional interpolation.

Note that details of the last two steps depend on the circuit simulator being used. Although methods specific to Advanced Design System (ADS) by Agilent are given here, most of the modern circuit simulators have features that enable one to accomplish this easily; so, little or no programming is required by the user.

The summarized approach is straightforward yet very powerful. Since sweeping parameters are completely automated, one can generate scalable passive libraries containing dozens of components for a given MMIC process in a matter of days using today's PC engineering workstations.

7.4 DIMENSIONAL ANALYSIS

Dimensional analysis has been extensively employed in other branches of engineering to study the functional dependencies of physical phenomena through experiments. It hasn't been used in electrical engineering much, mainly because the governing differential equations of electrical phenomena can be derived and solved instead. On the other hand, parameterized modeling of microwave lumped elements could benefit some from dimensional analysis in terms of reducing the number of parameters that need to be swept. Another application of the subject to microwave engineering is the dimensional modeling where scaled models of passive components can be built and measured at relatively low frequencies. Therefore, an introduction to dimensional analysis will be provided here. It should be stated, however, that dimensional analysis is a vast subject, and interested readers should refer to the literature for more information [70–72].

We will start our discussion by defining dimensional homogeneity. An equation is said to be dimensionally homogeneous if the form of the equation does not depend on the fundamental units of measurement. According to Bridgman's principle of absolute significance of relative magnitude, a number Q obtained by inserting the numerical values of base quantities into a formula is a physical quantity if the ratio of any two samples of it remains constant when the base unit sizes are changed [72]. It can also be shown that only a monomial form for number Q satisfies this principle:

$$Q = \alpha A^a B^b C^c \cdots \qquad (7.81)$$

where A, B, C, and so forth are the numerical values of base quantities. The coefficient α and exponents a, b, c, and so forth, are real numbers to distinguish one type of derived quantity from another. Thus, it can be deduced that all the summations in a physical equation must have the same dimensions. Similarly, special functions (e.g., natural logarithms, exponentials) must be applied only to dimensionless numbers. Dimensional analysis anchors on Buckingham's famous theorem, which states that if an equation is dimensionally homogeneous, then it can be reduced to a relationship among a complete set of dimensionless products. This is the heart of dimensional analysis.

The first step in dimensional analysis is to decide what variables enter the problem [70]. If a parameter does not have any significant effect on the phenomenon, then too many variables may appear in the final equation. On the other hand, if a crucial variable is skipped, then the result will be erroneous. It is important to note that even though some variables are practically constant, they may be essential because they combine with other parameters to form dimensionless products. In selecting the variables, it makes sense to let the first variable be the dependent variable. Then, the second variable is selected as the easiest variable to regulate experimentally. The third variable will be the next easiest to regulate experimentally, and so on. To introduce how the dimensionless products are formed, let's suppose that a physical system has m system variables and n fundamental units. Then, a unit of each variable can be expressed as follows:

$$[p_i] = u_1^{b_{1i}} \cdot u_2^{b_{2i}} \cdot u_3^{b_{3i}} \cdot \ldots \cdot u_n^{b_{ni}} \tag{7.82}$$

where b_{1i}, b_{2i}, and b_{ni} are the exponents of each fundamental unit. Then, a dimensionless product is defined by

$$\pi = p_1^{a_1} \cdot p_2^{a_2} \cdot p_3^{a_3} \cdot \ldots \cdot p_m^{a_m} \tag{7.83}$$

where a_1, a_2, and a_m are the exponents of each variable. Note that the symbol of π has nothing to do with the transcendental number π in mathematics. It merely represents the product of variables. The unit of π is then given by

$$[\pi] = u_1^{\sum a_i b_{1i}} \cdot u_2^{\sum a_i b_{2i}} \cdot u_3^{\sum a_i b_{3i}} \cdot \ldots \cdot u_n^{\sum a_i b_{ni}} \tag{7.84}$$

In order to have a dimensionless product, the following system of equations must be satisfied:

$$\begin{bmatrix} b_{11} & b_{12} & \cdots & b_{1m} \\ b_{21} & \ddots & \ddots & b_{2m} \\ \vdots & \ddots & \ddots & \vdots \\ b_{n1} & b_{n2} & \cdots & b_{nm} \end{bmatrix} \begin{bmatrix} a_1 \\ a_2 \\ \vdots \\ a_m \end{bmatrix} = \begin{bmatrix} 0 \\ 0 \\ \vdots \\ 0 \end{bmatrix} \tag{7.85}$$

Any nontrivial solution of this system forms a dimensionless product. Then, from Buckingham's theorem, one can construct a function representing the phenomenon using these dimensionless products as follows:

$$f(\pi_1, \pi_2, \pi_3, \ldots, \pi_q) = 0 \tag{7.86}$$

The number of dimensionless products, q, is equal to the number of variables, m, minus the dimensional matrix rank, r. In general, it is not necessary to consider all the rows in the dimensional matrix if the rank, r, is less than the number of rows, m. It would be sufficient to choose the rows whose rank is r.

Note that dimensional analysis does not provide the form of the function; it only provides a way of distilling system variables into appropriate dimensionless products that the function will depend on. The actual functional form must be determined in a separate process using inspection or numerical algorithms, such as least-squares. For this purpose, it is helpful to visually inspect the behavior of the function versus the resultant dimensionless products by plotting it. Since the dimensions of the variables in terms of fundamental physical units are important in dimensional analysis, a summary of them is provided in Table 7.3 for electrical and magnetic entities [70].

Table 7.3
Dimensions and Units of Electrical and Magnetic Entities

	Mass	*Length*	*Time*	*Charge*	*Name of Unit*
Mass	1	0	0	0	kg
Electric charge	0	0	0	1	coulomb
Permittivity, ε	−1	−3	2	2	farad/m
Permeability, μ	1	1	0	−2	ohm·sec/m
Electric current density, J	0	−2	−1	1	amp/m^2
Electric current	0	0	−1	1	amp
Electric displacement, D	0	−2	0	1	amp·sec/m^2
Electric field intensity, E	1	1	−2	−1	volt/m
Electric potential	1	2	−2	−1	volt
Electric resistance	1	2	−1	−2	ohm
Magnetic field intensity, H	0	−1	−1	1	amp/m
Magnetic induction, B	1	0	−1	1	weber/m^2
Flux of magnetic induction, Φ	1	2	−1	−1	weber
Inductance	1	2	0	−2	henry
Capacitance	−1	−2	2	2	farad
Electric energy	1	2	−2	0	joule
Electric power	1	1	−3	0	watt

Source: [70].

Example 7.4

The foregoing discussion can be best summarized by an example. For this purpose, we will apply dimensional analysis to the characteristic impedance of the microstrip line shown in Figure 7.33. This example is due to Mah et al. [73]. It is required to determine a function for the line characteristic impedance such as

$$f(Z_c, h_1, h_2, w, L, \varepsilon_1, \varepsilon_2, c) = 0 \qquad (7.87)$$

where Z_c and c are the characteristic impedance and speed of light in a vacuum, respectively. The other parameters are defined in Figure 7.33.

The first step of the dimensional analysis is to form the dimension matrix as follows:

$$
\begin{array}{c|cccccccc}
 & Z_c & h_1 & h_2 & w & L & \varepsilon_1 & \varepsilon_2 & c \\
\hline
M & 1 & 0 & 0 & 0 & 0 & -1 & -1 & 0 \\
L & 2 & 1 & 1 & 1 & 1 & -3 & -3 & 1 \\
T & -1 & 0 & 0 & 0 & 0 & 2 & 2 & -1 \\
Q & -2 & 0 & 0 & 0 & 0 & 2 & 2 & 0
\end{array}
\qquad (7.88)
$$

Then, the π term is written as

$$\pi = Z_c^a \cdot h_1^b \cdot h_2^c \cdot w^d \cdot L^e \cdot \varepsilon_1^f \cdot \varepsilon_2^g \cdot c^h$$

where the superscripts $a, b, c,...,h$ are constants yet to be determined. Using Table 7.3, a unit of π is written as follows:

$$[\pi] = M^{(a-f-g)} \cdot L^{(2a+b+c+d+e-3f-3g+h)} \cdot T^{(-a+2f+2g-h)} \cdot Q^{(-2a+2f+2g)}$$

Thus, to have π be dimensionless, the following system must be satisfied:

$$\begin{bmatrix} 1 & 0 & 0 & 0 & 0 & -1 & -1 & 0 \\ 2 & 1 & 1 & 1 & 1 & -3 & -3 & 1 \\ -1 & 0 & 0 & 0 & 0 & 2 & 2 & -1 \\ -2 & 0 & 0 & 0 & 0 & 2 & 2 & 0 \end{bmatrix} \begin{bmatrix} a \\ b \\ c \\ d \\ e \\ f \\ g \\ h \end{bmatrix} = \begin{bmatrix} 0 \\ 0 \\ 0 \\ 0 \end{bmatrix}$$

Note that the above system is underdetermined and has infinite solutions. Any of these solutions would form a dimensionless product. We select the solutions in such a way that each variable will appear at least once in the final equation. It is easy to show that the above system has a rank of three. Since the number of parameters is eight, we will have five dimensionless products. We can also delete the last row because it is linearly dependent on the first one. The remaining rows would still have the rank of three. By inspection, it can be seen that the following set of variables satisfies the above system:

$$\begin{array}{llllllll} a=0 & b=1 & c=-1 & d=0 & e=0 & f=0 & g=0 & h=0 \\ a=0 & b=-1 & c=0 & d=1 & e=0 & f=0 & g=0 & h=0 \\ a=0 & b=0 & c=0 & d=1 & e=-1 & f=0 & g=0 & h=0 \\ a=1 & b=0 & c=0 & d=0 & e=0 & f=1 & g=0 & h=1 \\ a=0 & b=0 & c=0 & d=0 & e=0 & f=-1 & g=1 & h=0 \end{array}$$

which results in the following dimensionless products:

$$\pi_1 = \frac{h_1}{h_2} \quad \pi_2 = \frac{w}{h_1} \quad \pi_3 = \frac{w}{L} \quad \pi_4 = Z_c \varepsilon_1 c \quad \pi_5 = \frac{\varepsilon_2}{\varepsilon_1}$$

Based on these, the equation that we have been seeking will have the following dependences:

$$f(\pi_1,\pi_2,\pi_3,\pi_4,\pi_5) = f\left(\frac{h_1}{h_2}, \frac{w}{h_1}, \frac{w}{L}, Z_c \varepsilon_1 c, \frac{\varepsilon_2}{\varepsilon_1}\right) = 0$$

or

$$Z_c = \frac{1}{c\varepsilon_1} f\left(\frac{h_1}{h_2}, \frac{w}{h_1}, \frac{w}{L}, \frac{\varepsilon_2}{\varepsilon_1}\right)$$

where f is a function that needs to be determined using subsequent analysis. It is very important to stress that dimensional analysis has helped us to reduce the number of variables that need to be swept from six to four in this example. This is a substantial reduction and shows the power of dimensional analysis. For instance, if 10 steps were to be used for each variable, then the overall CPU time would be reduced by a factor of one hundred after applying the dimensional analysis.

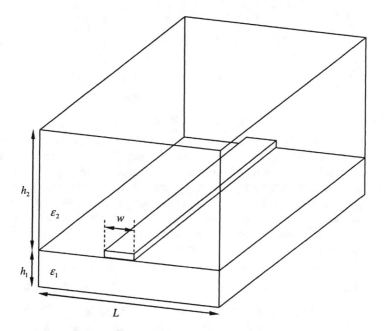

Figure 7.33 Model of a microstrip line in a metallic box showing variables for dimensional analysis of its characteristic impedance.

References

[1]   Daniel A. Daly et al., "Lumped Elements in Microwave Integrated Circuits," *IEEE Trans. Microwave Theory Tech.*, vol. MTT-15, pp. 713–721, Dec. 1967.

[2]   Martin Caulton, Stanley P. Knight, and Daniel A. Daly, "Hybrid Integrated Lumped-Element Microwave Amplifiers," *IEEE Trans. Microwave Theory Tech.*, vol. MTT-16, pp. 397–404, July 1968.

[3]   Martin Caulton et al., "Status of Lumped Element in Microwave Integrated Circuits — Present and Future," *IEEE Trans. Microwave Theory Tech.*, vol. MTT-19, pp. 588–599, July 1971.

[4]   Martin Caulton, "The Lumped Element Approach to Microwave Integrated Circuits," *Microwave Journal*, pp. 51–58, May 1970.

[5]   Robert A. Pucel, "Design Considerations for Monolithic Microwave Circuits," *IEEE Trans. Microwave Theory Tech.*, vol. MTT-29, pp. 513–534, June 1981.

[6]   Ewald Pettenpaul et al., "CAD Models of Lumped Elements on GaAs up to 18 GHz," *IEEE Trans. Microwave Theory Tech.*, vol. MTT-36, pp. 294–304, Feb. 1988.

[7]   Gary D. Alley, "Interdigital Capacitors and Their Applications to Lumped-Element Microwave Integrated Circuits," *IEEE Trans. Microwave Theory Tech.*, vol. MTT-18, pp. 1028–1033, Dec. 1970.

[8]   Roberto Sorrentino, "Numerical Methods for Passive Components," *IEEE MTT-S Digest*, pp. 619–622, 1988.

[9]   Ingo Wolff, and Godfrey Kibuuka, "Computer Models for MMIC Capacitors and Inductors," *Proc. 14th European Microwave Conf.*, pp. 853–858, 1984.

[10]   Spartak S. Gevorgian et al., "CAD Models for Multilayered Substrate Interdigital Capacitors," *IEEE Trans. Microwave Theory Tech.*, vol. MTT-44, pp. 896–904, June 1996.

[11]   Jyoti P. Mondal, "An Experimental Verification of a Simple Distributed Model of MIM Capacitors for MMIC Applications," *IEEE Trans. Microwave Theory Tech.*, vol. MTT-35, pp. 403–408, Apr. 1987.

[12]   Reza Esfandiari, Douglas W. Maki, and Mario Siracusa, "Design of Interdigital Capacitors and Their Applications to Gallium Arsenide Monolithic Filters," *IEEE Trans. Microwave Theory Tech.*, vol. MTT-31, pp. 57–64, Jan. 1983.

[13]   Michael B. Steer, John W. Bandler, and Christopher M. Snowden, "Computer-Aided Design of RF and Microwave Circuits and Systems," *IEEE Trans. Microwave Theory Tech.*, pp. 996–1005, Mar. 2002.

[14]   John W. Bandler et al., "Space Mapping Technique for Electromagnetic Optimization," *IEEE Trans. Microwave Theory Tech.*, vol. 42, pp. 2536–2544, Dec. 1994.

[15]   Ralph Levy, "Derivation of Equivalent Circuits of Microwave Structures Using Numerical Techniques," *IEEE Trans. Microwave Theory Tech.*, pp. 1688–1695, Sept. 1999.

[16]   Nitin Jain and Peter Onno, "Methods of Using Commercial Electromagnetic Simulators for Microwave and Millimeter-Wave Circuit Design and Optimization," *IEEE Trans. Microwave Theory Tech.*, pp. 724–746, May 1997.

[17]   James C. Rautio, "Planar Electromagnetic Analysis," *IEEE Microwave Magazine*, pp. 35–41, Mar. 2003.

[18]   Inder Bahl and Prakash Bhartia, *Microwave Solid State Circuit Design*, New York: John Wiley and Sons, 1988.

[19]   Inder Bahl, *Lumped Elements for RF and Microwave Circuits*, Norwood, MA: Artech House, 2003.

[20]   T. C. Edwards, *Foundations for Microwave Circuit Design*, New York: John Wiley and Sons, 1981.

[21]   Reinmut K. Hoffmann, *Handbook of Microwave Integrated Circuits*, Norwood, MA: Artech House, 1987.

[22] I. Kasa, *Microwave Integrated Circuits*, New York: Elsevier, 1991.

[23] Brian C. Wadell, *Transmission Line Design Handbook*, Norwood, MA: Artech House, 1991.

[24] Jeffrey Frey, and Kul Bhasin (Eds.), *Microwave Integrated Circuits*, Dedham, MA: Artech House, 1985.

[25] David Kincaid and Ward Cheney, *Numerical Analysis*, Belmont, CA: Thomson Publishing Group, 1991.

[26] J. Stoer and R. Bulirsch, *Introduction to Numerical Analysis*, 2nd ed., New York: Springer-Verlag, 1992.

[27] F. B. Hildebrand, *Introduction to Numerical Analysis*, 2nd ed., New York: Dover, 1974.

[28] Carl D. Meyer, *Matrix Analysis and Applied Linear Algebra*, SIAM Publications, 2000.

[29] William H. Press et al., *Numerical Recipes*, Cambridge: Cambridge University Press, 1989.

[30] Van Valkenburg, *Introduction to Modern Network Synthesis*, New York: John Wiley and Sons, 1960.

[31] Wai-Kai Chen, *Passive and Active Filters: Theory and Implementations*, New York: John Wiley and Sons, 1986.

[32] Wai-Kai Chen, *Theory and Design of Broadband Matching Networks*, Pergamon International Library, 1976.

[33] Ernst A. Guillemin, *Synthesis of Passive Networks*, New York: John Wiley and Sons, 1957.

[34] Mustafa Celik, Lawrence Pileggi, and Altan Odabasioglu, *IC Interconnect Analysis*, Boston, MA: Kluwer Academic Publishers, 1996.

[35] John R. Long and Miles A. Copeland, "The Modeling, Characterization, and Design of Monolithic Inductors for Silicon RF IC's," *IEEE J. Solid-State Circuits*, vol. 32, pp. 357–369, Mar. 1997.

[36] Sunderarajan S. Mohan et al., "Simple Accurate Expressions for Planar Spiral Inductances," *IEEE J. Solid-State Circuits*, vol. 34, pp. 1419–1424, Oct. 1999.

[37] Ali M. Niknejad and Robert G. Meyer, "Analysis, Design, and Optimization of Spiral Inductors and Transformers for Si RF IC's," *IEEE J. Solid-State Circuits*, vol. 33, pp. 1470–1481, Oct. 1998.

[38] Franz Sischka, "RF Measurements and Modeling with Special Emphasis on Test Structures," *ICMTS Digest*, Monterey, California, Mar. 2000.

[39] Jan De Geest et al., "Adaptive CAD-Model Building Algorithm for General Planar Microwave Structures," *IEEE Trans. Microwave Theory Tech.*, pp. 1801–1809, Sept. 1999.

[40] Steven D. Corey and Andrew T. Yang, "Automatic Netlist Extraction for Measurement-Based Characterization of Off-Chip Interconnect," *IEEE Trans. Microwave Theory Tech.*, pp. 1934–1940, Oct. 1997.

[41] James C. Rautio, "Synthesis of Lumped Models from N-Port Scattering Parameter Data," *IEEE Trans. Microwave Theory Tech.*, pp. 535–537, Mar. 1994.

[42] Tobias Mangold and Peter Russer, "Full-Wave Modeling and Automatic Equivalent-Circuit Generation of Millimeter-Wave Planar and Multilayer Structures," *IEEE Trans. Microwave Theory Tech.*, pp. 851–858, June 1999.

[43] Ian Timmins and Ke-Li Wu, "An Efficient Systematic Approach to Model Extraction for Passive Microwave Circuits," *IEEE Trans. Microwave Theory Tech.*, pp. 1565–1573, Sept. 2000.

[44] Peter Russer et al., "Lumped Element Equivalent Circuit Parameter Extraction of Distributed Microwave Circuits Via TLM Simulation," *IEEE MTT-S Digest*, pp. 887–890, 1994.

[45] M. Righi et al., "Lumped-Element Equivalent-Circuit Parameters Extraction of Coplanar MMIC Components Via TLM Simulation," *International Conference on Microwave and Optronics (MIOP)*, pp. 253–257, 1995.

[46] Andreas C. Cangellaris et al., "Electromagnetic Model Order Reduction for System-Level Modeling," *IEEE Trans. Microwave Theory Tech.*, vol. 47, pp. 840–850, June 1999.

[47] Matt Kamon, Frank Wang, and Jacob White, "Generating Nearly Optimally Compact Models from Krylov-Subspace-Based Reduced-Order Models," *IEEE Trans. on Circuits and Systems-II*, vol. 47, pp. 239–248, Apr. 2000.

[48] Ramachandra Achar and Michel S. Nakhla, "Simulation of High-Speed Interconnects," *Proceedings of the IEEE*, vol. 89, no. 5, pp. 693–728, May 2001.

[49] Tom Dhaene et al., "Adaptive Frequency Sampling Algorithm for Fast and Accurate S-parameter Modeling of General Planar Structures," *IEEE MTT-S Digest*, pp. 1427–1430, 1995.

[50] Jan Ureel et al., "Adaptive Frequency Sampling of Scattering Parameters Obtained by Electromagnetic Simulation," *IEEE AP-S Digest*, pp. 1162–1165, 1994.

[51] Robert Lehmensiek and Petrie Meyer, "Creating Accurate Multivariate Rational Interpolation Models of Microwave Circuits by Using Efficient Adaptive Sampling to Minimize the Number of Computational Electromagnetic Analyses," *IEEE Trans. Microwave Theory Tech.*, pp. 1419–1430, Aug. 2001.

[52] Tom Dhaene, Jan De Geest, and Daniel De Zutter, "EM-Based Multidimensional Parameterized Modeling of General Passive Planar Components," *IEEE MTT-S Digest*, pp. 1745-1748, 2001.

[53] Albert E. Ruehli and Andreas C. Cangellaris, "Progress in the Methodologies for the Electrical Modeling of Interconnects and Electronic Packages," *Proceedings of the IEEE*, vol. 89, no. 5, pp. 740–771, May 2001.

[54] G. J. Burke, E. K. Miller, and S. Chakrabarti, "Using Model-Based Parameter Estimation to Increase the Efficiency of Computing Electromagnetic Transfer Functions," *IEEE Trans. on Magnetics*, vol. 25, No. 4, pp. 2807–2809, July 1989.

[55] J. N. Brittingham, E. K. Miller, and J. L. Willows, *The Derivation of Simple Poles in a Transfer Function from Real-Frequency Information*, Lawrence Livermore Laboratory Report, No. UCRL-52050, 1976.

[56] J. N. Brittingham, E. K. Miller, and J. L. Willows, *The Derivation of Simple Poles in a Transfer Function from Real-Frequency Information: Results from Real EM Data*, Lawrence Livermore Laboratory Report, No. UCRL-52118, 1976.

[57] J. N. Brittingham, E. K. Miller, and J. L. Willows, "Pole Extraction from Real-Frequency Information," *IEEE Proceedings*, vol. 68, pp. 263–273, Feb. 1980.

[58] Krishna Naishadham and Xing Ping Lin, "Application of Spectral-Domain Prony's Method to the FDTD Analysis of Planar Microstrip Circuits," *IEEE Trans. Microwave Theory Tech.*, vol. 42, pp. 2391–2398, Dec. 1994.

[59] Krishnamoorthy Kottapalli et al., "Accurate Computation of Wide-Band Response of Electromagnetic Systems Utilizing Narrow-Band Information," *IEEE Trans. Microwave Theory Tech.*, pp. 682–687, Apr. 1991.

[60]   Sharath Narayana et al., "A Comparison of Two techniques for the Interpolation/Extrapolation of Frequency Domain Responses," *Digital Signal Processing,* No. 6, pp. 51–67, 1996.

[61]   Soren F. Peik, Rafaat R. Mansour, and Y. Leonard Chow, "Multidimensional Cauchy Method and Adaptive Sampling for an Accurate Microwave Circuit Modeling," *IEEE Trans. Microwave Theory Tech.,* vol. 46, pp. 2364–2371, Dec. 1998.

[62]   Yingbo Hua and Tapan K. Sarkar, "Matrix Pencil and System Poles," *Signal Processing,* no. 21, pp. 195–198, 1990.

[63]   Yingbo Hua, and Tapan K. Sarkar, "Generalized Pencil-of-Function Method for Extracting Poles of an EM System from Its Transient Response," *IEEE Trans. Microwave Theory Tech.,* vol. 37, pp. 229–234, Feb. 1989.

[64]   Tapan K. Sarkar and Odilon Pereira, "Using the Matrix Pencil Method to Estimate the Parameters of a Sum of Complex Exponentials," *IEEE Antennas and Propagation Magazine,* vol. 37, pp. 48–55, Feb. 1995.

[65]   Yingbo Hua and Tapan K. Sarkar, "Matrix Pencil Method for Estimating Parameters of Exponentially Damped/Undamped Sinusoids in Noise," *IEEE Trans. on Acoustics, Speech, and Signal Process.,* vol. 38, pp. 814–824, May 1990.

[66]   Mustafa Celik et al., "Pole-Zero Computation in Microwave Circuits Using Multipoint Pade Approximation," *IEEE Trans. on Circuits and Systems I,* vol. 42, pp. 6–13, Jan. 1995.

[67]   Gregory L. Creech et al., "Artificial Neural Networks for Fast and Accurate EM-CAD of Microwave Circuits," *IEEE Trans. Microwave Theory Tech.,* pp. 794–802, May 1997.

[68]   Fang Wang, Vieja Kumar Devabhaktuni, and Qi-Jun Zhang, "A Hierarchical Neural Network Approach to the Development of a Library of Neural Models for Microwave Design," *IEEE Trans. Microwave Theory Tech.,* pp. 2391–2403, Dec. 1998.

[69]   Fang Wang and Qi-Jun Zhang, "Knowledge-Based Neural Models for Microwave Design," *IEEE Trans. Microwave Theory Tech.,* pp. 2333–2343, Dec. 1997.

[70]   Henry L. Langhaar, *Dimensional Analysis and Theory of Models,* New York: John Wiley and Sons, 1951.

[71]   Thomas Szirtes, *Applied Dimensional Analysis and Modeling,* New York: McGraw-Hill, 1998.

[72]   Ain A. Sonin, *The Physical Basis of Dimensional Analysis,* Department of Mechanical Engineering, MIT, Boston, MA, 2001.

[73]   Misoon Y. Mah et al., "Design Methodology of Microstrip Lines Using Dimensional Analysis," *IEEE Microwave Guided Wave Lett.,* vol. 8, no. 7, pp. 248–250, July 1998.

About the Authors

Noyan Kinayman received his B.Sc. and M.Sc. from Middle East Technical University, Ankara, Turkey, in 1990 and 1993, respectively, and his Ph.D. from Bilkent University, Ankara, in 1997, all in electrical engineering.

He worked at Aselsan, Inc., Ankara, a military electronics company, as an electrical engineer between 1990 and 1994. Here, he developed computer-controlled and manual test stations for the testing of military communications equipment. Between 1994 and 1997, he was a research assistant with Bilkent University's electromagnetics group, where he studied novel algorithms for computer simulation of printed circuits. After obtaining his Ph.D., he joined the Corporate Research and Development Department at Microwave Associates (M/A-COM), in Lowell, Massachusetts, as a senior electrical engineer. Currently, he is working in the same department as a principal electrical engineer. His main responsibilities at M/A-COM are electromagnetic analysis, modeling, and microwave circuit design. He has designed various active and passive microwave circuits (e.g., low-noise amplifiers, power amplifiers, phase shifters) using SiGe technology.

Dr. Kinayman has authored or coauthored 25 technical publications in peer-reviewed international journals and conferences and holds four patents. He has cocreated the commercially available full-wave electromagnetic simulation software EMPLAN to simulate planar microstrip circuits. His main professional interests are electromagnetic theory, numerical solution of electromagnetic problems, and microwave circuit design.

M. I. Aksun received his B.Sc. and M.Sc. in electrical and electronics engineering at the Middle East Technical University, Ankara, Turkey, in 1981 and 1983, respectively. He received his Ph.D. in electrical and computer engineering at the University of Illinois, Urbana-Champaign, in 1990.

From 1990 to 1992, he was a postdoctoral fellow at the Electromagnetic Communication Laboratory at the University of Illinois, Urbana-Champaign. From 1992 to 2001, he was on the faculty of the Department of Electrical and Electronics Engineering at Bilkent University, Ankara, where he became a professor in 1999. In 2001, he joined the Department of Electrical and Electronics

Engineering at Koc University, Istanbul, Turkey, as a professor. Professor Aksun was appointed the acting dean of engineering in November 2003, and since May 2004, he has been the dean of the College of Engineering at Koc University.

His research interests include numerical methods for electromagnetics, microwave printed circuits in layered media, microstrip antennas, and indoor and outdoor propagation models.

Index

Recent Titles in the Artech House Microwave Library

The RF and Microwave Circuit Design Handbook, Stephen A. Maas

RF and Microwave Coupled-Line Circuits, Rajesh Mongia, Inder Bahl, and Prakash Bhartia

RF and Microwave Oscillator Design, Michal Odyniec, editor

RF Power Amplifiers for Wireless Communications, Steve C. Cripps

RF Systems, Components, and Circuits Handbook, Ferril Losee

Stability Analysis of Nonlinear Microwave Circuits, Almudena Suárez and Raymond Quéré

TRAVIS 2.0: Transmission Line Visualization Software and User's Guide, Version 2.0, Robert G. Kaires and Barton T. Hickman

Understanding Microwave Heating Cavities, Tse V. Chow Ting Chan and Howard C. Reader

For further information on these and other Artech House titles, including previously considered out-of-print books now available through our In-Print-Forever® (IPF®) program, contact:

Artech House Publishers
685 Canton Street
Norwood, MA 02062
Phone: 781-769-9750
Fax: 781-769-6334
e-mail: artech@artechhouse.com

Artech House Books
46 Gillingham Street
London SW1V 1AH UK
Phone: +44 (0)20 7596 8750
Fax: +44 (0)20 7630 0166
e-mail: artech-uk@artechhouse.com

Find us on the World Wide Web at:
www.artechhouse.com